D1657582

Edition date: January 2008

ISBN-10: 91-973730-4-4
ISBN-13: 978-91-973730-4-3
EAN: 9789197373043

UMETRICS AB
Box 7960
S-907 19 Umeå
Sweden

Tel. +46 (0)90 184800
Fax. +46 (0)90 184899

Email: info@umetrics.com

Home page: www.umetrics.com

Contents

4 Analysis of factorial designs 75

5 Experimental objective: Screening 101

14 Blocking the experimental design 279

15 The Taguchi approach to robust design 293

16 Models of variability: Distributions 309

0 Preface

0.1 Why conduct experiments?

How can we find optimum production conditions? What combination of factors will give the best car engine, at the lowest possible cost, with the lowest possible fuel consumption, and producing the minimum pollution? These and similar questions are commonly asked in business, research, and industry. Often in research, development, and production, half of the available experimental resources are dedicated to solving optimization problems. With the rapidly increasing cost of experiments, it is essential that these questions are answered as efficiently as possible. Design of Experiments, DOE, is used for this purpose – to ensure that the selected experiments produce the maximum amount of relevant information.

0.1.1 An argument for DOE based on cost and information

Experience shows there is rarely a one-to-one relationship between the number of experiments performed and the information obtained, since the quality of the experiments, in terms of the variations amongst them, affects their information content as well as their quality. Figure 0.1 (left plot) indicates, schematically, how the potential information in an experimental series may change as a function of the number of experiments (N). The information content increases rapidly to begin with, but then tends to taper off for larger values of N. A related and interesting observation is that the costs of acquiring information increase slightly more rapidly than they would if they were linearly related to the number of experiments (Figure 0.1, middle graph). This implies that when information and cost are both considered, there is a fairly small value of N that provides the optimal information per unit cost. This optimal value of N is what we wish to identify (and characterize) using DOE.

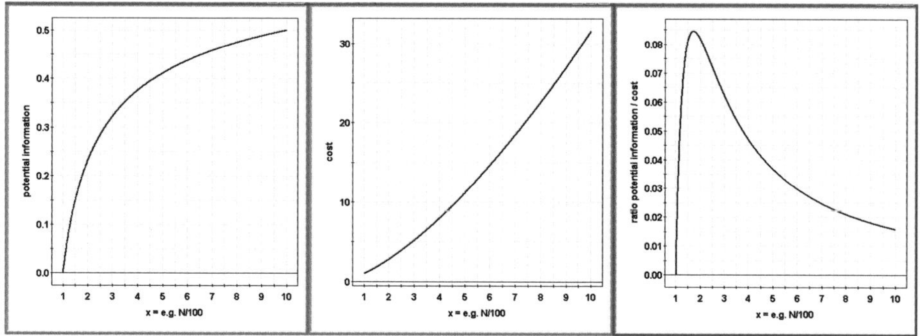

Figure 0.1: *(left) Relationship between potential information and the number of experiments (N). (middle) Relationship between cost and N. (right) Relationship between the potential information / cost ratio and N. The conclusion is that there is a fairly small N which corresponds to optimal information per unit cost.*

0.1.2 Why Design of Experiments (DOE) is used

DOE is used for a number of different purposes, including:

- Development of new products and processes
- Enhancement of existing products and processes
- Optimization of quality and performance of a product
- Optimization of an existing manufacturing procedure
- Screening important factors
- Minimization of production costs and pollution
- Robustness testing of products and processes

0.1.3 Sectors where DOE is used

DOE is used in diverse industrial sectors, including:

- The chemical industry
- The polymer industry
- The car manufacturing industry
- The pharmaceutical industry
- The biotech industry
- The food and dairy industry
- The pulp and paper industry
- The steel and mining industry
- The plastics and paints industry
- The telecom industry
- Marketing (where it is known as conjoint analysis)

0.2 The three primary experimental objectives

Basically, there are three main types of problems to which DOE is applicable. In this book, these problem areas will be referred to as *experimental objectives*. The first experimental objective is screening. Screening is used to identify the most influential factors, and to determine the ranges in which these should be investigated. This is a fairly straightforward aim, so screening designs require few experiments in relation to the number of factors. The second experimental objective is optimization. Here, the interest lies in defining which combination of the important factors will result in optimal operating conditions. Since optimization is more complex than screening, optimization designs require more experiments per factor. The third experimental objective is robustness testing. Here, the aim is to determine the sensitivity of a product or production procedure to small changes in the factor settings. Such small changes usually correspond to fluctuations in the factors occurring during a "bad day" for production, or the customer not following product usage instructions.

0.2.1 The main questions associated with the primary experimental objectives

- **Screening**
  Which factors are most influential?
  What are their appropriate ranges?
- **Optimization**
  How can we find the optimum operating conditions?
  Is there a unique optimum, or is a compromise necessary to meet conflicting demands imposed by the responses?
- **Robustness testing**
  How should we adjust our factors to guarantee robustness?
  Do we need to change our product specifications prior to claiming robustness?

0.3 The "intuitive" approach to experimental work

We may ask ourselves – *how is experimental work traditionally conducted*? Let us consider the optimization of a product or a process. This involves regulating the important factors so that the result becomes optimal. Traditionally this is achieved by changing the value of *one* factor at a time, until there is no further improvement. This is called the COST approach, and represents the intuitive way of performing experiments. An illustration of the COST approach is given in Figure 0.2.

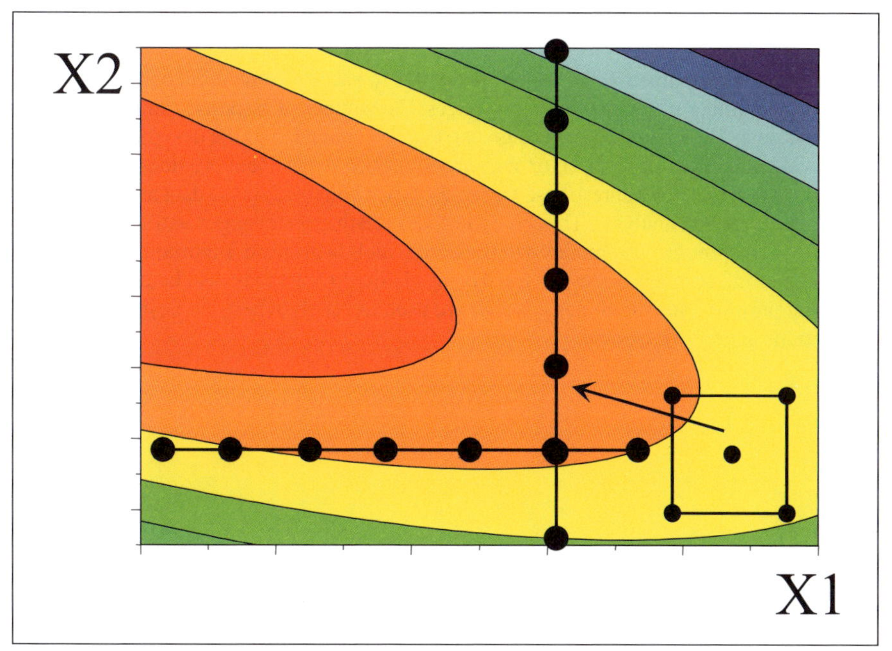

Figure 0.2: *An illustration of the COST approach and the DOE approach. Changing all factors at the same time according to a statistical experimental design (lower right corner) provides better information about the location of the optimum than the COST approach. Since the experiments here are laid out in a rectangular design a direction is indicated in which a better result is likely to be found.*

Unfortunately, the COST approach is inefficient. Early in the 20th century it was proven that changing one factor at a time does *not* necessarily provide information about the optimum conditions. This is particularly true when there are interactions between the factors. In such a case the COST approach can become bogged-down, usually some way from the real optimum. The problem is that the experimenter perceives that the optimum *has* been reached, simply because changing one factor at a time does not lead to any further improvement in the result. Changing a single factor at a time (COST) does not allow the experimenter to identify the real optimum, and produces different results with different starting points, see Figure 0.2. The two factors, x_1 and x_2, are varied separately, making it difficult to reach the optimum.

0.4 A better approach – DOE

If we are not to use the COST approach, what do we do instead? The solution is to construct a carefully selected *set* of experiments, in which *all* relevant factors are varied simultaneously. This is called statistical experimental design, or, Design Of Experiments (DOE). Such a set of experiments does not usually contain more than 10–20 runs, but this number can be adjusted to meet specific requirements. Figure 0.2 shows how two factors can be studied at the same time using a small experimental design. Since the experiments are distributed in a rectangular design space, it will be possible to identify a direction that will produce a better result. Analysis of the data acquired from the experiments performed will identify the optimal conditions, and reveal which factors influence the results. In other words, DOE provides a reliable basis for decision-making, thus providing a framework for changing all the important factors systematically, but requiring only a limited number of experiments.

0.5 Overview of DOE

Prior to conducting any experiments, the experimenter has to specify some input conditions (see Figure 0.3): the number of factors and their ranges, the number of responses, and the experimental objective. The experimental design can then be created, and the designated experiments carried out, either in parallel, or sequentially. Each experiment provides some results, i.e. numerical values for the response variables. Once collected, these data are investigated using regression analysis. This gives a model relating the changes in the factors to the changes in the responses. The model will indicate which factors are important, and how they combine in influencing the responses. The modelling results may also be converted into response contour plots, so-called maps, which are used to clarify where the best operating conditions are to be expected.

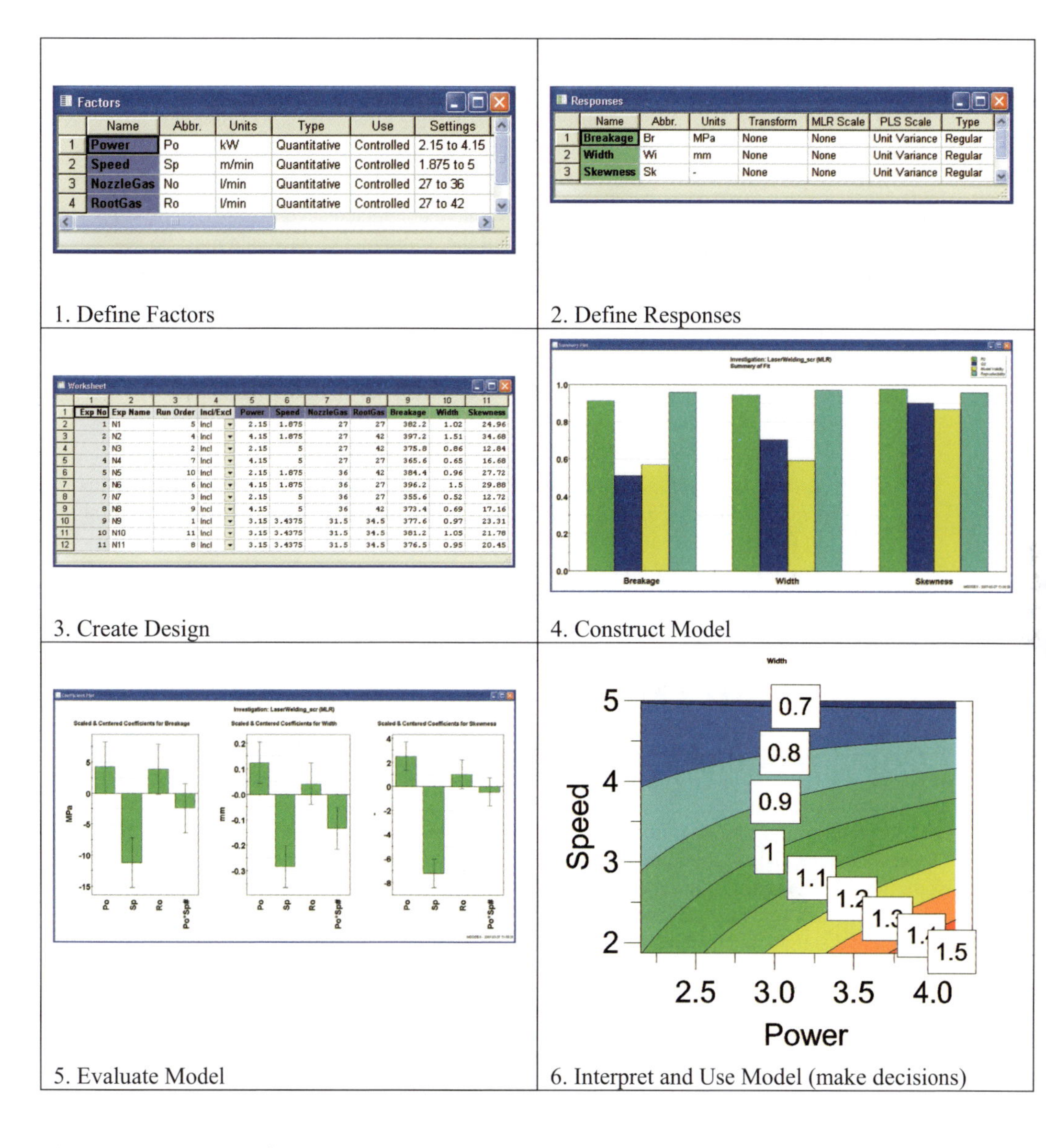

Factors

	Name	Abbr.	Units	Type	Use	Settings
1	Power	Po	kW	Quantitative	Controlled	2.15 to 4.15
2	Speed	Sp	m/min	Quantitative	Controlled	1.875 to 5
3	NozzleGas	No	l/min	Quantitative	Controlled	27 to 36
4	RootGas	Ro	l/min	Quantitative	Controlled	27 to 42

1. Define Factors

Responses

	Name	Abbr.	Units	Transform	MLR Scale	PLS Scale	Type
1	Breakage	Br	MPa	None	None	Unit Variance	Regular
2	Width	Wi	mm	None	None	Unit Variance	Regular
3	Skewness	Sk	-	None	None	Unit Variance	Regular

2. Define Responses

Worksheet

	1	2	3	4	5	6	7	8	9	10	11
1	Exp No	Exp Name	Run Order	Incl/Excl	Power	Speed	NozzleGas	RootGas	Breakage	Width	Skewness
2	1	N1	5	Incl	2.15	1.875	27	27	382.2	1.02	24.96
3	2	N2	4	Incl	4.15	1.875	27	42	397.2	1.51	34.68
4	3	N3	2	Incl	2.15	5	27	42	375.8	0.86	12.84
5	4	N4	7	Incl	4.15	5	27	27	365.6	0.65	16.68
6	5	N5	10	Incl	2.15	1.875	36	42	384.4	0.96	27.72
7	6	N6	6	Incl	4.15	1.875	36	27	396.2	1.5	29.88
8	7	N7	3	Incl	2.15	5	36	27	355.6	0.52	12.72
9	8	N8	9	Incl	4.15	5	36	42	373.4	0.69	17.16
10	9	N9	1	Incl	3.15	3.4375	31.5	34.5	377.6	0.97	23.31
11	10	N10	11	Incl	3.15	3.4375	31.5	34.5	381.2	1.05	21.78
12	11	N11	8	Incl	3.15	3.4375	31.5	34.5	376.5	0.95	20.45

3. Create Design

4. Construct Model

5. Evaluate Model

6. Interpret and Use Model (make decisions)

__Figure 0.3:__ An overview of some common steps in DOE.

0.6 The objectives of the DOE textbook

This DOE textbook is intended to show how:

- *To conduct experiments efficiently*. We demonstrate how to specify a problem definition, then plan a series of experiments that spans the experimental domain and is sufficiently informative to address the questions posed.
- *To analyze the data*. The mere fact that we have conducted several experiments does *not* guarantee that we know anything about the problem. Our raw data must be processed, i.e. analyzed, to extract the information they contain. This is accomplished by using good statistical tools.
- *To interpret the results*. The data analysis will result in a mathematical model describing the relationships between the factors and the responses. We will place a great deal of emphasis on interpreting such a model, i.e. on understanding its messages about the nature of the problem being investigated. This type of interpretation is greatly facilitated by the use of graphical tools.
- *To transform the model interpretation into concrete actions*. The aim of this last point is to encourage the user to make use of models obtained to produce firm experimental plans. In the ideal case, all problems are solved as a result of the first experimental series, and there is really no need for more experiments other than for verification. Often, however, more experiments are needed. It is then important to be able to decide how many, and what, additional experiments are most suitable in each situation.

In conclusion, the overall objective of this volume is to make the learner sufficiently confident to master these four aspects of DOE.

0.7 Organization of the book

This course book consists of 22 chapters, divided into four parts. The first part, chapters 1-4, provides a thorough introduction to the basic principles of DOE, focusing on two-level full factorial designs. The second part, chapters 5-7, is application-oriented and presents typical examples of screening, optimization, and robustness testing studies. The third part, chapters 8-18, is intended to equip the reader with insights into interesting additional topics and fundamental statistical concepts and methods. The final part, consisting of chapters 19-22, focuses on DOE applications within the pharmaceutical industry.

0.8 Summary

In this chapter, we have highlighted three important experimental objectives for which DOE is useful. These are screening, optimization and robustness testing. In essence, DOE designates a series of representative experiments to be carried out, in which all factors are varied systematically and simultaneously. Thus, DOE corresponds to a rigorous framework for planning experiments, performing experiments, analyzing data, evaluating the resulting model, and converting modelling results into informative maps representing the explored system.

Part I

Chapters 1-4

Basic principles of DOE

1 Introduction

1.1 Objective

In this chapter, we shall discuss where and how design of experiments, DOE, is used by industry today. Regardless of whether the experimental work takes place in the laboratory, the pilot plant, or the full-scale production plant, design of experiments is useful for three primary experimental objectives, screening, optimization and robustness testing. We shall discuss these three objectives and introduce one general example corresponding to each objective. These three examples will accompany us throughout the text-book.

We shall also describe another industrial application, the CakeMix application, which will help us to highlight some of the key elements involved in DOE. This example shows how changes in some important factors, that is, ingredients, can be linked to the changes in the response, that is, taste. In this way, an understanding is gained of how one must proceed to modify the amounts of the ingredients to improve the taste.

The chapter will focus on three critical problems, which are difficult to cope with using the COST-approach, but which are well handled with designed experiments. This problem discussion will automatically lead to the concept of variability. Variability is always present in experimental work, in terms both of the wanted systematic part - caused by changing important factors - and of the unwanted unsystematic noise. It is important to plan the experiments so as to allow estimation of the size of systematic and unsystematic variability. We will discuss the consequences of variability on the planning of the experimental design. The CakeMix application will be used to illustrate how variability is taken into account and interpreted.

Additionally, the aim is to describe the concept of mathematical models. We will discuss what such models are and what they look like. A mathematical model is an approximation of reality, which is never 100% perfect. However, when a model is informationally sound, the researcher is equipped with an efficient tool for manipulating the reality in a desired direction. We shall discuss the relationship between the investigation range of factors and model complexity, and conclude that quadratic polynomial models are sufficiently flexible for most practical cases.

Finally, the aim of this chapter is also to overview some of the most commonly employed DOE design families and point out when they are meaningful. The chapter ends with some arguments emphasizing the main benefits of adhering to the DOE methodology.

1.2 When and where DOE is useful

Design of experiments, DOE, is used in many industrial sectors, for instance, in the development and optimization of manufacturing processes. Typical examples are the production of wafers in the electronics industry, the manufacturing of engines in the car industry, and the synthesis of compounds in the pharmaceutical industry. Another main type of DOE application is the optimization of analytical instruments. Many applications are found in the scientific literature describing the optimization of spectrophotometers and chromatographic equipment.

Usually, however, an experimenter does not jump directly into an optimization problem; rather initial screening experimental designs are used in order to locate the most fruitful part of the experimental region in question. Other main types of application where DOE is useful is robustness testing and mixture design. The key feature of the latter application type is that all factors sum to 100%.

Areas where DOE is used in industrial research, development and production:

- optimization of manufacturing processes
- optimization of analytical instruments
- screening and identification of important factors
- robustness testing of methods
- robustness testing of products
- formulation experiments

1.3 What is DOE?

One question which we might ask ourselves at this stage is *what is design of experiments*? DOE involves making a set of experiments representative with regards to a given question. The way to do this is, of course, problem dependent, and in reality the shape and complexity of a statistical experimental design may vary considerably. A common approach in DOE is to define an interesting standard reference experiment and then perform new, representative experiments around it (see Figure 1.1). These new experiments are laid out in a symmetrical fashion around the standard reference experiment. Hence, the standard reference experiment is usually called the *center-point*.

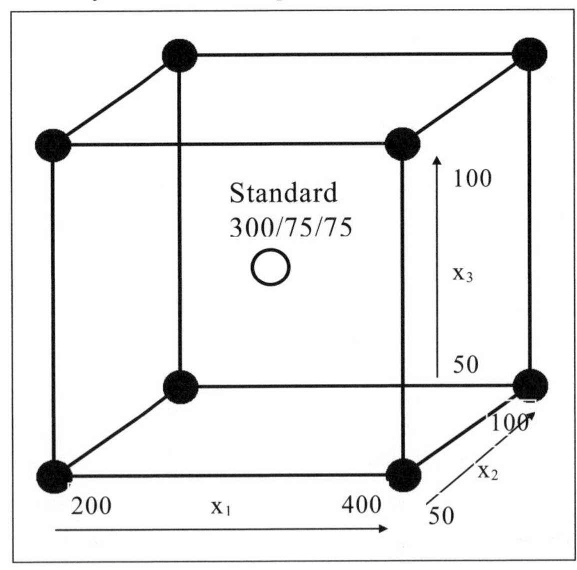

Figure 1.1: *A symmetrical distribution of experimental points around a center-point experiment.*

In the given illustration, the standard operating condition was used as the center-point. It prescribed that the first factor (x_1) should be set at the value 300, the second factor (x_2) at 75, and the third factor (x_3) at 75. In the next step, these three factors were varied according to the cubic pattern shown in Figure 1.1. This cubic pattern arises because the three factors are varied systematically around the center-point experiment. Thus, the first factor, x_1, is tested at a level slightly below the center-point, the value 200, and at a level slightly above the center-point, the value 400. A similar reasoning applies to factors x_2 and x_3.

Moreover, at a later stage in the experimental process, for instance, at an optimization step, already performed screening experiments may be used to predict a suitable reference experiment for an optimization design.

In the next three sections, we will introduce three representative DOE applications, which will accompany us in the course.

1.4 General Example 1: Screening

Screening is used at the beginning of the experimental procedure. The objective is (i) to explore many factors in order to reveal whether they have an influence on the responses, and (ii) to identify their appropriate ranges. Consider the laser welding material displayed in Figure 1.2. This is a cross-section of a plate heat-exchanger developed and manufactured by Alfa Laval Thermal.

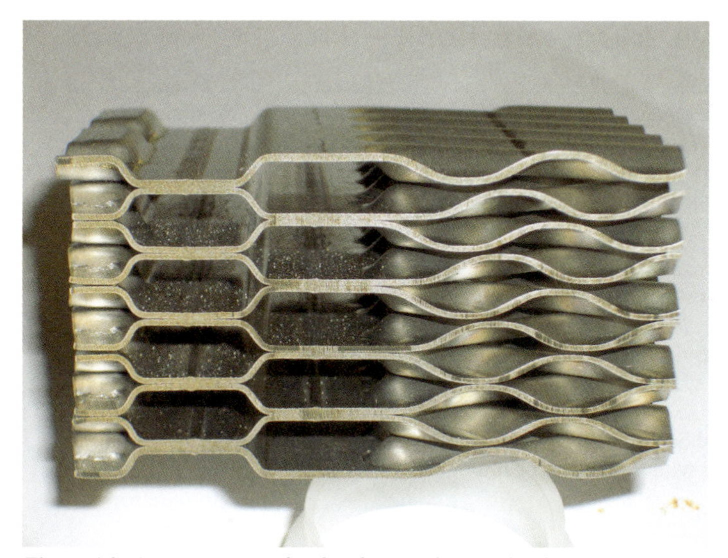

Figure 1.2: *A cross-section of a plate heat-exchanger developed and manufactured by Alfa Laval Thermal.*

In this application, the influence of four factors on the shape and the quality of the laser weld was investigated. The four factors were *power of laser*, *speed of laser*, *gas flow at nozzle*, and *gas flow at root* (underside) of the welding. The units and settings of low and high levels of these factors are seen in Figure 1.3. The experimenter measured three responses to characterize the shape and the quality of the weld, namely *breakage of weld*, *width of weld*, and *skewness of weld*. These are summarized in Figure 1.4. The aim was to obtain a persistent weld (high value of breakage), of a well-defined width and low skewness.

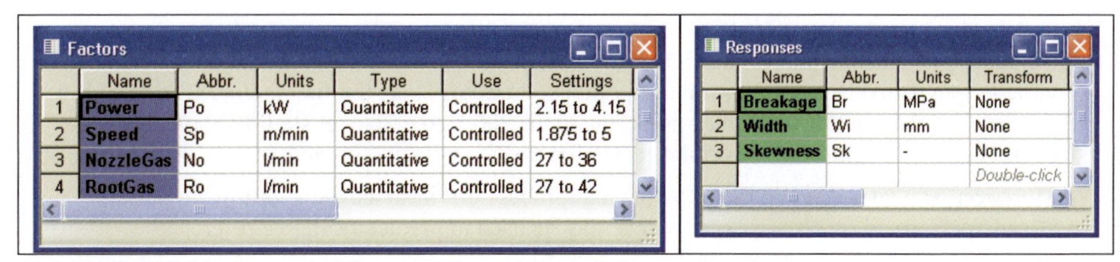

Factors

	Name	Abbr.	Units	Type	Use	Settings
1	Power	Po	kW	Quantitative	Controlled	2.15 to 4.15
2	Speed	Sp	m/min	Quantitative	Controlled	1.875 to 5
3	NozzleGas	No	l/min	Quantitative	Controlled	27 to 36
4	RootGas	Ro	l/min	Quantitative	Controlled	27 to 42

Responses

	Name	Abbr.	Units	Transform
1	Breakage	Br	MPa	None
2	Width	Wi	mm	None
3	Skewness	Sk	-	None
				Double-click

Figure 1.3: *(left) The four varied factors of General Example 1.*
Figure 1.4: *(right) The three measured responses of General Example 1.*

In the first stage, the investigator carried out eleven experiments. During the data analysis, however, it soon became apparent that it was necessary to upgrade the initial screening design with more experiments. Thus, the experimenter conducted another set of eleven experiments, selected to well supplement the first series. We will provide more details later.

In summary, with a screening design, the experimenter is able to extract a *yes* or *no* answer with regard to the influence of a particular factor. Information is also gained about how to modify the settings of the important factors, to possibly further enhance the result. Screening designs need few experiments in relation to the number of factors.

1.5 General Example 2: Optimization

Optimization is used after screening. The objective is (i) to predict the response values for all possible combinations of factors within the experimental region, and (ii) to identify an optimal experimental point. However, when several responses are treated at the same time, it is usually difficult to identify a single experimental point at which the goals for all responses are fulfilled, and therefore the final result often reflects a compromise between partially conflicting goals.

Our illustration of the optimization objective deals with the development of a new truck piston engine, studying the influence on fuel consumption of three factors, *air mass used in combustion*, *exhaust gas re-circulation*, and *timing of needle lift*. The settings of these factors are shown in Figure 1.5. Besides monitoring the *fuel consumption*, the investigator measured the levels of *NOx* and *Soot* in the exhaust gases. These responses are summarized in Figure 1.6. The goal was to minimize fuel consumption while at the same time not exceeding certain stipulated limits of NOx and Soot. The relationships between the three factors and the three responses were investigated with a standard 17 run optimization design. We will provide more details in Chapter 6.

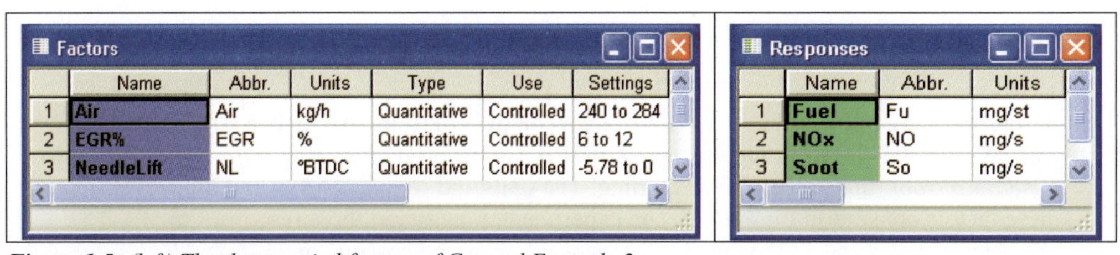

Factors

	Name	Abbr.	Units	Type	Use	Settings
1	Air	Air	kg/h	Quantitative	Controlled	240 to 284
2	EGR%	EGR	%	Quantitative	Controlled	6 to 12
3	NeedleLift	NL	°BTDC	Quantitative	Controlled	-5.78 to 0

Responses

	Name	Abbr.	Units
1	Fuel	Fu	mg/st
2	NOx	NO	mg/s
3	Soot	So	mg/s

Figure 1.5: *(left) The three varied factors of General Example 2.*
Figure 1.6: *(right) The three measured responses of General Example 2.*

In summary, with an optimization design the experimenter is able to extract detailed information regarding how the factors combine to influence the responses. Optimization designs require many experiments in relation to the number of investigated factors.

1.6 General Example 3: Robustness testing

The third objective is robustness testing, and it is applied as the last test just before the release of a product or a method. When performing a robustness test of a method – as in the example cited below – the objective is (i) to ascertain that the method is robust to small fluctuations in the factor levels, and, if non-robustness is detected, (ii) to understand how to alter the bounds of the factors so that robustness may still be claimed.

To portray a typical robustness test of an analysis method, we have selected an application taken from the pharmaceutical industry, which deals with a high-performance liquid chromatography (HPLC) system. Five factors, of which four were quantitative and one qualitative, were examined. These factors were *amount of acetonitrile* in the mobile phase, *pH*, *temperature*, *amount of the OSA counterion* in the mobile phase, and *type of stationary phase* (column). These five factors, summarized in Figure 1.7, were investigated using a design of 12 experiments. To describe the chromatographic properties of the HPLC system, three responses were recorded, that is, the *capacity factor* k_1 of analyte 1, the *capacity factor* k_2 of analyte 2, and the *resolution Res1* between these two analytes (Figure 1.8).

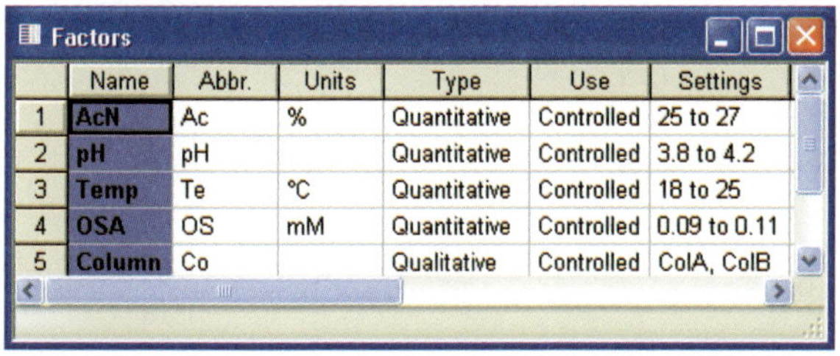

Factors

	Name	Abbr.	Units	Type	Use	Settings
1	AcN	Ac	%	Quantitative	Controlled	25 to 27
2	pH	pH		Quantitative	Controlled	3.8 to 4.2
3	Temp	Te	°C	Quantitative	Controlled	18 to 25
4	OSA	OS	mM	Quantitative	Controlled	0.09 to 0.11
5	Column	Co		Qualitative	Controlled	ColA, ColB

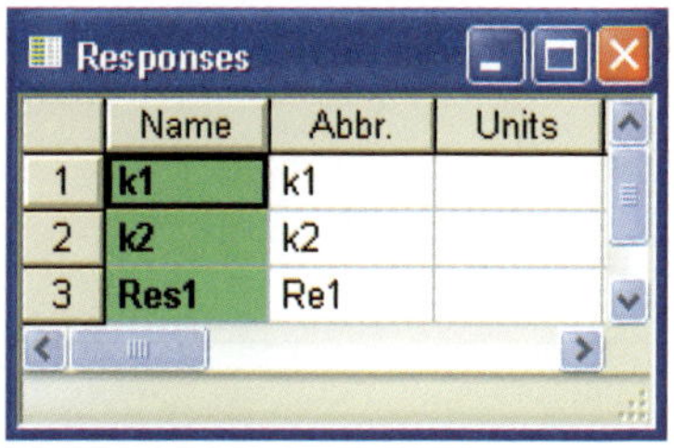

Responses

	Name	Abbr.	Units
1	k1	k1	
2	k2	k2	
3	Res1	Re1	

Figure 1.7: *(left) The five investigated factors of General Example 3.*
Figure 1.8: *(right) The three registered responses of General Example 3.*

In HPLC, the capacity factors measure the retention of compounds, and the resolution the separation between compounds. In the present case, the resolution response was the main interest and required to be robust. More information regarding this example will be given at a later stage (Chapter 7).

In summary, with a robustness testing design, it is possible to determine the sensitivity of the responses to small changes in the factors. Where such minor changes in the factor levels have little effect on the response values, the analytical system is determined to be robust.

1.7 A small example: The CakeMix data set

We will now concentrate on the CakeMix application, which is helpful in illustrating the key elements of DOE. This is an industrial pilot plant application in which the goal was to map a process producing a cake mix to be sold in a box, for instance, at a supermarket or shopping mall. On the box there will be instructions on how to use the cake mix, and these will include recommendations regarding baking temperature and time.

There are many parameters which might affect the production of a cake mix, but in this particular investigation we will only be concerned with the recipe. The experimental objective was screening, to determine the impact of three cake mix ingredients on the taste of the resulting cake. The first varied factor (ingredient) was *Flour*, the second S*hortening (fat)*, and the third *Eggpowder*. In reality, the investigated cake mix contained other ingredients, like sugar and milk, but to keep things simple only three ingredients were varied.

Firstly, the standard operating condition, the center-point, for the three factors was defined, and to do this a recommended cake mix composition was used. The chosen center-point corresponded to 300g Flour, 75g Shortening, and 75g Eggpowder. Secondly, the low and the high levels of each factor were specified in relation to the center-point. It was decided to vary Flour between 200 and 400g, Shortening between 50 and 100g, and Eggpowder between 50 and 100g. Thirdly, a standard experimental plan with eleven experiments was created. This experimental design is shown in Figure 1.9, and in this table each row corresponds to one cake.

Cake Mix Experimental Plan

Cake No	Flour	Shortening	Egg Powder	Taste
1	200	50	50	3.52
2	400	50	50	3.66
3	200	100	50	4.74
4	400	100	50	5.20
5	200	50	100	5.38
6	400	50	100	5.90
7	200	100	100	4.36
8	400	100	100	4.86
9	300	75	75	4.73
10	300	75	75	4.61
11	300	75	75	4.68

Factors		
	Levels (Low/High)	Standard condition
Flour	200 g / 400 g	300 g
Shortening	50 g / 100 g	75 g
Egg powder	50 g / 100 g	75 g
Response: Taste of the cake, obtained by averaging the judgment of a sensory panel.		

Figure 1.9: *The experimental plan of the CakeMix application.*

For each one of the eleven cakes, a sensory panel was used to determine how the cake tasted. The response value used was the average judgment of the members of the sensory panel. A high value corresponds to a good-tasting cake, and it was desired to get as high value as possible. Another interesting feature to observe is the repeated use of the standard cake mix composition in rows 9-11. Such repeated testing of the standard condition is very useful for determining the size of the experimental variation, known as the replicate error.

Apart from listing all the experiments of the design as a table, it is also instructive to make a graphical presentation of the design. In the CakeMix application, a cube is a good tool to visualize the design and thus better understand its geometry. This is shown in Figure 1.10.

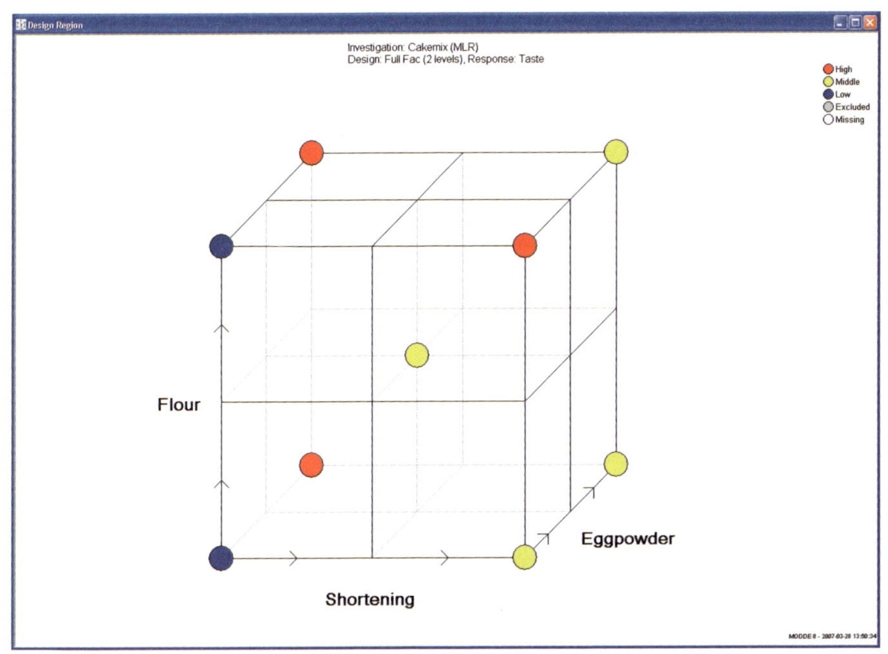

Figure 1.10: *A geometrical representation of the CakeMix experimental protocol.*

After the completion of an experimental plan, one must analyze the data to find out which factors influence the responses. Usually, this is done by fitting a polynomial model to the data. In the CakeMix application, the performed experimental design supports the model

$$y = \beta_0 + \beta_1 x_1 + \beta_2 x_2 + \beta_3 x_3 + \beta_{12} x_1 x_2 + \beta_{13} x_1 x_3 + \beta_{23} x_2 x_3 + \varepsilon,$$

where y is the response, x's the three ingredients, β_0 the constant term, β's the model parameters, and ε the residual response variation not explained by the model. The model concept, the philosophy of modeling, and model adequacy are further discussed in Sections 1.15 – 1.17.

The aim of the data analysis is to estimate numerical values of the model parameters, the so called regression coefficients, and these values will indicate how the three factors influence the response. Such regression coefficients are easy to overview when plotted in a bar chart, and the results for the cake mix data are displayed in Figure 1.11. We see that the strongest term is the two-factor interaction between Shortening and Eggpowder.

Normally, one uses a regression coefficient plot to *detect* strong interactions, but response contour plots to *interpret* their meaning. The response contour plot displayed in Figure 1.12 shows how Taste varies as a function of Shortening and Eggpowder, while keeping the amount of Flour fixed at its high level. Apparently, to obtain a cake with as high "taste" as possible, we should stay in the upper left-hand corner, i.e., use much Flour, much Eggpowder and little Shortening.

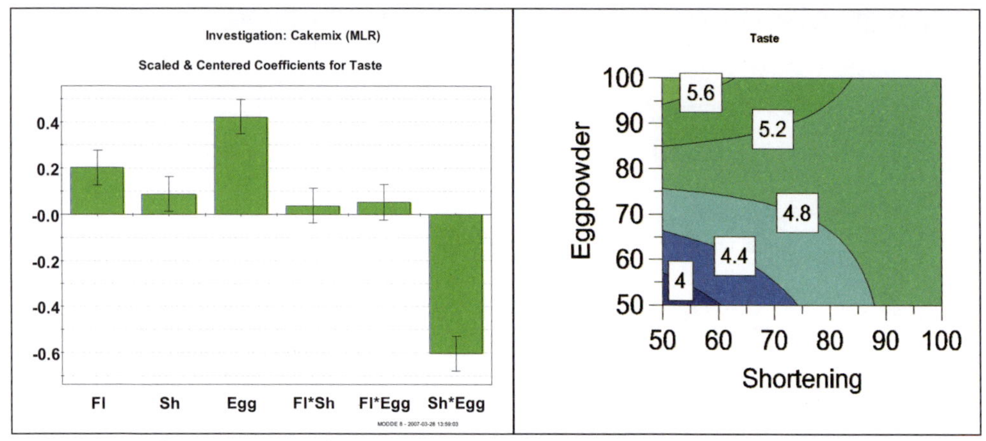

Figure 1.11: *(left) Regression coefficient plot of CakeMix regression model.*
Figure 1.12: *(right) Response contour plot of taste as a function of Eggpowder and Shortening.*

The type of response contour plot displayed in Figure 1.12 is useful for decision making – it suggests what to do next, that is, where to continue experimentally. Thus, we have converted the experimental data into an informative map with quantitative information about the modeled system. This is actually the essence of DOE, to plan informative experiments, to analyze the resulting data to get a good model, and from the model create meaningful maps of the system.

1.8 Three critical problems

There are three critical problems which DOE handles more efficiently than the COST approach. The first problem concerns the understanding of a system or a process influenced by many factors. In general, such systems are poorly studied by changing one factor at a time (COST), because interactions between factors cannot be estimated. The DOE approach, however, enables the estimation of such interactions. Secondly, systematic and unsystematic variability, the former of which is called *effects* and latter of which is called *noise*, are difficult to estimate and consider in the computations without a designed series of experiments. This second problem will be discussed later. Finally, the third critical problem is that reliable maps of the investigated system are hard to produce without a proper DOE foundation. It is very useful to inspect a reliable response contour plot of the investigated system to comprehend its behaviour. Unfortunately, such a contour plot may be misleading unless it is based on a set of designed experiments. For a response contour plot to be valid and meaningful, it is essential that the experiments have been positioned to well cover the domain of the contour plot. This is usually *not* the case with the COST approach.

1.9 Variability

We will now discuss the concept of variability. Consider Figure 1.13 in which the upper graph displays the yield of a product measured ten times under identical experimental conditions. Apparently, these data vary, despite being obtained under identical conditions. The reason for this is that every measurement and every experiment is influenced by noise. This happens in the laboratory, in the pilot-plant, and in the full-scale production. It is clear that each experimenter must know the size of the experimental noise in order to draw correct conclusions. Indeed, experimental designs are constructed in such a way that they permit a proper estimation of such noise.

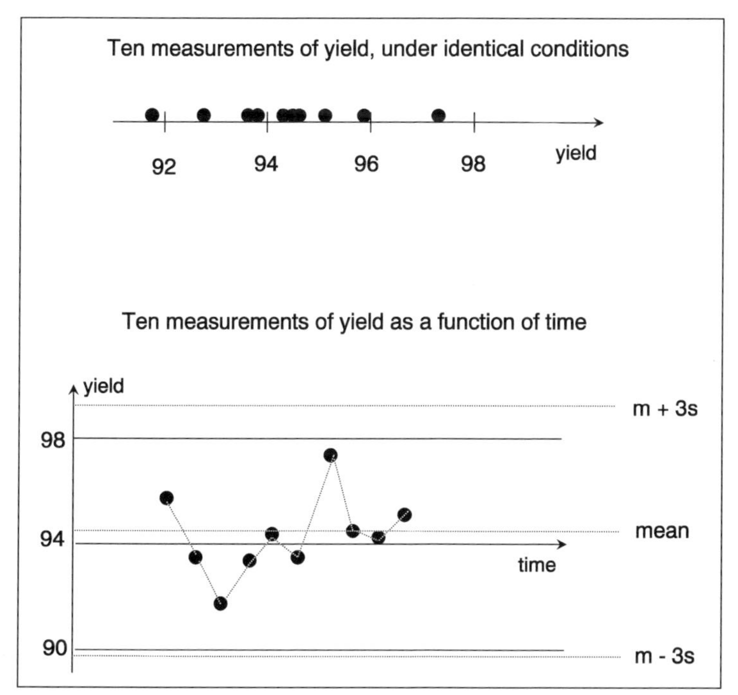

Figure 1.13: *The yield of a product measured ten times under identical experimental conditions.*

Moreover, since the ten measurements were sampled over time, one may use a time series plot showing the variation over time. Such a plot is displayed in the lower graph of Figure 1.13, together with some "control limits" indicating acceptable variation. We see that the data vary in a limited interval and with a tendency for grouping around a central value. The *size* of this interval, usually measured by the *standard deviation*, and the *location* of this central value, usually estimated by the *average*, may be used to characterize the properties of the variability. Once these quantities have been determined they may be used to monitor the behaviour of the system or the process. Under stable conditions, it can be expected that every process and system varies around its average, and stays within the specified control limits.

1.10 Reacting to noise

Let us consider the foregoing experimental system from another angle. It was decided to carry out two new experiments and change the settings of one factor, say, temperature. In the first experiment, the temperature was set to 35°C and in the second to 40°C. When the process was operated at 35°C the yield obtained was slightly below 93%, whereas at 40°C the yield became closer to 96%. This is shown in Figure 1.14. Is there any real difference between these two yields, i.e., does temperature have an effect on yield? An experimenter ignorant of the experimental variability of this system would perhaps conclude: Yes, the 5°C temperature increase induces an approximate 3% change in the yield. However, an experimenter who compares the 3% change in yield against the existing experimental variability, would not arrive at this conclusion. For him it would be obvious that the observed 3% change lies completely within the boundaries established by the ten replicated experiments. In this case, one cannot be sure about the size of the effect of the temperature. However, it is likely that the temperature *has* an effect on the yield; the question is how large. Thus, what we would like to accomplish is a partition of the 3% change into two components, the *real effect* of the temperature and the *noise*. As will soon be seen, this is best accomplished with DOE.

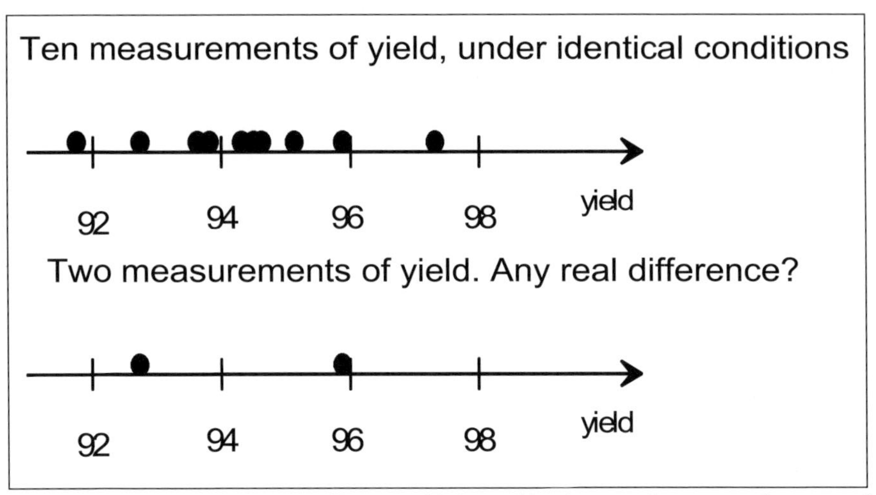

Figure 1.14: An illustration of the effect on yield obtained by changing the settings of one critical factor.

1.11 Focusing on effects

Using COST to investigate the "effect" of a factor often leads to a reaction to noise. Another problem with this approach is the number of experiments required, which is illustrated in Figure 1.15. In the left-hand part, five experiments are laid out in a COST fashion to explore the relationship between one factor and one response. This is an informationally inefficient distribution of the experiments. With a better spread, as few as five to seven runs are sufficient for investigating the effects of *two* factors. This is displayed in the center part of the figure. When arrayed in a similar manner, as few as nine to eleven experiments are sufficient for investigating *three* factors, as seen in the right-hand part of the figure. Such square and cubic arrangements of experiments are informationally optimal and arise when DOE is used.

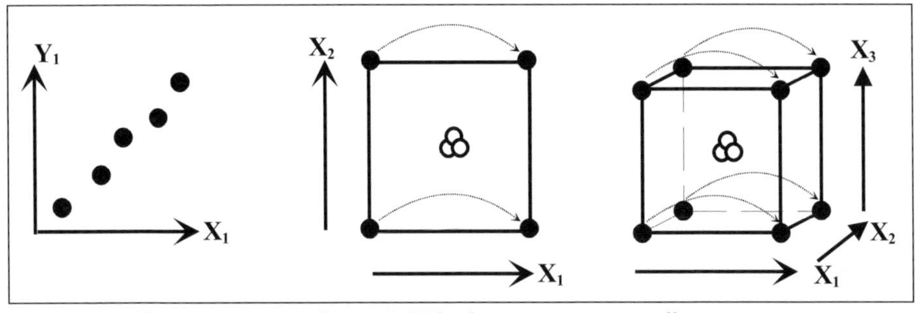

Figure 1.15: The averaging procedure in DOE leading to more precise effect estimates.

The squared arrangement of four experiments has the advantage that we can make *two* assessments of the effect of each factor. Consider the factor x_1. It is possible to estimate the effect of this factor both at the low level of x_2 *and* at the high level of x_2, that is, to study the change in the response when moving along the bottom and top edges of the square. This is schematically illustrated with the two curved, dotted arrows. Each one of these two arrows provides an estimate of the effect of x_1. These estimates are then averaged. The formation of this average implies that the estimated effect of x_1 is sharpened. Analogously, in the three-factor case, the effect of x_1 is computed as the average of four assessments, obtained by moving along the four curved arrows of the cube, and hence the effect of x_1 is well estimated.

Furthermore, with DOE it is possible to obtain an estimate of the noise. This is accomplished by considering the part of the variation which the mathematical model leaves unexplained, the so called residuals. In summary, with DOE it is possible not only to sharpen the estimate of the real effect, thanks to averaging, but also to estimate the size of the noise, e.g. the standard deviation of the residuals. This leads to a focusing on the *real effects* of the factors, and not on some coincidental noise effect. In addition, DOE always needs fewer experiments than COST.

1.12 Graphical display of real effects and noise

Given that it is possible to estimate both *real effects* of factors and experimental *noise*, one may wonder *how do we use such estimates in our daily work?* We will exemplify this using the CakeMix example. Figure 1.16 shows the regression coefficients of the interaction model and their confidence intervals. The first three coefficients, also called linear terms, reveal the *real effects* of the three ingredients. The last three coefficients, also called interaction terms, show if there are interactions among the factors. The uncertainty of these coefficients is given by the confidence intervals, and the size of these depends on the size the *noise. Hence, real effects are given by the coefficients, and the noise is accounted for by the confidence intervals.*

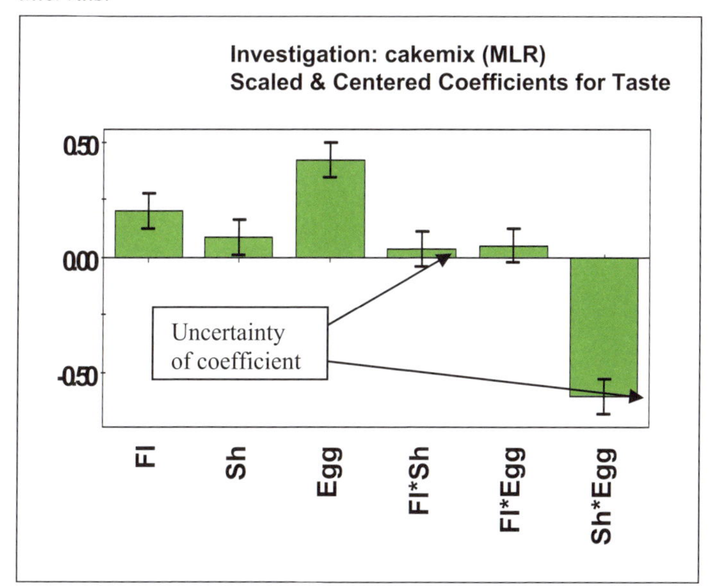

Figure 1.16: *Regression coefficients of the CakeMix interaction model.*

We now turn to the interpretation of the model in Figure 1.16. The linear terms are the easiest ones to interpret, and directly reveal the importance of each factor. We can see that Eggpowder has the strongest impact on the taste. Its coefficient is +0.42, which is interpreted in the following way: When the amount of Eggpowder is increased from the standard condition, 75g, to its high level, 100g, and keeping the other factors fixed at their standard condition, the taste of the cake will increase by 0.42. The latter value is expressed in the same unit as the taste response. The second-most influential factor is the amount of Flour. When Flour is increased from 300g to 400g the taste is modelled to increase by 0.2 unit. The third ingredient, Shortening, has comparatively little impact on taste. Interestingly, the most important term in the model is the interaction between Shortening and Eggpowder. This term will be interpreted later.

In summary, we have here exemplified the partition of *observed effects* into *real effects* and *noise*, and shown how these complement each other in the model evaluation.

1.13 Consequence of variability

An important consequence of variability is that it matters a lot *where* the experiments are performed. Consider the top graph of Figure 1.17. In this graph we have plotted an arbitrary response, y, against an arbitrary factor, x. Two experiments have been carried out, and the variability around each point is depicted with error bars. With only two experiments it is possible to calculate a simple mathematical model, a line. When these two experiments are very close to each other, that is, the investigation range of the factor is small, the slope of the line will be poorly determined. In theory, the slope of this line may vary considerably, as indicated in the plot. This phenomenon arises because the model is unstable, that is, it is based on ill-positioned experiments.

However, when these two experiments are far away from each other (upper right-hand graph of Figure 1.17), the slope of the line is well determined. This is because the investigation range of the factor is considerably larger than the experimental variability, and hence there is a strong enough "signal" for the factor to be modelled. Furthermore, with a quantitative factor, it is also favourable to put in an extra point in between, a so called center-point, to make sure that our model is OK. This is shown in the bottom left-hand graph of Figure 1.17. With this third level of the factor, it is possible to determine whether there is a linear or non-linear relationship prevailing between the factor and the response.

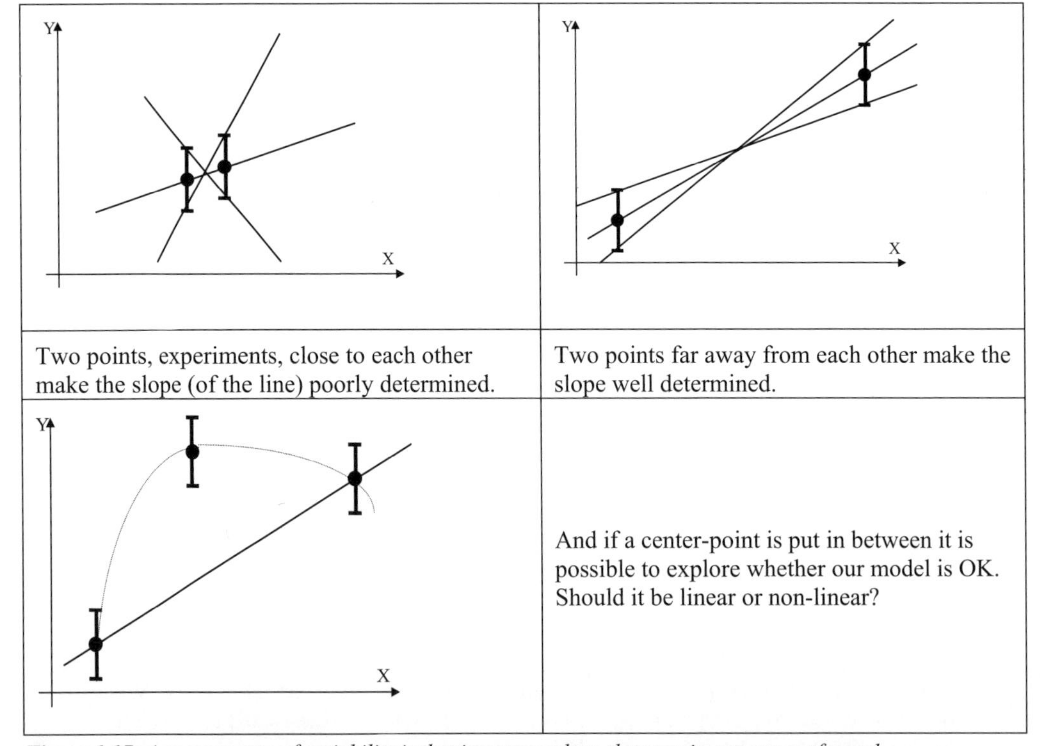

Figure 1.17: *A consequence of variability is that it matters where the experiments are performed.*

1.14 Connecting factors and responses

In DOE, there are two fundamental types of variables, *factors* and *responses* (see Figure 1.18). The responses inform us about properties and general conditions of the studied system or process. Putting it loosely, one may say that they reveal whether the system behaves in a healthy or unhealthy manner. Typical responses might be taste of cake, stability of weld, fuel consumption of truck engine, resolution of analytical peaks in liquid chromatography, and so on. The factors, on the other hand, are our tools for manipulating the system. Since they exert an influence on the system, the nature of which we are trying to map, it is usually possible to force the system towards a region where it becomes even healthier. Typical factors might be amount of flour in a cake mix recipe, gas flow at nozzle of a welding equipment, exhaust gas re-circulation in a truck engine, pH of mobile phase in liquid chromatography, and so on.

Once factors have been chosen and responses measured, it is desirable to get an understanding of the relationships between them, that is, we want to connect the information in the factor changes to the information in the response values. This is conveniently done with a mathematical model, usually a polynomial function. With such a model it is possible to extract clues like: *to maximize the third response factor 1 should be set high and factor 2 be set low; to minimize the fourth response just lower all factors; the first response is not explicable with the factors studied; there is a non-linear relationship between response 2 and factor 3*, and so on. Such insights are invaluable when it comes to specifying further experimental work.

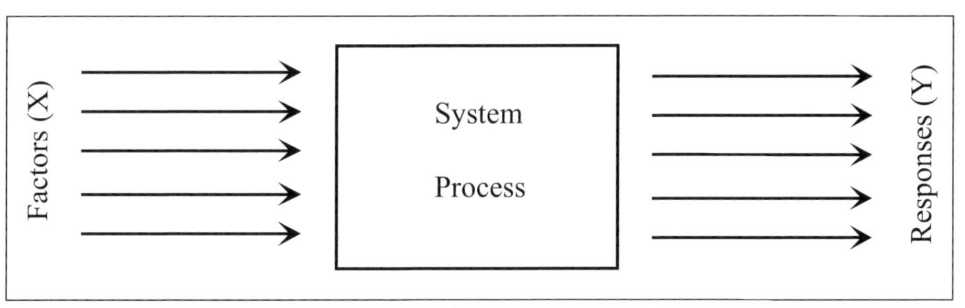

Figure 1.18: *The measured responses describe the properties of the investigated system. By changing the most influential factors the features of the system might be altered according to a desired response profile.*

1.15 The model concept

It is of utmost importance to recognize that a model is an *approximation*, which simplifies the study of the reality. A model will never be 100% perfect, but still be very useful. To understand the model concept, some simple examples are given in Figure 1.19. Consider the electric toy train, something which most children have played with. Such a toy train mimics the properties of a full-sized train, though in a miniature format, and may thus be regarded as a model. Another example is found in geography. For instance, a geographic map of Iceland, with volcanoes and glaciers marked, is a useful and necessary model of reality for tourists. Also, the response contour plot of taste in the CakeMix application is a model, which suggests how to compose a cake mix to get a tasty cake. These three examples well document what is meant with a model. Models are not reality, but approximate representations of some important aspects of reality. Provided that a model is sound – there are tools to test this – it constitutes an excellent tool for understanding important mechanisms of the reality, and for manipulating parts of the reality according to a desired outcome.

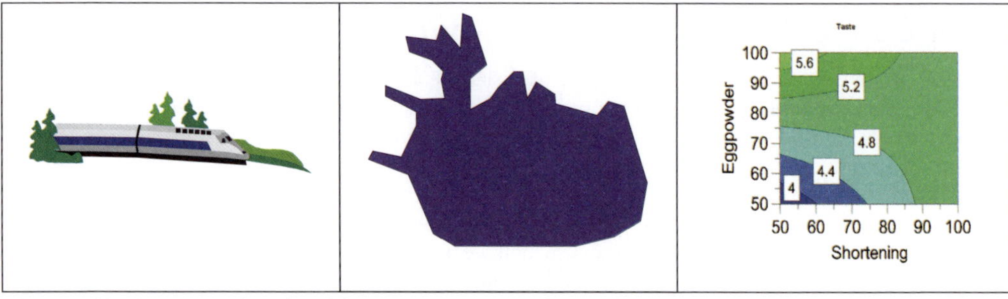

Figure 1.19: *Three examples of models, a toy train, a map of Iceland, and a response contour plot of taste.*

1.16 Empirical, semi-empirical and theoretical models

In this course book, we will work with *mathematical* models. Certain classes of mathematical models are discernible, i.e., *empirical models*, *semi-empirical models*, and *theoretical models* (see Figure 1.20). A theoretical model, also called a *hard* model, is usually derived from a well-established and accepted theory within a field. Consider, for example, the Schrödinger equation, which is a typical representative of this class of models. Theoretical models are often regarded as fundamental "laws" of natural science even though the label "models" would be more appropriate in experimental disciplines. In most cases, however, the mechanism of a system or a process is usually not understood well enough, or maybe too complicated, to permit an exact model to be postulated from theory. In such circumstances, an empirical model based on experiments, also called a *soft* model, describing how factors influence the responses in a local interval, might be a valuable alternative.

Further, in DOE, the researcher often has *some* prior knowledge that *certain* mathematical operations might be beneficial in the model building process. For instance, we all know that it is not particularly meaningful to work with the concentration of the hydrogen ion in water solution. Rather, it is more tractable to work with the negative logarithm of this concentration, a quantity normally known as pH. Another example might be that the experimenter knows that it is not the factors A and B that are interesting per se, but that the ratio B/A is what counts. When we enter this kind of prior knowledge into an empirical investigation, we are conducting *partial* empirical modeling, and consequently the prefix semi- is often added. Because of this, it is often stated that DOE involves semi-empirical modeling of the relationships between factors and responses.

Empirical	Semi-empirical	Fundamental
$y = a + bx + \varepsilon$	$y = a + b\log x + \varepsilon$	$H\psi = E\psi$

Figure 1.20: *An overview of empirical, semi-empirical and fundamental mathematical models.*

1.17 Semi-empirical modeling – Taylor series expansions

Taylor series expansions are often used in semi-empirical modeling. Consider Figure 1.21 in which the "true" relationship between one response and one factor, the target function y = f(x), is plotted. We will try to find an alternative function, y = P(x), which can be used to approximate the true function, y = f(x). In a limited factor interval, Δx, any continuous and differentiable function, f(x), can be arbitrarily well approximated by a polynomial, P(x), a Taylor series of the form: $y = P(x) = b_0 + b_1x + b_2x^2 + ... + b_px^p + e$, where e represents a residual term. In this polynomial function, the degree, p, gives the complexity of the equation.

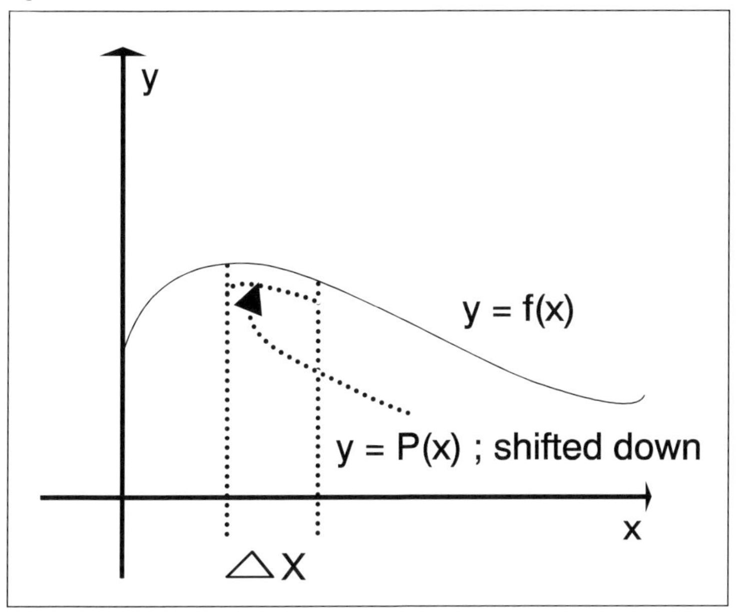

Figure 1.21: *The "true" relationship, y = f(x), between a response, y, and a factor, x, approximated by an alternative function, y = P(x), in a limited factor interval Δx.*

Because there exists an important relationship between the factor interval, Δx, and the degree of the polynomial model, p, it is worthwhile to carry out a closer inspection of Figure 1.21. Let us assume that the degree p has been fixed. Then, for this degree, p, the approximation of y = f(x) by y = P(x) gets better the smaller the factor interval, Δx, becomes. Analogously, for a fixed size of the factor interval, Δx, the approximation is better the higher the degree, p, is. Hence, what is important to always bear in mind in modeling is this trade-off between model complexity and investigation ranges of factors. Of course, this kind of reasoning can be generalized to functions of many factors and many responses.

1.18 Conceptual basis of semi-empirical modeling

A semi-empirical model is a *local* model, which describes in detail the situation within the investigated interval. This is a desirable feature. We do not want a *global* model only providing superficial knowledge of larger aspects of reality. A small example will explain why. Imagine a tourist wanting to find the exact position of the Eiffel Tower in Paris. Surely, this person would not look at a globe of our planet, because it represents too global a model, with little or no useful information of where to find the Eiffel Tower. Not even a map of France is local enough, but a map of the central parts of Paris would do the job well. Such a town map of Paris represents a local model of a small aspect of reality, and provides detailed information of where to find the great tourist attraction.

Models of complicated systems or processes in science and technology function in the same way. Local models pertaining to narrowly defined investigation regions provide more detail than models covering large regions of seemingly endless character. This is one reason why it is important to carefully specify the investigation range of each factor. Interestingly, our long experience in technology, chemistry, biology, medicine, and so on, shows that within such a restricted experimental area nature is fairly smooth and not extensively rugged (see Figure

1.22). This means that when several factors are explored, a smooth, waving response surface may typically be encountered. Such a smooth response surface can be well approximated by a simple polynomial model, usually of quadratic degree.

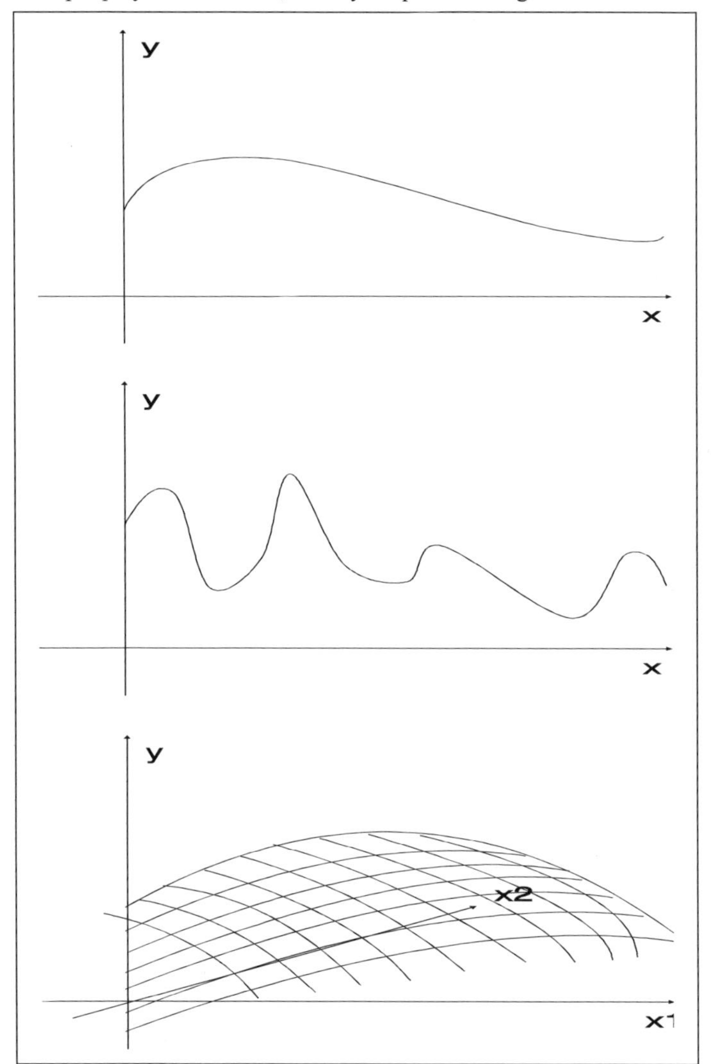

Figure 1.22: *Our long experience (technology, chemistry, biology, medicine,...) shows that nature is fairly smooth and not rugged. With several factors a smooth response surface is usually applicable. Such a smooth surface can be well approximated by a quadratic polynomial model.*

1.19 Configuration of experiments depends on the objective

As we have seen, the DOE concept may be viewed as a framework for experimental planning. We shall here briefly overview a few basic designs of this framework, which are used to deal with the three major experimental objectives, and point out their common features and differences. Figure 1.23 provides a summary of the designs discussed.

The first row of Figure 1.23 shows complete, or full, factorial designs for the investigation of two and three factors. These are screening designs, and are called *full* because all possible corners are investigated. The snowflake in the interior part depicts replicated center-point experiments carried out to investigate the experimental error. Usually, between 3-5 replicates are made. The second row in the figure also shows a screening design, but one in which only a fraction of all possible corners have to be carried out. It belongs to the *fractional factorial* design family, and this family is extensively deployed in screening. Fractional factorial designs are also used a lot for robustness testing. The last row of Figure 1.23 displays designs originating from the *composite* design family, which are used for optimization. These are called composite designs because they consist of the building blocks, corner (factorial) experiments, replicated center-point experiments, and axial experiments, the latter of which are denoted with open circles.

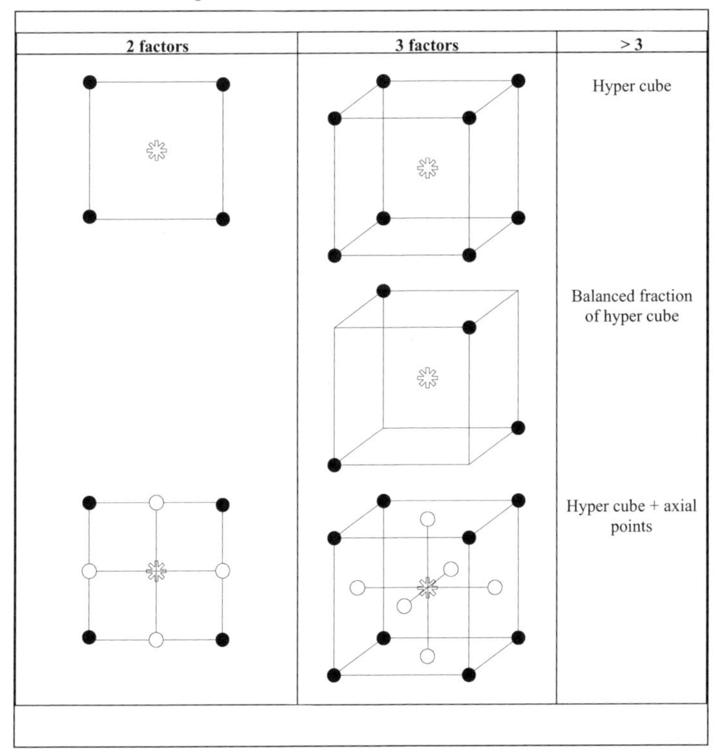

Figure 1.23: *Examples of full factorial, fractional factorial, and composite designs used in DOE.*

1.20 Benefits of DOE

The great advantage of using DOE is that it provides an *organized* approach, with which it is possible to address both simple and tricky experimental problems. The experimenter is encouraged to select an appropriate experimental objective, and is then guided to devise and perform a *set* of experiments, which is adequate for the selected objective. Although the experimenter may feel some frustration about having to perform a series of experiments, experience shows that DOE requires fewer experiments than any other approach. Since these few experiments belong to an experimental plan, they are mutually connected and thereby linked in a logical and theoretically favorable manner. Thus, by means of DOE, one obtains more useful and more precise information about the studied system, because the joint influence of all factors is assessed. After checking the model adequacy, the importance of the factors is evaluated in terms of a plot of regression coefficients, and interpreted in a response contour plot. The latter type of plot constitutes a map of the system, with a familiar geometrical interpretation, and with which it is easy to decide what the next experimental step ought to be.

1.21 Questions for Chapter 1

- Which are the three main experimental objectives?
- What are the main questions addressed using these objectives?
- Which are the three critical problems that DOE addresses better than the COST-approach?
- What is variability?
- Is it important to consider variability?
- Is it possible to separate observed effects into real effects and noise with COST? With DOE?
- Why is it important that the range of a factor is made sufficiently large?
- Why is it useful to also use center-point experiments in DOE?
- Why is it necessary to obtain a model linking factors and responses together?
- What is a model?
- Is a model always a 100% perfect representation of reality?
- What is a theoretical model?
- What is a semi-empirical model?
- What can you say about the trade-off between model complexity and size of factor ranges?
- Why is a local model preferable to a global model?

1.22 Summary and discussion

Design of experiments is useful in the laboratory, the pilot plant and full-scale production, and is used for any experimental objective, including screening, optimization, and robustness testing. We have introduced three general examples - the laser welding case, the truck engine study, and the HPLC robustness problem - which will be used to illustrate these three objectives. In addition, the CakeMix application was outlined for the purpose of overviewing some of the key elements involved in DOE. By conducting an informative set of eleven experiments, it was possible to create a meaningful response contour plot, showing how to modify the cake mix recipe to achieve even better tasting cakes.

It was shown that DOE efficiently deals with three critical problems, where the COST approach is inefficient or unsuccessful. These three problems concern (i) how to monitor a system simultaneously influenced by many factors, (ii) how to separate observed effects into real effects and noise, and (iii) how to produce reliable maps of an explored system.

Furthermore, in this chapter we have introduced the concept of variability. Emphasis has been placed on the importance of well resolving the real effect of a factor and get it “clear” of the noise, to avoid reacting to just noise. This was demonstrated using the CakeMix application. Here, the calculated regression coefficients relate to the real effects, while the confidence intervals inform about the noise.

In order to understand how factors and responses relate to each other, and to reveal which factors are influential for which responses, it is favorable to calculate a polynomial model. It is important to recall that such a model is a simplification of some small aspects of reality, and that it will never be 100% perfect. However, with a sufficiently good model, we have an efficient tool for manipulating a small part of reality in a desired direction.

In DOE, we work with semi-empirical modeling, and the models that are calculated have a local character, because they are applicable in a confined experimental region. Because we are investigating reality in a local interval, where nature is often smooth, simple low-order polynomials up to quadratic degree are usually sufficient.

2 Problem formulation

2.1 Objective

The problem formulation step is of crucial importance in DOE, because it deals with the specification of a number of important characteristics critically influencing the experimental work. In the problem formulation, the experimenter must specify the experimental objective, define factors and responses, choose a regression model, and so on, and any mistakes at this step may result in difficulties at later stages. The aim of this chapter is to overview how a correct problem formulation is carried out. A secondary aim is to outline six important stages in the experimental process. These are (i) familiarization, (ii) screening, (iii) finding the optimal region, (iv) optimization, (v) robustness testing, and (vi) mechanistic modelling.

The chapter also provides an expanded discussion regarding the factor and response definitions in the problem formulation. To this end, we will address qualitative factors at two or more levels, and some complications that arise in the design generation. For instance, with qualitative factors it is no longer possible to define pure center-point experiments, and the balancing of a design is also rendered more difficult. Moreover, we will discuss the situation that arises when it is impossible to make experiments at certain factor combinations, for instance, when parts of the experimental region are inaccessible for experimentation. Simple linear constraints may then be defined for the quantitative factors involved, and used to regulate where the experiments of the design are positioned. The topic of selecting an appropriate metric for factors and responses will also be discussed. Selecting the correct metric involves considering whether a transformation is needed, and which kind of transformation is the optimal one.

The choice of the regression model is an integral part of the problem formulation. The aim of the last part of this chapter is therefore to describe the semi-empirical polynomial models that are commonly chosen in DOE. We will examine some properties of linear, interaction and quadratic models, and pay special attention to their geometric features. We will also review how these models are associated with appropriate experimental designs.

2.2 Introduction

The problem formulation is of central importance in DOE, regardless of whether the application concentrates on screening, optimization, or robustness testing. The objective of carrying out the problem formulation, or problem definition, is to make completely clear, for all involved parties, the intentions underlying an experimental investigation. There are a number of things to discuss and agree about, and it is necessary to consider six points. These six items regard (i) the experimental objective, (ii) the factors, (iii) the responses, (iv) the model, (v) the design, and (vi) the worksheet.

The experimental objective defines what kind of investigation is required. One should ask *why is an experiment done*? And *for what purpose*? And *what is the desired result*? The factors are the variables that are changed to give different results on the measured responses. The fourth point, the model, means that one has to specify a polynomial model that corresponds to the chosen experimental objective. Next, an experimental design is created which supports the selected model. Thus, the composition of the design follows from the experimental objective and the polynomial model, but also depends on the shape of the experimental region and the number of experiments.

When a design has been proposed, it is important to sit down and go through the proposed experiments to make sure that all of them look reasonable, can be performed, and have the potential of fulfilling the experimental objective. Subsequently, the experimental worksheet is created. In general, such a worksheet is similar to the actual design. However, in the worksheet, information of practical experimental character is normally appended, such as a proposed random run order, and supplementary experiments already carried out.

In Section 2.3, we will discuss the selection of the experimental objective (item (i)), and in Sections 2.4 – 2.8 the other five important points (items (ii)-(vi)).

2.3 Problem formulation, Step 1: The experimental objective

We will now describe the first step of the problem formulation, the selection of the experimental objective. Basically, the experimental objective may be selected from six DOE stages:

- (i) familiarization,
- (ii) screening,
- (iii) finding the optimal region,
- (iv) optimization,
- (v) robustness testing, and
- (vi) mechanistic modelling.

We will highlight the merits and requirements of these stages. However, we will *not* enter into any details regarding mechanistic modelling, as it rarely occurs in industrial practice. In summary, for each one of these stages/objectives, the experimental design will vary depending on the type of factors, the number of factors, possible constraints (restrictions), and the selected model.

2.3.1 Familiarization

The familiarization stage is used when faced with an entirely new type of application or equipment. Many researchers, who are already familiar with their applications, may skip this step and go directly to the next step, screening. In the familiarization stage, one normally spends only a limited amount of the available resources, say 10% at maximum. With the term *resources* we refer to money, time, personnel, equipment, starting materials, and so on. The basic idea is to obtain a brief feeling for what is experimentally possible, and how one should proceed to exert some kind of influence on the investigated system or process. In other words, it is desirable to find out whether the process is possible at all, and, if it is possible and working properly, roughly how good it is.

Familiarization may be accomplished using chemical and technological insight, intuition, and very simple designs. Such a simple design is usually selected from the factorials family and is predominantly constructed in the two factors believed to be most important. As seen in Figure 2.1, this factorial design may be supplemented with some center-point runs. Usually, such a design is constructed only to verify that similar results are obtained for the replicated center-points, and that different results are found in the corners of the design.

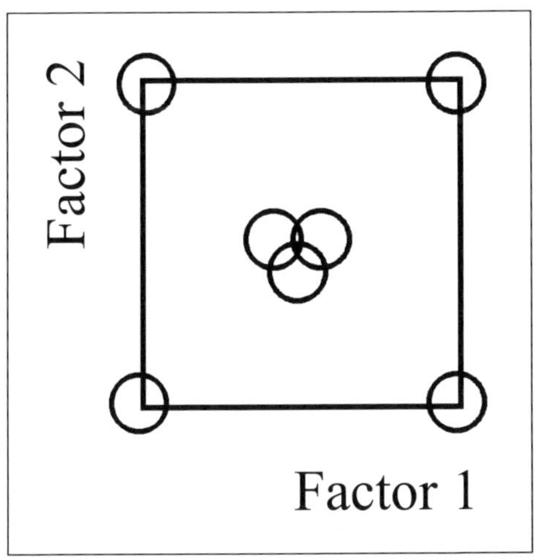

Figure 2.1: *A simple factorial design useful in familiarization. Usually, such a design is constructed only to verify that similar results are obtained for the replicated center-points, and that different results are found in the corners of the design.*

In addition, another main aim of this stage concerns the development of response measurement techniques. One should always choose responses that are relevant to the goals set up in the problem formulation, and many responses are often necessary. In this stage, we will therefore have to verify that our response measurement techniques work in reality.

2.3.2 Screening

The second experimental stage is screening, which in this text-book is highlighted as one of the three major experimental objectives (screening, optimization, and robustness testing). With screening one wants to find out a little about *many* factors, that is, which factors are the dominating ones, and what are their optimal ranges? Interestingly, the 80/20 rule, also known as the Pareto principle after the famous Italian economist Vilfredo Pareto, applies well to this stage. In the context of DOE, the Pareto principle states that 80% of the effects on the responses are caused by 20% of the investigated factors. This is illustrated in Figure 2.2, in which approximately 20% of the investigated factors have an effect exceeding the noise level. These strong factors are the ones we want to identify with the screening design. Another way to see this is exemplified in Figure 2.3. Prior to putting the screening design into practice, the examined factors are ascribed the same chance of influencing the responses, whereas after the design has been performed and evaluated, only a few important factors remain. With the screening objective, simple linear or interaction polynomials are sufficient, and the designs employed are primarily of the fractional factorial type.

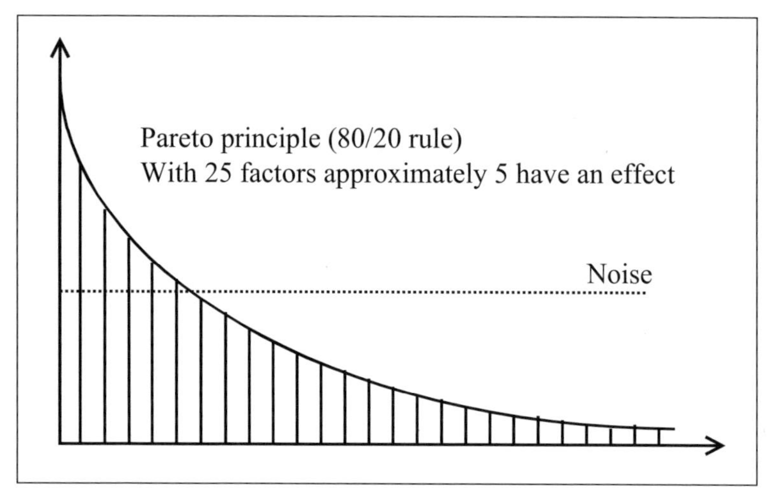

Figure 2.2: *The Pareto principle in DOE suggests that 20% of the factors have an effect above the noise level.*

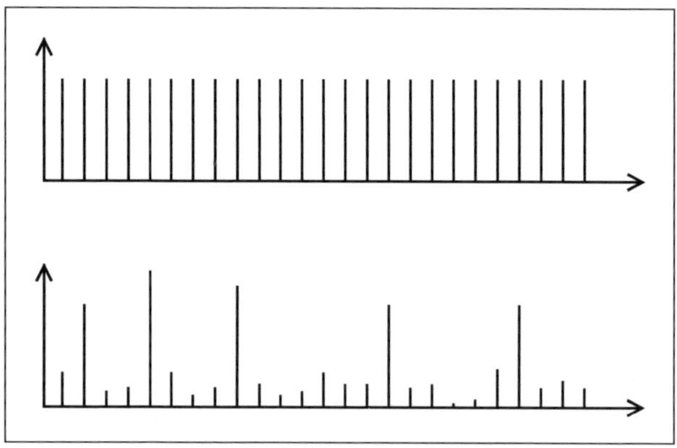

Figure 2.3: *Prior to the screening examination all factors are ascribed the same chance of influencing the responses, and after the screening only a few dominating ones remain.*

2.3.3 Finding the optimal region

Occasionally, after the screening stage, one has to conclude that the experimental region explored is unlikely to contain the optimum. In such a situation, it is warranted to move the experimental region into a domain more likely to include the desired optimal point. The question which arises, then, is *how do we move the experimental region to an appropriate location*? This is illustrated in Figure 2.4. The tricky part lies in the maneuvering of the experimental region up to the top of the mountain.

In reality, a re-positioning of the experimental region is accomplished by using the outcome of the existing screening design, that is, the polynomial model, which dictates in which direction to move. This is exemplified in Figure 2.5. The interesting direction, which lies perpendicular to the level curves (parallel lines) of the response contour plot, is used for moving according to the principles of *steepest ascent* or *steepest descent*, depending on whether a maximum or minimum is sought. The example shows a steepest descent application. Alternatively, the software optimizer might be used to accomplish this experimental objective (see Chapters 5 and 6).

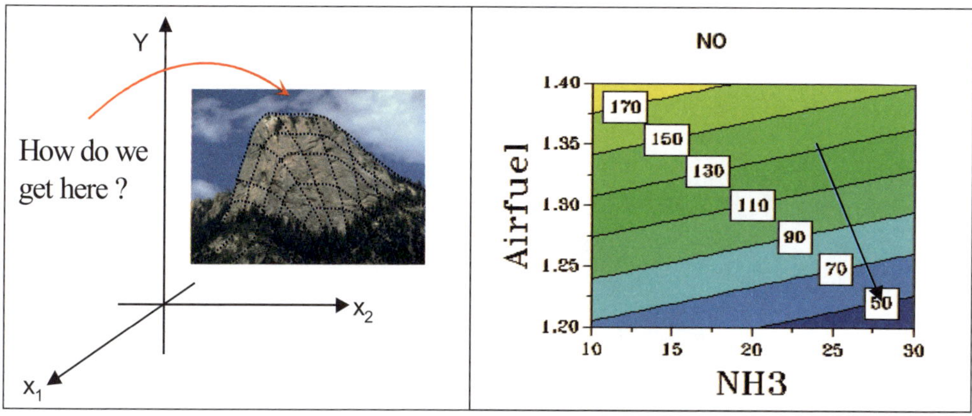

Figure 2.4: *(left) After the screening stage, one is often interested in moving the experimental region so that it includes the optimum.*
Figure 2.5: *(right) To accomplish a re-positioning of the experimental region, one uses the results from the screening design to move in the direction of steepest ascent for maximum or steepest descent for minimum.*

We emphasize that these gradient techniques are greatly facilitated by the use of graphical tools, such as response contour plots. The usual result of this stage is that one obtains a new experimental point which is suited for anchoring an optimization design, but sometimes it is found that an entirely new screening design is necessary. Screening is further discussed in Chapter 5.

2.3.4 Optimization

The next experimental stage is optimization. Now, our perspective is that the most important factors have been identified, as a result of the screening phase, and that the experimental region is appropriately positioned so that it presumably contains the optimal point. Before we discuss optimization in more detail, we shall contrast screening and optimization. At the screening stage, *many* factors are investigated in *few* runs, but in optimization the reverse is true, because *few* factors are explored in comparatively *many* experiments. In addition, we also have to re-phrase our main question. In screening, we ask *if* a factor is relevant, and the expected answer is either yes or no. In optimization, we ask *how* a factor is important, and expect an answer declaring whether there is a positive or negative relation between that factor and a response. This answer must also reveal the nature of such a relationship, that is, is it linear, quadratic, or maybe even cubic?

Hence, we may say that in optimization the main objective is to extract in-depth information about the *few* dominating factors. This is achieved with designs of the *composite* family, which support quadratic polynomial models by encoding between three and five levels of each factor. A quadratic model is flexible and may closely approximate the "true" relation between the factors and the responses. A convenient way of overviewing the implication of a fitted model is to display the modelling results in terms of a response surface plot. Figure 2.6 shows an example of a response surface plot. Hence, this approach is known as *response surface modelling*, or RSM for short.

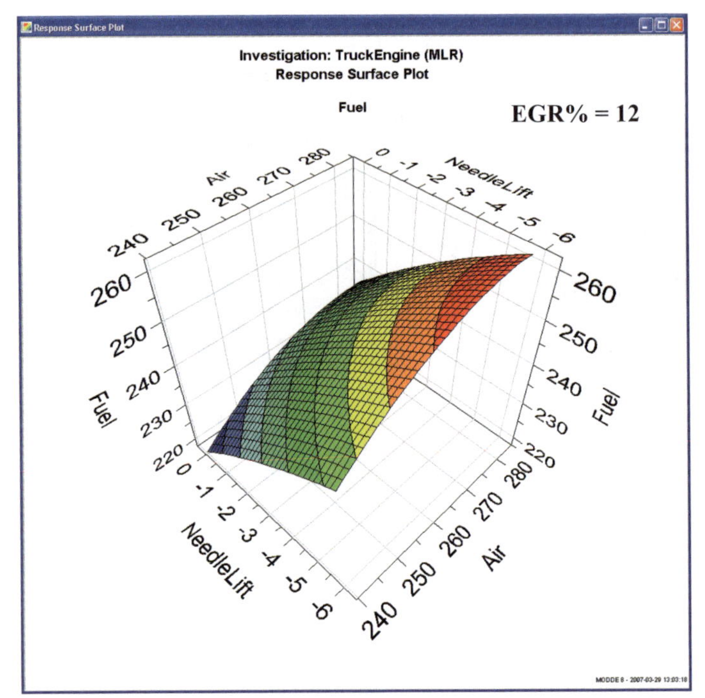

Figure 2.6: *In optimization, the main objective is to extract in-depth information about the* ***few*** *dominating factors. A quadratic RSM model is usually used, closely approximating the "true" relation between the factors and the response(s):* $y = \beta_0 + \beta_1 x_1 + \beta_2 x_2 + \beta_{11} x_1^2 + \beta_{22} x_2^2 + \beta_{12} x_1 x_2 + \ldots + \varepsilon$.

When an RSM model has been fitted it can be used for two primary purposes. The first is the prediction of response values for any factor setting in the experimental region. The second is the identification of the factor setting corresponding to the optimal point, i.e., optimization. With few responses, say two or three, finding the optimum is an easy task. But with more responses, and with sometimes conflicting demands, the identification of an optimal point may be cumbersome, and the final result is often a compromise between various objectives. Optimization is further discussed in Chapter 6.

2.3.5 Robustness testing

Robustness testing is an important, but often overlooked, stage in the experimental process. It is usually carried out before the release of an almost finished product, or analytical system, as a last test to ensure quality. Accordingly, we highlight robustness testing as one of the three primary experimental objectives. With robustness testing one wants to identify those factors which *might* have an effect on the result and regulate them in such a manner that the outcome is within given specifications. Here, we think of factors that we normally vary in a statistical design, such as, temperature, pressure, gas flow, and so on. Likewise, we want to reveal those factors which in principle have *little* effect, but which may still cause an undesired spread around the ideal result. Typical examples of such factors are ambient temperature, humidity, variability in raw material composition, etc. We want to understand their impact and try to adjust all factors so that the variation in the responses is minimized. There are different ways to tackle a robustness testing problem. Usually, a fractional factorial design or Plackett-Burman design is used. This topic will be pursued in more detail in Chapter 7.

2.3.6 Mechanistic modelling

Mechanistic modelling is the last of the stages in the experimental process that we will describe. This stage is attempted when there is a need to establish a theoretical model within a field. One or more semi-empirical models are then utilized in trying to build such a theoretical model. However, it is important to remember that the semi-empirical models are *local* approximations of the underlying fundamental relationships between the factors and the responses. As such, they will never be 100% perfect, which is a source of uncertainty when trying to formulate an exact mechanistic model. In this conversion process, the coefficients of the semi-empirical models are used to get an idea of appropriate derivative terms in the mechanistic model, and it is also crucial that the shape of the former conforms with the shape of the latter, more fundamental relationship. One may use one or many semi-empirical models to attempt to falsify, or maybe prove, a number of competing theoretical models. However, the success of the mechanistic modelling relies heavily on the use of a correct problem formulation, a correct experimental design, and a correct data analysis. The extent to which mechanistic modelling is approached in this way in industry is limited.

2.4 Problem formulation, Step 2: Definition of factors

We have now completed the description of the six experimental stages in DOE. The experimenter should select one of these stages as part of the problem formulation. Subsequently, the next step in the problem formulation involves the specification of which factors to change. We will now conduct a general discussion related to the specification of the factors.

The factors are the variables which, due to changes in their levels, will exert an influence on the system or the process. In the broadest sense, factors may be divided into *controllable* and *uncontrollable* factors (see Figure 2.7). Controllable factors are the easiest ones to handle and investigate, and the experimenter is usually alerted when such a factor changes. Uncontrollable factors are factors which are hard to regulate, but which may have an impact on the result. Typical examples of the latter are ambient temperature and humidity. We strongly recommend that such uncontrollable factors be kept track of.

Another way to categorize factors is according to the system of *process factors* and *mixture factors*. Process factors are factors which can be manipulated independently of one another, and which may be investigated using factorial and composite designs. Typical examples are the power of the laser welding equipment, the air mass used in combustion in a truck engine, the pH in the mobile phase of a liquid chromatography system, and so on. Mixture factors are factors which display the amounts of the ingredients, or constituents, of a mixture, and they add to 100%. Hence, mixture factors can *not* be varied independently of one another, and therefore require special designs other than factorial and composite designs.

A third, and perhaps the most common way, of dividing factors is to consider them as either *quantitative* or *qualitative*. A quantitative factor is a factor which may change according to a continuous scale. The three process factors just mentioned are good examples from this category. A qualitative factor is a categorical variable, which can only assume certain discrete values. A typical qualitative factor might be the kind of flour used in making a cake mix, which could, for instance, assume the three levels flour supplier A, flour supplier B, and flour supplier C. In the MODDE software, all these types of factors can be specified.

Controlled

Quantitative (Continuous)

	low	center	high
Temperature	35°C	40°C	45°C
Amount	2g	4g	6g
Speed	200 rpm	250 rpm	300 rpm
pH	5	6	7

Qualitative (Discrete)

Catalyst	Pd	Pt	
Flour supplier	A	B	C
Type of car	Volvo	Saab	Fiat

Uncontrolled

Outside/Inside Temperature

Outside/Inside Moisture

Figure 2.7: *In the specification of quantitative factors, it is mandatory to specify a low and a high investigation level. It is recommended to experiment also at a center level, so called center-points, located half-way between the low and high settings. In the specification of qualitative factors, all levels of each factor must be given. It is practical to define between two and five levels for a qualitative factor. In MODDE, it is possible to explore as many as 12 levels of a qualitative factor, however, this requires many experiments. It is not possible to define true center-point experiments with qualitative factors. It is also favorable in the factor specification to define which factors are uncontrollable. This is done in order to keep track of their values at the time of the experimentation. In the data analysis it may be investigated whether the uncontrollable factors actually have an effect on the responses.*

Apart from deciding whether a factor is quantitative, qualitative, or uncontrollable, and so on, the specification of factors also includes defining the investigation range of each factor. For a quantitative factor it is common practice to define a low and a high level. But, since it is favorable to use replicated experiments in the interior part of a design, defining a center level, located half-way between the low and high levels, is a good practice. This is exemplified in Figure 2.7.

When the low and the high levels of a quantitative factor have been set, they define the investigation *range* of that particular factor. The range of a quantitative factor predominantly depends on three criteria: experimental feasibility, experimental objective, and experimental noise. With experimental feasibility we mean that it is important to make clear which settings are *relevant* for the problem at hand. For instance, when taking a bath, the temperature of the bathing water is of great significance. Theoretically, this temperature may be varied between 0° and 100° C, but it is only relevant to consider the temperature range 35-45°C.

Also, the experimental objective influences the range of a factor. In screening, one normally maps larger intervals as it is not certain beforehand where the best settings are found. And in optimization, it is possible to narrow the ranges, since one then has a good idea of where to encounter the optimal point. Finally, it is important to consider the experimental error. It is crucial to make the investigation range large enough to allow the effect of each factor to be captured.

For a qualitative factor, which can only assume certain discrete levels, one must specify the exact number of levels for each factor. It is practical to examine a qualitative factor at two-to-five levels, but with many more levels, the number of required experiments increases drastically.

2.5 Problem formulation, Step 3: Specification of responses

The next step in the problem formulation is the specification of responses. Note that some thought may have been given to this point during the familiarization stage. In this step, it is important to select responses that are relevant according to the problem formulation. Many responses are often necessary to map well the properties of a product or the performance characteristics of a process. Also, with modern regression analysis tools, it is not a problem to handle many responses at the same time.

There are three types of responses in MODDE, regular, derived and linked. A *regular* response is a standard response measured and fitted in the current investigation. A *derived* response is an artificial response computed as a function of the factors and/or regular or linked responses. A *linked* response is a response that is invoked in the current application, but was defined in another project. The option of linking responses to the ongoing investigation makes it possible to fit separate models to fundamentally different responses and to optimize them together.

When specifying regular responses one must first decide whether a response is of a quantitative or qualitative nature (see Figure 2.8). Examples of the former category are the breakage of a weld, the amount of soot released when running a truck engine, and the resolution of two analytical peaks in liquid chromatography. Quantitative responses are easier to handle than qualitative, because interpretation of regression models is rendered easier.

Sometimes, however, the experimenter is forced to work with qualitative responses. The general advice here is that one should strive to define a qualitative response in as many levels as possible. Let us say that we are developing a product, and that we are unable to measure its quality with a quantitative measurement technique. The only workable option is a visual quality inspection. Then it is preferable to try to grade the quality in as many levels as is found reasonable. A classification of the product quality according to the five levels worthless/bad/OK/good/excellent is more tractable than a mutilated yes/no type of answer (Figure 2.8). Human quality inspection requires a lot of training, but for instance a well qualified sensory panel can grade the taste and flavor of cheese according to a nine-grade scale.

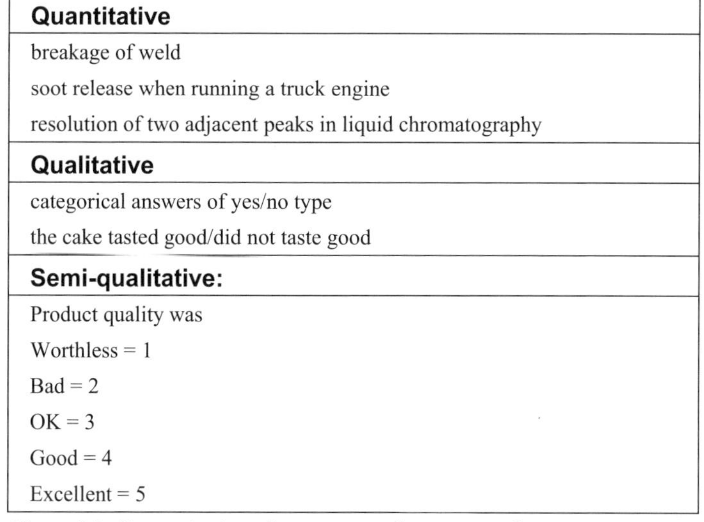

Quantitative
breakage of weld soot release when running a truck engine resolution of two adjacent peaks in liquid chromatography
Qualitative
categorical answers of yes/no type the cake tasted good/did not taste good
Semi-qualitative:
Product quality was Worthless = 1 Bad = 2 OK = 3 Good = 4 Excellent = 5

Figure 2.8: *Categorization of responses and some examples.*

2.6 Problem formulation, Step 4: Selection of model

The next step in the problem formulation is the selection of an appropriate regression model. This selection is an integral part of the problem formulation. We distinguish between three main types of polynomial models, which are frequently used. These are *linear*, *interaction*, and *quadratic* polynomial models (Figure 2.9).

Linear
$y = \beta_0 + \beta_1 x_1 + \beta_2 x_2 + ... + \varepsilon$
Interaction
$y = \beta_0 + \beta_1 x_1 + \beta_2 x_2 + \beta_{12} x_1 x_2 + ... + \varepsilon$
Quadratic
$y = \beta_0 + \beta_1 x_1 + \beta_2 x_2 + \beta_{11} x_1^2 + \beta_{22} x_2^2 + \beta_{12} x_1 x_2 + ... + \varepsilon$

Figure 2.9: *In DOE, we distinguish between three main types of polynomial models, that is, linear, interaction and quadratic regression models.*

Being the most complex model, a quadratic polynomial requires more experiments than the others. An interaction model requires fewer experiments, and a linear model fewer still. The choice of which model to use is *at this stage* of the research not completely free, because part of the choice was made already in the first step of the problem formulation, that is, when the experimental objective was selected. If optimization was selected, only a quadratic model will do. If screening was selected, either a linear or an interaction model is pertinent. We recommend an interaction model, if the number of experiments required is practical. If robustness testing was selected, a linear model might be the appropriate choice. The properties of these models (displayed in Figure 2.9) are discussed in Section 2.17.

Moreover, one may occasionally end up in a situation where a cubic model is necessary. This is especially the case when modelling the performance of living systems, e.g., growth rate of algae as a function of availability of nutritious agents, water temperature, amount of sunlight, and other factors. However, it should be made clear that cubic models are *rarely* relevant in industrial practice, because they are too complex and demand too many experiments.

2.7 Problem formulation, Step 5: Generation of design

Subsequent to the selection of the regression model, the next stage in the problem formulation is the generation of an appropriate experimental design. However, one cannot select an arbitrary experimental design and hope that it will work for a given problem. *The chosen model and the design to be generated are intimately linked.* The MODDE software will consider the number of factors, their levels and nature (quantitative, qualitative, …), and the selected experimental objective, and propose a recommended design, which will well suit the given problem. It is possible to override this proposed design, but this choice should only be exercised by experienced DOE practitioners.

In general, the designs recommended by MODDE will be of the types displayed in Figure 2.10. The upper two rows of the figure show factorial and fractional factorial designs, which are screening designs, and support linear and interaction models. These designs have two investigation levels for each factor, plus an optional number of center-point experiments, here symbolized by the snow-flake. The last row displays composite designs, which are useful for optimization. Composite designs support quadratic models, because every factor will be explored at three or five levels.

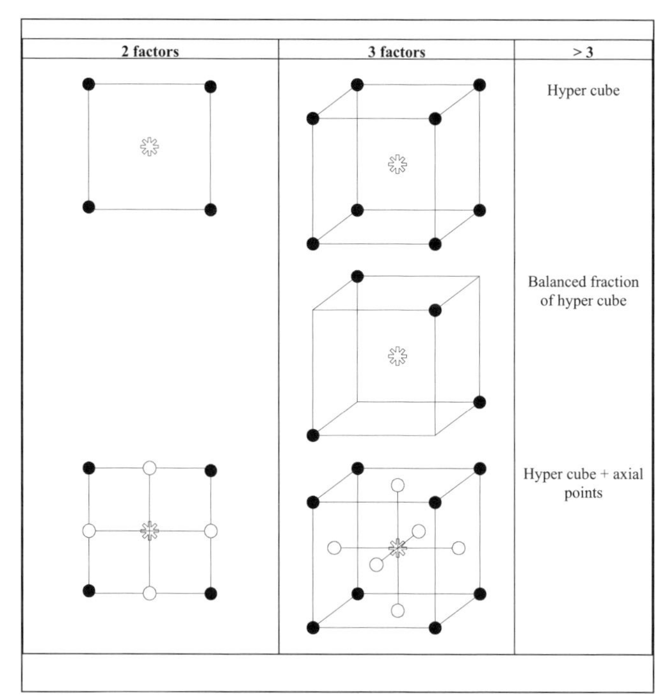

Figure 2.10: *Examples of full factorial, fractional factorial, and composite designs used in DOE.*

It is important to observe that these factorial, fractional factorial, and composite designs have a regular geometry, because the experimental region is regular. As soon as the experimental region becomes irregular, for instance, as a result of one corner being inaccessible experimentally, other types of designs displaying irregular geometries have to be employed. The handling of such irregularities in the factor definition and design generation is discussed in Section 2.13.

2.8 Problem formulation, Step 6: Creation of worksheet

In the problem formulation, the last stage is the creation of the experimental worksheet. The worksheet is, in principle, very similar to a table containing the selected experimental design. However, in the worksheet it is possible to add vital information linked to the actual execution of the experiments. An example worksheet is shown in Figure 2.11.

Worksheet

	1	2	3	4	5	6	7	8	9	10	11
1	Exp No	Exp Name	Run Order	Incl/Excl	Factor1	Factor2	Factor3	Factor4	Factor5	Response1	Response2
2	1	N1	1	Incl	10	50	25	25			
3	2	N2	4	Incl	20	50	25	25			
4	3	N3	10	Incl	10	100	25	25			
5	4	N4	8	Incl	20	100	25	25			
6	5	N5	11	Incl	10	50	75	25			
7	6	N6	6	Incl	20	50	75	25			
8	7	N7	7	Incl	10	100	75	25			
9	8	N8	3	Incl	20	100	75	25			
10	9	N9	2	Incl	15	75	50	25			
11	10	N10	9	Incl	15	75	50	25			
12	11	N11	5	Incl	15	75	50	25			
13	12	N12	12	Incl	12	80	65	25	25	46	53
14	13	N13	13	Incl	20	70	65	25	23	49	60
15	14	N14	14	Incl	15	75	25	25	24	40	89

Figure 2.11: *An example worksheet with extra information. Remember to go through the proposed experiments to make sure that all of them look reasonable, can be performed, and have the potential of fulfilling the experimental objective.*

One important component in the worksheet is the proposed run order, which tells in which randomized order to conduct the experiments. Another appealing feature is the possibility of incorporating a descriptive name for each trial. Such a name might be linked to a page number of the experimenter's hand written laboratory notebook, or a filename on his computer hard disk, or some other reference system, describing conditions and events occurring during the experimental work.

Furthermore, in the worksheet, one may include additional factors beyond the ones that are actually varied. In the example worksheet, the first three factors are controlled, quantitative and manipulable. We see that the fourth factor is constant. It is not part of the design, but is included in the worksheet to help the experimenter. Also, the uncontrollable fifth factor is included in the worksheet, because it is important to record the values of an uncontrollable factor, and during the data analysis to examine whether this factor had an influence on the responses. Finally, it is possible to append to the worksheet supplementary experiments already carried out. This is done in order to consider such information in the data analysis. In the example worksheet, the last three rows illustrate such existing experimental information.

2.9 Summary of the problem formulation phase

Now, we have completed a detailed discussion of the basic aspects of the problem formulation step in DOE. In the next few sections, we expand the discussion regarding (i) the factor and response definitions and (ii) the properties of the various models encountered in the problem formulation.

2.10 Qualitative factors at two levels

A qualitative factor is a categorical variable which is varied in discrete steps, that is, it can only assume certain distinct levels. As an illustration, consider General Example 3 in which the fifth factor, the type of stationary phase in the HPLC system, is a qualitative factor. This factor is said to have two levels, or settings, column type A and type B (see Figure 2.12). Clearly, there is no center level definable in this case, because a hybrid A/B column does not exist. This is graphically illustrated in Figure 2.13, using the factor sub-space defined by the three factors AcN, pH and column type.

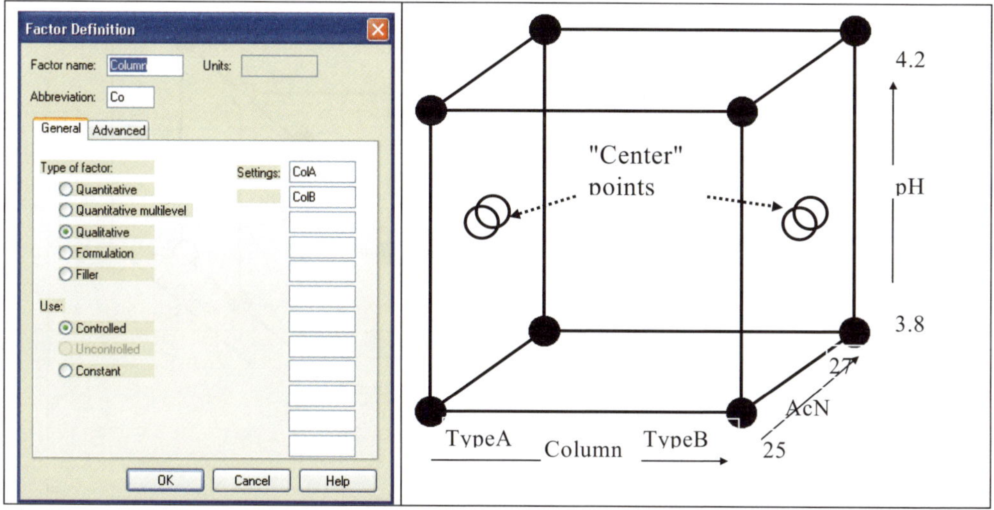

Figure 2.12: *(left) Factor definition of the factor Column of General Example 3.*
Figure 2.13: *(right) A geometrical representation of the design underlying General Example 3.*

All the 12 performed experiments are plotted in Figure 2.13, and we can see that in this three-dimensional subspace they pattern a regular two-level factorial design. Evidently, the experimenter has chosen to make replicated pseudo center-points, located at the centers of the left and right surfaces of the cube. In this way, at least the quantitative factors are varied in three levels, which is desirable when it comes to the data analysis. Another alternative might be to replicate the whole cube once, that is, doing one extra experiment at each corner, but this is more costly in terms of the number of experiments.

Thus, in summary, it is not possible to define technically correct center-points when dealing with two-level qualitative factors, but it is often possible to define centers-of-surfaces which will encode reasonable alternative experiments. In addition, qualitative factors at two levels are easily handled within the standard two-level factorial and fractional factorial framework.

2.11 Qualitative factors at many levels

When a qualitative factor is explored at three or more levels, the factor definition is as simple as with two-level qualitative variables. What becomes more challenging, however, is the task of creating a good experimental design with few runs. Full factorial designs with qualitative factors at many levels are unrealistic. Imagine a situation where two qualitative factors and one quantitative factor are varied. Factor A is a qualitative factor with four levels, Factor B is also qualitative with three settings, and Factor C a quantitative factor changing between –1 and +1. A full factorial design would in this case correspond to 4*3*2 = 24 experiments, which is depicted by all filled and open circles in Figure 2.14.

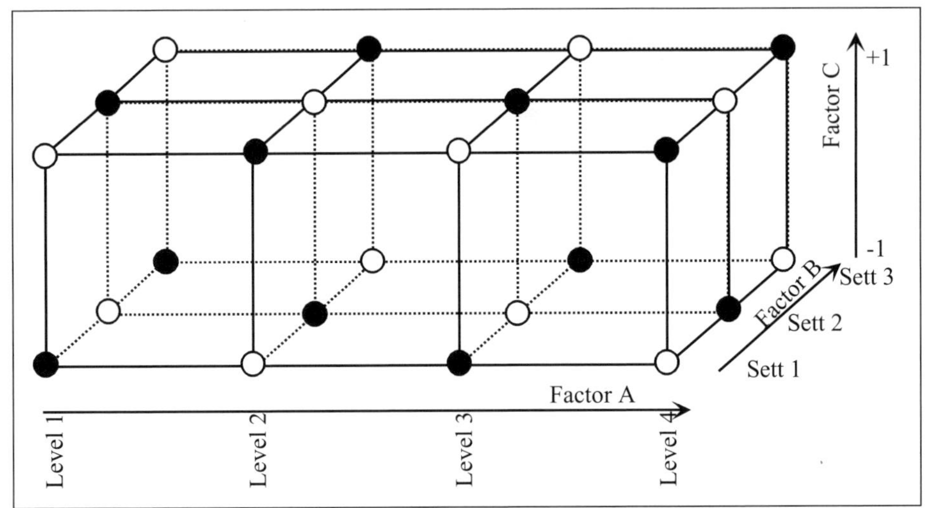

*Figure 2.14: A full 4*3*2 factorial design in 24 experiments.*

With a screening objective and a linear model, 24 runs is unnecessarily high, and an alternative design with fewer experiments is therefore legitimate. Such an alternative design in only 12 experiments is given by the solid circles, or equivalently, by the set of unfilled circles. These subset selections were made using a theoretical algorithm, a D-optimal algorithm, for finding those experiments with the best spread and best balanced distribution. A balanced design has the same number of runs for each level of a qualitative factor. This feature is not absolutely necessary, but often convenient and makes the design a lot easier to understand and evaluate.

With simple applications and few qualitative factors at few levels, the generation of a suitable design may be done by hand using simple geometrical tools like squares and cubes. However, with more complex applications involving several qualitative factors at many levels, the construction of an appropriate design is by no means an easy task. This latter task demands a D-optimal selection of experiments. D-optimal designs are discussed in Chapter 10.

2.12 Qualitative factors in screening, optimization and robustness testing

We now know that qualitative factors at more than two levels require different types of experimental design, other than the regular two-level factorials and fractional factorials. Further, because no center-points are determinable for such factors, replicated experiments cannot be located at the interior part of a design, but must be positioned elsewhere. Also, we would stress that qualitative factors with more than two levels are mainly encountered in screening and robustness testing. This is because in screening the idea is to uncover which setting of a qualitative factor is most favorable, and in robustness testing to reveal how sensitive a response is to changes in such settings. In optimization, however, usually the advantageous level of a qualitative factor will have been found, and there is no need to manipulate such a factor. Thus, most optimization studies will be done by keeping qualitative factors fixed and changing quantitative factors until an optimal point, or a region of decent operability, is detected.

2.13 Regular and irregular experimental regions

We will now consider the concept of the *experimental region*, which is an important aspect of the problem formulation. The experimental region arises as a result of the factor definition, and corresponds to the factor space in which experiments are going to be performed. It is usually easy to comprehend the geometry of such a region by making plots. The simplest experimental regions are those of regular geometry, such as, squares, cubes, and hypercubes, which are linked to quantitative and qualitative factors without experimental restrictions (Figure 2.15).

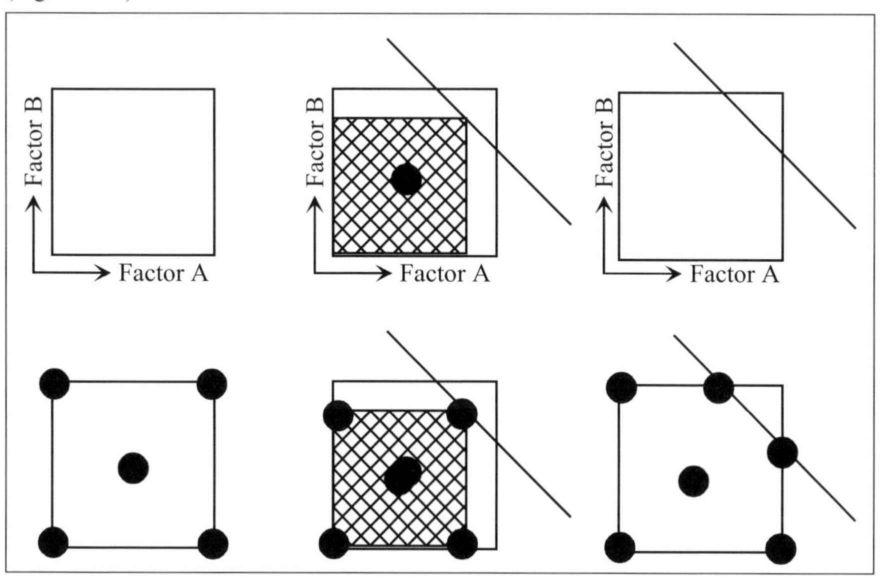

Figure 2.15: *Regular and irregular experimental regions, and some ways to tackle them.*

The left-hand column of Figure 2.15 shows a regular experimental region in two factors, and how it might be mapped with a suitable experimental design. Designs yielding such regular regions are of the factorial, fractional factorial, and composite design types. Sometimes, when the experimenter has a particular requirement concerning the number of experimental runs, other designs like Plackett-Burman and D-optimal designs may be called upon. These are discussed later.

With irregular experimental regions, on the other hand, laying out an appropriate design is not as straightforward. This is exemplified in the middle and right-hand columns of Figure 2.15, where the upper right-hand corner of the experimental region is inaccessible for experimentation. This often happens in industrial practice, e.g., because of unwanted process complications, excessive costs for raw materials, high energy consumption, different process mechanisms, highly toxic by-products, and so on.

In principle, there are two ways of handling an irregular region. The first is to shrink the factor ranges so that the region becomes regular. This is shown by the hatched area in the figure. However, this shrinking is made at the expense of certain parts of the region now being overlooked. The second way is to make use of a D-optimal design, which is well adapted to spreading experiments in an irregular region. This is demonstrated in the right-hand column. Observe that since an irregular region is more complex than a regular one, the former demands more runs than the latter.

We note that in the problem formulation, an irregular experimental region may be defined by specifying linear *constraints,* depicted by the oblique line, among the factors involved. How to specify such factor constraints is discussed below.

2.14 Specifying constraints on factors

A common problem in industry is that experimentation may not be allowed in some portions of the experimental region. It may, for example, not be possible in an experiment to have high temperature and simultaneously low pH, as is illustrated in Figure 2.16. In this case, the lower right corner is going to be ignored, and this must be specified in the problem formulation. This is accomplished by specifying a *constraint*, a linear function, of the two quantitative factors Temperature and pH, and this is depicted by the line chopping off the corner. When an appropriate constraint has been defined it ensures that an irregular region will be considered in the subsequent design generation.

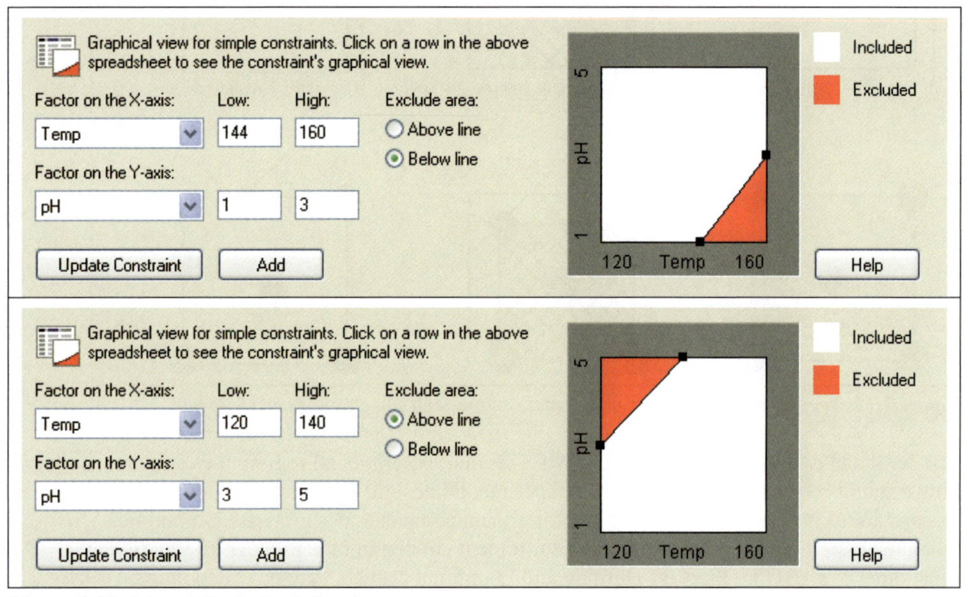

Figure 2.16: *(above) Exclusion below line.*
Figure 2.17: *(below) Exclusion above line.*

In more general terms, a constraint may be used to specify which portion of an experimental region that should either be *included* or *excluded*. MODDE includes a functionality for setting simple linear constraints of quantitative factors. Figure 2.16 exemplifies an exclusion below the line, and Figure 2.17 above the line. It is possible to define more than one constraint in an application, and by using the two example constraints one may generate the D-optimal design shown in Figure 2.18.

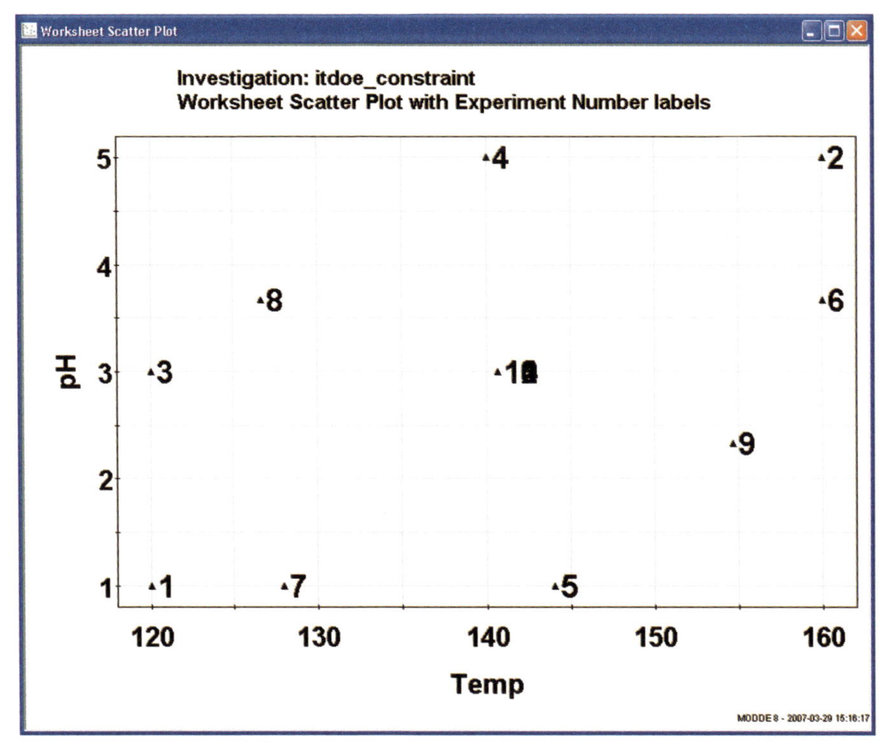

Figure 2.18: A D-optimal design mapping the irregular region arising from the two linear constraints of Figures 2.16 and 2.17.

2.15 Metric of factors

Another feature regarding the factors that must be treated within the problem formulation is how to set the pertinent *metric* of a factor. The metric of a factor is the numerical system chosen for expressing the measured values, e.g., no transformation or the logarithm transformation. Probably the most well-known transformed factor is pH, the negative logarithm of the H^+ concentration in an aqueous solution. In fact, pH is one of the five varied factors in General Example 3, the robustness testing study. Another common factor transformation is the square root.

As seen in Figure 2.19, a correctly used factor transformation may simplify a response function, and hence the regression model, by linearizing the non-linearity.

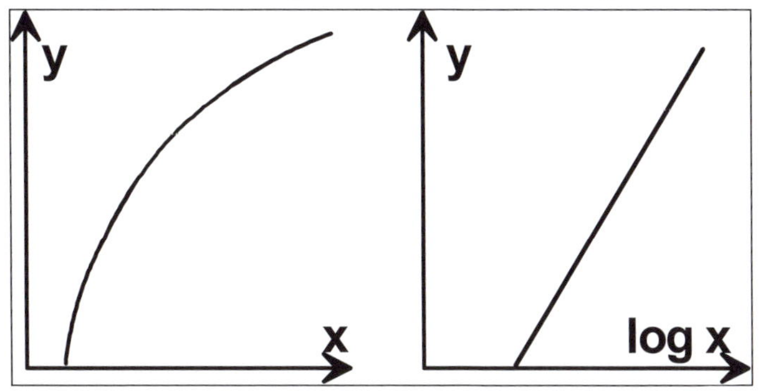

Figure 2.19: A correctly chosen factor transformation may linearize a response function.

It is extremely important that a factor transformation is decided upon *before* the worksheet is created and the experiments are executed, because otherwise the design will get distorted. This is evident in the next two figures. The two factors were initially defined as untransformed (Figure 2.20), but in the data analysis of the resulting data, one factor was log-transformed, an operation completely ruining the symmetry of the original design (Figure 2.21).

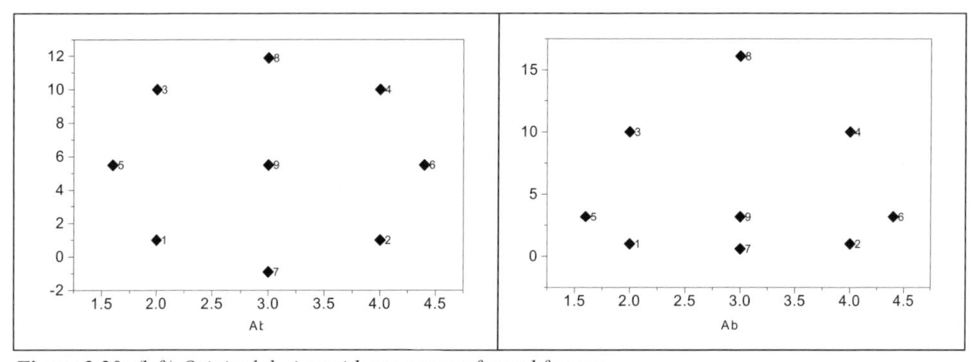

Figure 2.20: *(left) Original design with two untransformed factors.*
Figure 2.21: *(right) Design distortion due to a log-transformation of one of the factors.*

To carry out a factor transformation, one specifies low and high levels as usual, and then the design is created in the transformed unit. For practical purposes, all factor values are then re-expressed in the original unit in the worksheet.

Furthermore, it is necessary to ask the question *when is it relevant to transform a factor*? This is not a simple question, but our general advice is that as soon as it is rational to think in powers of 10 for a factor, that factor should probably be log-transformed. Typical examples are concentrations and amounts of raw materials. Other clues may come from the way a factor is expressed, say, a volume in m^3, where the third-root transformation might be useful. Factor transformations are common in analytical and biological systems.

2.16 Metric of responses

Contrary to the transformation of factors, the transformation of responses is *not* a critical part of the problem formulation. The transformation of a response does *not* affect the shape and geometry of the selected experimental design, and is always performed in the data analysis stage after the completion of all experiments. A properly selected transformation may (i) linearize and simplify the response/factor relationship, (ii) stabilize the variance, and (iii) remove outliers. The linearization capability is graphically illustrated in Figure 2.22. In Chapter 4, we will describe a procedure for discovering the need for response transformation. The MODDE software supports seven kinds of response transformations, including the no transformation option.

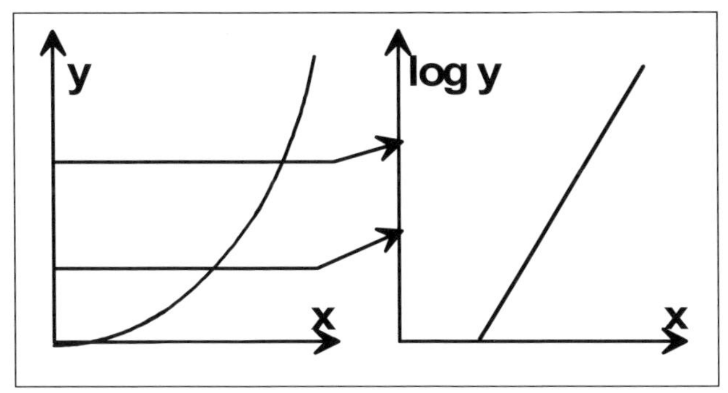

Figure 2.22: A non-linear relationship between y and x, may be linearized by a suitable transformation of y.

2.17 Overview of models

As was discussed in Section 2.6, the selection of a proper regression model is an important part of the problem formulation. It is important to remember that a model is a mathematical representation of the investigated system or process. This model can be viewed as a map which is valid only in the investigated region. It may be risky if a model is used for outside orientation. This was discussed in Chapter 1, where the main emphasis was placed on models as a conceptual phenomenon. In this chapter we will highlight the *geometric* properties of such models.

We recall that with DOE it is possible to establish semi-empirical models, that is, empirical models with some theoretical input, reflecting the underlying true relationship between factors and responses. Linear and interaction models are normally used in screening. Similar models are employed in robustness testing, although linear models are prevalent. Unlike screening and robustness testing, optimization investigations use quadratic models, or RSM models, which are more flexible and adaptable to complex response functions. Other models, such as partly cubic or cubic are sometimes seen in optimization modelling. Regardless of which model is selected for use in DOE, it must be borne in mind that all of them are approximations, simplifications, of a complicated reality. We must avoid being gullible and making over-interpretations of the meaning of a model.

2.17.1 Geometry of linear models

We will start by examining the geometry of linear models and use General Example 3 as an illustration. This example is a robustness testing application, in which five factors are explored in twelve experiments. The experimental plan supports a linear model. In this linear model, each of the five factors occurs as a linear term only. This is shown in Figures 2.23 and 2.24.

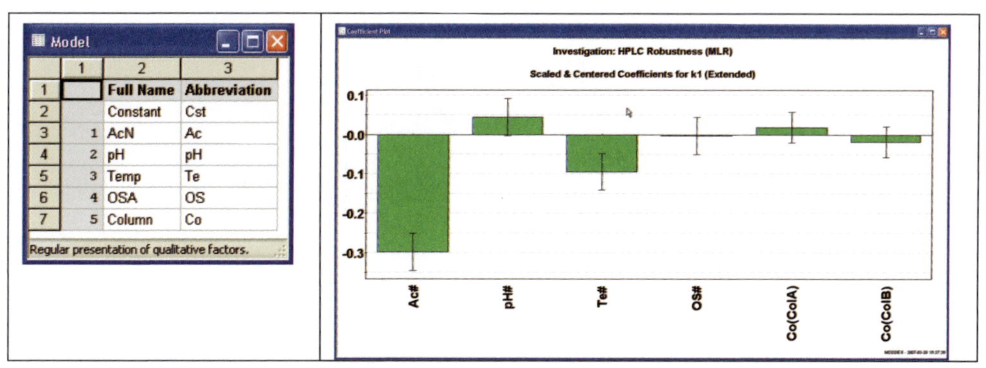

	1	2	3
1		Full Name	Abbreviation
2		Constant	Cst
3	1	AcN	Ac
4	2	pH	pH
5	3	Temp	Te
6	4	OSA	OS
7	5	Column	Co

Regular presentation of qualitative factors.

Figure 2.23: *(left) The linear model of General Example 3.*
Figure 2.24: *(right) Regression coefficients of the linear model of k_1.*

Let us take a closer look at how two of the five factors, acetonitrile and temperature, influence the first response, the k_1 capacity factor. One illustrative way of doing this is to construct either a response contour plot or a response surface plot of k_1, the latter of which is shown in Figure 2.25. We can see that the linear model forms a *plane* in the space defined by the response and the two factors. With more than two factors, the linear model corresponds to fitting a *hyperplane* in the space defined by the response and the factors $x_1, x_2, \ldots, x_k$. Another way to inspect a linear model is to make a so called *main effect* plot of each single factor. Such a plot is shown in Figure 2.26 and this graph applies to the acetonitrile factor. This line may be interpreted as the front edge of the surface seen in the surface plot (Figure 2.25). The interpretation of Figure 2.26 suggests that when the amount of acetonitrile in the mobile phase is increased from 25 to 27%, the k_1 capacity factor decreases by almost 0.6 unit, from 2.3 to 1.7.

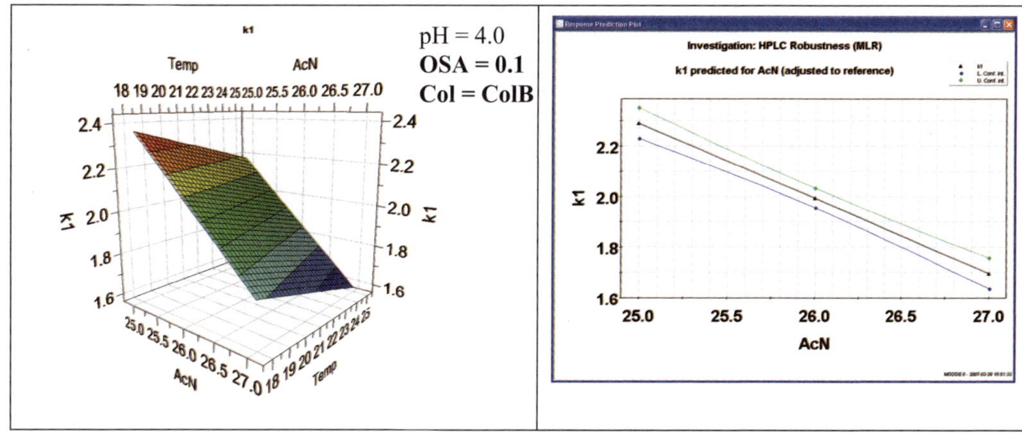

Figure 2.25: *(left) Response surface plot of k_1. The varied factors are acetonitrile and temperature.*
Figure 2.26: *(right) Main effect plot of acetonitrile with regards to k_1.*

In addition, there is another feature visible in Figure 2.24, which needs to be highlighted. In this coefficient plot, each quantitative factor is represented by one bar, whereas the impact of the last factor, type of HPLC column, is depicted by two bars. This has to do with the fact that the latter is a qualitative factor in two levels. Since there is no unique center-point for this qualitative factor, one obtains as many bars in the coefficient plot as there are levels of the qualitative factor. The interpretation of the HPLC model indicates that higher k_1 capacity factors are acquired with column type A than type B.

2.17.2 Geometry of interaction models

An interaction model is more complex than a linear model, and may therefore fit more intricate response functions. To understand the geometry of an interaction model, we shall use General Example 1, which is a screening application carried out in two steps. In the first step of this application, the four factors Power, Speed, NozzleGas and RootGas were varied using 11 experiments. The details pertaining to the data analysis are given later in the course book (Chapter 5), and for now we shall focus on the second response, the Width of the weld. It was found that one of the main effects, NozzleGas, had no influence on Width. Further, only one two-factor interaction, the Power*Speed term, was meaningful. Hence, the model obtained is not a full interaction model, but a partial interaction model. Figure 2.27 shows the model composition and Figure 2.28 the model shape.

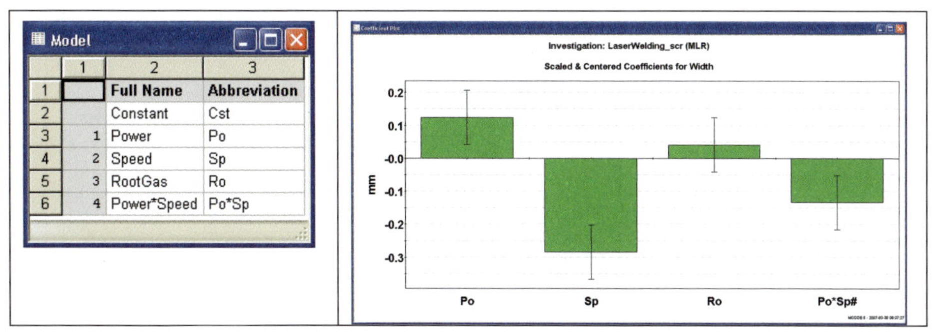

	1	2	3
1		Full Name	Abbreviation
2		Constant	Cst
3	1	Power	Po
4	2	Speed	Sp
5	3	RootGas	Ro
6	4	Power*Speed	Po*Sp

Figure 2.27: *(left) The interaction model of General Example 1.*
Figure 2.28: *(right) Regression coefficients of the interaction model of Width.*

Unlike in a linear model, the surface in the response surface plot of an interaction model (Figure 2.29) is no longer an undistorted plane but a *twisted* plane. Because of the presence of the two-factor interaction, the surface is twisted. This two-factor interaction may be further explored by means of the interaction plot (Figure 2.30), which may be thought of as displaying the front and rear edges of the surface in Figure 2.29. We can see that when changing Power from low to high level, we get a larger impact on the response when the second factor Speed is at its low level rather than its high level. This is the verbal definition of a two-factor interaction: *the effect of one factor depends on the level of the other factor.*

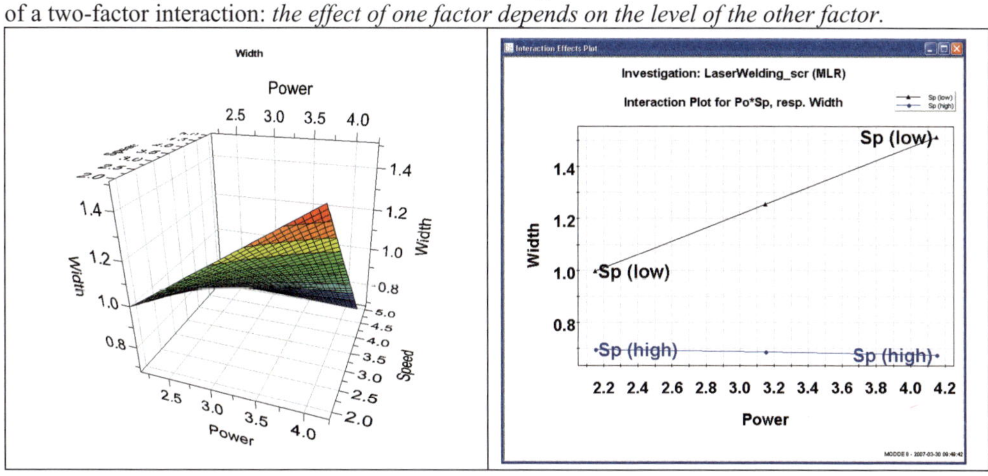

Figure 2.29: *(left) The twisted response surface of Width. The varied factors are Power and Speed.*
Figure 2.30: *(right) Interaction effect plot of Power and Speed with regards to Width.*

2.17.3 Geometry of quadratic models

In order to comprehend the geometry of a quadratic model we shall use General Example 2. This is an optimization investigation of the combustion in a truck engine. Three factors were varied according to a composite design incorporating 17 runs. This design supports a full quadratic polynomial model. Figure 2.31 shows the best model obtained when simultaneously considering the three responses Fuel, NOx, and Soot. Figure 2.32 displays the model appearance with regards to the Fuel response. Apparently the factors Air and NeedleLift dominate the model with regard to Fuel, and a rather large squared term for Air is detected.

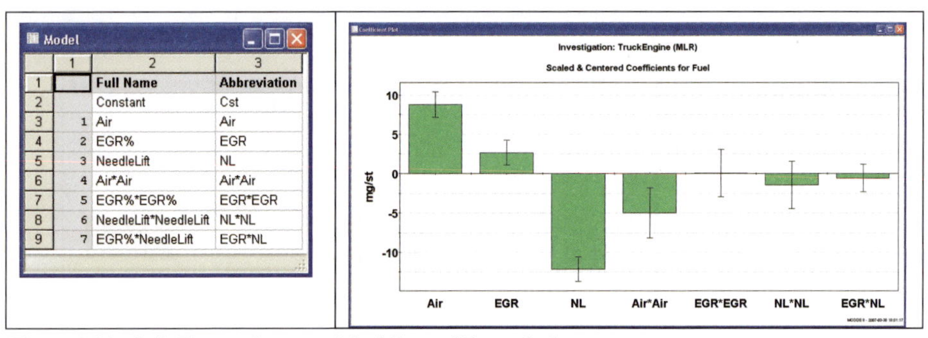

	1	2	3
1		Full Name	Abbreviation
2		Constant	Cst
3	1	Air	Air
4	2	EGR%	EGR
5	3	NeedleLift	NL
6	4	Air*Air	Air*Air
7	5	EGR%*EGR%	EGR*EGR
8	6	NeedleLift*NeedleLift	NL*NL
9	7	EGR%*NeedleLift	EGR*NL

Figure 2.31: *(left) The quadratic model of General Example 2.*
Figure 2.32: *(right) Regression coefficients of the quadratic model of Fuel.*

Further, Figure 2.33 shows the response surface plot of Fuel obtained when varying Air and NeedleLift. This surface is now *curved*, and the possibility of modelling curvature is due to the presence of the quadratic terms. In order to diminish the volume of consumed fuel, the truck engine should be run with a high value of NeedleLift and a low value of Air, according to the model interpretation. Another way of examining the model curvature is to make main effect plots of the separate factors. One such plot, pertaining to the combination Air and Fuel, is provided in Figure 2.34. This plot also indicates that in order to decrease Fuel, the amount of Air used in combustion ought to be lowered.

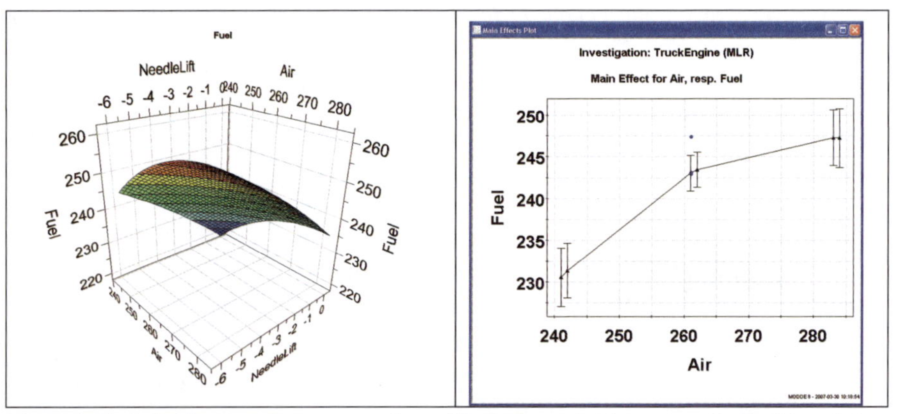

Figure 2.33: *(left) The curved response surface of Fuel. The varied factors are NeedleLift and Air.*
Figure 2.34: *(right) Main effect plot of Air with regard to Fuel.*

2.17.4 Generation of adequate designs

In the problem formulation it is often a question of making an "educated guess" when selecting an appropriate model. As a rule-of-thumb, one can select a linear model for robustness testing, a linear or an interaction model for screening, and a quadratic model for optimization. This model selection then strongly influences the design selection. It is instructive to overview how models and designs are linked, and this is graphically summarized in Figure 2.35.

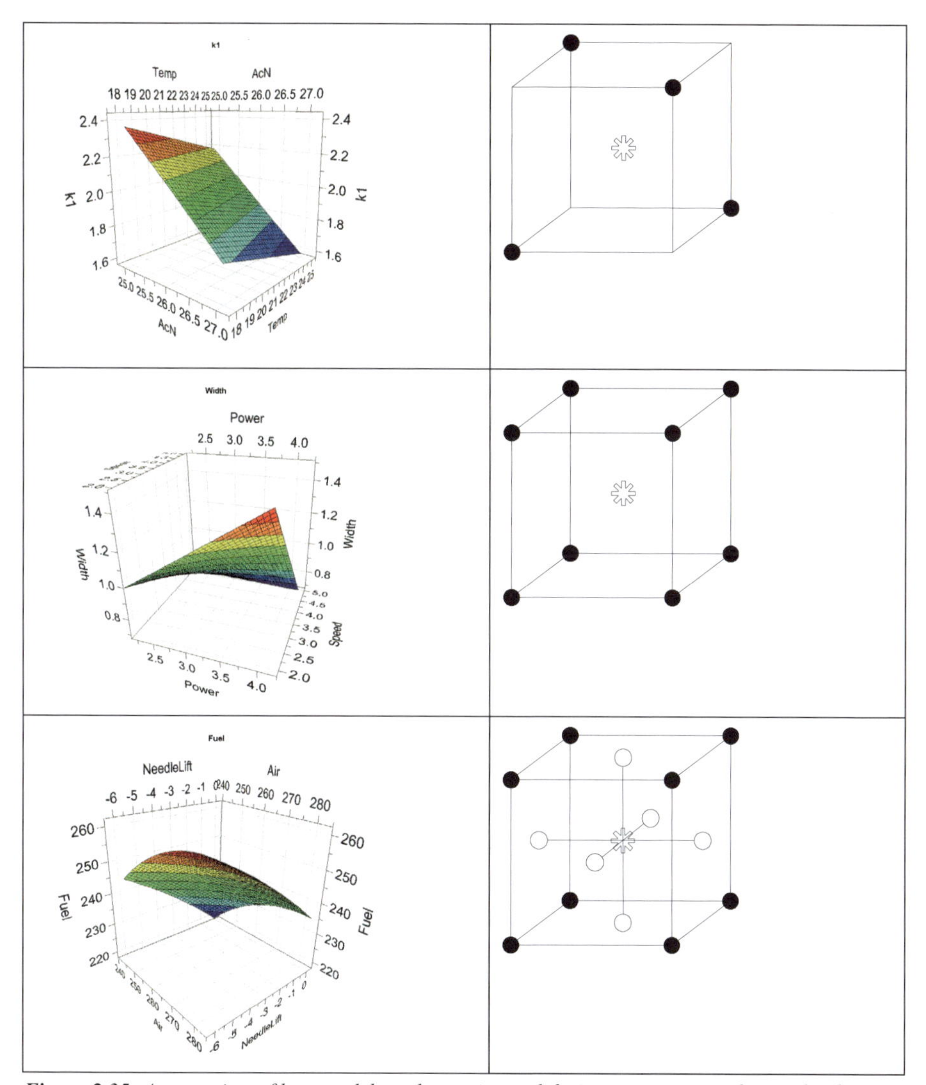

Figure 2.35: An overview of how models and experimental designs are connected to each other.

The first row of Figure 2.35 shows the *plane* surface of the linear robustness testing model, and a simplified version of the corresponding underlying fractional factorial design. For display purposes, the design is drawn in only three factors, although the application itself comprises five factors. The fractional factorial design encodes only a limited number of experiments, because not all corners are investigated. In the second row of the figure, the *twisted* plane of the interaction model of the screening application is shown, together with a complete factorial design in three factors. We can see that an interaction model requires more experiments than a linear model. As far as interaction models are concerned, they may sometimes also be supported by designs drawn from the fractional factorials family. This is actually the case in the current screening example. Finally, the third row of Figure 2.35 displays the relationship between the *curved* surface of the optimization application and the underlying composite design. The quadratic optimization model demands most experiments.

In reality, it is the data analysis that will reveal whether the initial "educated guess" of model was appropriate. Should it turn out that the experimenter did underestimate the model complexity, it is possible to modify the model and augment the design. For instance, the plotted fractional factorial design may be expanded to the full factorial design, which, in turn, may be augmented to the composite design. With few factors the expansion of a design is easy, with many factors more taxing. In this context, one may talk about *updating* and *upgrading* of models. By updating we mean the addition of a *limited number* of well-identified model terms, for instance, a two-factor interaction term to a linear model. By upgrading we mean the addition of a *set* of model terms, converting, for instance, a linear model to a quadratic model. This will be discussed later (see Chapter 10).

2.18 Questions for Chapter 2

- Which six points are important in the problem formulation?
- Which are the six stages in the experimental process?
- Of these six stages, three are more important as experimental objectives than the others. Which?
- What is important regarding the responses at the familiarization stage?
- What is the objective in the screening stage?
- What kind of polynomial model is sufficient for screening?
- How are gradient techniques applied in the search for the optimal experimental region?
- What is the main objective in the optimization stage?
- What is an RSM model? How can it be interpreted?
- What are the main differences between screening and optimization?
- When is robustness testing relevant?
- What are controllable and uncontrollable factors?
- Why is it important to monitor uncontrollable factors?
- What are process and mixture factors?
- What are the differences between quantitative and qualitative factors?
- What is the minimum number of levels which must be defined for a quantitative factor?
- How many levels are reasonable for investigating the impact of a qualitative factor?
- Why are quantitative responses more useful than qualitative ones?
- When is a linear model applicable? Interaction model? Quadratic model?
- How many levels of each factor is specified by a composite design?
- Which kind of model is supported by a composite design?
- What is the geometric property of a linear model?

- What is the geometric property of an interaction model?
- What is the geometric property of a quadratic model?
- Which type of design is typically linked to a linear model? Interaction model? Quadratic model?
- Why is it impossible to define replicated center-point experiments when working with qualitative factors?
- What are the alternative approaches for creating replicated experiments?
- What are the problems to consider when working with qualitative factors at many levels?
- What is a balanced design in the context of qualitative factors?
- How many levels of qualitative factors are normally investigated in screening? In optimization? In robustness testing?
- What is the difference between regular and irregular experimental regions?
- When may irregular experimental regions occur?
- Which type of design is useful for irregular regions?
- What is a constraint?
- What is meant by the *metric* of factors and responses?
- Which technique is usually used for changing the metric of factors?
- Why does the transformation of factors have a decisive impact in the problem formulation?

2.19 Summary and discussion

In this chapter, a detailed account of the problem formulation step in DOE was given. The problem formulation is composed of six steps, (i) selection of experimental objective, (ii) specification of factors, (iii) specification of responses, (iv) selection of regression model, (v) generation of design, and (vi) creation of worksheet. To understand which are the relevant experimental objectives, a discussion regarding six stages in DOE was carried out. The six highlighted stages were (i) familiarization, (ii) screening, (iii) finding the optimal region, (iv) optimization, (v) robustness testing, and (vi) mechanistic modelling.

In this chapter, we have also scrutinized the geometric properties of linear, interaction and quadratic polynomial models, which are the kind of models typically used in DOE. A linear model needs a fractional factorial design, an interaction model either a full or a fractional factorial design, and a quadratic model a composite design. When making use of response surface plotting, a linear model corresponds to an undistorted flat or inclined plane, an interaction model to a twisted plane, and a quadratic model to a curved surface.

Additionally, this chapter has been devoted to a deeper treatment of factor and response specifications in the problem formulation. Qualitative factors at two or more levels have been addressed, and it has been shown how replicated experiments and balancing of D-optimal designs may be achieved. Further, some attention was given to the experimental region concept, and regular and irregular regions were contrasted. In this context, the concept of linear constraints among quantitative factors was described. Factor constraints are useful for demarcating portions of the experimental region where experiments are undesired. Lastly, parts of this chapter explained the metric of factors and responses, and the need to specify the factor metric *prior* to the design generation was strongly underlined.

3 Factorial designs

3.1 Objective

Full factorial designs form the basis for all classical experimental designs used in screening, optimization, and robustness testing. Hence, a good understanding of these designs is important, because this will make it easier to understand other related, and more commonly used, experimental designs. In this chapter we will introduce two-level factorial designs. The aim is to consider their use and construction.

The aim of this chapter is also to define the main effect of a factor and the interaction effect between two factors, and to illustrate how these may be displayed graphically. An additional aim is to elucidate some elementary ways of analyzing data from full factorials by means of simple arithmetic. This is done in order to increase the understanding of basic concepts. We will also briefly discuss the relationship between effects and regression coefficients. In order to facilitate this discussion, a brief introduction to least squares analysis is given.

3.2 Introduction to full factorial designs

In this chapter, we shall consider two-level full factorial designs. Such factorial designs support interaction models and are used in screening. They are important for a number of reasons:

- they require relatively few runs per investigated factor
- they can be upgraded to form composite designs, which are used in optimization
- they form the basis for two-level *fractional* factorial designs, which are of great practical value at an early stage of a project
- they are easily interpreted by using common sense and elementary arithmetic

Factorial designs are regularly used with two-to-four factors, but with five or more factors the number of experiments required tends to be too demanding. Hence, when many factors are screened, fractional factorial designs constitute a more appealing alternative.

3.3 Notation

To perform a general two-level full factorial design, the investigator has to assign a low level and a high level to each factor. These settings are then used to construct an orthogonal array of experiments. There are some common notations in use to represent such factor settings. Usually, the low level of a factor is denoted by –1 or just –, and the high level by +1 or simply +. As a consequence, the center level, usually chosen for replication, will be denoted by 0. As seen in Figure 3.1, these alternatives are called standard and extended notation. Both these notations are said to operate in a coded, -1 to +1, unit.

Item	Low	High	Center
Standard notation	–	+	0
Extended notation	–1	+1	0
Example; temperature	100°C	200°C	150°C
Example; pH	7	9	8
Example; Catalyst (A, B)	A	B	N/A

Figure 3.1: *Standard and extended notations for factor settings.*

For simple systems, such as the two factor situation sketched in Figure 3.2, it may be convenient to display the coded unit together with the original factor unit. In this example, we easily see that when using catalyst A and raising the temperature from 100°C to 200°C, the yield is enhanced.

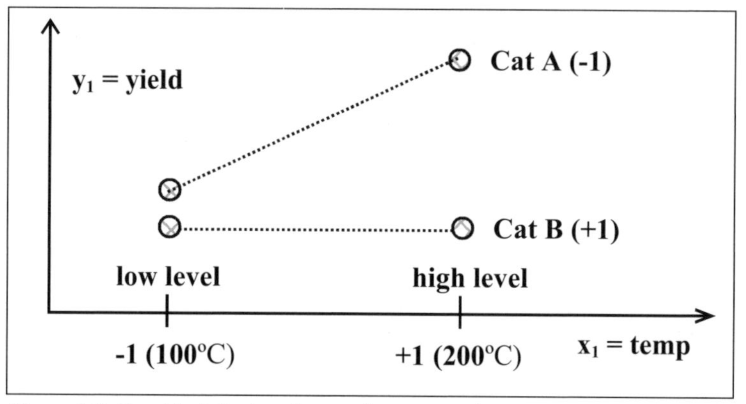

Figure 3.2: *Graphical display of a two-factor example.*
1. The low level of a factor will be denoted by -1 or just -;
2. The high level of a factor will be denoted by +1 or just +;
3. Consequently, the center is denoted by 0.

3.4 The 2^2 full factorial design – construction & geometry

The 2^2 full factorial design is the simplest of its kind, and the 2^2 nomenclature is understood as a *two-level* design in *two* factors. We shall consider a simple example, known as "ByHand", to illustrate the principles of a two-level factorial design. This example stems from the field of organic chemistry and deals with the reduction of an enamine. Enamines are reduced by formic acid to saturated amines. The experimenter decided to vary two factors (Figure 3.3). One factor, x_1, was the molar ratio of the two reacting compounds, formic acid and enamine. As seen in Figure 3.3, this ratio was varied between 1 and 1.5. The second factor, x_2, was the reaction temperature, which was varied between 25°C and 100°C. In order to monitor the success of the reaction, three responses were measured. We will only focus on one response, here called y_3, the formation of the desired product, which should be maximized (Figure 3.3).

Factors		Levels		
		- (1)	0	+ (1)
X_1	Amount formic acid/enamine (mole/mole)	1.0	1.25	1.5
X_2	Reaction temperature (°C)	25	62.5	100
Response				
Y_3	The desired product %			

Figure 3.3: *The two factors and the response considered in the ByHand example.*

With two factors and two levels of each, there are four possible factor-combinations, that is, low-low, high-low, low-high, and high-high, which correspond to the four first rows of the design table shown in Figure 3.4. In addition, three replicated experiments carried out at the center of the experimental region have been added. Such experiments are typically termed *center-points*, since they are located midway between the low and high levels.

	Factors		Factors		Response
	Original unit		Coded unit		%
Exp. no	x_1	x_2	x_1	x_2	y_3
1	1	25	-	-	80.4
2	1.5	25	+	-	72.4
3	1	100	-	+	94.4
4	1.5	100	+	+	90.6
5	1.25	62.5	0	0	84.5
6	1.25	62.5	0	0	85.2
7	1.25	62.5	0	0	83.8

Figure 3.4: *The 2^2 factorial design of the ByHand example.*

Geometrically, the experimental design created may be interpreted as a square, and hence the experimental region is said to be of regular geometry (Figure 3.5). The important point is that each row in the experimental design (Figure 3.4) corresponds to one experiment, and may be interpreted as a point in the two-dimensional factor space (Figure 3.5).

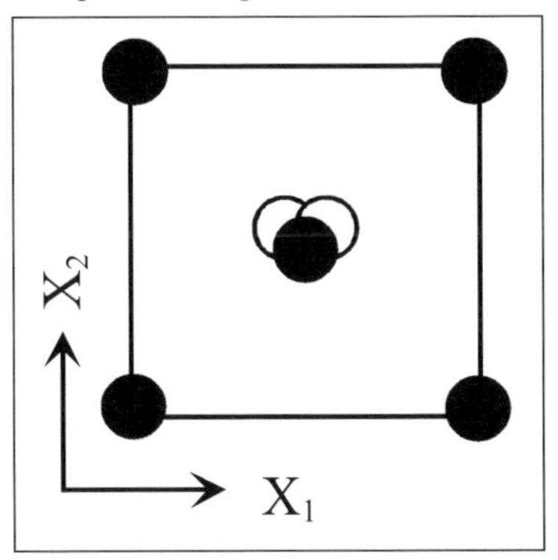

Figure 3.5: *A geometrical representation of the factorial design used in the ByHand example.*

3.5 The 2^3 full factorial design – construction & geometry

The two-level full factorial design in three factors, denoted 2^3, is constructed analogously to the full factorial design in two factors. As an illustration, we shall use the CakeMix application. In the construction of the 2^3 design, the experimenter first has to assign to each investigated factor a low and a high level. Then the actual *design matrix* is created as follows (Figure 3.6). The first and leftmost column of the design matrix is created by writing minus and plus signs alternatingly in eight rows, simply because the factorial part of the design will have $2^3 = 8$ rows. Subsequently, the second column is created by successive pairs of minus and plus signs continuing until the first eight rows of the table have entries. Finally, the third column of the design matrix is generated by doubling the number of signs in a sequence, and then alternately laying out such sequences. This implies that the third column starts with four minus signs followed by four plus signs. The design that we have just created is said to be written in *standard order*.

Design Matrix				Experimental matrix			
Exp No	Flour	Shortening	Egg	Flour	Shortening	Egg	Taste
1	-	-	-	200	50	50	3.52
2	+	-	-	400	50	50	3.66
3	-	+	-	200	100	50	4.74
4	+	+	-	400	100	50	5.2
5	-	-	+	200	50	100	5.38
6	+	-	+	400	50	100	5.9
7	-	+	+	200	100	100	4.36
8	+	+	+	400	100	100	4.86
9	0	0	0	300	75	75	4.68
10	0	0	0	300	75	75	4.73
11	0	0	0	300	75	75	4.61

Figure 3.6: *The 2^3 factorial design of the CakeMix example.*

In addition to the eight experiments of the factorial part, it is recommended to incorporate replicated experiments, usually carried out as center-points. In the CakeMix case, three center-points were added, denoted with zeroes in the design matrix (rows 9-11 in Figure 3.6). In order to facilitate the experimental work, the *design matrix* may also be converted to the *experimental matrix* by inserting the original factor settings. This is also shown in Figure 3.6. Every row in the design table, that is, each experiment, represents a point in the three-dimensional experimental space. Obviously, the experimental region in the CakeMix case is a cube of regular geometry (Figure 3.7).

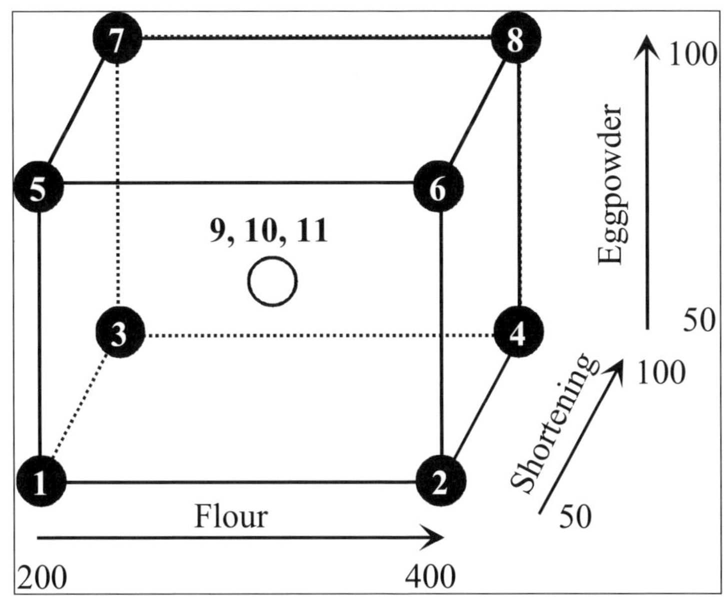

Figure 3.7: The three-dimensional experimental region of the CakeMix example.

3.6 The 2^4 and 2^5 full factorial designs

We will now consider the construction of the 2^4 and 2^5 full factorial designs. Fortunately, the procedure described in the foregoing paragraph is useful for this purpose. First we consider the 2^4 design. The trick is to compute the number of necessary rows, which in this case is $2^4 = 16$, and then lay out the first and leftmost column by means of a series of alternating minus and plus signs (Figure 3.8). To complete the second and third columns we adhere to the already described procedure and fill them with 16 entries. Finally, the fourth column is created by first writing eight minus signs and then eight plus signs. This design matrix is indicated by the blue area in Figure 3.8.

	X 1	X 2	X 3	X 4	X 5
1	-	-	-	-	-
2	+	-	-	-	-
3	-	+	-	-	-
4	+	+	-	-	-
5	-	-	+	-	-
6	+	-	+	-	-
7	-	+	+	-	-
8	+	+	+	-	-
9	-	-	-	+	-
10	+	-	-	+	-
11	-	+	-	+	-
12	+	+	-	+	-
13	-	-	+	+	-
14	+	-	+	+	-
15	-	+	+	+	-
16	+	+	+	+	-
17	-	-	-	-	+
18	+	-	-	-	+
19	-	+	-	-	+
20	+	+	-	-	+
21	-	-	+	-	+
22	+	-	+	-	+
23	-	+	+	-	+
24	+	+	+	-	+
25	-	-	-	+	+
26	+	-	-	+	+
27	-	+	-	+	+
28	+	+	-	+	+
29	-	-	+	+	+
30	+	-	+	+	+
31	-	+	+	+	+
32	+	+	+	+	+

Figure 3.8: *The 2^4 (blue area) and 2^5 factorial designs.*

Geometrically, the 2^4 design corresponds to a regular hypercube with four dimensions. In fact, the laser welding application may be used as an illustration of the 2^4 design. Although this application was carried out in two steps, the combined factorial parts form the 2^4 design. This is shown in Figure 3.9, but note that for clarity the performed center-points have been omitted.

	1	2	3	4	5	6	7	8
1	Exp No	Exp Name	Run Order	Incl/Excl	Effect	Speed	NozzleGas	RootGas
2	1	N1	3	Incl	2.15	1.875	27	27
3	2	N2	16	Incl	4.15	1.875	27	27
4	3	N3	12	Incl	2.15	5	27	27
5	4	N4	14	Incl	4.15	5	27	27
6	5	N5	9	Incl	2.15	1.875	36	27
7	6	N6	5	Incl	4.15	1.875	36	27
8	7	N7	6	Incl	2.15	5	36	27
9	8	N8	1	Incl	4.15	5	36	27
10	9	N9	10	Incl	2.15	1.875	27	42
11	10	N10	11	Incl	4.15	1.875	27	42
12	11	N11	8	Incl	2.15	5	27	42
13	12	N12	15	Incl	4.15	5	27	42
14	13	N13	13	Incl	2.15	1.875	36	42
15	14	N14	4	Incl	4.15	1.875	36	42
16	15	N15	2	Incl	2.15	5	36	42
17	16	N16	7	Incl	4.15	5	36	42

Figure 3.9: *The combined factorial parts of the laser welding experiments form a 2^4 factorial design.*

Further, the 2^5 design is constructed similarly to the 2^4 design, except that the design matrix now has 32 rows and five columns. This design matrix is shown in Figure 3.8, and geometrically corresponds to a regular five-dimensional hypercube. The 2^5 full factorial design is not used to any great extent in industrial practice, because of the large number of experimental trials. Instead, there exists an efficient fractional factorial design in 16 runs, which is almost as good as the full factorial counterpart. Also observe that to the 2^4 and 2^5 designs three to five center-points are normally added.

3.7 Pros and cons of two-level full factorial designs

With two-level full factorial designs it is possible to estimate interaction models, which well serve to fulfill the objectives underlying screening. This type of design consists of a *set* of experimental runs in which each factor is investigated at both levels of all the other factors. In other words, the two-level full factorial design corresponds to a *balanced* and *orthogonal* arrangement of experiments. This arrangement enables the effect of one factor to be assessed independently of all the other factors. With k investigated factors the two-level full factorial design has $N = 2^k$ runs. In Figure 3.10, we have tabulated the number of runs for k = 2 to k = 10. It must be observed that this compilation does *not* take any replicated experiments into account.

No of investigated factors (k)	No of runs Full factorial	No of runs Fractional factorial
2	4	---
3	8	4
4	16	8
5	32	16
6	64	16
7	128	16
8	256	16
9	512	32
10	1024	32

Figure 3.10: *An overview of the number of experiments encoded by two-level full and fractional factorial designs in $k = 2$ to $k = 10$ factors.*

As seen in Figure 3.10, with more than five factors the number of experiments increases dramatically, so that only the full factorials with two to four factors are realistic choices. The rightmost column lists the more manageable number of experiments required by designs of the two-level fractional factorial family. For example, it is realistic to screen ten factors in 32 experiments, plus some optional replicates. In summary, two-level full factorial designs are experimentally practical and economically defendable only when considering few factors. With more than four factors, a switch to fractional factorial designs is more favorable.

3.8 Main effect of a factor

We will now describe the main effect of a factor. Consider the graph in Figure 3.11. This is a graphical representation of how a response, y_1, may change due to changing values of the factor, x_1. The response might be the taste of a cake as in the CakeMix case, and the factor the amount of flour consumed. The main effect of a factor is defined as the change in the response due to varying one factor from its low level to its high level, and keeping the other factors at their average. Thus, in our case, the main effect of flour is the change in taste when increasing the amount of flour in the cake mix recipe from 200g to 400g, and keeping shortening and egg powder at their center level of 75g.

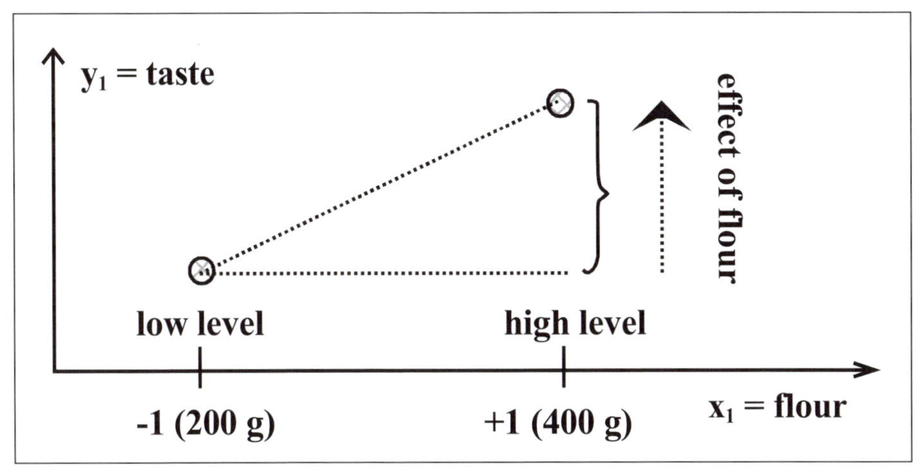

Figure 3.11: Illustration of the main effect of a factor, e.g., flour in the cake mix recipe.

Moreover, we can see in Figure 3.12 that the main effect of flour amounts to approximately 0.4 unit, from taste 4.5 to 4.9. Since the experimental plan in the CakeMix case is a two-level full factorial design, all factors are varied simultaneously and in a systematically balanced manner. This allows the estimation of a factor's main effect independently of the other factors.

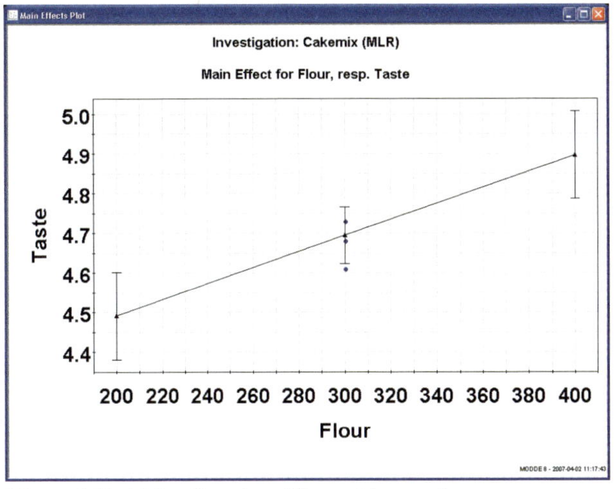

Figure 3.12: Main effect plot of flour with regard to taste.

3.9 Computation of main effects in the 2^2 case

With a 2^2 factorial design two main effects may be calculated. To illustrate graphically and arithmetically how this is carried out, we shall revisit the ByHand example introduced in Section 3.4. Recall that this example has two factors, the molar ratio of formic acid/enamine and the reaction temperature, and that focus was given to the response reflecting the yield of the desired product. We may ask ourselves, *what is the main effect of the molar ratio on the yield of the desired product*? One manner in which to compute and geometrically understand this main effect is given in Figure 3.13. There are two estimates of the impact of changing the molar ratio from 1.0 to 1.5. One estimate, called Δ1,2, indicates that the yield will decrease by 8%. Another estimate, denoted Δ3,4, suggests that the yield will drop only 3.8%.

To compute the *main effect* of the molar ratio we simply average these two estimates, that is, the main effect of the molar ratio is –5.9%, which is plotted in Figure 3.14.

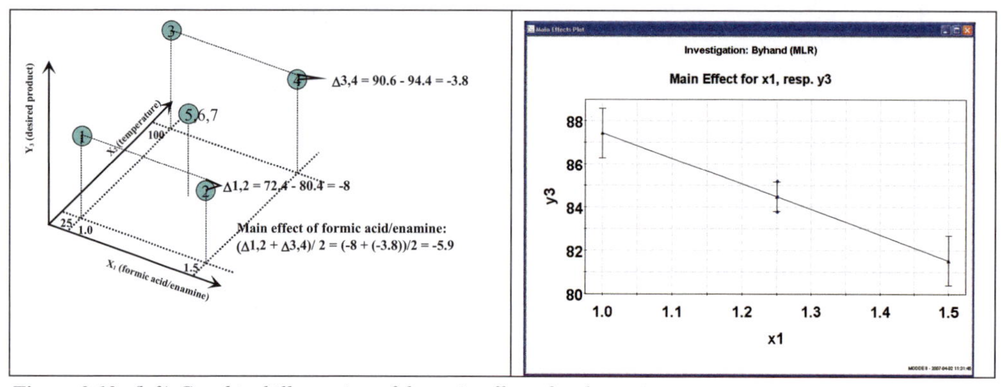

Figure 3.13: *(left) Graphical illustration of the main effect of molar ratio.*
Figure 3.14*: (right) Main effect plot of molar ratio.*

If we want to compute the main effect of the other factor, the reaction temperature, we proceed similarly but along the temperature axis. This is portrayed in Figure 3.15. We can see that the two estimates, called Δ1,3 and Δ2,4, of the temperature effect, are 14.0 and 18.2, respectively. The average of these two values, 16.1, corresponds to the main effect of the temperature, that is, the change in yield when raising the temperature from 25ºC to 100ºC. This main effect is plotted in Figure 3.16.

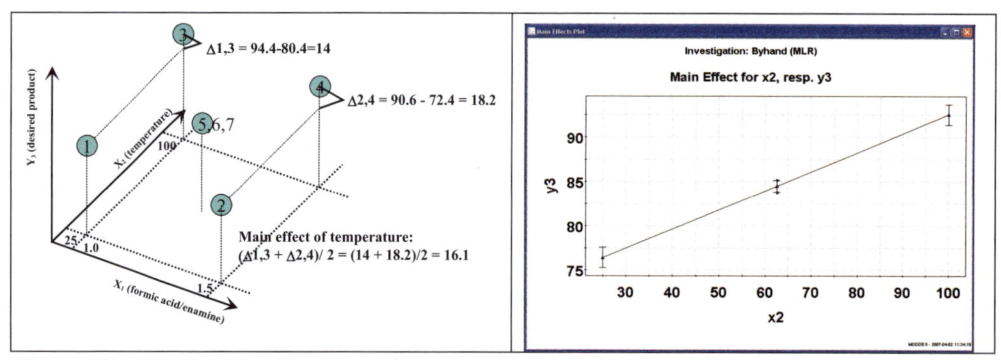

Figure 3.15: *(left) Graphical illustration of the main effect of temperature.*
Figure 3.16: *(right) Main effect plot of temperature.*

3.10 A second method of understanding main effects

A second way of understanding and computing the main effects of the molar ratio and the reaction temperature is to make pair-wise comparisons of the edges of the design square. Figure 3.17 shows how this may be done to derive the main effect of the molar ratio. By forming the average of the right side, 81.5, and subtracting from this value the average of the left side, 87.4, one obtains –5.9, which is the estimate of the main effect of the molar ratio. Compare with the main effect plotted in Figure 3.14. Analogously, the main effect of the reaction temperature may be formed by comparing the rear edge with the front edge. As seen in Figure 3.18, the rear edge average is 92.5 and the front edge average 76.4, which gives the main effect of 16.1. Compare with the temperature main effect plotted in Figure 3.16. Notice

(1) that all four corner experiments are used to supply information on each of the main effects, and (2) that each effect is determined with the precision of a two-fold replicated difference.

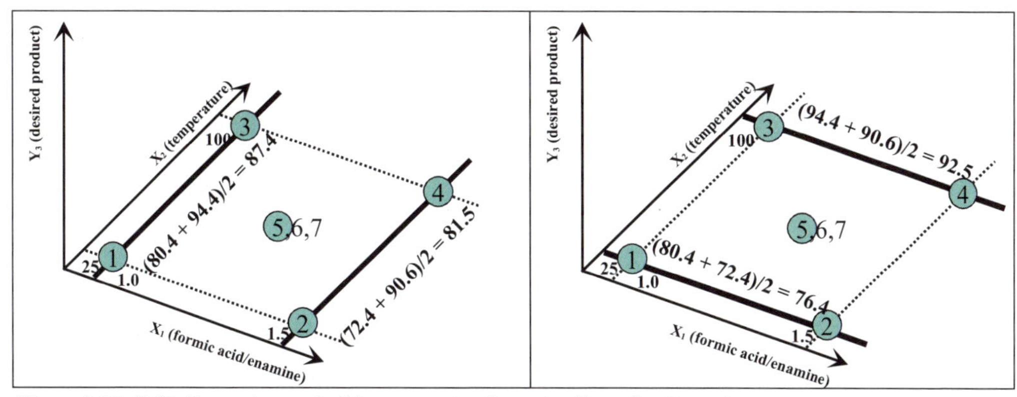

Figure 3.17: *(left) Alternative method for computing the main effect of molar ratio.*
Figure 3.18: *(right) Alternative method for computing the main effect of reaction temperature.*

3.11 A quicker by-hand method for computing effects

We will now examine a quicker method for computing the effects of factors, which is based on using the columns of the computational matrix. Figure 3.19 gives the computational matrix of the ByHand example.

	Experimental matrix		Computational matrix				Response
Exp. no	x1	x2	mean	x1	x2	x1*x2	y3
1	1	25	+	-	-	+	80.4
2	1.5	25	+	+	-	-	72.4
3	1	100	+	-	+	-	94.4
4	1.5	100	+	+	+	+	90.6
5	1.25	62.5	+	0	0	0	84.5
6	1.25	62.5	+	0	0	0	85.2
7	1.25	62.5	+	0	0	0	83.8

Figure 3.19: *Experimental and computational matrices of the ByHand example. Calculations below refer to the computational matrix.*
1st column gives the mean: (+80.4+72.4+94.4+90.6+84.5+85.2+83.8)/7 = 84.5;
2nd column gives the molar ratio, x_1, main effect: (-80.4+72.4-94.4+90.6)/2 = -5.9;
3rd column gives the reaction temperature, x_2, main effect: (-80.4-72.4+94.4+90.6)/2 = 16.1;
*4th column gives the x_1*x_2 two-factor interaction: (+80.4-72.4-94.4+90.6)/2 = 2.1*

The first column of the computational matrix does not provide any information related to factor effects, but is used to compute the average response according to (+80.4+72.4+94.4+90.6+84.5+85.2+83.8)/7 = 84.5. The molar ratio, x_1, main effect is calculated from the second column according to (-80.4+72.4-94.4+90.6)/2 = -5.9. Observe that the replicated center-point rows of the computational matrix do not contribute to this computation. Analogously, the reaction temperature, x_2, main effect is calculated according to (-80.4-72.4+94.4+90.6)/2 = 16.1. Finally, the fourth column, derived by multiplying the x_1 and x_2 columns, is used to encode the two-factor interaction molar ratio*reaction temperature, x_1*x_2, according to (+80.4-72.4-94.4+90.6)/2 = 2.1. Interestingly, by using the same graphical interpretation as in the preceding paragraph, this two-factor interaction may be interpreted as the difference between the response averages of the black and gray diagonals in Figure 3.20, that is, 85.5-83.4 = 2.1.

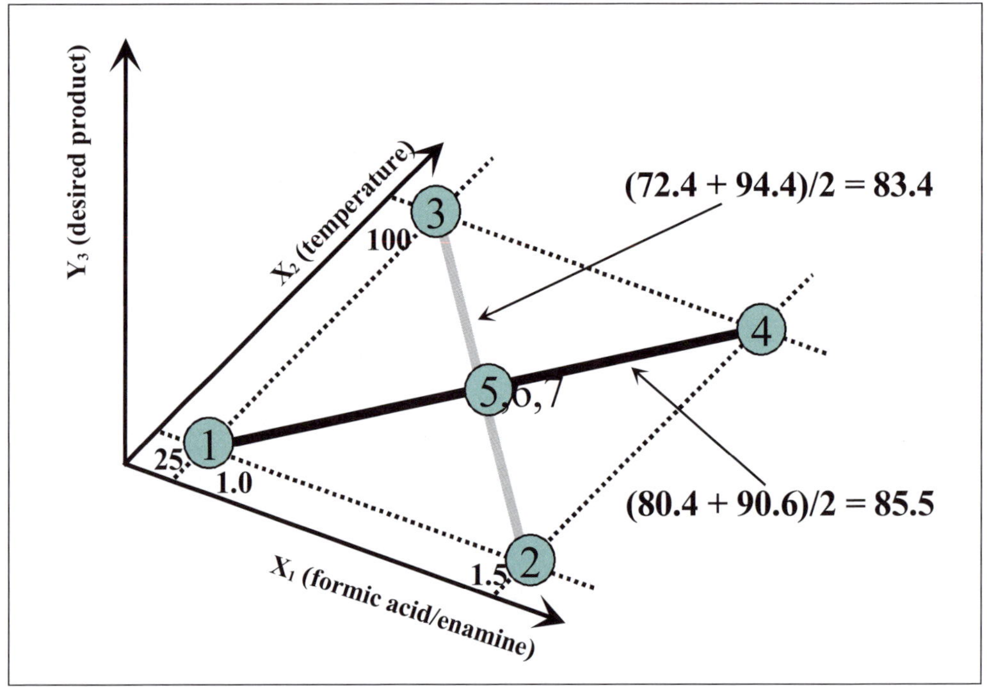

Figure 3.20: *The molar ratio*reaction temperature two factor interaction is interpreted as the difference between the black and gray diagonal averages.*

3.12 Plotting of main and interaction effects

Main effects and two-factor interaction effects have different impacts on the appearance of a semi-empirical model. Figure 3.21 shows the response surface obtained when fitting an interaction model to the yield of the desired product. The two main effects make the surface slope and the two-factor interaction causes it to twist. This is one way of interpreting these effects.

In addition, it is possible to create interaction plots specifically exploring the nature of interactions. Such plots are provided in Figure 3.22 and 3.23, and these may be thought of as representing the edges of the response surface plot shown in Figure 3.21. Figure 3.22 shows that when increasing the molar ratio, the yield of the desired product diminishes. However, the influence of the molar ratio is greater when the reaction temperature is set to its low level than to its high level. In other words, *the effect of the molar ratio depends on the level of the reaction temperature*. In a similar way, we see in Figure 3.23 that when the reaction temperature is raised the yield is increased. This influence is slightly more pronounced when the molar ratio is high.

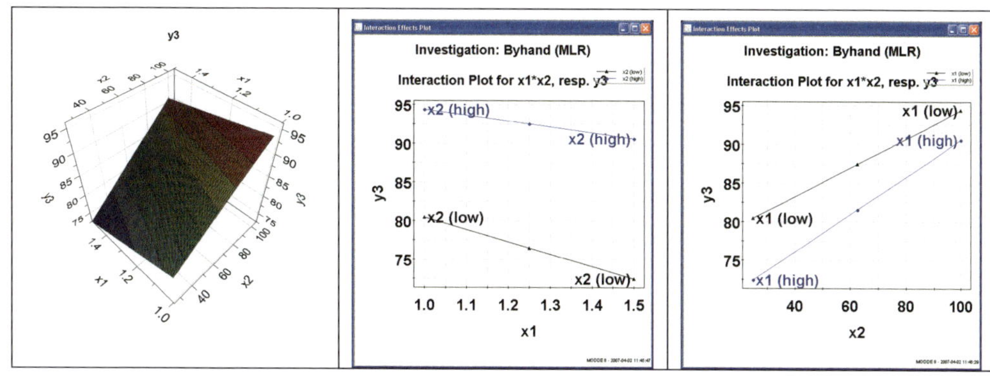

Figure 3.21: *(left) Response surface plot of yield.*
Figure 3.22: *(middle) Interaction plot of molar ratio*reaction temperature.*
Figure 3.23: *(right) Interaction plot of molar ratio*reaction temperature.*

Interestingly, the interaction plot may be used to uncover the strength of an interaction. Figures 3.24 to 3.26 show what the interaction plot may look like when dealing with almost no interaction (Figure 3.24), a mild interaction (Figure 3.25), and a strong interaction (Figure 3.26).

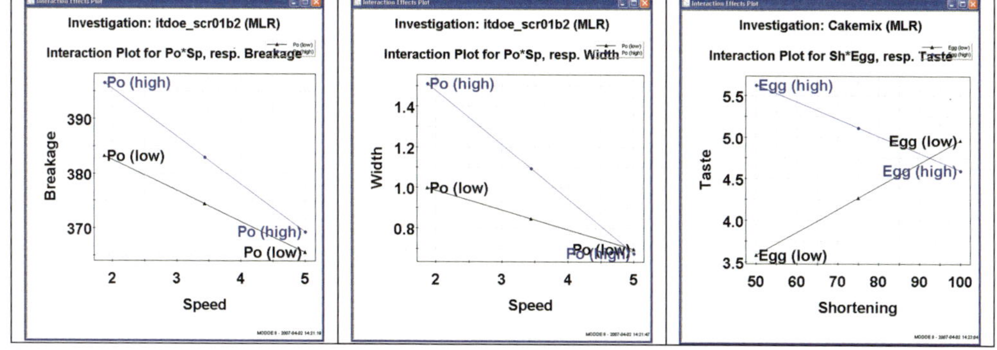

Figure 3.24: *(left) Almost no interaction.*
Figure 3.25: *(middle) Mild interaction.*
Figure 3.26: *(right) Strong interaction.*

3.13 Interpretation of main and interaction effects in the 2^3 case

We have now examined main and interaction effects in the 2^2 case, and seen how these may be interpreted by means of simple geometrical concepts. The geometric interpretation of main and interaction effects in a 2^3 factorial design is no more difficult. Consider the CakeMix application. Both the design matrix and the experimental matrix are shown in Figure 3.27, and the geometry of the experimental design is plotted in Figure 3.28. Notice that for clarity only the corner experiments are plotted.

	Design Matrix			Experimental Matrix			
Exp. No.	Flour	Shortn.	Egg	Flour	Shortn.	Egg	TASTE
1	-	-	-	200	50	50	3.52
2	+	-	-	400	50	50	3.66
3	-	+	-	200	100	50	4.74
4	+	+	-	400	100	50	5.2
5	-	-	+	200	50	100	5.38
6	+	-	+	400	50	100	5.9
7	-	+	+	200	100	100	4.36
8	+	+	+	400	100	100	4.86
9	0	0	0	300	75	75	4.68
10	0	0	0	300	75	75	4.73
11	0	0	0	300	75	75	4.61

Figure 3.27: *Design matrix and experimental matrix of the CakeMix application.*

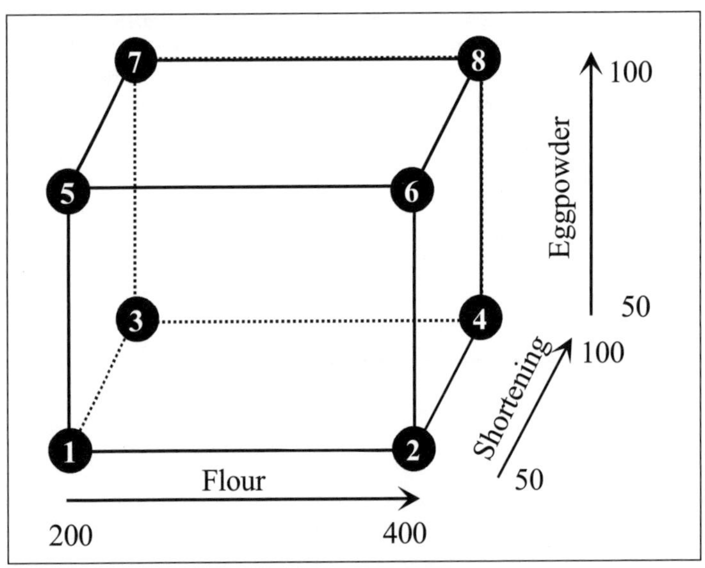

Figure 3.28: *Geometry of the CakeMix design.*

With the experimental design displayed in Figure 3.27, three main effects, Flour, Shortening, and Eggpowder, and three interaction effects, Flour*Shortening, Flour*Eggpowder, and Shortening*Eggpowder, are determinable. Figure 3.29 shows how the main effect of flour is interpretable as the difference between the gray plane average and hatched plane average, that is, right side of cube minus left side of cube. Similarly, Figures 3.30 – 3.34 illustrate the geometric interpretation of the remaining five effects. They are all interpretable as the difference between the averages of two planes, the gray plane minus the hatched plane.

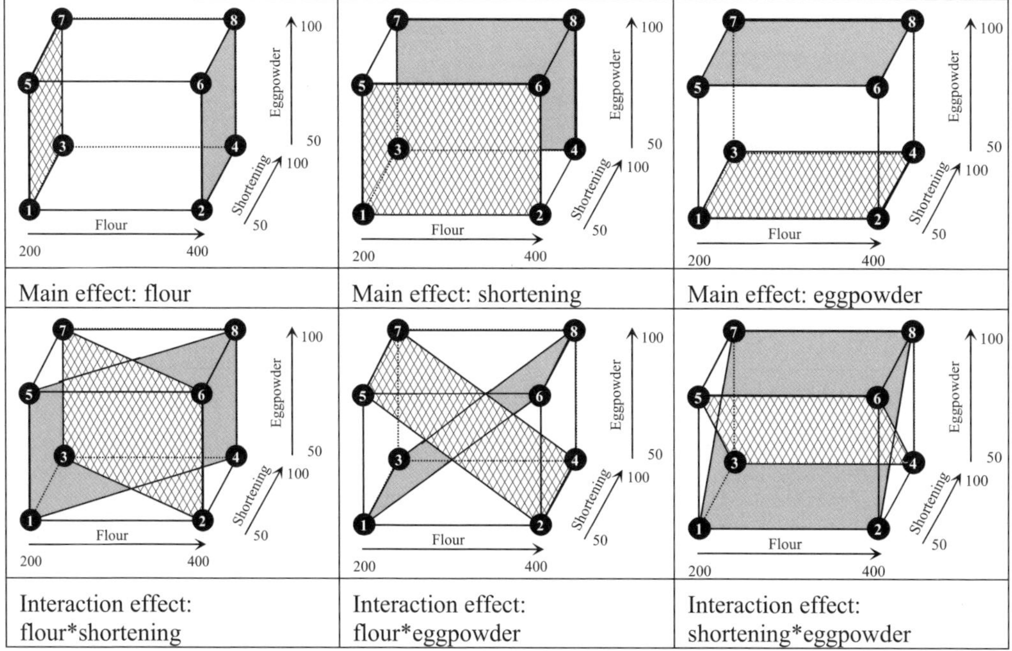

Figure 3.29: *(upper left) Main effect of flour.*
Figure 3.30: *(upper middle) Main effect of shortening.*
Figure 3.31: *(upper right) Main effect of eggpowder.*
Figure 3.32: *(lower left) Interaction effect of flour*shortening.*
Figure 3.33: *(lower middle) Interaction effect of flour*eggpowder.*
Figure 3.34: *(lower right) Interaction effect of shortening*eggpowder.*

3.14 Computation of effects using least squares fit

It would be tiresome if factor main effects and two-factor interactions had to be calculated by any of the by-hand principles outlined in this chapter. Fortunately, in reality, we may analyze DOE data by calculating a regression model using least squares fit. An introduction to least squares analysis is given in Section 3.15. The rationale for the by-hand methods is that they assist in gaining an understanding of the main and interaction effects concepts. Their big disadvantage, however, is that they are sensitive to slight errors in the factor settings, such as when, for example, the reaction temperature became 27°C instead of the wanted value 25°C. Experimental designs with such failing data are hard to analyze with the by-hand methods just described. Least squares analysis is much less sensitive to this problem.

In fact, least squares analysis has a number of advantages, notably, (i) the tolerance of slight fluctuations in the factor settings, (ii) the ability to handle a failing corner where experiments could not be performed, (iii) the estimation of the experimental noise, and (iv) the production of a number of useful model diagnostic tools. An important consequence of least squares analysis is that the outcome is *not* main and interaction effect estimates, but a regression model consisting of *coefficients* reflecting the influence of the factors. *Such a regression coefficient has a value half of that of the corresponding effect estimate.* Figures 3.35 and 3.36 show the relationship between effects and coefficients in the ByHand example. This relationship is discussed in more detail in Section 3.18. Also notice that the effect plot (Figure 3.36) is sorted according to the size of the effect, whereas the coefficient plot (Figure 3.35) is unsorted.

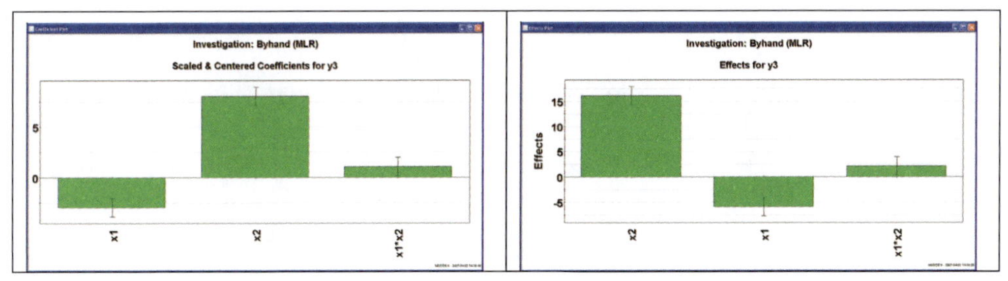

Figure 3.35: *(left) Regression coefficient plot of the ByHand example.*
Figure 3.36: *(right) Effect plot of the ByHand example.*

3.15 Introduction to least squares analysis

Consider Figure 3.37. In this graph the relationship between a single factor, called x_1, and a single response, called y_1, is plotted. Our goal is to obtain a model which can be used to predict y_1 from x_1, a common goal in, for instance, calibration studies. Because the relationship is linear the task will be to calculate the "best" straight line going through the swarm of points. However, prior to this we must first define the criterion which we are going to use for deciding when the "best" model has been found. Consider the dotted line. This line represents the best linear relationship according to a modelling criterion known as least squares. It was found with a technique known as linear regression (LR). Linear regression is based on seeking the model that minimizes the vertical deviation distances between all the experimental points and the line. An example of such a deviation, technically known as a residual, is given by the double-headed arrow. Notice that one experiment gives rise to one residual.

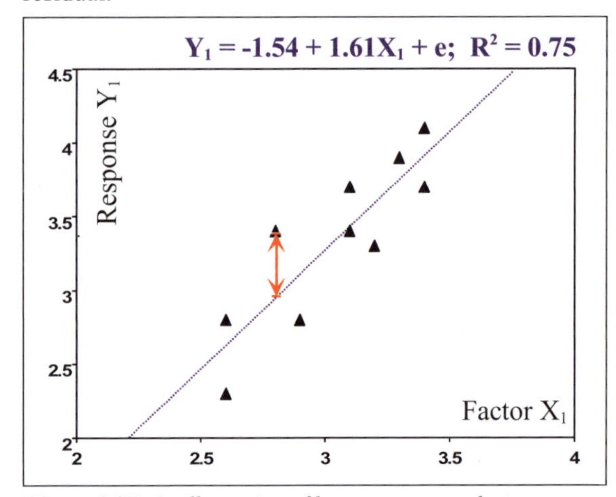

Figure 3.37: *An illustration of least squares analysis.*

Since some of these residuals will be positive and some negative, it is sensible to seek to minimize *the sum of the squares of the residuals*. Minimization of the residual sum of squares is a widely used criterion for finding the "best" straight line, and also explains the frequent use of the term *least squares analysis*. We can see in Figure 3.37 that the least squares line obtained has the equation $y_1 = -1.54 + 1.61x_1 + e$. The value –1.54 is called the intercept and gives the intersection of the line with the y_1-axis, that is, when $x_1 = 0$. The value 1.61 is called the gradient and indicates the slope of the line. Finally, the last term, e, represents the modelling residuals, that is, the discrepancies between the model (the line) and the reality. The R^2 parameter may be interpreted as a measure of the goodness of the model.

It may vary between 0 and 1, and is a measure of how closely the line can be made to fit the points. When R^2 equals 1 the fit is perfect and all the points are situated on the line. Hence, the lower the value of R^2 the more scattered the points are. An R^2 of 0.75 indicates a rough, but stable and useful relationship, as is also evident from the figure.

3.16 Least squares analysis applied to the CakeMix data

Least squares analysis is not restricted to a one-factor-situation, but is applicable to many factors. When least squares analysis is applied to the modelling of several factors it is commonly known as *multiple* linear regression, MLR. MLR is explained in the statistical appendix. We are now going to apply MLR to the CakeMix data. Recall that the 2^3 full factorial design, augmented with three center-points, supports the interaction model

$$y = \beta_0 + \beta_1 x_1 + \beta_2 x_2 + \beta_3 x_3 + \beta_{12} x_1 x_2 + \beta_{13} x_1 x_3 + \beta_{23} x_2 x_3 + \varepsilon.$$

Also recall that in this equation y denotes taste, x_1 flour, x_2 shortening, and x_3 eggpowder. The model obtained was

$$y = 4.695 + 0.203x_1 + 0.088x_2 + 0.423x_3 + 0.038x_1x_2 + 0.053x_1x_3 - 0.602x_2x_3 + e.$$

Unlike the previous example, in which a straight line was fitted to the experiments, the fitted regression model now corresponds to a twisted hyperplane in a four-dimensional space spanned by three factors and one response. We know that the hyperplane is twisted because of the large x_2x_3 two-factor interaction. This hyperplane is fitted such that the sum of the squared residuals, that is, the distances from the experimental points to the twisted hyperplane, is as small as possible.

In this context, a popular way of displaying the performance of the regression model is to make a scatter plot of the relationship between measured and calculated response values. Such a plot is displayed in Figure 3.38. Apparently, the fit is excellent, because all points are located close to the 1:1 line and $R^2 = 0.99$. However, we can see that there are some small deviations between the measured and calculated response values, i.e., small *vertical* distances from the points onto the regression line. An average estimate of these deviations is given by the residual standard deviation, RSD.

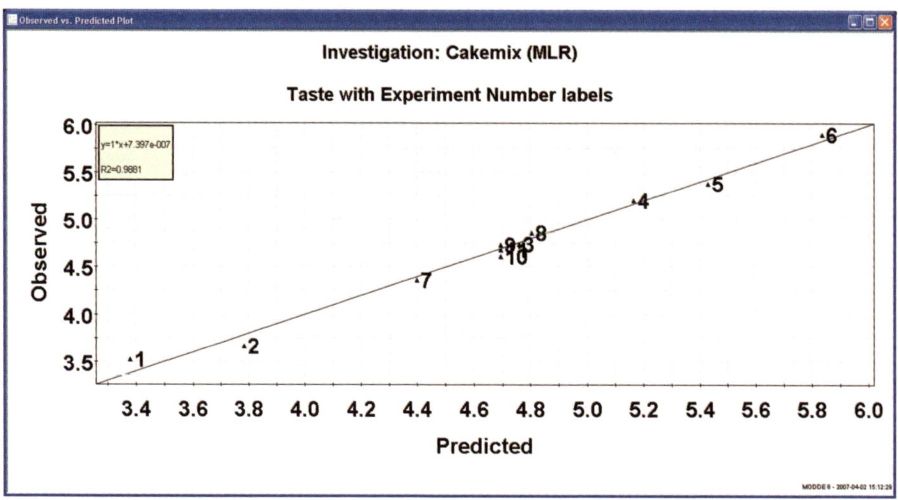

Figure 3.38: *The relationship between calculated and observed response values of the CakeMix application.*

3.17 The proper way of expressing regression coefficients

In the two examples that we have discussed so far in this chapter, the regression coefficients obtained were not listed according to the same principles. With the one factor/one response setup in the first example, the derived equation was $y_1 = -1.54 + 1.61x_1 + e$. Here the regression coefficient, 1.61, is *unscaled* and refers to the original measurement scale of the factor x_1. As a consequence, the constant term, -1.54, represents the estimated value of y_1 when x_1 is zero.

In the CakeMix case the regression model was $y = 4.695 + 0.203x_1 + 0.088x_2 + 0.423x_3 + 0.038x_1x_2 + 0.053x_1x_3 - 0.602x_2x_3 + e$. In this equation, the regression coefficients are *scaled and centered*. This means that they are no longer expressed according to the original measurement scales of the factors, but have been re-expressed to relate to the coded –1/+1 unit. Therefore, the constant term, 4.695, relates to the estimated taste at the design center-point, that is, when the factors have the value zero *in the coded unit*. The constant term does *not* relate to the *natural zero*, that is, zero grams of flour, shortening, and eggpowder, as this is a totally irrelevant cake mix composition.

The two kinds of regression coefficients are compared in Figures 3.39 and 3.40 using the CakeMix example. We can see that the regression coefficients differ dramatically. Figure 3.40 shows another disadvantage of using unscaled coefficients, namely that they are extremely difficult to interpret. According to the unscaled coefficients, shortening is more important than flour for improving the taste, whereas in reality the reverse is true. It is therefore common practice in DOE to express the regression results in terms of scaled and centered coefficients, as this enhances the interpretability of the model.

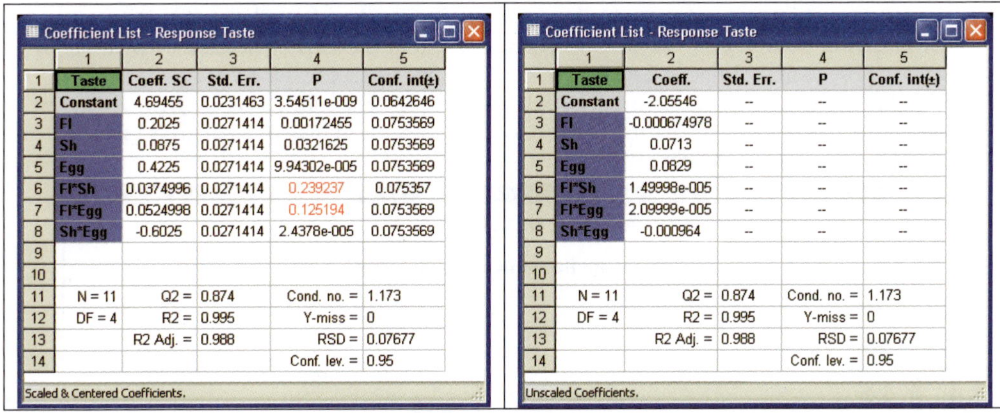

Coefficient List - Response Taste

	1	2	3	4	5
1	Taste	Coeff. SC	Std. Err.	P	Conf. int(±)
2	Constant	4.69455	0.0231463	3.54511e-009	0.0642646
3	Fl	0.2025	0.0271414	0.00172455	0.0753569
4	Sh	0.0875	0.0271414	0.0321625	0.0753569
5	Egg	0.4225	0.0271414	9.94302e-005	0.0753569
6	Fl*Sh	0.0374996	0.0271414	0.239237	0.075357
7	Fl*Egg	0.0524998	0.0271414	0.125194	0.0753569
8	Sh*Egg	-0.6025	0.0271414	2.4378e-005	0.0753569
9					
10					
11	N = 11	Q2 =	0.874	Cond. no. =	1.173
12	DF = 4	R2 =	0.995	Y-miss =	0
13		R2 Adj. =	0.988	RSD =	0.07677
14				Conf. lev. =	0.95

Scaled & Centered Coefficients.

Coefficient List - Response Taste

	1	2	3	4	5
1	Taste	Coeff.	Std. Err.	P	Conf. int(±)
2	Constant	-2.05546	--	--	--
3	Fl	-0.000674978	--	--	--
4	Sh	0.0713	--	--	--
5	Egg	0.0829	--	--	--
6	Fl*Sh	1.49998e-005	--	--	--
7	Fl*Egg	2.09999e-005	--	--	--
8	Sh*Egg	-0.000964	--	--	--
9					
10					
11	N = 11	Q2 =	0.874	Cond. no. =	1.173
12	DF = 4	R2 =	0.995	Y-miss =	0
13		R2 Adj. =	0.988	RSD =	0.07677
14				Conf. lev. =	0.95

Unscaled Coefficients.

***Figure 3.39**: (left) Scaled and centered regression coefficients of the CakeMix application.*
***Figure 3.40:** (right) Unscaled regression coefficients of the CakeMix application.*

3.18 Use of coefficient and effect plots

It is convenient to display regression coefficients in a bar chart. Figure 3.41 presents the scaled and centered coefficients of the CakeMix model. On each bar, the corresponding 95% confidence interval is superimposed. We will here concentrate on how to use confidence intervals, whereas their mathematical background is dealt with in Chapter 16. Confidence intervals are error-bars which indicate the uncertainty of each coefficient. Their size depend on three factors, (i) the quality of the experimental design, (ii) the goodness of the regression model, and (iii) the number of degrees of freedom. In principle, this means that the narrowest limits are obtained with (i) a perfect design with no geometrical defects, (ii) a model with low RSD, and (iii) enough experiments.

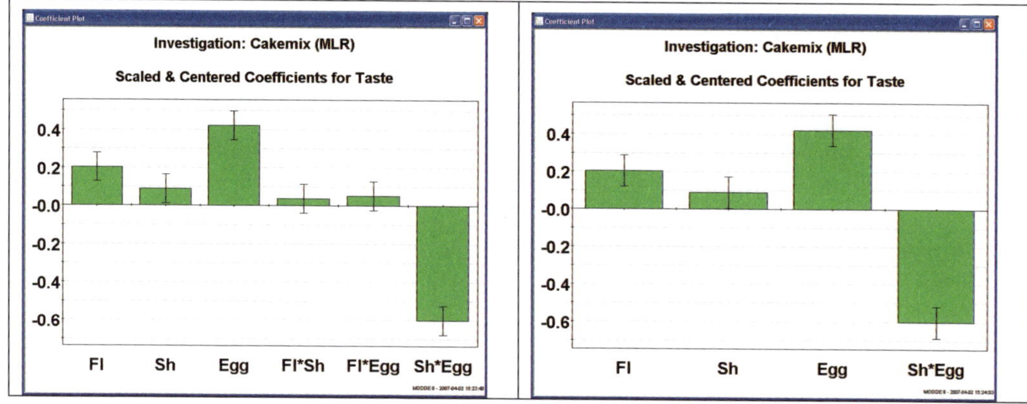

Figure 3.41: *(left) Regression coefficients of the initial CakeMix model.*
Figure 3.42: *(right) Regression coefficients of the refined CakeMix model.*

We can conclude that in the CakeMix case two coefficients, those pertaining to the two-factor interactions flour*shortening and flour*eggpowder, are statistically insignificant as their confidence intervals include zero. Hence, they can be removed and the model refitted. The results after refining the model are displayed in Figure 3.42. The numerical values of the four remaining coefficients and their confidence intervals have not changed to any appreciable extent. We also observe that the main effect of shortening is a borderline case according to the confidence interval assessment. However, this term is allowed to stay in the pruned model because it contributes to a highly significant two-factor interaction. This is because there exists a hierarchy among model terms, and a term of lower order should not be deleted from the model if it participates in the formation of a higher order term.

Now, we turn to Figures 3.43 and 3.44, which show plots of effects instead of coefficients, prior to, and after, model refinement. Note that these effect plots are sorted according to the numerical size, and that the effect is twice as large as the coefficient.

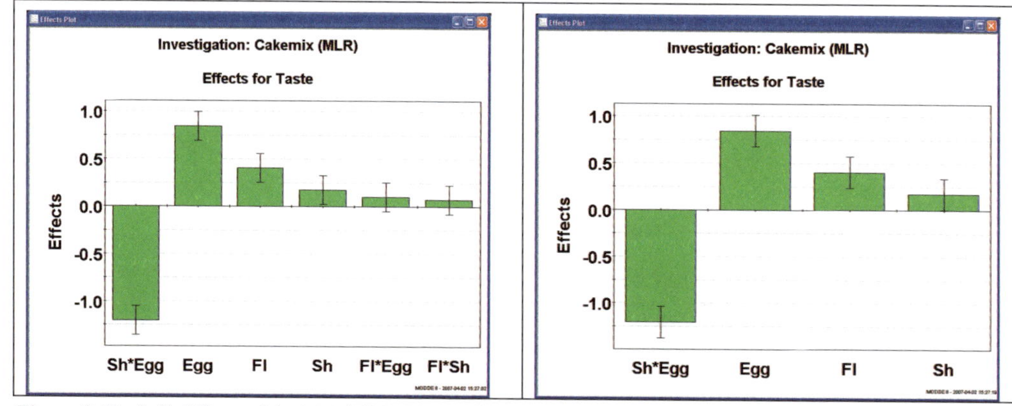

Figure 3.43: *(left) Effects of the initial CakeMix model.*
Figure 3.44: *(right) Effects of the refined CakeMix model.*

3.19 Other effect plots

Factor effects may also be displayed by plots other than bar charts. We will now overview some of these plotting alternatives using the refined model of the CakeMix application as an illustration. The three main effects of this model can be plotted separately, as seen in Figures 3.45 – 3.47. In comparison with the shortening*eggpowder two-factor interaction, displayed in Figure 3.48, the main effects are smaller in magnitude and of lesser importance. This is the same interpretation as before. Traditionally, a third way of plotting factor effects has been in use: the creation of normal probability plots (N-plots) of effects. But since such plots are less useful than those described here, N-plots of effects are not recommended.

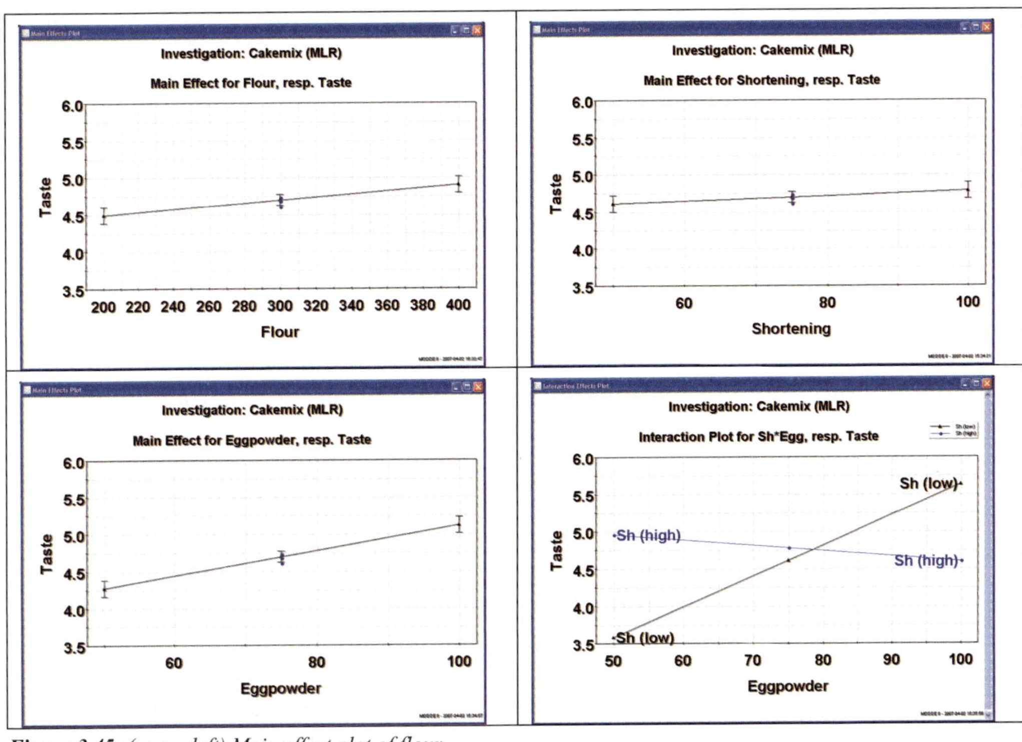

Figure 3.45: (upper left) Main effect plot of flour.
Figure 3.46: (upper right) Main effect plot of shortening.
Figure 3.47: (lower left) Main effect plot of eggpowder.
Figure 3.48: (lower right) Interaction effect plot of shortening with eggpowder.

3.20 Questions for Chapter 3

- What is a two-level factorial design?
- How many runs are included in the factorial part of the 2^2 design? 2^3? 2^4? 2^5?
- What is the geometry of the 2^3 design?
- How many center-point experiments are usually incorporated in a factorial design?
- How many factors are practical for two-level full factorial designs?
- When is it recommended to switch to fractional factorial designs?
- How is the main effect of a factor defined?

- What is the meaning of a two-factor interaction effect?
- How are main and two-factor interaction effects interpretable in a 2^2 design? In a 2^3 design?
- A two-factor interaction effect may be categorized in three levels of strength. Which?
- Which is the quickest by-hand method for computing effects?
- Give four reasons why it is better to use least squares fit for the data analysis?
- What is least squares analysis?
- What is the relationship between effects and coefficients?
- What is a residual?
- What does R^2 signify?
- What are unscaled coefficients?
- What are scaled and centered coefficients?
- What does the constant term represent in the two previous questions?
- How can you use confidence intervals to create a better model?

3.21 Summary and discussion

In this chapter we have focused on defining, and gaining an understanding of, main and two-factor interaction effects in simple full factorial designs, such as the 2^2 and 2^3 designs. In the 2^2 case, the main effect may be geometrically understood as the difference between the average response values of two opposite edges of the square experimental region, and the interaction effect as a similar difference between the two diagonals of the square. In the 2^3 case, the main effect corresponds to the difference between the average response values of two sides of the experimental cube, and the interaction effect to a difference between two diagonal planes inserted in the three-dimensional factor space. Our reasoning here may be generalized to k factors, but this is beyond the scope of this course book.

In addition, we have introduced three methods in which simple by-hand arithmetic is used for computing main and two-factor interaction effects. The main advantage of these methods is that they give an understanding of the concepts involved. However, in reality, least squares fit of a regression model to the data is a better approach. The use of least squares fit results in regression coefficients, with half the numerical values of the corresponding effects.

Based on the principle of least squares analysis, multiple linear regression represents a very useful method for the analysis of DOE data. MLR fits a model to the data such that the sum of the squared y-residuals is minimized. The outcome is a model consisting of regression coefficients, which are utilized to interpret the influence of the factors. Such regression coefficients are usually expressed in either an unscaled, or a scaled and centered format. Using the scaled and centered representation, a regression coefficient indicates the change in the response when the factor is raised from its zero level to its high level. In this situation, the constant term expresses the estimated average response at the design center. This is the recommended option in DOE as it will facilitate model interpretation. Using the alternative, unscaled version, a regression coefficient is expressed in the original measurement unit of the factor. In this case, the constant term relates to the situation at the natural zero, and not the coded zero. Another means of model interpretation is offered through the computed effects, which are twice as large in numerical value as the coefficients.

This chapter has summarized various plotting alternatives for displaying coefficients and effects. In addition, we have described how to use confidence intervals. Confidence intervals indicate the uncertainty in the coefficients or the effects, and are useful for identifying the most important factors. The narrowest confidence intervals are obtained with (i) a perfect design with no geometrical defects, (ii) a model with low RSD, and (iii) enough experiments.

4 Analysis of factorial designs

4.1 Objective

The analysis of experimental data originating from full factorial designs involves three basic steps. These steps are (i) evaluation of raw data, (ii) regression analysis and model interpretation, and (iii) use of regression model. The aim of this chapter is two-fold, i.e., first to give an introductory description of these three steps, and then to describe four common causes of poor models.

The diagnostic tools used in the data analysis will be grouped according to three ambition levels, i.e., basic level, recommended level and advanced level. On the basic level, we will (i) introduce a useful plot for the evaluation of raw data, (ii) introduce the powerful diagnostic tool R^2/Q^2, and (iii) demonstrate how a regression model may be used for decision making.

On the recommended level, the objective is to review and deepen the understanding of the diagnostic tools available. Thus, we will outline four additional tools, which are useful for evaluating the raw data. These tools are called condition number, scatter plot, histogram of response, and descriptive statistics of response. Regarding the regression analysis and model interpretation, two additional model diagnostic tools will be explained, i.e., the lack of fit test and the normal probability plot of the response residuals. Finally, regarding the use of the regression model, a procedure for converting the information in response contour plots and response surface plots into predictions for new experiments will be exemplified.

On the advanced level, we will provide an introduction to the partial least squares projections to latent structures, PLS, method. PLS has certain features which are appealing in the analysis of more complicated designs, notably its ability to cope with several correlated responses in one single regression model. The PLS model represents a different mathematical construction, which has the advantage that a number of new diagnostic tools emerge. These tools are useful for model interpretation in more elaborate applications, and are more informative than other diagnostic tools.

Finally, the objective is to describe four common causes of poor models, describe how to pinpoint these causes, and illustrate what to do when they have been detected. Fortunately, in most cases, the measures needed to solve such problems are not sophisticated.

4.2 Introduction to the analysis of factorial designs

The analysis of experimental data generated through DOE consists of three primary stages. The first stage, *evaluation of raw data*, focuses on a general appraisal of regularities and peculiarities in the data. In most cases, important insights can be obtained even at this stage. Such insights should be used in order to enhance the subsequent regression analysis. There are a number of useful diagnostic tools available, which will be described. The second stage, *regression analysis and model interpretation*, involves the actual calculation of the model linking the factors and the response(s) together, and the interpretation of this model. It is of crucial importance to derive a model with optimal predictive capability, and we will here describe the usefulness of the R^2/Q^2 diagnostic tool. As far as the model interpretation is concerned, we will mainly consider the plotting of coefficients. Finally, in the third stage, *use of regression model*, the model obtained is utilized to predict the best point at which to conduct verifying experiments or in which to anchor a subsequent design. In this respect, the analyst may use response contour plots, response surface plots and/or an interactive optimization routine.

4.3 Basic level of data analysis

4.3.1 Evaluation of raw data – Replicate plot

The replicate plot is a useful graphical tool for evaluating raw data. In such a graph, the measured values of a response are plotted against the unique number of each experiment. We will use the ByHand and CakeMix applications to illustrate this plotting principle. Replicate plots pertaining to the ByHand application are shown in Figure 4.1 for the three responses *side product* (y_1), *unreacted starting material* (y_2), and *desired product* (y_3).

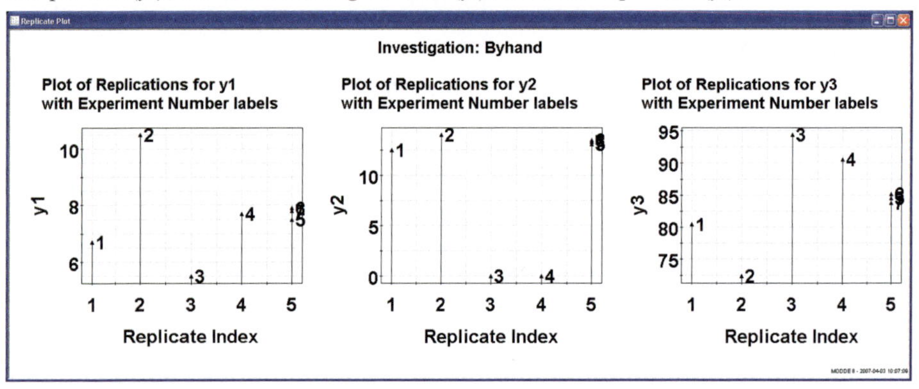

Figure 4.1: *(left) Replicate plot of side product (y_1). (middle) Replicate plot of unreacted starting material (y_2). (right) Replicate plot of desired product (y_3).*

In a plot of replications, any experiment with a unique combination of the factors will appear isolated on a bar, whereas experiments with identical factors settings, that is, replicates, will show up on the same bar. Hence, in the ByHand application there are three replicates with the experiment numbers 5, 6 and 7 (Figure 4.1). Since the variation in these three replicates is much smaller than the variation in the entire investigation series, we can conclude that the replicate error will not complicate the data analysis.

In a similar manner, Figure 4.2 represents the variation in *taste* in the CakeMix application, and apparently the replicate error is small in this case as well. Conversely, Figure 4.3 exemplifies a situation in which the replicate error is so large that a good model cannot be obtained. While not going into any details of this application, we can see that the replicates, numbered 15-20, vary almost as much as the other experiments.

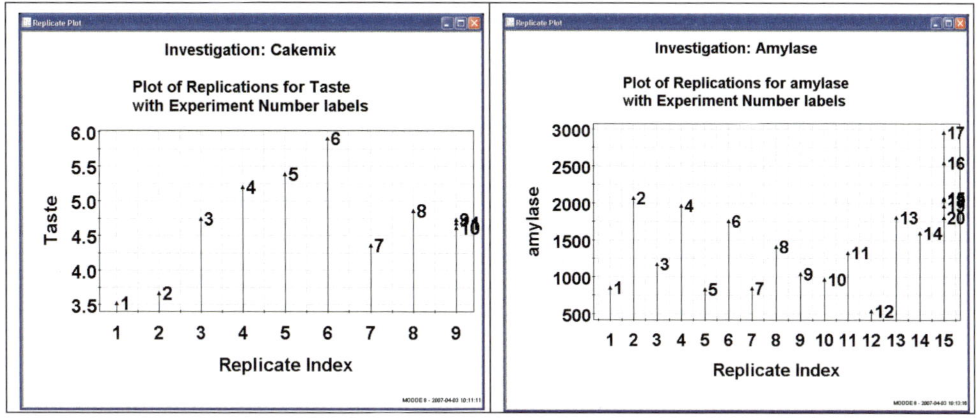

Figure 4.2: *(left) Replicate plot of taste.*
Figure 4.3: *(right) Replicate plot indicating a too large a replicate error.*

4.3.2 Regression analysis – The Summary of Fit plot

The replicate plot represents a minimum level of evaluation of raw data. Since the replicate plots did not show any problems in the ByHand and CakeMix applications, we can move on to the next stage, the regression analysis and model interpretation. When fitting a regression model the most important diagnostic tool consists of the two companion parameters R^2 and Q^2. A plot of these for the CakeMix model is shown in Figure 4.4.

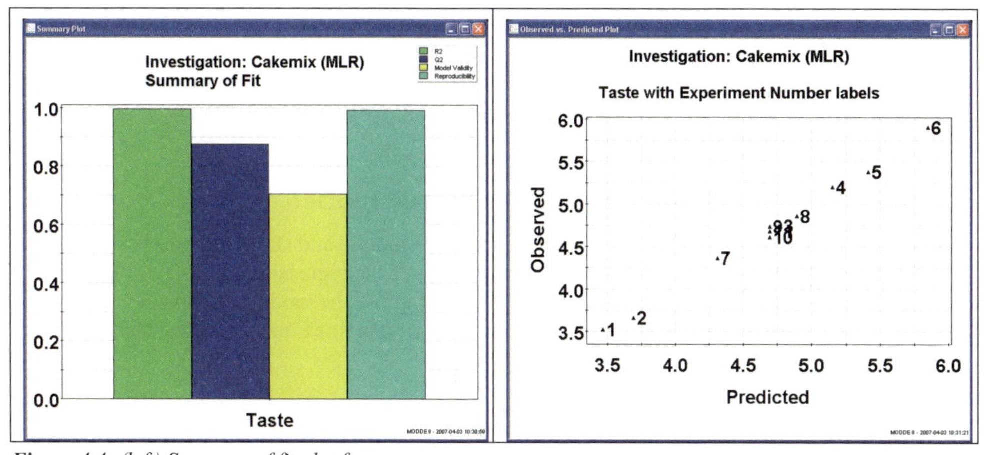

Figure 4.4: *(left) Summary of fit plot for taste.*
Figure 4.5: *(right) Observed/Predicted plot for taste.*

The leftmost bar in Figure 4.4 is R^2 and it amounts to 0.99. This parameter is called the *goodness of fit*, and is a measure of how well the regression model can be made to fit the raw data. R^2 varies between 0 and 1, where 1 indicates a perfect model and 0 no model at all. When R^2 is 1 all points are situated on the diagonal in Figure 4.5. The main disadvantage of R^2 is that it can be made arbitrarily close to 1, by, for instance, including more terms in the model. Hence, R^2 alone is not a sufficient indicator for probing the validity of a model.

A much better indication of the usefulness of a regression model is given by the Q^2 parameter. Q^2 is the second bar from the left in Figure 4.4 and it equals 0.87. This parameter is called the *goodness of prediction*, and estimates the predictive power of the model. This is a more realistic and useful performance indicator, as it reflects the final goal of modelling – predictions of new experiments. Like R^2, Q^2 has the upper bound 1, but its lower limit is minus infinity. For a model to pass this diagnostic test, both R^2 and Q^2 should be high, and preferably not separated by more than 0.2 – 0.3. A substantially larger difference constitutes a warning of an inappropriate model. Generally speaking, a $Q^2 > 0.5$ should be regarded as good, and $Q^2 > 0.9$ as excellent, but these limits are application dependent.

The third bar in the summary of fit plot (Figure 4.4) is called *model validity*. It reflects whether the model is appropriate in a general sense, i.e., if the right type of model (linear, interaction, quadratic, ...) was chosen from the beginning in the problem formulation. The higher the numerical value the more valid the model is, and a value above 0.25 suggests a valid model. Formally, the model validity statistic is based on the lack of fit test carried out as part of the analysis of variance (ANOVA) evaluation. The ANOVA evaluation is explained below (see Section 4.4.5).

Finally, the rightmost bar in the summary of fit plot is called the *reproducibility* diagnostic tool. This performance indicator is a numerical summary of the variabilities plotted in the replicate plot (see Figure 4.2). The higher the numerical value the smaller the replicate error is in relation to the variability seen across the entire design. Conversely, if the value of the reproducibility bar is small, below 0.5, you have a large pure error and poor control of the experimental procedure.

In summary, for a model to be judged as good, the model performance indicators should comply with the following reference values:

- Difference $R^2 - Q^2$ $< 0.2\text{-}0.3$
- Q^2 > 0.5
- Model validity > 0.25
- Reproducibility > 0.5

For the relevant equations regarding how to calculate these parameters, we refer to the statistical appendix. Also observe that model validity and reproducibility are only available when replicated experiments have been performed.

4.3.3 Summary of Fit plot pointing to a poor model

Unfortunately, the regression modelling is not always as simple and straightforward as in the previous section. We will now make a closer inspection of the regression analysis of the ByHand data, where things become a little more challenging. The model performance parameters are displayed in Figure 4.6. In this case, we obtain three quartets of bars, simply because the data-set contains three response variables.

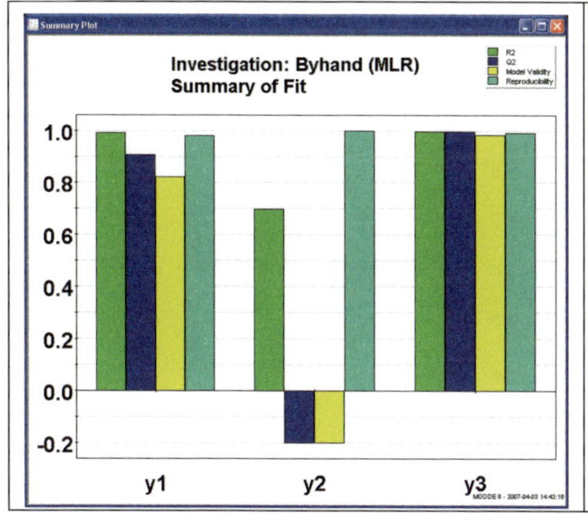

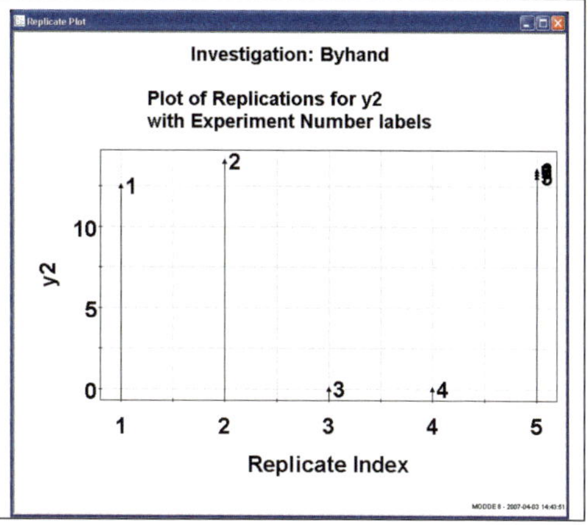

Figure 4.6: *(left) Summary of fit plot of the ByHand model.*
Figure 4.7: *(right) Replicate plot of y_2 (unreacted starting material).*

According to Figure 4.6, y_1 *(side product)* and y_3 (*desired product*) are well fitted and predicted by the interaction model, since both R^2 and Q^2 are high, and closer together than 0.2 – 0.3. In contrast, the situation is far from ideal for the second response, y_2 (*unreacted starting material*), which has a fairly high R^2 but negative Q^2 and model validity. The negative Q^2 and model validity indicate a poor model devoid of any predictive power and with significant lack of fit. Now, we have to figure out why we get this model behaviour. Actually, we have already examined a plot hinting at the underlying reason - the replicate plot of y_2, which is re-displayed in Figure 4.7.

Since the response values of the replicated center-points are amongst the highest, we must have a non-linear dependence between y_2 and the two factors, molar ratio of formic acid/enamine and temperature. For a linear or an interaction model to be valid, one would expect to see the replicated center-points in the middle part of the response interval. Clearly, this is not the case and hence we can conclude that the response/factor relationship is curved. Such a curvature can only be adequately represented by a model with quadratic terms. Unfortunately, the full factorial design used here does not allow the estimation of quadratic terms. Hence, the fitted interaction model is of too low a complexity. This means that the predictive ability will break down, as manifested by the negative Q^2-value. In Section 4.6, we will see that this problem is rather easily addressed.

4.3.4 Model interpretation – Coefficient plot

Model interpretation also plays an important role in the data analysis. This section will describe how the coefficient plot can be used for model interpretation, and, eventually, model pruning. Consider the CakeMix data set. According to the model performance indicators plotted in Figure 4.8, the fitted interaction model is excellent. In this case, it is logical to interpret the model, and this is accomplished with the coefficient plot shown in Figure 4.9. According to Figure 4.9, there are two small and insignificant two-factor interactions, Fl*Sh and Fl*Egg. These terms may be omitted and the model refitted to the data.

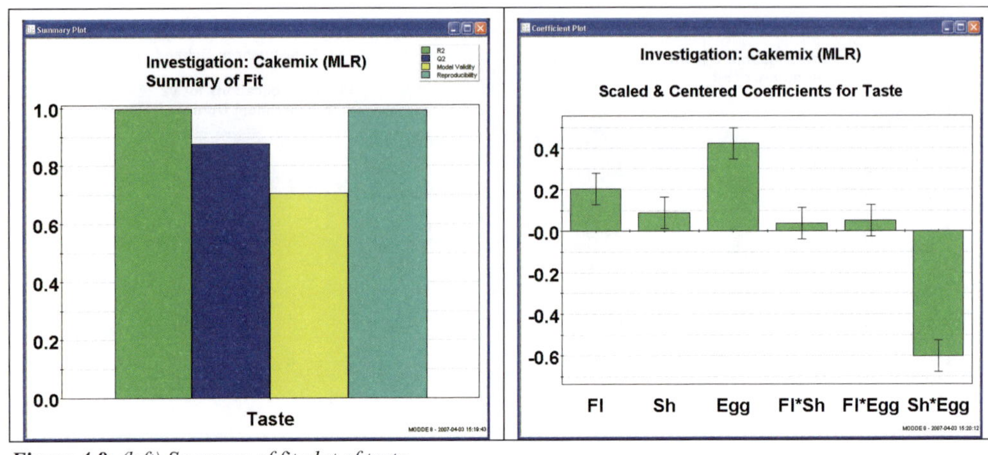

Figure 4.8: *(left) Summary of fit plot of taste.*
Figure 4.9: *(right) Regression coefficients of the model for taste.*

Figures 4.10 and 4.11 show the outcome of this model refinement. Notice, that due to the orthogonality of the experimental design, the numerical values of the remaining coefficients have not changed. We may also observe that R^2 has undergone a tiny decrease, and that Q^2 increased from 0.87 to 0.94. The increase in Q^2 is not large, but appreciable, and indicates that we now have a simpler model with better predictive ability. The interpretation of the refined model indicates that in order to improve the taste one should concentrate on increasing the amounts of eggpowder and flour. The large two-factor interaction will be examined shortly.

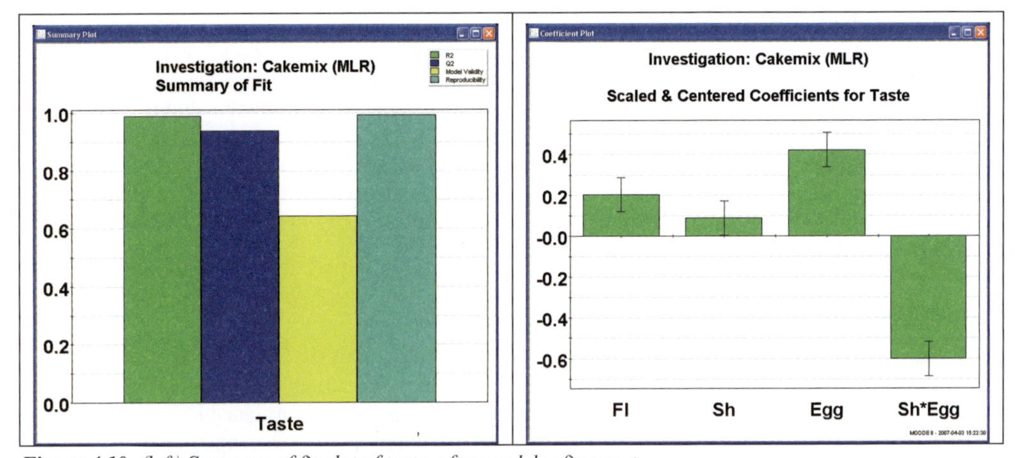

Figure 4.10: *(left) Summary of fit plot of taste, after model refinement.*
Figure 4.11: *(right) Regression coefficients of taste, after model refinement.*

4.3.5 Use of model – Response contour plot

When it is believed that the optimal regression model has been acquired it is pertinent to carry out the third stage of the data analysis, *use of model*. Here, the aim is to gain a better understanding of the modelled system, to decide if it is necessary to continue experimenting, and, if so, to locate good factor settings for doing this. First it must be decided which response contour plots are most meaningful. In the CakeMix data set, the strong two-factor interaction between shortening and eggpowder regulates this choice. Figure 4.12 shows a response contour plot created with the factors shortening and eggpowder as axes, and flour fixed at its high level.

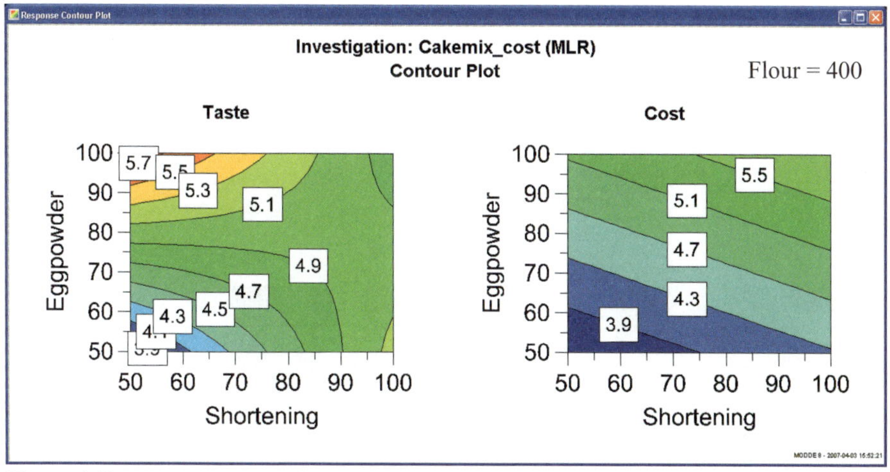

Figure 4.12: Response contour plots of taste and cost of ingredients.

The response contour plot in Figure 4.12 is twisted as a result of the strong two-factor interaction. Obviously, to improve the taste, we should position new (verifying) experiments in the upper left-hand corner.

In reality, however, maximizing one response variable, like taste of cake or quality of product, is usually not the single operative goal, and sometimes the most important goal is to minimize the production cost. We may transfer this reasoning to the CakeMix study. Since shortening is a cheaper ingredient than eggpowder, it may be argued that the lower right-hand corner represents a more relevant region of operability (Figure 4.12). This corner offers a reasonable compromise between high taste and low cost.

4.3.6 Model interpretation and use in the case of several responses

Since the CakeMix study contains only one response, the taste, only one model and hence one set of regression coefficients needs to be contemplated. This makes the model interpretation quite simple. With M responses M models are fitted with MLR, producing M sets of regression coefficients for model interpretation. Thus, there are three sets of regression coefficients to consider in the ByHand case. These regression coefficients are given in Figure 4.13.

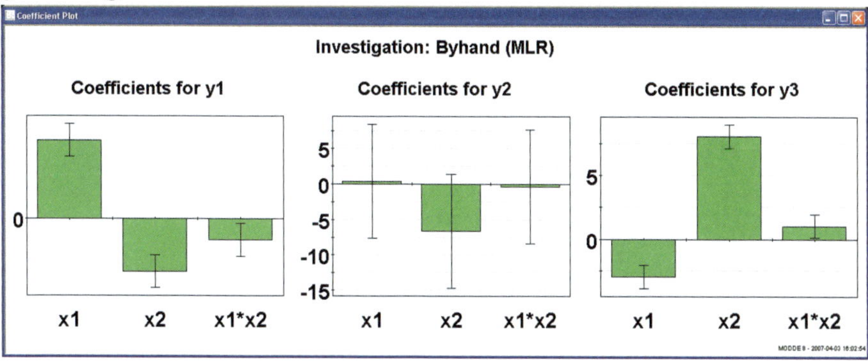

Figure 4.13: (left) Regression coefficients of side product (y_1). (middle) Regression coefficients of unreacted starting material (y_2). (right) Regression coefficients of desired product (y_3).

At first glance, there are two striking features in the sets of regression coefficients (Figure 4.13). The most conspicuous feature is the huge confidence intervals of the model for y_2, *unreacted starting material*. But, really, this comes as no surprise, since we know that this response exhibits a quadratic behaviour, which is poorly modelled by the interaction model (see Figure 4.7). The other feature is the weakness of the two-factor interaction, which is barely significant for y_1, *side product*, and y_3, *desired product*. For simplicity and clarity, however, we will keep these models.

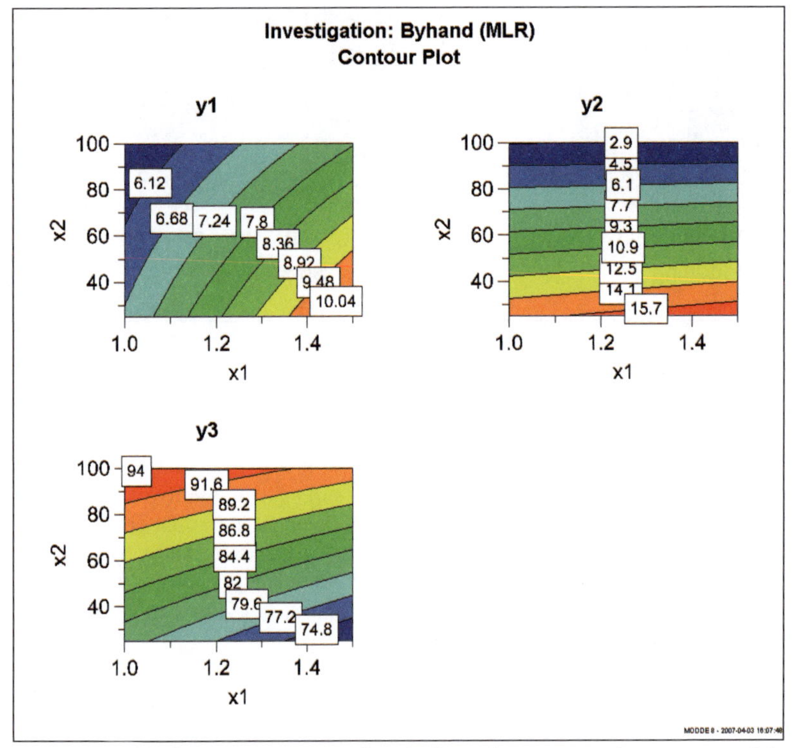

Figure 4.14: *Triple response contour plot of side product, unreacted starting material, and desired product.*

A convenient way of surveying these models is to construct three response contour plots and place them near each other. Such a triple-contour plot is given Figure 4.14. The experimental goal was to decrease the side product, decrease the unreacted starting material, and increase the desired product. Figure 4.14 reveals that these objectives are in no conflict and are predicted to be best met in the upper left-hand corner, that is, with low molar ratio and high temperature. It is appropriate to carry out new experiments in that region. Moreover, in the interpretation of this triple-contour plot, it must be borne in mind that the model underlying the y_2 contour plot is weak, and this will result in uncertainties in the prediction phase.

4.4 Recommended level of data analysis

We will now review and deepen the understanding of the diagnostic tools related to (i) evaluation of raw data, (ii) regression analysis and model interpretation, and (iii) use of regression model. Thus, we will outline additional diagnostic tools, which are useful for the routine user of DOE.

4.4.1 Evaluation of raw data – Condition number

The *condition number* is a tool that can be used to evaluate the performance of an experimental design *prior to its execution*. Formally, the condition number is the ratio of the largest and the smallest singular values of the X-matrix, that is, the matrix of the factors extended with higher order terms. Informally, the condition number may be regarded as the ratio of the longest and shortest design diagonals. A schematic drawing of this is shown in Figure 4.15.

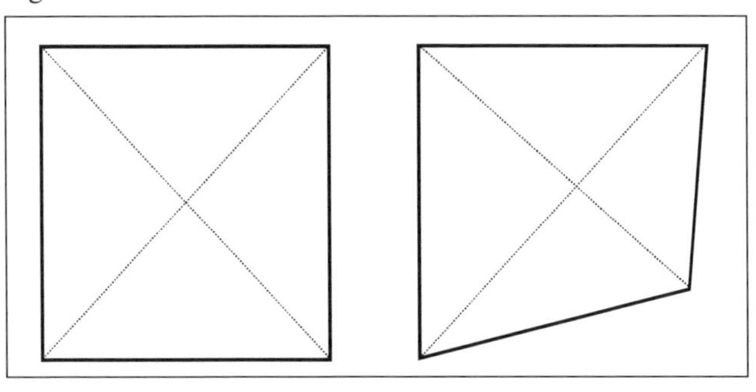

Condition Number	Objective: Screening & Robustness testing	Objective: Optimization
Good Design	< 3	< 8
Questionable design	3-6	8-12
BAD design	>6	>12

Figure 4.15: *A schematic illustration of the condition number. This number may be interpreted as the ratio between the longest and shortest design diagonals.*

For the left-hand design, which is symmetrical, the ratio of the two diagonals is 1. For the right-hand design, which is unsymmetrical, the ratio is > 1. From this, we see that the condition number is a measure of the *sphericity* of the design, or, its *orthogonality*. All two-level factorial designs, full and fractional, without center-points have a condition number of 1, and then all the design points are situated on the surface of a circle or a sphere.

Interestingly, by means of the condition number it is possible to formulate guidelines for the assessment of the performance of an experimental design. Such guidelines, applicable to designs in quantitative factors, are given in the lower part of Figure 4.15. With qualitative factors and mixture factors, the condition numbers are generally much higher. The listed reference values permit the evaluation of the design prior to carrying out its experiments. Hence, when a screening design has a condition number < 3 or an optimization design < 8, the designs as such are very good.

Furthermore, whenever an experimenter wishes to make a change in a design, we would advise that the condition numbers before and after the change be computed, in order to get a feeling for their magnitude. If the condition number changes by less than 0.5 we regard it as a small change, and the modification is justifiable from a theoretical point of view. If the condition number changes by more than 3.0 we regard it as a large change, and such a modification of the design would require serious reconsideration.

4.4.2 Evaluation of raw data – Scatter plot

The scatter plot is a useful tool in connection with the condition number assessment, particularly in investigations of limited size. When a design has been evaluated and its condition number found, one may understand its geometry by making scatter plots of the factors. We will exemplify this with the ByHand application. Figure 4.16 shows the scatter plot of the two factors *molar ratio of formic acid/enamine* and *temperature*. As seen, the design is symmetrical. It is a screening design having a condition number of 1.3. Observe that it is the existence of the three center-points that makes the condition number slightly exceed 1. Apparently, this is a good design for screening. However, were the ByHand application to be modified by excluding one corner from the experimentation, a design like that illustrated in Figure 4.17 would be obtained. This D-optimal design has a condition number of 1.6, and is therefore capable of doing a good job as a screening design, despite its skewed geometry.

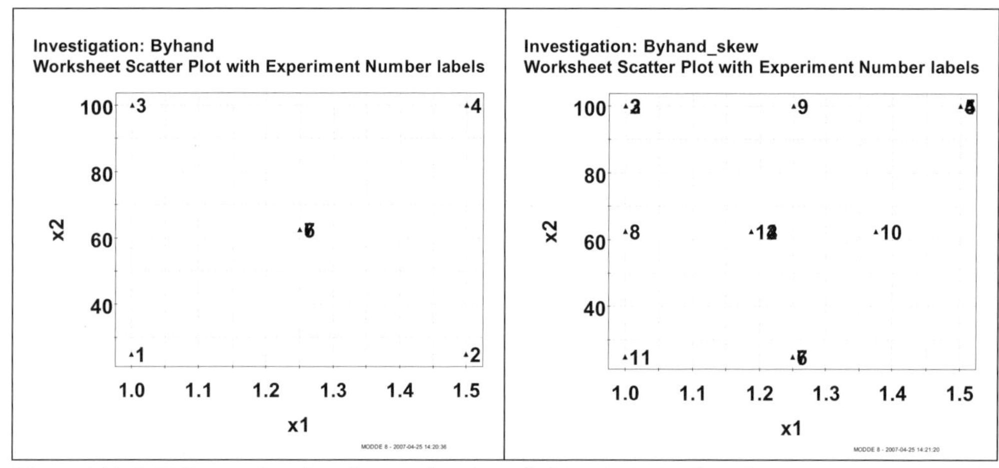

Figure 4.16: *(left) Scatter plot of two factors when the underlying design is of regular geometry.*
Figure 4.17: *(right) Scatter plot of two factors when the underlying design is of irregular geometry.*

The scatter plot can also be used for investigating relationships between factors and responses. In a response/factor plot we are interested in overviewing whether the relationship is linear or curved (non-linear). Figure 4.18 applies to the ByHand case and shows the relationship between the response *desired product* and the factor *temperature*. Clearly, in this case one should expect a linear relationship. In contrast, Figure 4.19 indicates a curved relationship between the response *unreacted starting material* and the factor *temperature*. Such a curved relationship will not be adequately explained by a linear or an interaction model; it needs a quadratic model. By obtaining this kind of insight prior to the regression analysis, the analyst is well prepared. Furthermore, this information also helps us to understand why we obtained such a poor model for this response (unreacted starting material) in Section 4.3.3. Recall that Q^2 and model validity were negative.

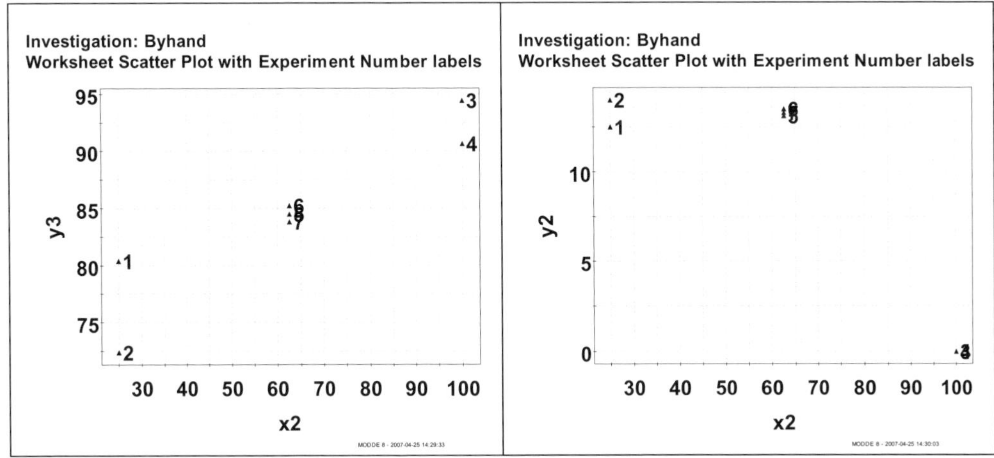

Figure 4.18: *(left) Scatter plot of one response and one factor when the relationship is linear.*
Figure 4.19: *(right) Scatter plot of one response and one factor when the relationship is curved.*

In summary, scatter plots of raw data make it possible to map factor/factor and response/factor relationships and gain useful insights, which can be used to guide the subsequent regression analysis. However, it must be realized that this plotting approach works best for uncomplicated investigations with few factors and responses, and is impractical with more than four factors and/or four responses.

4.4.3 Evaluation of raw data – Histogram of response

We will now address two tools that are useful for evaluating the statistical properties of *response* data. The first tool is *histogram of response* and the second tool is *descriptive statistics of response*. A short background to the use of these tools is warranted.

In regression analysis, it is advantageous if the data of a response variable are normally distributed, or nearly so. This improves the efficiency of the data analysis, and enhances model validity and inferential reliability. The histogram plot is useful for studying the distributional shape of a response variable. We will create some histograms to illustrate this point. Figure 4.20 shows a histogram of the response *desired product* of the ByHand application. This is a response which is approximately normally distributed, and which may be analyzed directly. As seen in the next histogram (Figure 4.21), the same statement holds true for the response *taste* of the CakeMix application.

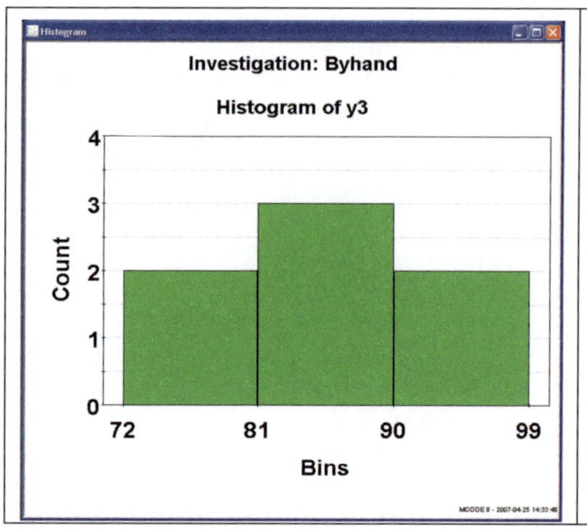

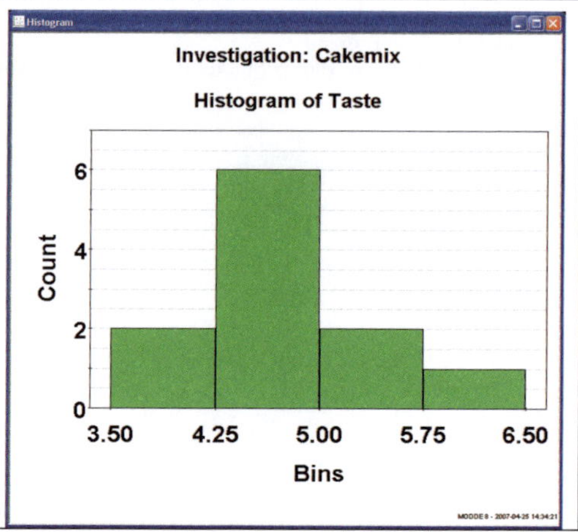

Figure 4.20: *(left) Histogram of an approximately normally distributed response.*
Figure 4.21: *(right) Histogram of an approximately normally distributed response.*

However, the third histogram (Figure 4.22), pertaining to General Example 3, shows that the response *skewness of weld* is not approximately normally distributed. This histogram has a heavy tail to the right, and indicates that one measurement is not as the others. It is much larger. It is not recommended to apply regression analysis to a response with this kind of distribution, as that would correspond to assigning the extreme measurement an undue influence in the modelling. Fortunately, it is easy to solve this problem. A simple logarithmic transformation of the response is all that is needed. Indeed, after a logarithmic transformation the extreme measurement is much closer, and hence more similar, to the majority of the data points (Figure 4.23).

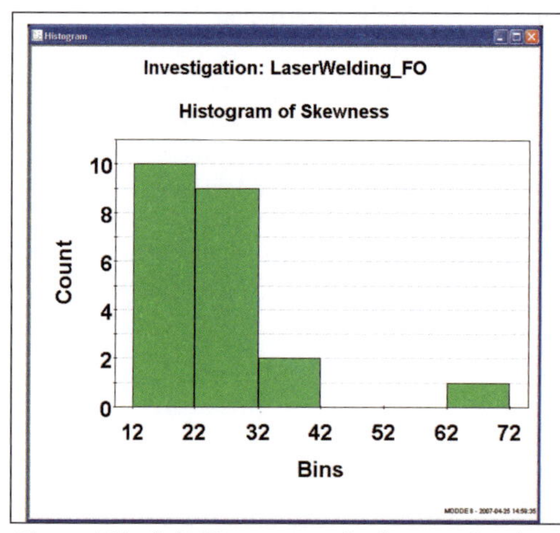

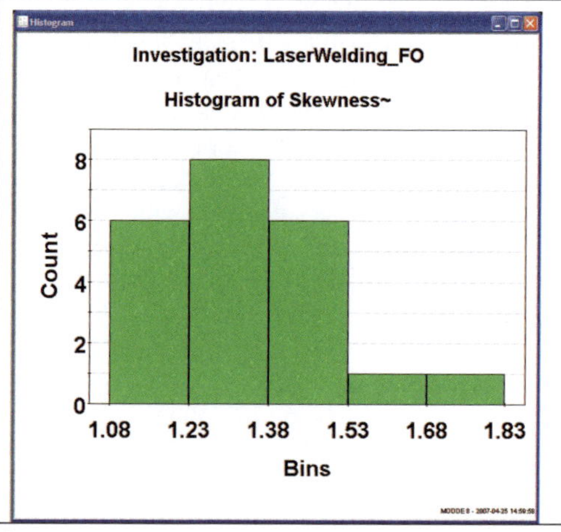

Figure 4.22: *(left) Histogram with a heavy tail to the right.*
Figure 4.23: *(right) Histogram of a log-transformed variable.*

In summary, the histogram is a powerful graphical tool for determining if a transformation of responses is needed. Also notice, that in DOE it is less common to transform a factor after the design has been executed. The reason for this is that the symmetry and balance of the design then degrade substantially (see discussion in Chapter 2).

4.4.4 Evaluation of raw data – Descriptive statistics of response

Another tool for investigating properties of responses is called descriptive statistics of response. It is often used in conjunction with histograms, especially when there are only a few measured values. If there are few measured values it is difficult to define appropriate data intervals for the histogram, and the histogram will then have a rugged appearance. To facilitate the understanding of the descriptive statistics tool we will compare its results with the histograms. Figures 4.24 – 4.26 show histograms of three responses which are (i) nearly normally distributed, (ii) positively skewed, and (iii) negatively skewed.

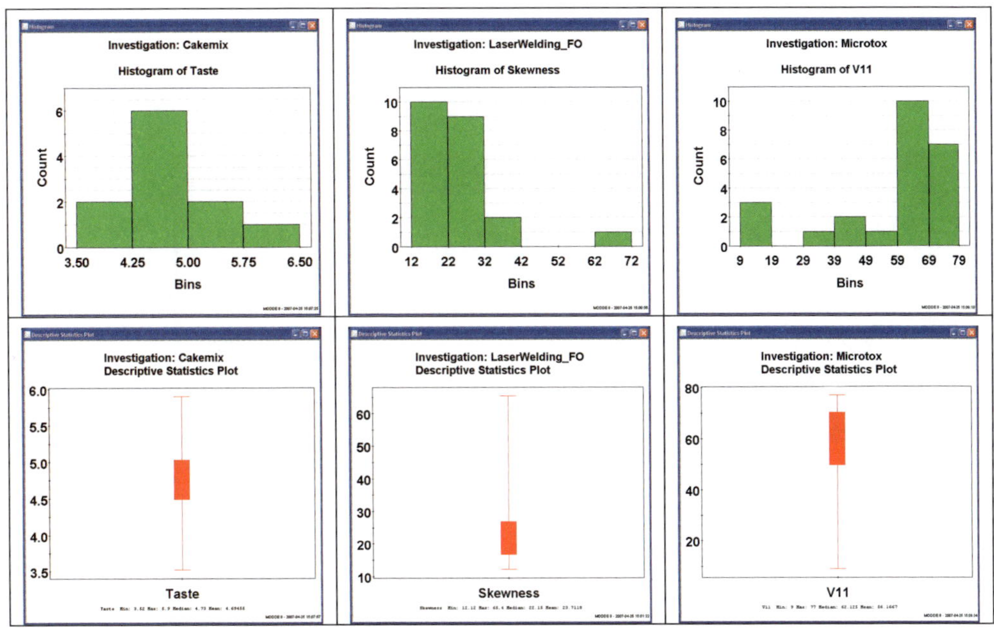

Figure 4.24: *(upper left) Histogram of a nearly normally distributed response.*
Figure 4.25: *(upper middle) Histogram of a positively skewed response.*
Figure 4.26: *(upper right) Histogram of a negatively skewed response.*
Figure 4.27: *(lower left) Descriptive statistics corresponding to Figure 4.24.*
Figure 4.28: *(lower middle) Descriptive statistics corresponding to Figure 4.25.*
Figure 4.29: *(lower right) Descriptive statistics corresponding to Figure 4.26.*

In MODDE, the descriptive statistics tool comprises a type of graph called a Box-Whisker plot (Figures 4.27 – 4.29). The Box-Whisker plot is made up of a rectangular body, the box, and two attached antennae, the whiskers. Whenever the two whiskers attached to the box are of similar length, the distribution of the data is roughly normal. Figures 4.27 – 4.29 display Box-Whisker plots corresponding to the histograms of Figures 4.24 – 4.26, and we may see how the Box-Whisker plot assumes different shapes depending on the type of data distribution.

We will now give some details regarding the Box-Whisker plot. The lower and upper short horizontal lines denote the 5 and 95 percentiles of the distribution. In the box, the lowest long horizontal line depicts the lower quartile and the upper line the upper quartile. To summarize the use of the Box-Whisker plot one may say that when the whiskers are symmetrical, the response is approximately normally distributed. Whether to use a Box-Whisker plot or a histogram plot of a response much depends on personal preference.

In summary, we have now introduced four tools for evaluating raw data: condition number scatter plot, histogram of response and descriptive statistics of response. Together with the replicate plot these form a good basis for probing anomalies and errors in the input data. When the raw data are understood, the next phase is regression modelling and model interpretation.

4.4.5 Regression analysis – Analysis of variance (ANOVA)

In Section 4.3.2, we introduced the summary of fit diagnostic tool. We will now introduce two other diagnostic tools: first the analysis of variance, ANOVA, and its lack of fit test (current section), and then later the normal probability plot of residuals (next section). ANOVA is described in Chapter 17, and for the moment we will only focus on how to *use* this tool as a diagnostic test. ANOVA is concerned with estimating different types of variability in the response data, and then comparing such estimates with each other by means of F-tests. Figure 4.30 shows the tabulated output typically obtained from performing an ANOVA. In this case, the *taste* response data of the CakeMix application have been analyzed.

ANOVA Table - Response Taste

	1	2	3	4	5	6	7
1	Taste	DF	SS	MS (variance)	F	p	SD
2	Total	11	247.205	22.4731			
3	Constant	1	242.426	242.426			
4							
5	Total Corrected	10	4.77829	0.477829			0.691252
6	Regression	4	4.72142	1.18035	124.525	**0.000**	1.08644
7	Residual	6	0.056873	0.00947883			0.0973593
8							
9	Lack of Fit	4	0.0496064	0.0124016	3.41329	**0.239**	0.111362
10	(Model Error)						
11	Pure Error	2	0.00726666	0.00363333			0.0602771
12	(Replicate Error)						
13							
14		N = 11	Q2 =	0.937	Cond. no. =	1.173	
15		DF = 6	R2 =	0.988	Y-miss =	0	
16			R2 Adj. =	0.980	RSD =	0.09736	

The selected response 'Taste'

Figure 4.30: *Typical output from an ANOVA evaluation.*

In ANOVA, two F-tests are made, and to evaluate these one examines the probability values, p, here highlighted by frames. The first test assesses the significance of the regression model, and when $p < 0.05$ this test is satisfied. We can see that the CakeMix model is statistically significant as p is approaching 0 (it equals 6.68E-06). The second test compares the model error and the replicate error. When a sufficiently low model error is obtained the model shows good fit to the data, that is, the model has no *lack of fit*. Hence, this latter test is known as the lack of fit test, and it is satisfied when $p > 0.05$. In the CakeMix case, p is 0.239, which is larger than the reference value, and therefore we conclude that the model has no lack of fit. As mentioned in Section 4.3.2, the outcome of the lack of fit test, the p-value, is used when computing the model validity value displayed in the Summary of Fit plot. When the p-value of the lack of fit test is 0.05 the model validity parameter is 0.25. Notice, however, that lack of fit cannot be assessed unless replicated experiments have been performed.

4.4.6 Regression analysis – Normal probability plot of residuals

Our third important diagnostic tool is the normal probability plot, N-plot, of response residuals. This is a good tool for finding deviating experiments, so called outliers. An example N-plot is shown in Figure 4.31. It displays the residuals of the NOx response of

General Example 2. The vertical axis in this plot gives the normal probability of the distribution of the residuals. The horizontal axis corresponds to the numerical values of the residuals. However, the residuals are not expressed in the original unit of the NOx response, rather each entry has been divided by the standard deviation of the residuals. In this way, the scale of the horizontal axis is expressed as standard deviations, SDs.

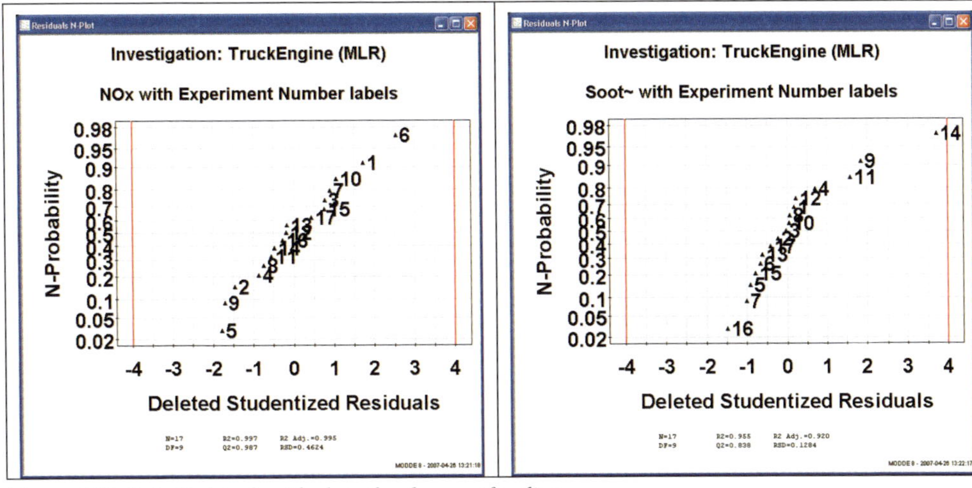

Figure 4.31: *(left) N-plot of residuals in the absence of outliers.*
Figure 4.32: *(right) N-plot of residuals when a suspect point is present.*

To detect possibly deviating experiments one proceeds as follows. A straight line going through the majority of the points is fitted with the eye. This line must pass through the point (0, 50%). Any point *not* falling close to this straight line is a suspect outlier. Note, that for an N-plot of residuals to be meaningful around 12-15 experiments are needed. Otherwise it is hard to draw a straight line. It is also favorable if the number of degrees of freedom exceeds three.

In the NOx case, all the residuals are approximately normally distributed, and no deviating experiment is detectable. In the same application, however, we have another response, Soot, for which the situation is different. This is shown in Figure 4.32. In particular, experiment #14 seems to have a much larger residual than the others. This is an experiment that ought to be checked more closely.

Interestingly, it is possible to formulate warning and action limits in the N-plot. The warning limit is ± 3 SDs and the action limit is ± 4 SDs. This means that all experimental points that are found inside ± 3 SDs are good and should be kept in the model. Then there is a "gray-zone" between ± 3 and ± 4 SDs in which the analyst has to start paying attention to suspected outliers. All experiments outside ± 4 SDs are considered as statistically significant outliers and may be deleted. However, we emphasize that a critical and cautious attitude towards removing outliers from the model will be rewarding in the long run. It is always best to re-run the suspicious experiment.

In summary, we have now outlined three model diagnostic tools, the summary of fit plot, the lack of fit test, and the N-plot of residuals. Any regression model that is to be used for predictive purposes should ideally pass these three diagnostic tools. Also note, that the summary of fit plot provides information about the ANOVA lack of fit test in terms of the model validity parameter.

4.4.7 Use of model – Making predictions

The last stage in the analysis of DOE data is to use the regression model for making predictions. In Sections 4.3.5 and 4.3.6, it was demonstrated how graphical tools, such as the response contour plot, could be used for this purpose. At this stage of the course, we feel that it is appropriate to broaden the scope a little. Consider the response contour plot of the taste response given in Figure 4.33. This plot only indicates a *point estimate* of the taste, and does not provide the uncertainty involved. To assess the prediction uncertainty, there is the option in MODDE of transferring the most interesting factor settings into a prediction spreadsheet and computing the confidence intervals of the predicted response values.

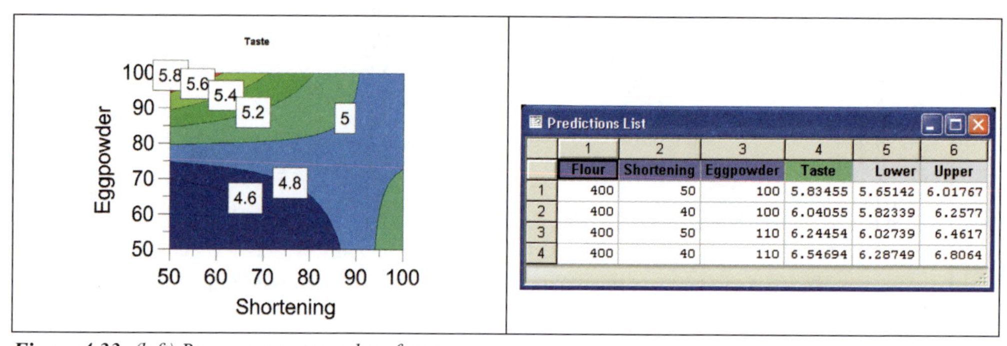

	1	2	3	4	5	6
	Flour	Shortening	Eggpowder	Taste	Lower	Upper
1	400	50	100	5.83455	5.65142	6.01767
2	400	40	100	6.04055	5.82339	6.2577
3	400	50	110	6.24454	6.02739	6.4617
4	400	40	110	6.54694	6.28749	6.8064

Figure 4.33: *(left) Response contour plot of taste.*
Figure 4.34: *(right) Prediction spreadsheet providing 95% confidence intervals around predicted taste values.*

Assuming that the predominant objective is that of maximizing the taste, we show a prediction spreadsheet in Figure 4.34 in which predictions are made for factor settings expected to be relevant for the goal. The first predicted point corresponds to the upper left-hand corner of the response contour plot. We can see that the 95% confidence interval indicates that at this point cakes with a taste value of 5.83 ± 0.18 are obtainable. The other three points (Figure 4.34) correspond to extrapolated factor settings outside the model calibration domain. Because they are extrapolations greater prediction uncertainties are associated with these points. However, all three predicted points unanimously indicate that even better tasting cakes are obtainable outside the explored experimental region. Hence, the process operator should select one of these proposed recipes and carry out the verifying experiment.

We note that this manual and graphically based approach works best in small designs with few runs and responses, and becomes less practical when working with many responses and sometimes conflicting goals. Fortunately, as we shall se later, it is possible to automate this kind of search for better factors settings.

4.5 Advanced level of data analysis

In this section our aim is to provide an introduction to the partial least squares projections to latent structures, PLS, method. PLS has certain features which are appealing in the analysis of more complicated designs, notably its ability to cope with several correlated responses in one single regression model. The PLS model represents a different mathematical construction, which has the advantage that a number of new diagnostic tools emerge. These tools are useful for model interpretation in more elaborate applications, and are more informative than other diagnostic tools. A detailed account of PLS is given in Chapter 18 and in the statistical appendix. For the moment the intention is to position PLS in the DOE framework

4.5.1 Introduction to PLS regression

PLS is a pertinent choice if (a) there are several correlated responses in the data set, (b) the experimental design has a high condition number, above 10, and (c) there are small amounts of missing data in the response matrix. The nice feature with PLS is that all the diagnostic tools that we have described so far are retained. In addition to R^2/Q^2, ANOVA and N-plot, PLS provides other diagnostic tools known as *scores* and *loadings*.

Consider the ByHand application. As seen in Figure 4.35, PLS experiences the same problems as MLR in the modelling of the second response, but accounts well for y_1 and y_3. The loading plot in Figure 4.36 is informative in the model interpretation. In this plot, one can see the correlation structure between all factors and responses at the same time. Points that lie close to each other and far from the origin in the loading plot indicate variables (factors and responses) that are highly correlated. For instance, the inspection of the plot in Figure 4.36 indicates that factor x_2 and response y_3 are strongly correlated. This means that x_2 has a strong positive influence on y_3 – that an increase in x_2 is likely to result in an increased value of y_3. On the other hand, factor x_1 is plotted relatively distant from this response, indeed on the opposite side of the origin – indicating that these two variables are less well correlated. We therefore expect this factor to have a lesser, and negative, influence on y_3.

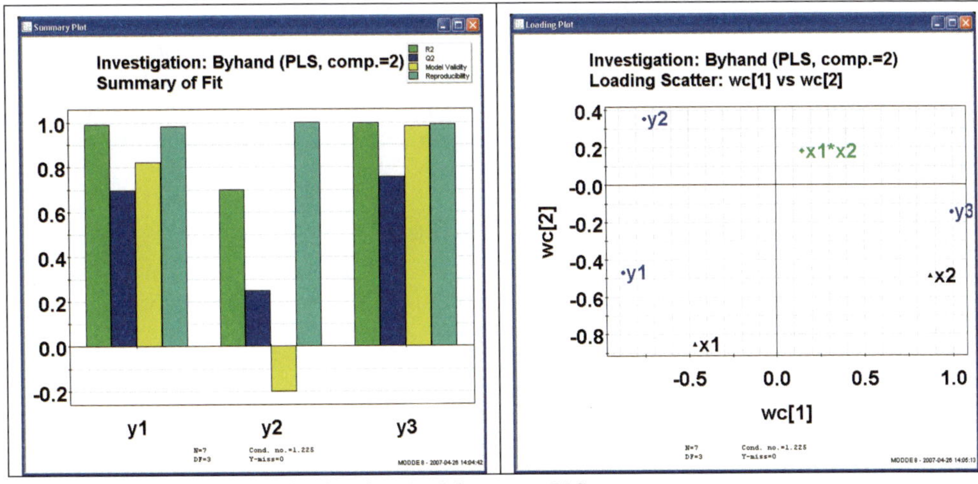

Figure 4.35: *(left) Summary of fit plot of ByHand data using PLS.*
Figure 4.36: *(right) Loading plot of the PLS model.*

It is possible to convert the PLS solution into an expression based on regression coefficients. This is shown in Figure 4.37. Notice that in this plot the coefficients are expressed a little differently from usual, in that they have been normalized; that is, the raw coefficients (scaled and centered) have been divided by the standard deviation of the respective response. This gives us the best overview of the effect of each factor on each response. We can see that the information in Figure 4.37 corroborates that in Figure 4.36; namely that x_2 has a large positive coefficient for y_3, while the corresponding coefficient for x_1 is small and negative.

However, these coefficient plots do not reflect the correlation structure among the responses. Fortunately, such information is evident from the loading plot (Figure 4.36), which indicates that y_1 and y_2 are positively correlated, but mutually negatively correlated with y_3. That this interpretation makes sense can be checked with the correlation matrix shown in Figure 4.38. It can be seen that the response correlation coefficients are 0.52 for y_1/y_2, -0.80 for y_1/y_3 and -0.79 for y_2/y_3. Any correlation coefficient larger than |0.5| indicates a strong correlation. Hence, the three ByHand responses may well be modelled together with PLS.

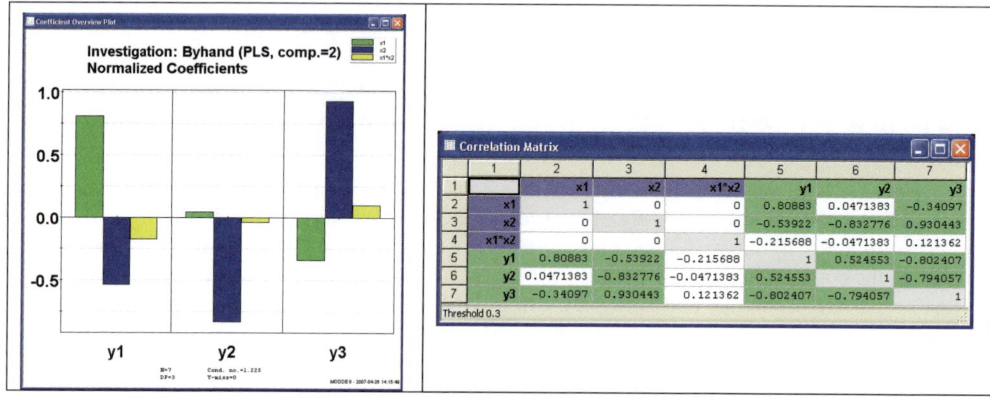

	1	2	3	4	5	6	7
1		x1	x2	x1*x2	y1	y2	y3
2	x1	1	0	0	0.80883	0.0471383	-0.34097
3	x2	0	1	0	-0.53922	-0.832776	0.930443
4	x1*x2	0	0	1	-0.215688	-0.0471383	0.121362
5	y1	0.80883	-0.53922	-0.215688	1	0.524553	-0.802407
6	y2	0.0471383	-0.832776	-0.0471383	0.524553	1	-0.794057
7	y3	-0.34097	0.930443	0.121362	-0.802407	-0.794057	1

Threshold 0.3

Figure 4.37: *(left) Regression coefficient overview of the PLS model.*
Figure 4.38: *(right) Correlation matrix of ByHand data.*

4.6 Causes of poor model

Now, that we have completed the treatment of the data analytical cycle used in DOE, we shall see that regression analysis of DOE data may not always run smoothly. Our objective is to describe four common causes of poor models, describe how to pinpoint these causes, and illustrate what to do when they have been detected. Fortunately, in most cases, the measures needed to solve such problems are not sophisticated.

4.6.1 Cause 1 – Skew response distribution

Failure to recognize that a response variable has a skewed distribution is a common reason for poor modelling results. The need for response transformation may be detected with a histogram plot or a Box-Whisker plot of the raw data. Consider General Example 2 and the third response called Soot. Figure 4.39 shows the histogram plot for Soot and Figure 4.40 the corresponding Box-Whisker plot. It is seen that this response is too skewed to be modellable, and, in fact, the modelling of this response is quite poor in comparison with the other two responses of this investigation. This is indicated in Figure 4.41 by the comparatively low Q^2 of 0.71. Recall that this is an optimization and hence rather high demands should be put on Q^2. Since the Soot response is skewed with a tail to the right the first choice of transformation is the logarithmic transformation.

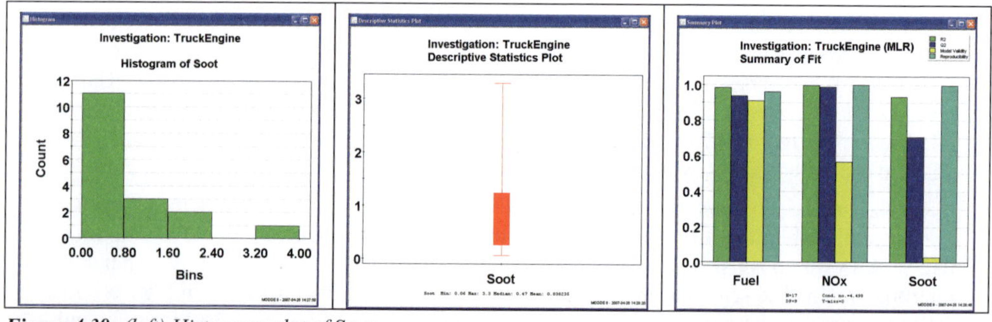

Figure 4.39: *(left) Histogram plot of Soot.*
Figure 4.40: *(middle) Box-Whisker plot of Soot.*
Figure 4.41: *(right) Summary of fit plot for General Example 2.*

The next triplet of figures, Figures 4.42 – 4.44, show the results after log-transforming Soot. Both the histogram plot and the Box-Whisker plot have improved substantially. Also, the goodness of prediction parameter, Q^2, has increased from its previous value of 0.71 to 0.84, which constitutes a strong indication that the deployed transformation was sensible. The revised model is also much better according to the model validity parameter.

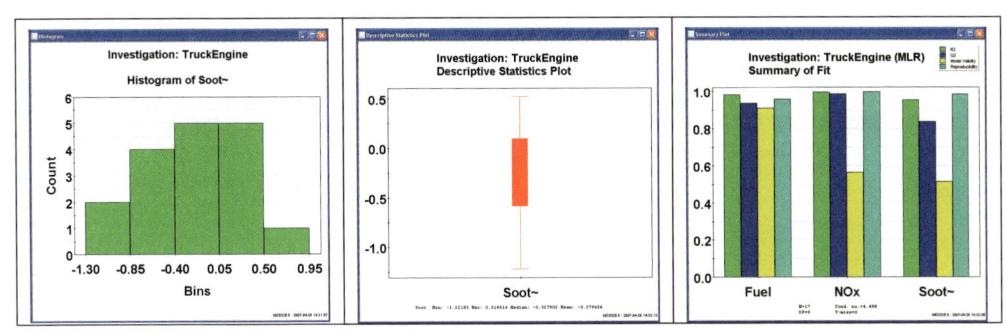

Figure 4.42: *(left) Histogram plot of Soot after transformation.*
Figure 4.43: *(middle) Box-Whisker plot of Soot after transformation.*
Figure 4.44: *(right) Summary of fit plot for General Example 2 after transformation.*

Often, non-normally distributed responses are skewed to the right, that is, the majority of the measured values are small except for a few cases which have very large numerical values. Typical examples of responses that adhere to this kind of distribution are variables expressed as amounts, levels, and concentrations of substances, and retention times in chromatography. Such responses share another feature: they all have a natural zero, i.e., non-negativity among the numerical values. In the case of a positively skewed distribution, the logarithmic transformation is the most common method of repairing a poor model.

In other cases one may encounter responses which are skewed to the left, that is, containing predominantly high numerical values and only a few small values. The histogram plot and the Box-Whisker plot are useful for detecting a negatively skewed response, and an example was given in Section 4.4.4. Typical examples of such responses are variables which are expressed as percentages, and where almost all measured values are close to 100% except for a few which are somewhat lower. In order to make a response with a negative skew more normally distributed, we may use a modification of the logarithmic transformation called NegLog. With this transformation each measured value is subtracted from the maximum value, and then the negative logarithm is formed. There are also many other transformations in use, but this is an advanced topic which will not be addressed further in the course.

In summary, one important reason for poor models is non-normally distributed responses. A skewed response distribution is easily detected by making a histogram or a Box-Whisker plot. However, one should observe that other diagnostic tools may also be used to reveal a response which needs to be transformed.

4.6.2 Benefits of response transformation

A properly selected response transformation brings a number of benefits for regression modelling. Notably, it may (i) simplify the response function by linearizing a non-linear response-factor relationship, (ii) stabilize the variance of the residuals, and (iii) make the distribution of the residuals more normal, which effectively implies that outliers are eliminated. As an illustration, we shall consider a screening application aimed at producing a long-lasting device for service in an aircraft construction. Ten factors were varied using a screening design consisting of 32 experiments and the measured response was the lifetime in hours. The design employed supports a linear model.

As displayed in Figure 4.45, when fitting the linear model to the data the statistics $R^2 = 0.88$ and $Q^2 = 0.71$ were obtained. These results are reasonably satisfactory. However, a closer scrutiny of the model in terms of some residual plots provides a strong warning that something is not right with the computed model. Figures 4.46 – 4.48 provide plots displaying some of the modelling problems that hopefully might be relieved with a suitable response transformation. In Figure 4.46, the observed and fitted response values are plotted against each other. From the very strong curvature seen in this plot we can conclude that the response function between the ten factors and the lifetime response is non-linear.

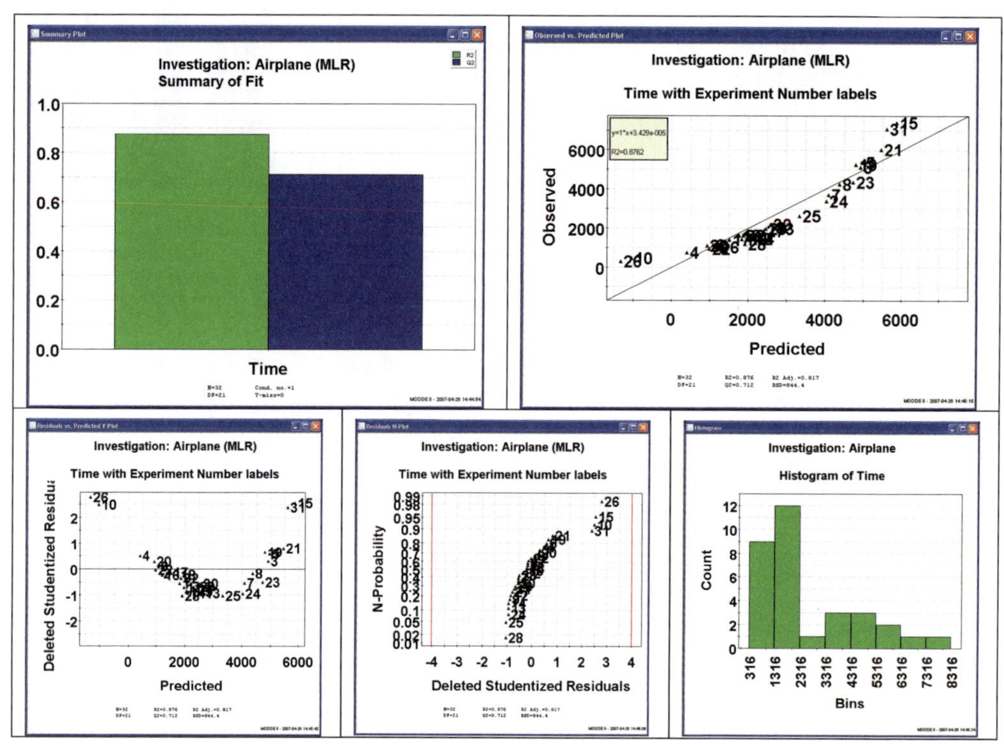

Figure 4.45: *(upper left) Summary of fit plot – before transformation.*
Figure 4.46: *(upper right) Observed versus predicted data – before transformation.*
Figure 4.47: *(lower left) Residuals versus predicted data – before transformation.*
Figure 4.48: *(lower middle) N-plot of residuals – before transformation.*
Figure 4.49: *(lower right) Histogram plot of response – before transformation.*

Further, in Figure 4.47 the modelling residuals are plotted against the fitted response values. In this kind of plot one does not want to see any systematic structure. In our case, however, a strong boomerang-like data distribution is seen, suggesting some structure to the variance of the residuals. This is unwanted. Similarly, in Figure 4.48, the normal probability plot of residuals shows a group of four deviating experiments. Recall that in this plot, the desired result is that all residuals are situated on a straight line which goes through the point (0, 50%). Because of the non-linear appearance of the residuals in the N-plot, it is logical to conclude that the response is unsymmetrically distributed. That this is indeed the case is evidenced by the histogram plot displayed in Figure 4.49.

Consequently, the lifetime response was log-transformed and the linear model refitted, the results of which are illustrated in Figures 4.50 – 4.54. Figure 4.50 shows that R^2 now becomes 0.99 and Q^2 0.98, which are significant improvements justifying the transformation undertaken. In addition, the response transformation has (i) simplified the response function by making the response-factor relationship linear (Figure 4.51), (ii) made the size of the residual independent of the estimated response value (Figure 4.52), and (iii) made the distribution of the residuals more nearly normal (Figure 4.53). The histogram of the log-transformed response is plotted in Figure 4.54 and shows that the response is much more symmetrically distributed after the transformation.

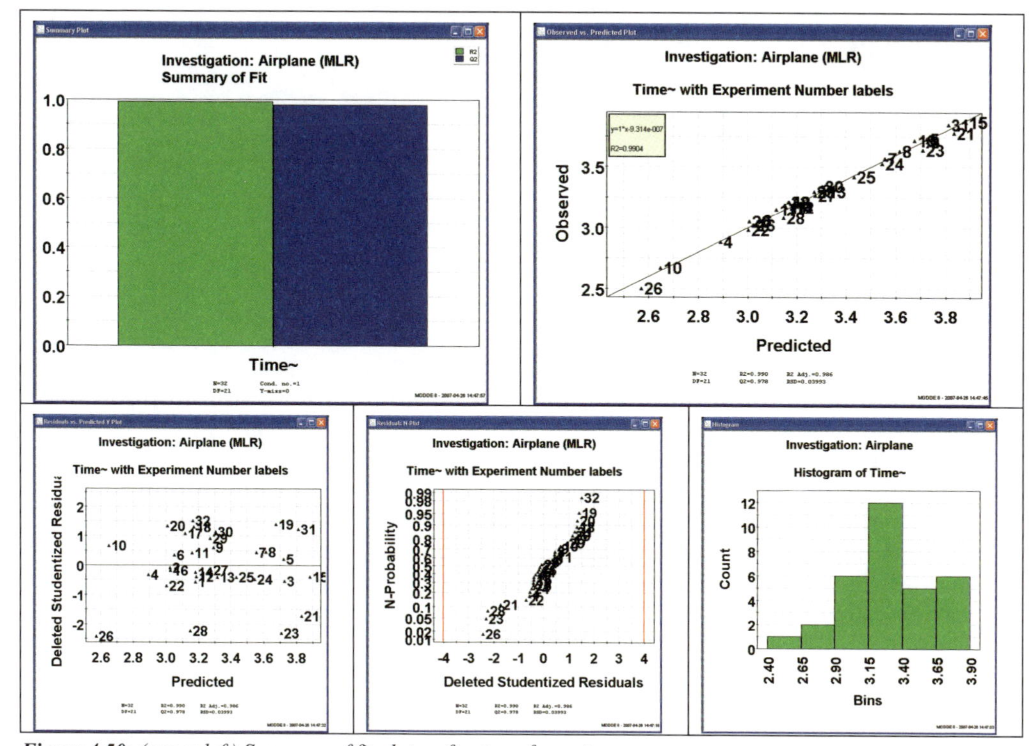

Figure 4.50: *(upper left) Summary of fit plot – after transformation.*
Figure 4.51: *(upper right) Observed versus predicted data – after transformation.*
Figure 4.52: *(lower left) Residuals versus predicted data – after transformation.*
Figure 4.53: *(lower middle) N-plot of residuals – after transformation.*
Figure 4.54: *(lower right) Histogram plot of response - after transformation.*

In summary, the need for transforming a response is frequently found from a histogram, but may be found in many other kinds of plot, as well. And, as seen in the outlined example, a carefully selected transformation may make the regression modelling simpler and more reliable.

4.6.3 Cause 2 – Curvature

Another common reason for obtaining a poor screening model is *curvature*. Curvature is a problem in screening because the normally used linear and interaction models are unable to fit such a phenomenon. Fortunately, problems related to curvature are easily detected and fixed. Let us consider the ByHand data set. We remember from Section 4.3.3 that a poor regression model was obtained concerning the second response, y_2, *unreacted starting material*. One clue to understanding this deficiency was found in the summary of fit plot (Figure 4.55). The model for y_2 has low model validity. It therefore exhibits significant lack of fit (Figure 4.56).

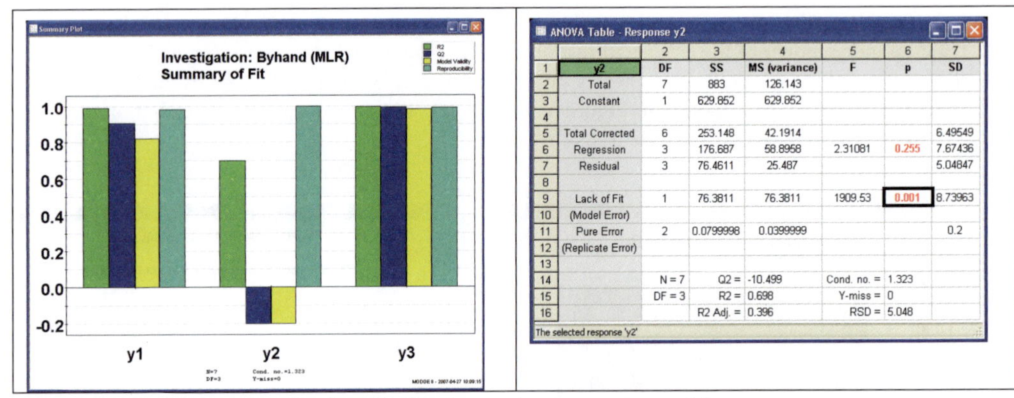

	1	2	3	4	5	6	7
1	y2	DF	SS	MS (variance)	F	p	SD
2	Total	7	883	126.143			
3	Constant	1	629.852	629.852			
4							
5	Total Corrected	6	253.148	42.1914			6.49549
6	Regression	3	176.687	58.8958	2.31081	0.255	7.67436
7	Residual	3	76.4611	25.487			5.04847
8							
9	Lack of Fit	1	76.3811	76.3811	1909.53	0.001	8.73963
10	(Model Error)						
11	Pure Error	2	0.0799998	0.0399999			0.2
12	(Replicate Error)						
13							
14		N = 7	Q2 =	-10.499	Cond. no. =	1.323	
15		DF = 3	R2 =	0.698	Y-miss =	0	
16			R2 Adj. =	0.396	RSD =	5.048	

Figure 4.55: *(left) Summary of fit plot of ByHand example – interaction model.*
Figure 4.56: *(right) ANOVA table of the second response – interaction model.*

In a screening application, lack of fit frequently indicates curvature. We know from previous observations that curvature exists between y_2 and the factor temperature. Lack of fit means that a regression model has some model error, that is, it contains some imperfections. Another way of formulating this deficiency in fitting ability is to state that the model error is *too large* in relation to the replicate error. It is reasonable to attempt to modify the model so that the model error decreases. Ideally, the model error should be comparable with the replicate error.

In the ByHand case we can modify the regression model, and test whether the introduction of the squared term of temperature is beneficial. Figure 4.57 shows the summary of fitting a modified model with the term $Temp^2$, and Figure 4.58 the corresponding ANOVA table. Evidently, the introduction of a square term is necessary for the second response. However, this operation is not advantageous for the first and third responses. The new ANOVA table shows that the lack of fit has vanished, indicating that the square term is meaningful.

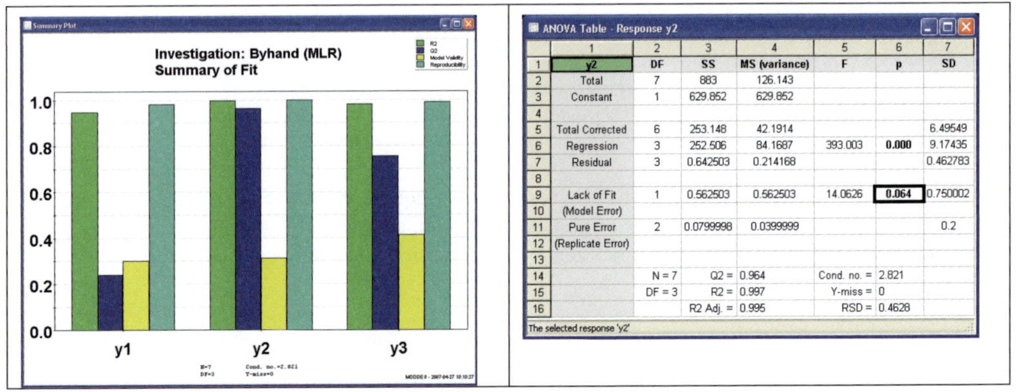

	1	2	3	4	5	6	7
1	y2	DF	SS	MS (variance)	F	p	SD
2	Total	7	883	126.143			
3	Constant	1	629.852	629.852			
4							
5	Total Corrected	6	253.148	42.1914			6.49549
6	Regression	3	252.506	84.1687	393.003	0.000	9.17435
7	Residual	3	0.642503	0.214168			0.462783
8							
9	Lack of Fit	1	0.562503	0.562503	14.0626	0.064	0.750002
10	(Model Error)						
11	Pure Error	2	0.0799998	0.0399999			0.2
12	(Replicate Error)						
13							
14		N = 7	Q2 =	0.964	Cond. no. =	2.821	
15		DF = 3	R2 =	0.997	Y-miss =	0	
16			R2 Adj. =	0.995	RSD =	0.4628	

Figure 4.57: *(left) Summary of fit plot of ByHand example – modified model.*
Figure 4.58: *(right) ANOVA table of the second response – modified model.*

Now, however, some words of caution are appropriate, because what we have just done is theoretically dubious. *One cannot reliably estimate a quadratic term with a screening design.* For a rigorous assessment of a quadratic term an optimization design is mandatory. Despite this, the model refinement performed suggests that the quadratic term is useful, and hence it seems motivated to expand the screening design with extra experiments, so that it becomes a valid optimization design. How this may be accomplished will be discussed later (see Chapter 5).

4.6.4 Cause 3 – Bad replicates

A third common cause resulting in a poor screening model is when replicated experiments spread too much. Bad replicates are easily detected either by a replicate plot, as pointed out in Section 4.3.1, or by studying the ANOVA table. Consider the replicate plot in Figure 4.59. It is obvious from this picture that the six replicates, numbered 15 – 20, vary almost as much as the other 14 experiments. Such a large variation between replicates will destroy any regression model.

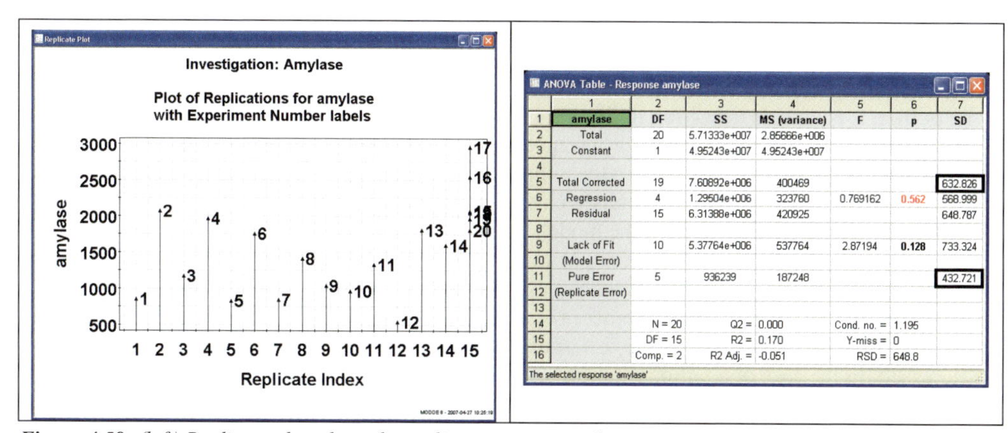

	1	2	3	4	5	6	7
1	amylase	DF	SS	MS (variance)	F	p	SD
2	Total	20	5.71333e+007	2.85666e+006			
3	Constant	1	4.95243e+007	4.95243e+007			
4							
5	Total Corrected	19	7.60892e+006	400469			632.826
6	Regression	4	1.29504e+006	323760	0.769162	0.562	568.999
7	Residual	15	6.31388e+006	420925			648.787
8							
9	Lack of Fit	10	5.37764e+006	537764	2.87194	0.128	733.324
10	(Model Error)						
11	Pure Error	5	936239	187248			432.721
12	(Replicate Error)						
13							
14		N = 20	Q2 =	0.000	Cond. no. =	1.195	
15		DF = 15	R2 =	0.170	Y-miss =	0	
16		Comp. = 2	R2 Adj. =	-0.051	RSD =	648.8	

Figure 4.59: *(left) Replicate plot where the replicate error is too large.*
Figure 4.60: *(right) ANOVA table corresponding to the model based on Figure 4.59.*

The replicate plot gives a graphical appraisal of the relationship between the replicate error and the variation across all samples. A more quantitative assessment is retrievable through ANOVA (Figure 4.60). Focus on the two standard deviation estimates marked by frames. The upper SD estimates the variation in the entire experimental design. The lower SD estimates the variation among the replicates. Clearly, these are of comparable magnitude, which is unsatisfactory. In conclusion, bad replicates are easy to detect, but harder to fix. Bad replicates means that you have a serious problem with your experimental set-up.

4.6.5 Cause 4 – Deviating experiments

Deviating experiments, or *outliers*, may degrade the predictive ability and blur the interpretation of a regression model. Hence, it is of great importance to detect outliers. The normal probability plot, N-plot, of residuals, introduced and explained in Section 4.4.6, is an excellent graphical tool for uncovering deviating experiments. An illustration of this point is given in Figure 4.61.

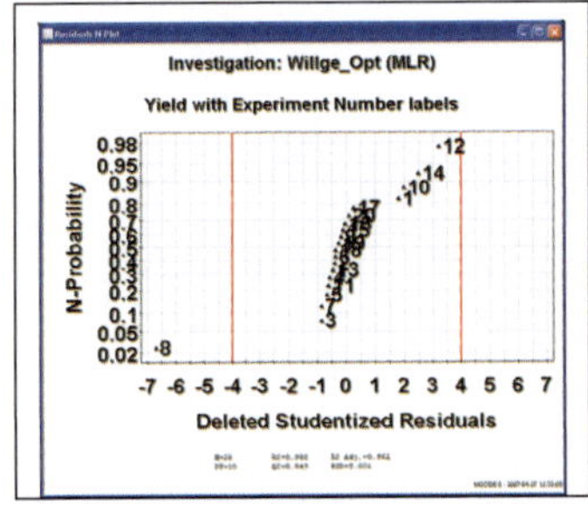

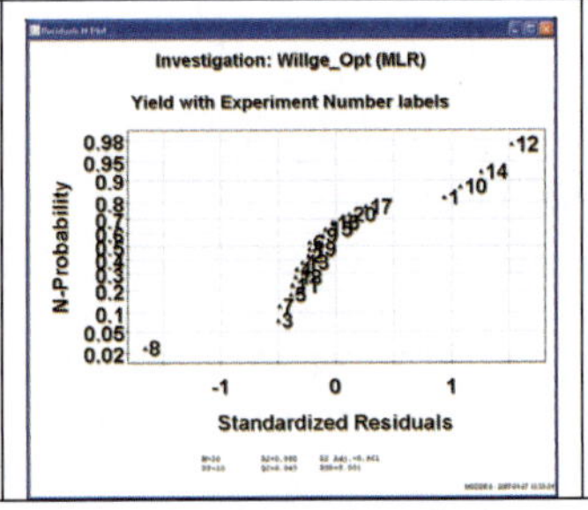

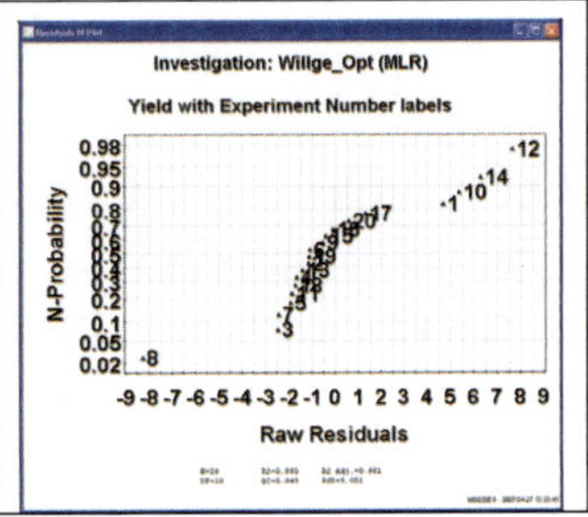

Figure 4.61: *(left) N-plot of residuals – Deleted studentized residuals.*
Figure 4.62: *(middle) N-plot of residuals – Standardized residuals.*
Figure 4.63: *(right) N-plot of residuals – Raw residuals.*

Without going into any details of this example, we may observe that experiment number 8 is an outlier. It is outside the action limit of ± 4 standard deviations. This deviating result could be caused by several phenomena, for instance, an incorrect experimental value in the worksheet, a failing experiment, or a too simple model. What to do with such a strong outlier is a non-trivial question. From a purely technical and statistical perspective, it is clear that the outlier should be eliminated from the model, as it degrades its qualities. However, from a scientific and ethical viewpoint, a more cautious attitude is recommended. It may well be the case that the deviating experiment is the really interesting one, and may indicate a substantially improved product or process operating condition. Automatic deletion of all outliers is not recommended, as the risk of obtaining spurious and meaningless models is increased. In the case of a failing experiment, we recommend that the initial experiment is removed from the model, but should be replaced by a re-tested run.

Moreover, the N-plot displayed in Figure 4.61 enables us to point out another important feature relating to this diagnostic test. We see that, besides experiment number 8, there are four other experiments which fall off the imaginary straight line. However, these are of lesser interest. What we are seeking in this kind of plot are the lonely and remote outliers. What happens here is that the model makes such strong efforts to model the behaviour of #8 that it looses its power to explain the other four experiments.

Finally, we will draw your attention to the kind of residuals being displayed. In MODDE, there are three alternatives, called *deleted studentized residuals, standardized residuals*, and *raw residuals*. These are displayed, for the same application, in Figures 4.61 – 4.63. The raw residual is the difference between the observed and the fitted response value, expressed in the original response metric. The standardized residual is the raw residual divided by the residual standard deviation, RSD. This division makes the x-axis of the N-plot be interpretable in terms of standard deviations, but does not in any other sense change the appearance of the N-plot. The deleted studentized residual is the raw residual divided by an alternative standard deviation, computed when that particular experiment was left out of the analysis and thus not influencing the model. One can think of the deleted studentized residual as a way of mounting an amplifier on this graphical device, simply resulting in a sharper and more efficient outlier diagnostic tool. Deleted studentized residuals require at least three degrees of freedom, and are available for MLR, but not PLS.

4.7 Questions for Chapter 4

- What are the three basic stages of data analysis?
- What is the purpose of the replicate plot?
- Which requirements must be met by the R^2/Q^2 diagnostic tool?
- When does this diagnostic tool indicate a model of low quality?
- What are the basic tools for model interpretation and model use?
- What does the condition number indicate?
- What are the condition number limits for good screening, robustness testing and optimization designs if only quantitative factors are present?
- How can the scatter plot be used in conjunction with the condition number?
- What can you deduce from histograms and Box-Whisker plots of responses?
- Why is it important to remove skewed data distributions prior to regression modelling?
- What is analysis of variance?
- What can you test with the lack of fit test?
- What kind of information is retrievable from a normal probability plot of residuals?
- Which three features of PLS make it useful for the evaluation of DOE data?
- How is the PLS loading plot useful?
- Which are the four common reasons for poor models?
- What precautionary measures may be used to reduce the impact of these four causes?

4.8 Summary and discussion

The data analysis of full factorial designs can be said to comprise a sequence of three stages. The first stage is *evaluation of raw data*, in which the replicate plot is an informative tool. This plot shows the size of the replicate error in relation to the variation across the entire investigation. One example was shown, in which the replicate error was too large to permit a good model to be derived. The second stage is *regression analysis and model interpretation*. In connection with this step, we introduced a powerful diagnostic tool: the simultaneous use of the four parameters R^2 (denoted *goodness of fit*), Q^2 (denoted *goodness of prediction*), model validity and reproducibility. For a model to be valid, both R^2 and Q^2 must be high, as close to one as possible, and *R^2 must not exceed Q^2 by more than 0.2-0.3*. Furthermore, $Q^2 > 0.5$, model validity > 0.25 and reproducibility > 0.5 are necessary conditions for declaring a model to be valid and reliable.

We also showed how the model in the CakeMix application was slightly improved by the removal of two small two-factor interactions, as a result of which Q^2 increased from 0.87 to 0.94. This is not a big increase, but as the model was also simplified, it well serves to illustrate the modelling message we are trying to convey, that of maximizing Q^2. In addition, attention was given to the use of the coefficient plot for model interpretation and the response contour plot for model usage. It was also shown how triple plots of coefficients and response contour plots could be used to graphically evaluate the ByHand application and its three responses.

In the continued treatment of measures needed to properly evaluate the raw data, we outlined four additional tools. These are condition number, scatter plot, histogram of response, and descriptive statistics of response. Together with the replicate plot, these tools can help uncover anomalies in the input data. A careful evaluation of the raw data is beneficial for the subsequent regression modelling, and may result in substantially shorter analysis times. In order to facilitate the regression modelling, we also introduced ANOVA and its lack of fit test, and the normal probability plot of residuals. With the former tool it is possible to pinpoint inadequate models, and with the latter deviating experiments.

On the advanced level of data analysis, we introduced the PLS regression method. PLS is an attractive choice when one is working with (i) several correlated responses, (ii) a design with a high condition number, or (iii) a response matrix with a moderate amount of missing data. Because PLS fits only one model to all responses, the model interpretation in terms of loadings is powerful. In addition, the interpretation of loadings may be supplemented with other tools with which we are more familiar, such as, a coefficient overview plot, a table of the correlation matrix, and scatter plots of raw data. PLS is further described in Chapter 18.

Finally, in the evaluation of data from full factorial designs, one can envision four common causes that may create problems in the regression modelling. These causes relate to (i) skew response distribution, (ii) curvature, (iii) bad replicates, and (iv) deviating experiments. We have described pertinent tools for detecting such troublesome features, and also indicated what to do when such problems are encountered.

Part II

Chapters 5-7

The three main experimental objectives

5 Experimental objective: Screening

5.1 Objective

Chapters 1-4 have outlined a general framework for DOE, by describing what are the key steps in the generation of an experimental design and the analysis of the resulting data. The objective of this chapter is to make use of this framework in the presentation and discussion of a DOE screening application. We will describe in detail an application dealing with the laser welding of a plate heat exchanger. In addition, we will introduce the family of fractional factorial designs, as these are the designs used most frequently in screening. Our objective is to provide a comprehensive overview of what are fractional factorial designs, and highlight their advantages and limitations. Further, the analysis of the welding data also involves the repeated use of a number of data evaluation and regression analysis tools, which were outlined in the foregoing chapters. Of necessity, there will be some switching between the repetition of previous concepts and their application, and new theory related to fractional factorial designs.

As far as the theoretical concepts related to two-level fractional factorial designs are concerned, we will discuss confounding of estimated effects, and give examples of more complex confounding patterns resulting from heavily reduced fractional factorial designs. It is possible to regulate and manipulate the confounding pattern of a fractional factorial design. In order to understand how this may be accomplished, we must understand the concept of a generator. A generator is a column of alternating minus and plus signs of the computational matrix that is used to introduce a new factor. One or more generators of a design control the actual fraction of experiments that is selected and regulate the confounding pattern. With two or more generators it is beneficial to concatenate them in one single expression. This expression is called the defining relation. We will show how to derive such an expression and how to use it for overviewing a confounding pattern. Another relevant property of a fractional factorial design, its resolution, is obtainable through the defining relation.

5.2 Background to General Example 1

Some years ago, Alfa Laval introduced a new type of plate heat exchanger, based on the principle of replacing polymer gaskets with an all-welded gasket-free plate design. The all-welded design enabled the apparatus to endure higher operating temperatures and pressures. Such plate heat exchangers are used in a wide variety of applications, for instance, in offshore oil production for cooling of hydrocarbon gas, in power plants for preheating of feed water, and in petrochemical industries for solvent recovery, reactor temperature control, and steam applications. A small cross-section of such a plate heat exchanger is shown in Figure 5.1.

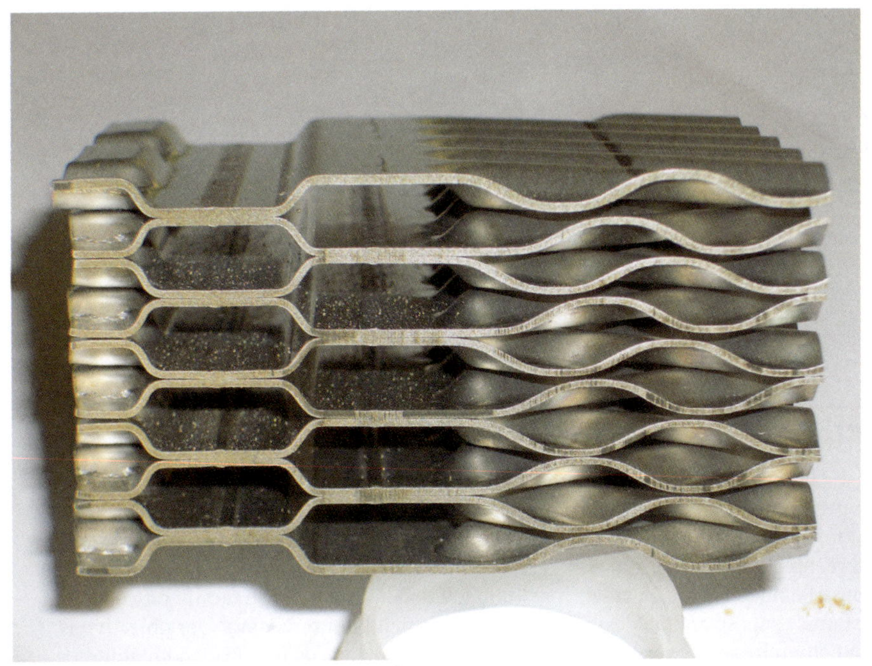

Figure 5.1: *Cross-section of plate heat exchanger.*

The example that we will present here relates to one step in the process of fabricating a plate heat exchanger, a laser welding step involving the metal nickel. The investigator, Erik Vännman, studied the influence of four factors on the shape and quality of the resulting weld. These factors were *power of laser, speed of laser*, *gas flow at nozzle* of the welding equipment, and *gas flow at root*, that is, the underside of the welding equipment. The units and settings of low and high levels of these factors are found in Figure 5.2.

In this example, the weld takes the shape of an hour-glass, where the width of the "waist" is one important response variable. To characterize the shape and the quality of the weld, the following three responses were monitored: (i) breakage of weld, (ii) width of weld, and (iii) skewness of weld. These are displayed in Figure 5.3.

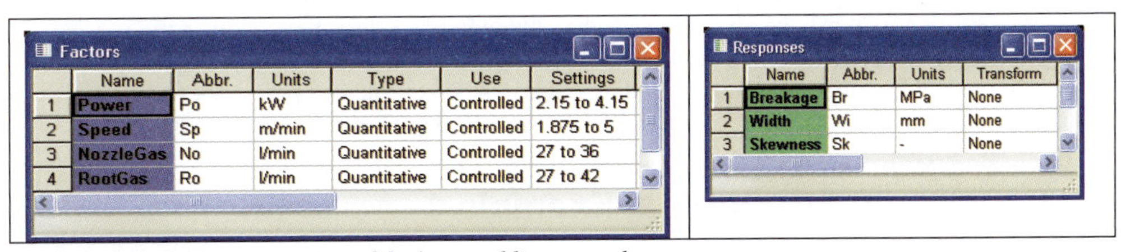

Factors

	Name	Abbr.	Units	Type	Use	Settings
1	Power	Po	kW	Quantitative	Controlled	2.15 to 4.15
2	Speed	Sp	m/min	Quantitative	Controlled	1.875 to 5
3	NozzleGas	No	l/min	Quantitative	Controlled	27 to 36
4	RootGas	Ro	l/min	Quantitative	Controlled	27 to 42

Responses

	Name	Abbr.	Units	Transform
1	Breakage	Br	MPa	None
2	Width	Wi	mm	None
3	Skewness	Sk	-	None

Figure 5.2: *(left) Factor overview of the laser welding example.*
Figure 5.3: *(right) Response overview of the laser welding example.*

Besides understanding which factors influence the welding process, the main goal was to obtain a robust weld (high value of breakage), of a well-defined width and low skewness. The desired response profile was a high value of breakage with the target set at 385 MPa, a width of weld in the range 0.7-1.0 mm, and a low value of skewness with 20 specified as the target. In the first stage, the investigator carried out eleven experiments using a medium-resolution two-level fractional factorial design. However, as we shall see during the analysis of this data-set, it was necessary to upgrade the initial screening design with more experiments. Indeed, the experimenter conducted a further set of eleven experiments, selected to complement the first series. Towards the end of this chapter we shall analyze both sets of eleven experiments together.

5.3 Problem Formulation (PF)

5.3.1 Selection of experimental objective

The problem formulation (PF) process is important in DOE, and it was discussed in Chapter 2. We will now re-iterate the steps of the problem formulation by reviewing the laser welding application. The first step in the problem formulation corresponds to choosing which experimental objective to use. Recall that the experimental objective may be selected from six stages of DOE, viz., (i) familiarization, (ii) screening, (iii) finding the optimal region, (iv) optimization, (v) robustness testing, and (vi) mechanistic modelling. In the laser welding application, the selected experimental objective was screening. With screening we want to find out a little about many factors, that is, which factors dominate and what are their optimal ranges? Typically, screening designs involve the study of between four and ten factors, but applications with as many as 12-15 screened factors are not uncommon. In the laser welding case there are four factors and this facilitates the overview of the results.

5.3.2 Specification of factors

The one thing that an experimenter always wants to avoid is to complete an experimental design, and then suddenly to be struck by the feeling that something is wrong, that some crucial factor has been left out of the investigation. If we forget an important factor in the initial design it takes a large number of extra experiments to map this factor. Thus, in principle, with DOE one can only *eliminate* factors from an investigation. In this regard, the Ishikawa, or fishbone, system diagram is a very helpful method to overview all factors *possibly* influencing the results. An example diagram is shown in Figure 5.4.

In the Ishikawa diagram a baseline is drawn. To this line is attached a number of new intersecting sub-lines corresponding to each main category of factors. Each such sub-line may be further decomposed with additional sub-lines. In the welding application we can discern four major types of factors, factors related to metal materials, factors related to weld gas, general equipment factors, and equipment factors of a more specific character. Now, let us focus on the category of metal material factors. Factors that are important in this group are the type of metal material, the thickness, the surface properties and pre-treatment procedures. The other categories of factors may be worked out in a similar fashion.

The final system diagram resulting from this kind of "mental" screening of factors is helpful for overviewing the interesting factors. Correctly used, it will diminish the risk of neglecting important factors. As discussed in Chapter 22, the fishbone diagram is regarded as an important tool in Process Analytical Technology (PAT) and Quality By Design (QBD).

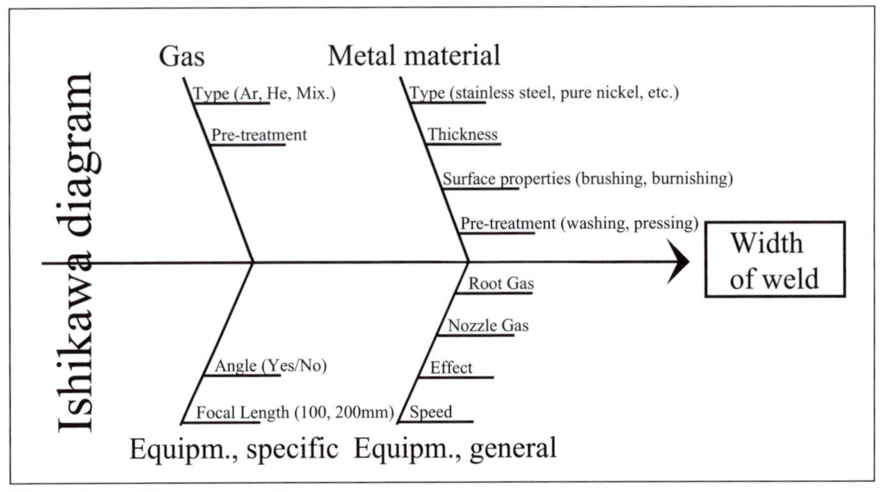

Figure 5.4: *Ishikawa system diagram.*

When all factors have been listed their ranges must be defined. This is accomplished by determining the low and high investigation values for each factor. Usually, comparatively large ranges are investigated in screening, because one does not want to run the risk of overlooking a meaningful factor. The factors selected in General Example 1 were *power of laser, speed of laser*, *gas flow at nozzle*, and *gas flow at root*. Their units and settings of low and high levels are found in Figures 5.5 – 5.8. As seen, these factors are specified according to their respective untransformed metric.

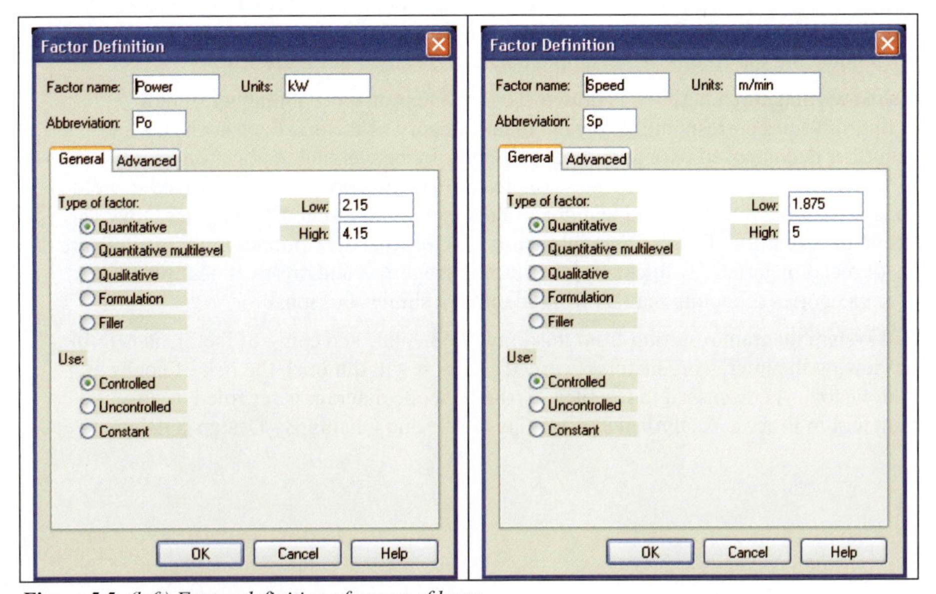

Figure 5.5: *(left) Factor definition of power of laser.*
Figure 5.6: *(right) Factor definition of speed of laser.*

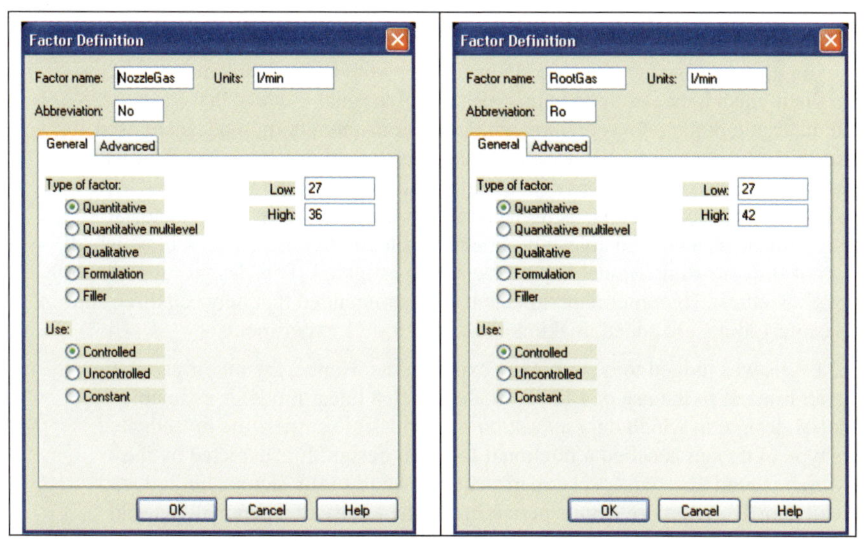

Figure 5.7: *(left) Factor definition of nozzle gas.*
Figure 5.8: *(right) Factor definition of root gas.*

5.3.3 Specification of responses

The next step in the problem formulation is to specify the responses. It is important to select responses that are relevant to the experimental goals. In the laser welding application, the quality and the shape of the resulting weld were the paramount properties. Hence, the investigator registered three responses, the breakage of the weld, the width of the weld, and the skewness of the weld. It is logical to choose three responses, because product quality is typically a multivariate property, and many responses are often needed to get a good appreciation of quality. Also, with modern regression analysis tools, it is no problem to handle many responses at the same time.

Furthermore, when specifying the responses one must first decide whether a response is of a quantitative or qualitative nature. Quantitative responses are preferable to qualitative ones, because the interpretation of the resulting regression model is simplified. We see in Figures 5.9 – 5.11 that the three weld responses are quantitative and untransformed. The first response is expressed in MPa, the second in mm, whereas the third is dimensionless. The overall goal was a response profile with a high value of breakage with the target set at 385 MPa, a width of the weld in the range 0.7-1.0 mm, and a low value of skewness with 20 specified as the target.

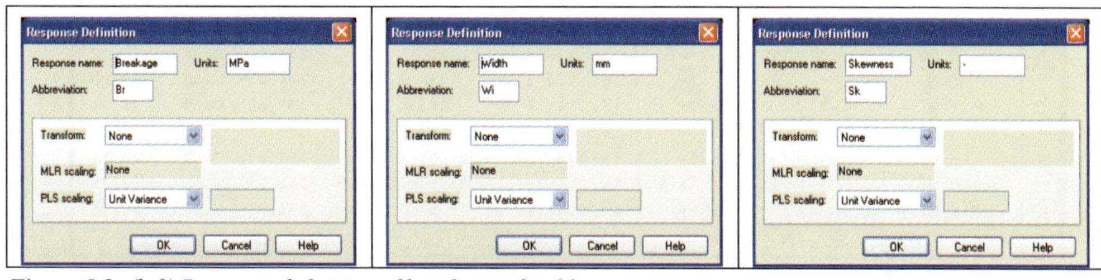

Figure 5.9: *(left) Response definition of breakage of weld.*
Figure 5.10: *(middle) Response definition of width of weld.*
Figure 5.11: *(right) Response definition of skewness of weld.*

5.3.4 Selection of regression model

The selection of an appropriate regression model is the next step in the problem formulation. Recall, that we distinguish between three main types of polynomial models, that is, linear, interaction and quadratic models. In screening, either linear or interaction models are used. This means that in the laser welding application, one could use either model, and the final choice must depend on the expected clarity in the information and on the number of experiments allowed. With four factors, as in the laser welding study, the 2^4 full factorial design in 16 experiments is a conceivable experimental protocol, that is, a design in which all possible corners of the four-dimensional hypercube are investigated. This design supports an interaction model. To these 16 corner experiments it is recommended that between three and five replicated center-points are added, making a total of 19 – 21 experiments.

Initially, 19 – 21 runs was judged to be too many experiments. Hence, the investigator decided to restrict himself to the use of a linear model. Such a linear model is estimable with a reduced factorial design, in which only a *fraction* of all possible corners are investigated. Therefore, this type of design is called a fractional factorial design. It is depicted by the notation 2^{4-1}, which is read as a two-level experimental design in four factors, but reduced one step. We will shortly give more details pertaining to the family of fractional factorial designs. For now, suffice it to say that the 2^{4-1} design encodes eight experiments, that is, eight corners out of the 16 theoretically possible. To these eight experiments, the investigator appended three replicated center-points.

5.3.5 Generation of design and creation of worksheet

The last two stages of the problem formulation deal with the generation of the statistical experimental design and the creation of the associated worksheet. We will treat these two steps together in describing the laser welding application. We must realize that the chosen regression model and the design to be generated are intimately linked. Since the experimenter selected a linear model, the 2^{4-1} fractional factorial design is an excellent choice of design. This is a standard design, which prescribes eight corner experiments.

In Figure 5.12, the four worksheet columns with numbers between 5 and 8 represent the identified design. The first eight rows are the corner experiments and the last three rows are the replicated center-points. Before we leave this part of the problem formulation, we will consider the column entitled RunOrder. It contains a proposed randomized order in which to run the eleven experiments. It is recommended that all experiments be run in a randomized order, to prevent any systematic time trend from influencing the experimental values.

Worksheet

	1	2	3	4	5	6	7	8	9	10	11
1	Exp No	Exp Name	Run Order	Incl/Excl	Power	Speed	NozzleGas	RootGas	Breakage	Width	Skewness
2	1	N1	5	Incl	2.15	1.875	27	27	382.2	1.02	24.96
3	2	N2	4	Incl	4.15	1.875	27	42	397.2	1.51	34.68
4	3	N3	2	Incl	2.15	5	27	42	375.8	0.86	12.84
5	4	N4	7	Incl	4.15	5	27	27	365.6	0.65	16.68
6	5	N5	10	Incl	2.15	1.875	36	42	384.4	0.96	27.72
7	6	N6	6	Incl	4.15	1.875	36	27	396.2	1.5	29.88
8	7	N7	3	Incl	2.15	5	36	27	355.6	0.52	12.72
9	8	N8	9	Incl	4.15	5	36	42	373.4	0.69	17.16
10	9	N9	1	Incl	3.15	3.4375	31.5	34.5	377.6	0.97	23.31
11	10	N10	11	Incl	3.15	3.4375	31.5	34.5	381.2	1.05	21.78
12	11	N11	8	Incl	3.15	3.4375	31.5	34.5	376.5	0.95	20.45

Figure 5.12: *Experimental data of the first eleven laser welding experiments.*

We have now completed all steps of the problem formulation. Now the design must be executed and the response data entered into the worksheet. In the three right-most columns of the worksheet, numbered 9-11, we can see the response values found.

5.4 Fractional Factorial Designs

5.4.1 Introduction to fractional factorial designs

We have now completed the problem formulation part of the laser welding application, and carried out all experiments. This means that it is time for the data analysis. However, before we continue with that story, we will make a temporary stop and get more closely acquainted with the family of fractional factorial designs. For this purpose, we consider the 2^7 full factorial design in 128 runs. With 128 experiments it is possible to estimate 128 model parameters, distributed as 1 constant term, 7 linear terms, 21 two-factor interactions, 35 three-factor interactions, 35 four-factor interactions, 21 five-factor interactions, 7 six-factor interactions, and 1 seven-factor interaction (Figure 5.13).

With the 2^7 full factorial design the following model parameters may be estimated:

constant	linear	2-fact. int.	3-fact. int.	4-fact. int.	5-fact. int.	6-fact. int.	7-fact. int.
1	7	21	35	35	21	7	1

Figure 5.13: *Estimable terms of the 2^7 full factorial design.*

Now, the fact that all these parameters can be estimated does not in any way guarantee that they are all of appreciable size and meaningful. Rather, there tends to be a certain hierarchy among model terms, making some terms more important than others. Looking at the absolute magnitude of model terms, we find that linear terms tend to be larger than two-factor interactions, which, in turn, tend to be larger than three-factor interactions, and so on. Consequently, it is often so that higher-order interactions tend to become negligible and can therefore be disregarded. Our long experience of applying DOE to chemical and technological problems suggests to us that three-factor interactions and interactions of higher order usually are negligible. This means that there tends to be a redundancy in a 2^k full factorial design, that is, an excess number of parameters which *can* be estimated but which lack relevance. This is the entry point for fractional factorial designs. They exploit this redundancy, by trying to reduce the number of necessary design runs.

5.4.2 A geometric representation of fractional factorial designs

The question which arises is *how is this decrease in the number of necessary experiments accomplished*? We begin by considering the geometry of the 2^3 full factorial design, as displayed in Figure 5.14. It represents the CakeMix application, except for the fact that all center-points are omitted from the drawing. With this full factorial design three factors are investigated in eight runs. Interestingly, the 2^3 full factorial design may be split into two balanced fractions, one shown in Figure 5.15 and one shown in Figure 5.16.

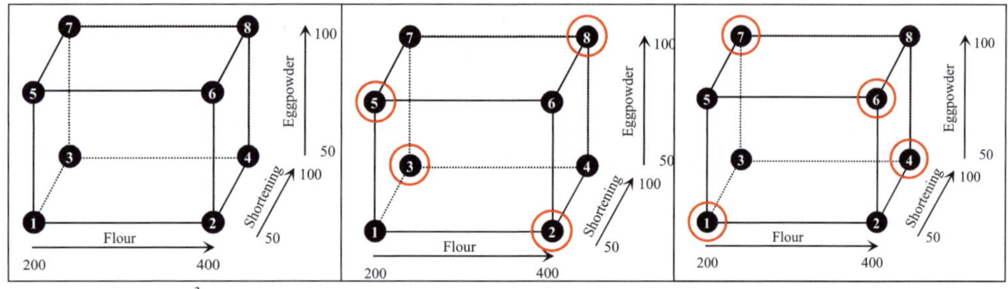

Figure 5.14: *(left) 2^3 full factorial design in eight runs.*
Figure 5.15: *(middle) First half-fraction of 2^3 design denoted 2^{3-1}.*
Figure 5.16: *(right) Second half-fraction of 2^3 design also denoted 2^{3-1}.*

The two half-fractions shown in Figures 5.15 and 5.16 are encoded by the so called 2^{3-1} fractional factorial design, and imply that three factors can be explored in four runs. From a design point of view these two fractions are equivalent, but in reality *one* may be preferred to the other because of practical experimental considerations. The fractional factorial design used in the laser welding application was created in an analogous fashion, only in this case the 16 possible corners were divided into half-fractions each comprising eight experiments. These half-fractions are encoded by the 2^{4-1} fractional factorial design, and one of them was selected as the design to be performed.

In fact, by forming fractions like this, any two-level full factorial design may be converted into a fractional factorial design with a balanced distribution of experiments. However, with five or more factors, the parent two-level full factorial design may be reduced by more than one step. For instance, the 2^5 full factorial design may be reduced one step to become the 2^{5-1} fractional factorial design, or two steps to become the 2^{5-2} fractional factorial design. Using the former fractional factorial design, the 32 theoretically possible corner experiments are divided into two fractions of 16 experiments, and one such half-fraction is then selected as the working design. In the latter case, the 32 corners are divided into four fractions of eight experiments, and one of these quarter-fractions is then selected as the working design. Thus, in principle, it is possible to screen 5 factors in either 32, 16, or 8 experiments, plus some additional center-points. Which design to prefer will be discussed shortly.

5.4.3 Going from the 2^3 full factorial design to the 2^{4-1} fractional factorial design

We will now briefly describe the technique behind the fractionation of two-level full factorial designs. Consider the computational matrix of the 2^3 design shown in Figure 5.17. This design consists of eight corner experiments, making it possible to calculate a regression model of seven coefficients and one constant term. These seven regression coefficients represent three factor main effects, three two-factor interactions, and one three-factor interaction. Thus, we have eight experiments in the design and estimate eight model terms, which means that all possible effects are estimated.

Run #	const.	x1	x2	x3	x1x2	x1x3	x2x3	x1x2x3
1	+	−	−	−	+	+	+	−
2	+	+	−	−	−	−	+	+
3	+	−	+	−	−	+	−	+
4	+	+	+	−	+	−	−	−
5	+	−	−	+	+	−	−	+
6	+	+	−	+	−	+	−	−
7	+	−	+	+	−	−	+	−
8	+	+	+	+	+	+	+	+

Figure 5.17: *Computational matrix of the 2^3 design.*

However, as was previously pointed out, three-factor interactions are usually negligible. This implies that in the 2^3 case the right-most column of the computational matrix encoding the $x_1x_2x_3$ three-factor interaction is not utilized fully. This is because it is set to represent a model term likely to be near-zero in numerical value. Hence, there is a slot "open" in the computational matrix for depicting a more prominent model term. As is shown in Figure 5.18, this free column may be assigned to another factor, say x_4. Thus, the column previously labeled $x_1x_2x_3$ now shows how to vary x_4 when doing the experiments. This way of introducing a fourth factor means that we have converted the 2^3 full factorial design to a 2^{4-1} fractional factorial design. Thus, we can explore four factors by means of eight experiments, and not 16 experiments, as is the case with the 2^4 full factorial design. We will soon see what price we have to pay for this experiment reduction.

Run #	const.	x1	x2	x3	x1x2	x1x3	x2x3	x4= x1x2x3
1	+	−	−	−	+	+	+	−
2	+	+	−	−	−	−	+	+
3	+	−	+	−	−	+	−	+
4	+	+	+	−	+	−	−	−
5	+	−	−	+	+	−	−	+
6	+	+	−	+	−	+	−	−
7	+	−	+	+	−	−	+	−
8	+	+	+	+	+	+	+	+

Figure 5.18: *How to introduce a fourth factor.*

5.4.4 Confounding of effects

The price we have to pay when reducing the number of experiments is that our factor effects can no longer be computed completely free of one another. The effects are said to be *confounded*, that is, to a certain degree mixed up with each other. The extent to which the effects are confounded in the 2^{4-1} fractional factorial design, and, indirectly the laser welding application, is delineated in Figure 5.19. This figure contains a table of the confounding pattern, and there are two important observations to make. The first observation relates to the fact that the 16 possible effects are evenly allocated as two effects per column. The second observation is that the main effects are confounded with the three-factor interactions, and that the two-factor interactions are mutually confounded. This is a comparatively simple confounding situation, because all the effects of primary concern in screening, *the main effects*, are aliased with the negligible three-factor interactions. This means that when we calculate our regression model, and use the resulting regression coefficient plot for interpreting the model, we will be able to identify the most important factors with high certainty. This is addressed graphically in the ensuing section.

	$x_1x_2x_3x_4$	$x_2x_3x_4$	$x_1x_3x_4$	$x_1x_2x_4$	$x_1x_2x_3$	x_3x_4	x_2x_4	x_2x_3
	constant	x_1	x_2	x_3	x_4	x_1x_2	x_1x_3	x_1x_4
1	+	−	−	−	−	+	+	+
2	+	+	−	−	+	−	−	+
3	+	−	+	−	+	−	+	−
4	+	+	+	−	−	+	−	−
5	+	−	−	+	+	+	−	−
6	+	+	−	+	−	−	+	−
7	+	−	+	+	−	−	−	+
8	+	+	+	+	+	+	+	+

Figure 5.19: *Confounding pattern of the 2^{4-1} fractional factorial design.*

5.4.5 A graphical interpretation of confoundings

We recall from the foregoing discussion that in the case of the 2^{4-1} fractional factorial design, the main effects are confounded with three-factor interactions and the two-factor interactions are confounded with each other. This means that when we try to interpret a regression model based on this design, we must bear in mind that the model terms are not completely resolved from one another. A schematic drawing of a regression model based on a 2^{4-1} fractional factorial design is shown in Figure 5.20. The confounding pattern of this model is listed underneath the coefficients.

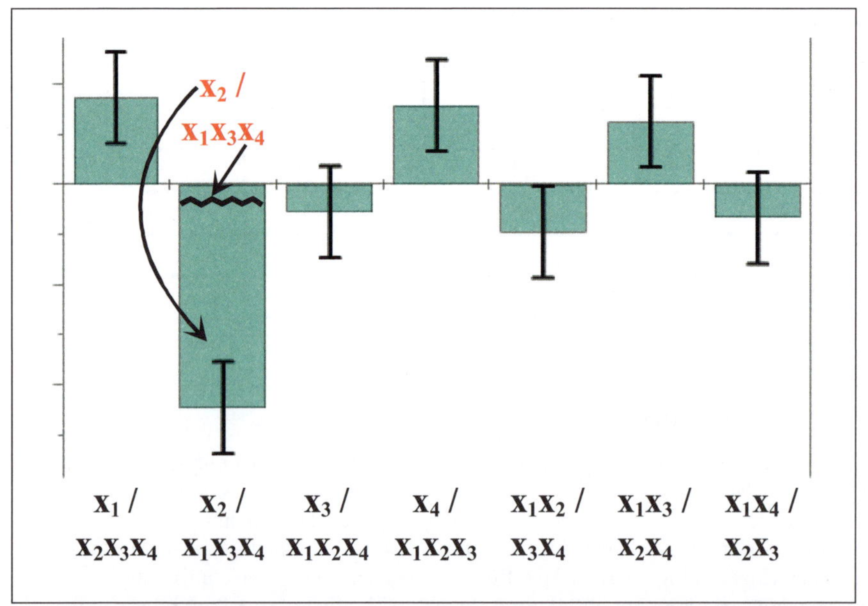

Figure 5.20: *Hypothetical regression coefficients of a model based on a 2^{4-1} design.*

Now, let us focus on the largest coefficient, the second one from the left. This regression coefficient represents the *sum* of the impact of the factor x_2 and the $x_1x_3x_4$ three-factor interaction. With the available experiments it is impossible to resolve these two terms. Fortunately, however, the three-factor interaction can be assumed to be of negligible relevance, and hence its contribution to the displayed regression coefficient is small. Thus, the regression coefficient that we see is a good approximation of the importance of x_2. A similar reasoning can of course be applied to the other main effects and three-factor interactions.

Concerning the confounding of the two-factor interactions, the situation is more complex and the interpretation harder. Sometimes it is possible to put forward an educated guess as to which two-factor interaction is likely to dominate in a confounded pair of two-factor interactions. Such reasoning is based on assessing which of the main effects are the largest ones. Factors with larger main effects normally contribute more strongly to a two-factor interaction than do factors with smaller main effects. Thus, regarding the set of regression coefficients given in Figure 5.20, it is reasonable to anticipate that in the confounded pair x_1x_3/x_2x_4, probably the x_2x_4 two-factor interaction contributes more strongly to the model. But the only way to be absolutely sure of this hypothesis is to do more experiments.

5.4.6 The concept of a design generator

Our aim is now to describe in more detail how it is possible to reduce the number of experiments, that is, how to select less than the full set of corner experiments. In doing so, we will start by introducing the concept of a generator. Consider the 2^3 full factorial design displayed in Figure 5.21. This is the CakeMix application. Now, let us assume that, for some reason, it is impossible to test all eight corners. The available resources for testing sanction only four corner experiments, plus an additional three replicated center-points. This means that we would like to construct the 2^{3-1} fractional factorial design in four corner experiments. Actually, as is evident from Figures 5.22 and 5.23, with this design, either of two possible half-fractions of four corner experiments can be selected.

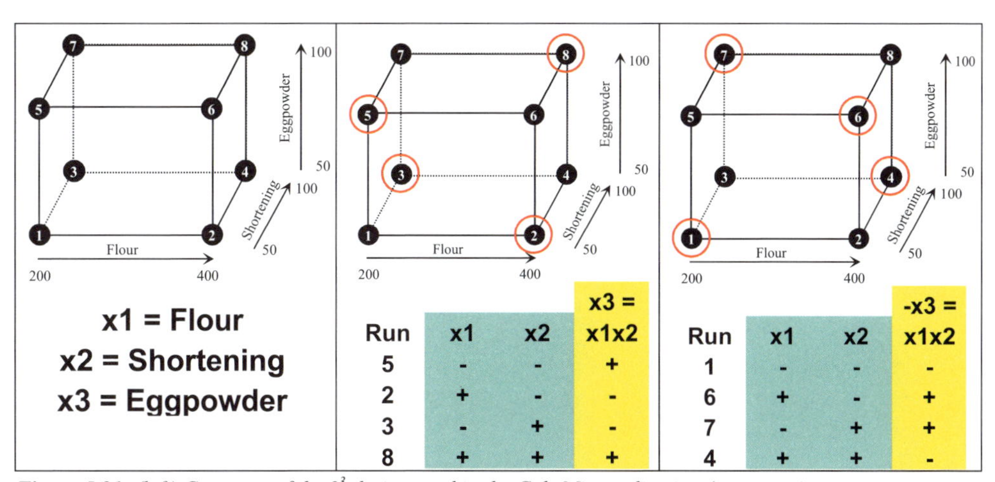

Run	x1	x2	x3 = x1x2
5	-	-	+
2	+	-	-
3	-	+	-
8	+	+	+

Run	x1	x2	-x3 = x1x2
1	-	-	-
6	+	-	+
7	-	+	+
4	+	+	-

Figure 5.21: *(left) Geometry of the 2^3 design used in the CakeMix application (center-points are not displayed).*
Figure 5.22: *(middle) Geometry of the first 2^{3-1} design created with the generator $x_3 = x_1x_2$.*
Figure 5.23: *(right) Geometry of the second 2^{3-1} design created with the generator $-x_3 = x_1x_2$.*

To understand how the two half-fractions can be selected it is instructive to consider the precursor 2^2 full factorial design, which is depicted by the two columns labeled x_1 and x_2 in Figure 5.22. These columns can be used to estimate the main effects of these two factors.

Attached to these two columns, we find the right-most column which may be used to estimate the x_1x_2 two-factor interaction. If we want to introduce a third factor, x_3, at the expense of some other term, the term we have to sacrifice is the two-factor interaction. It is shown in Figure 5.22 how x_3 is introduced in the design. Now we have created the first version of the 2^{3-1} fractional factorial design, and it corresponds to the four encircled experiments 2, 3, 5, and 8. It was the insertion of $x_3 = x_1x_2$ that generated this selection of four corners. Hence, the expression $x_3 = x_1x_2$ is called the *generator* of this 2^{3-1} design.

Alternatively, one may choose to incorporate x_3 into the 2^{3-1} design as $-x_3 = x_1x_2$ and obtain another selection of experiments. As seen in Figure 5.23, this alternative selection corresponds to the complementary half-fraction consisting of the runs 1, 4, 6, and 7. It was the new generator $-x_3 = x_1x_2$ that produced the complementary half fraction. From a DOE perspective these two complementary half-fractions are equivalent, but it may well be the case that practical considerations make one half-fraction preferable to the other.

In summary, we have just shown how two different generators may be used to construct two versions of the 2^{3-1} fractional factorial design, each version encoding one half-fraction of four corner experiments. This illustrates that it is the generator that dictates which specific fraction will be selected, and thereby, indirectly, controls the confounding pattern. The 2^{3-1} fractional factorial design is ideal for illustrating the generator concept. It is not very useful in reality, however, because it results in the confounding of main effects with two-factor interactions, which in a screening design is undesirable.

5.4.7 Generators of the 2^{4-1} fractional factorial design

Let us now consider four factors instead of three. With four factors we may use the 2^4 full factorial design corresponding to 16 runs. These 16 factor combinations are listed in the left-hand part of Figure 5.24. Often, however, we may not want to do all these experiments, and hence it is relevant to attempt a sub-set selection. Indeed, it is possible to make a sub-set selection, and an optimal selection of experiments is then encoded by the 2^{4-1} fractional factorial design. There are two versions of the 2^{4-1} design. One version is obtained when the generator $x_4 = x_1x_2x_3$ is used. This design represents the half-fraction of experiments listed in the upper right-hand part of Figure 5.24. The other version of the 2^{4-1} design is achieved when the alternative generator - $x_4 = x_1x_2x_3$ is employed. The latter generator results in the selection of the complementary half-fraction shown in the lower right-hand part of Figure 5.24.

										$x_1x_2x_3$ =
Run #	x_1	x_2	x_3	x_4		Run #	x_1	x_2	x_3	x_4
1	-	-	-	-		1	-	-	-	-
2	+	-	-	-		10	+	-	-	+
3	-	+	-	-		11	-	+	-	+
4	+	+	-	-		4	+	+	-	-
5	-	-	+	-		13	-	-	+	+
6	+	-	+	-		6	+	-	+	-
7	-	+	+	-		7	-	+	+	-
8	+	+	+	-		16	+	+	+	+
										$x_1x_2x_3$=
							x_1	x_2	x_3	$-x_4$
9	-	-	-	+		9	-	-	-	+
10	+	-	-	+		2	+	-	-	-
11	-	+	-	+		3	-	+	-	-
12	+	+	-	+		12	+	+	-	+
13	-	-	+	+		5	-	-	+	-
14	+	-	+	+		14	+	-	+	+
15	-	+	+	+		15	-	+	+	+
16	+	+	+	+		8	+	+	+	-

Figure 5.24: *Overview of how the 2^4 design relates to the two 2^{4-1} designs.*

In summary, we have seen that the application of generators in the 2^{3-1} and 2^{4-1} design cases is similar. In the former case, the generators are $\pm x_3 = x_1x_2$, and in the latter $\pm x_4 = x_1x_2x_3$. Thus, when only two generators exist, either will return a half-fraction of experiments.

5.4.8 Multiple generators

Now, we shall consider another fractional factorial design for which more than one pair of generators have to be contemplated. Consider a case with five factors. Five factors may be screened with either the 2^{5-1} or 2^{5-2} fractional factorial designs. With the former design, only two generators are defined, and hence using one of them must result in the selection of a half-fraction of 16 experiments.

On the other hand, when using the 2^{5-2} design only eight runs are encoded. When creating this design one must use the 2^3 full factorial design as starting point. The extended design matrix of the 2^3 design is shown in Figure 5.25. The best confounding pattern of the 2^{5-2} design is obtained when using the four generators $\pm x_4 = x_1x_2$ and $\pm x_5 = x_1x_3$. Here, the first pair of generators, $\pm x_4 = x_1x_2$, means that the fourth factor is introduced in the x_1x_2 column. Similarly, the second pair of generators, $\pm x_5 = x_1x_3$, implies that the fifth factor is inserted in the x_1x_3 column. It is easily realized that with two pairs of generators, four combinations of them are possible, i.e., the forth and fifth factors may be introduced in the design as $+x_4/+x_5$, $+x_4/-x_5$, $-x_4/+x_5$, or $-x_4/-x_5$. Each such combination prescribes the selection of a unique quarter-fraction comprising eight experiments. From a DOE perspective, these four quarter-fractions are equivalent, but the experimentalist may prefer one of them to the others.

				$\pm x_4 =$	$\pm x_5 =$		
	x_1	x_2	x_3	x_1x_2	x_1x_3	x_2x_3	$x_1x_2x_3$
1	-	-	-	+	+	+	-
2	+	-	-	-	-	+	+
3	-	+	-	-	+	-	+
4	+	+	-	+	-	-	-
5	-	-	+	+	-	-	+
6	+	-	+	-	+	-	-
7	-	+	+	-	-	+	-
8	+	+	+	+	+	+	+

Figure 5.25: *How the fourth and fifth factors may be introduced in a precursor 2^3 design.*

5.4.9 The defining relation

So far, we have mainly confronted fractional factorial designs dealing with three, four, or five factors, with comparatively few generators. However, when screening many more factors, one has to work with designs founded on several generators. Since it is not very practical to keep track of many generators in terms of isolated generator expressions, it would seem appropriate to replace such isolated expressions with some single relation tying them all together. Indeed, this is possible with what is called the *defining relation* of a design. The defining relation of a design is a formula derived from all its generators, that allows the calculation of the confounding pattern. Will we illustrate this using the 2^{4-1} fractional factorial design.

In order to derive the defining relation of the 2^{4-1} design, we must first understand two computation rules pertaining to column-wise multiplication of individual column elements. This is illustrated in Figure 5.26.

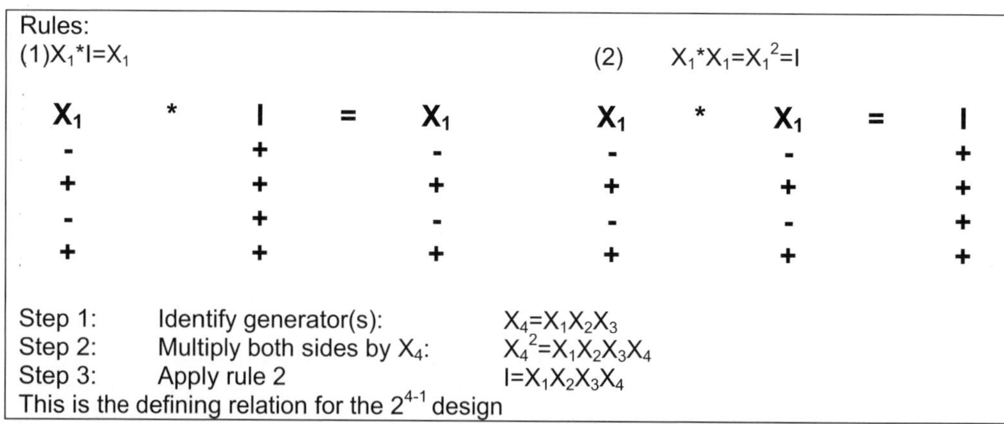

Figure 5.26: *How to derive the defining relation of the 2^{4-1} fractional factorial design.*

The first rule says that when multiplying the elements in any column, say column x_1, by the elements of the identity column, I, one obtains the same column x_1. The second rule says that when multiplying the elements in any column by a column of identical elements, one gets I. We will now put these rules into operation. The first step in the derivation of the defining relation is to identify the generator(s). In the case of the 2^{4-1} design used in the laser welding application, the generator is $x_4 = x_1x_2x_3$. Then, in the second step, both sides of this equality is multiplied by x_4. Finally, in step 3, upon applying rule 2, it is seen that $I = x_1x_2x_3x_4$. This is the defining relation of the 2^{4-1} fractional factorial design.

5.4.10 Use of the defining relation

The defining relation is very useful in that it immediately yields the confounding pattern. Figure 5.27 shows how to use the defining relation of the 2^{4-1} design, that is, the design underlying the first series of experiments in the laser welding case. To compute the confounding for a given term, one may proceed as follows: Step 1: Identify the defining relation; Step 2: Identify the term of interest, for instance x_1, and multiply both sides of the defining relation by that term; Step 3: Apply rule 1 to the left-hand side and rule 2 to the right-hand side; Step 4: Apply rule 1 to the right-hand side. By computing the confoundings of all other terms in this fashion, we get the results summarized in Figure 5.27. For comparative purposes, we show in Figure 5.28 how MODDE lists the confounding pattern of the initial laser welding design. For clarity, three-factor interactions have been omitted, as they are assumed to be of limited utility.

Defining relation	$I = x_1x_2x_3x_4$
What is x_1 confounded with? Step 1: $I = x_1x_2x_3x_4$ Step 2: $x_1I = x_1^2x_2x_3x_4$ Step 3: $x_1 = Ix_2x_3x_4$ Step 4: $x_1 = x_2x_3x_4$	$x_1 = x_2x_3x_4$
What is x_2 confounded with?	$x_2 = x_1x_3x_4$
What is x_3 confounded with?	$x_3 = x_1x_2x_4$
What is x_4 confounded with?	$x_4 = x_1x_2x_3$
What is x_1x_2 confounded with?	$x_1x_2 = x_3x_4$
What is x_1x_3 confounded with?	$x_1x_3 = x_2x_4$
What is x_1x_4 confounded with?	$x_1x_4 = x_2x_3$
What is x_2x_3 confounded with?	$x_2x_3 = x_1x_4$
What is x_2x_4 confounded with?	$x_2x_4 = x_1x_3$
What is x_3x_4 confounded with?	$x_3x_4 = x_1x_2$

Confoundings

	1	2	3
1	Term	Generator	Confounded with
2	Po	a	
3	Sp	b	
4	No	c	
5	Ro	abc	
6	Po*Sp		No*Ro
7	Po*No		Sp*Ro
8	Po*Ro		Sp*No
9	Sp*No		Po*Ro
10	Sp*Ro		Po*No
11	No*Ro		Po*Sp

Figure 5.27: *(left) How to compute the confounding for each individual term estimable with the initial laser welding model.*
Figure 5.28: *(right) An overview of the confounding pattern of the 2^{4-1} design (laser welding data).*

5.4.11 The defining relation of the 2^{5-2} fractional factorial design

The defining relation of the 2^{4-1} fractional factorial design is comparatively simple, and is not the most useful example for demonstrating the utility of this kind of expression. The key importance of the defining relation is better appreciated when considering the 2^{5-2} design. Recall that for this design two pairs of generators, $\pm x_4 = x_1x_2$ and $\pm x_5 = x_1x_3$, apply. By processing these generators according to the procedure previously outlined, it is possible to derive the defining relation $I = x_1x_2x_4 = x_1x_3x_5 = x_2x_3x_4x_5$, which is shown in Figure 5.29. In this expression, the last term, or "word", was obtained through multiplication of the generator-pairs. As seen in Figure 5.29, this defining relation may now be used to derive the confounding pattern of the 2^{5-2} fractional factorial design. Figure 5.29 lists the complete confounding pattern and Figure 5.30 a somewhat simplified confounding pattern as it is rendered in MODDE. We emphasize that in reality the experimenter does not have to pay a great deal of attention to the confounding pattern or the defining relation, as this is automatically taken care of by the software.

$I = x_1x_2x_4 = x_1x_3x_5 = x_2x_3x_4x_5$

$\mathbf{x_1} = x_2x_4 = x_3x_5 = x_1x_2x_3x_4x_5$

$\mathbf{x_2} = x_1x_4 = x_1x_2x_3x_5 = x_3x_4x_5$

$\mathbf{x_3} = x_1x_2x_3x_4 = x_1x_5 = x_2x_4x_5$

$\mathbf{x_4} = x_1x_2 = x_1x_3x_4x_5 = x_2x_3x_5$

$\mathbf{x_5} = x_1x_2x_4x_5 = x_1x_3 = x_2x_3x_4$

$\mathbf{x_1x_2} = x_4 = x_2x_3x_5 = x_1x_3x_4x_5$

$\mathbf{x_1x_3} = x_2x_3x_4 = x_5 = x_1x_2x_4x_5$

$\mathbf{x_1x_4} = x_2 = x_3x_4x_5 = x_1x_2x_3x_5$

$\mathbf{x_1x_5} = x_2x_4x_5 = x_3 = x_1x_2x_3x_4$

$\mathbf{x_2x_3} = x_1x_3x_4 = x_1x_2x_5 = x_4x_5$

$\mathbf{x_2x_4} = x_1 = x_1x_2x_3x_4x_5 = x_3x_5$

$\mathbf{x_2x_5} = x_1x_4x_5 = x_1x_2x_3 = x_3x_4$

$\mathbf{x_3x_4} = x_1x_2x_3 = x_1x_4x_5 = x_2x_5$

$\mathbf{x_3x_5} = x_1x_2x_3x_4x_5 = x_1 = x_2x_4$

$\mathbf{x_4x_5} = x_1x_2x_5 = x_1x_3x_4 = x_2x_3$

Confoundings

	1	2	3	4
1	Term	Generator	Confounded with	Confounded with
2	x1	a	x2*x4	x3*x5
3	x2	b	x1*x4	
4	x3	c	x1*x5	
5	x4	ab	x1*x2	
6	x5	ac	x1*x3	
7	x1*x2		x4	
8	x1*x3		x5	
9	x1*x4		x2	
10	x1*x5		x3	
11	x2*x3		x4*x5	
12	x2*x4		x1	x3*x5
13	x2*x5		x3*x4	
14	x3*x4		x2*x5	
15	x3*x5		x1	x2*x4
16	x4*x5		x2*x3	

Figure 5.29: *(left) The complete confounding pattern of the 2^{5-2} design.*
Figure 5.30: *(right) A simplified confounding pattern of the 2^{5-2} design, as given in MODDE.*

5.4.12 Resolution

The defining relation is useful in that it directly gives the confounding pattern of a fractional factorial design. In addition, this expression is of central importance when it comes to uncovering the *resolution* of a design. The concept of design resolution has to do with the complexity of the confounding pattern. A design of high resolution conveys factor estimates with little perturbation from other effect estimates. On the other hand, a design of low resolution experiences complicated confounding of effects.

In order to understand the resolution of a design, we need to consider the defining relation of the 2^{5-2} design. We remember that the relevant defining relation was $I = x_1x_2x_4 = x_1x_3x_5 = x_2x_3x_4x_5$. This expression is said to consist of three words. Two of the words have the length three and one the length four. The key point is that the resolution of a design is defined as the *length of the shortest word in the defining relation*. Because the shortest word in the defining relation of the 2^{5-2} design has length three, this design is said to be of resolution III.

The vast majority of screening designs used have resolutions in the range of III to V (Figure 5.31). We now give a summary of their confounding properties. In a resolution III design, main effects are confounded with two-factor interactions, and two-factor interactions with each other. This is a complex confounding pattern that should be avoided in screening. However, resolution III designs are useful for robustness testing purposes (see Chapter 7). In contrast to this, the confounding pattern of a resolution IV design is more tractable in screening. Here, main effects are unconfounded with two-factor interactions, but the two-factor interactions are still confounded with each other. With a resolution IV design the experimentalist has access to a design with an appropriate balance between the number of factors and the number of experiments. Resolution IV designs are recommended for screening. Further, with a design of resolution V, even the two-factor interactions themselves are unconfounded. This means that resolution V designs are almost as good as full factorial designs. Thus, although a resolution V design would be a technically correct choice for screening, it requires unnecessarily many runs.

Consider the summary displayed in Figure 5.31. This table provides an overview of the most commonly used fractional factorial designs and their respective resolutions. It is organized as follows: the vertical direction gives the number of corner experiments and the horizontal direction the number of treated factors. For each design, the resolution is given using roman numerals.

Factors/ Runs	3	4	5	6	7	8
4	2^{3-1} Res III $+/-X_3=X_1*X_2$	--	--	--	--	--
8	2^3	2^{4-1} Res IV $+/-X_4=X_1*X_2*X_3$	2^{5-2} Res III $+/-X_4=X_1*X_2$ $+/-X_5=X_1*X_3$	2^{6-3} Res III $+/-X_4=X_1*X_2$ $+/-X_5=X_1*X_3$ $+/-X_6=X_2*X_3$	2^{7-4} Res III $+/-X_4=X_1*X_2$ $+/-X_5=X_1*X_3$ $+/-X_6=X_2*X_3$ $+/-X_7=X_1*X_2*X_3$	--
16		2^4	2^{5-1} Res V $+/-X_5=X_1*X_2*X_3*X_4$	2^{6-2} Res IV $+/-X_5=X_1*X_2*X_3$ $+/-X_6=X_2*X_3*X_4$	2^{7-3} Res IV $+/-X_5=X_1*X_2*X_3$ $+/-X_6=X_2*X_3*X_4$ $+/-X_7=X_1*X_3*X_4$	2^{8-4} Res IV $+/-X_5=X_2*X_3*X_4$ $+/-X_6=X_1*X_3*X_4$ $+/-X_7=X_1*X_2*X_3$ $+/-X_8=X_1*X_2*X_4$
32			2^5	2^{6-1} D-opt Res VI $+/-X_6=X_1*X_2*X_3*X_4*X_5$	2^{7-2} D-opt Res IV $+/-X_6=X_1*X_2*X_3*X_4$ $+/-X_7=X_1*X_2*X_4*X_5$	2^{8-3} Res IV $+/-X_6=X_1*X_2*X_3$ $+/-X_7=X_1*X_2*X_4$ $+/-X_8=X_2*X_3*X_4*X_5$

Figure 5.31: *Overview of the resolution of some common fractional factorial designs.*

5.4.13 Summary of fractional factorial designs

The topics discussed so far in this chapter – confounding pattern, generators, defining relation, and resolution of fractional factorial designs – are intimately connected (see Figure 5.32). It has been our intention to provide an account of these fundamental design concepts. In the design generation phase, everything starts with the selection of one or more generators. The chosen generators control which fraction of experiments is to be investigated, and also regulate the confounding pattern. When several generators are in operation these may be concatenated through the defining relation. The major advantage of the defining relation is that it is an expeditious route towards an understanding of the confounding pattern. Since the complexity of a confounding pattern may vary considerably, it is desirable to somehow quantify the resolution of a design. This has been agreed upon by defining that the resolution of a fractional factorial design equals the length of the shortest word in the defining relation.

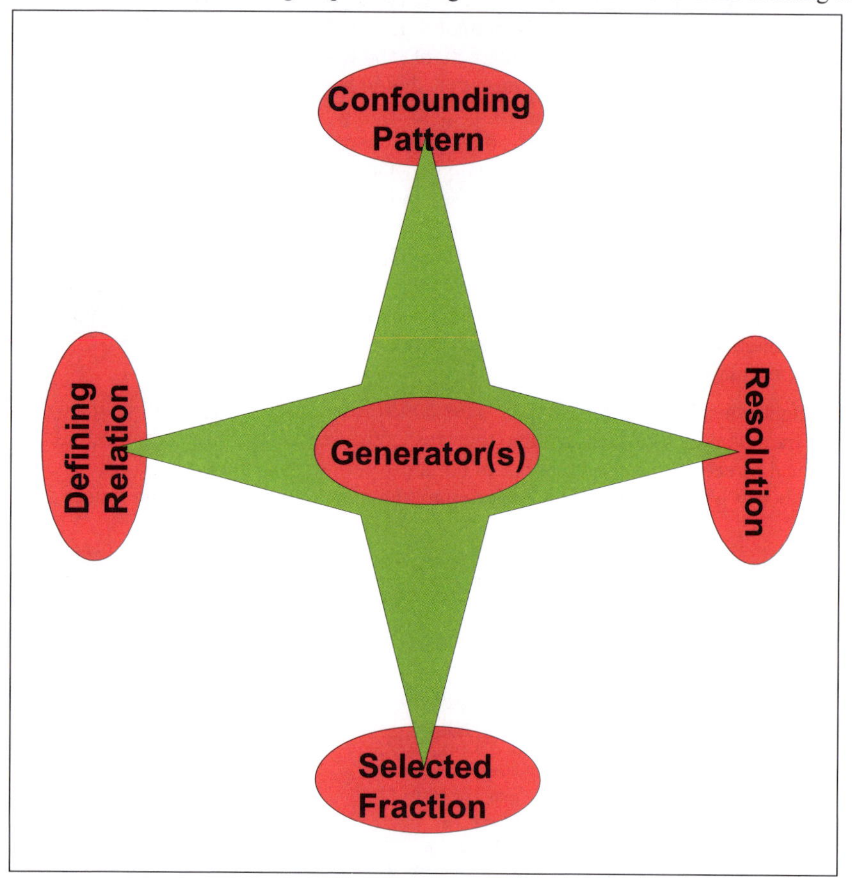

Figure 5.32: *The selected generators control the confounding pattern and the selected fraction of experiments. Indirectly, this means that the selected generators also influence the shape of the defining relation and the resolution of the design.*

In summary, we wish the reader to be familiar with these concepts and to have a basic understanding of how they work. In a real situation, any standard DOE software will take care of the technical details, and really the only thing an experimenter should worry about is to make sure that a resolution IV design is selected in screening. Occasionally, there may be no resolution IV design available, and then a resolution V design should be identified.

This concludes the introductory discussion concerning fractional factorial designs. The main advantage of fractional factorial designs is that many factors may be screened in drastically fewer runs. Their main drawback is the confounding of effects. However, to some extent, the confounding pattern may be regulated in a desired direction. With this discussion in mind we will now return to the laser welding application, and the data analysis of this data-set.

5.5 Laser welding application – Part I

5.5.1 Evaluation of raw data – I

We start the evaluation of the raw data by inspecting the replicate error of each response. The three relevant plots of replications are displayed in Figure 5.33. As seen, the replicate errors are small for all three responses, which means that we have good data to work with.

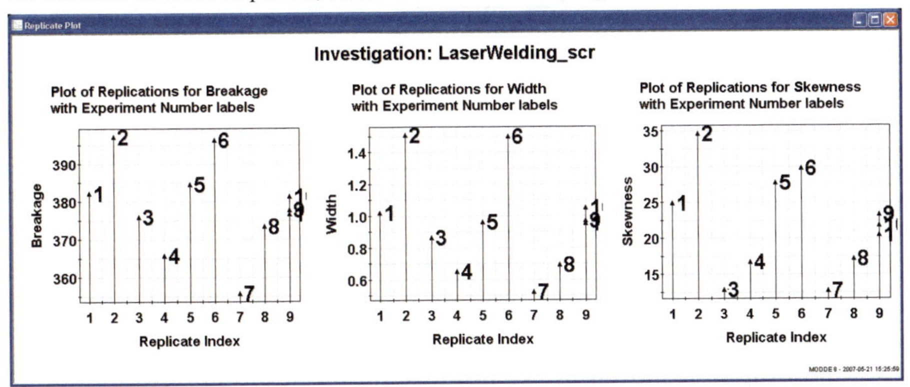

Figure 5.33: (left) Replicate plot of Breakage. (middle) Replicate plot of Width. (right) Replicate plot of Skewness.

Further, in the evaluation of the raw data, it is mandatory to check the data distribution of the responses, to reveal any need for response transformations. We may check this by making a histogram of each response. Such histograms are rendered in Figure 5.34, and they inform us that it is pertinent to work in the untransformed metric of each response.

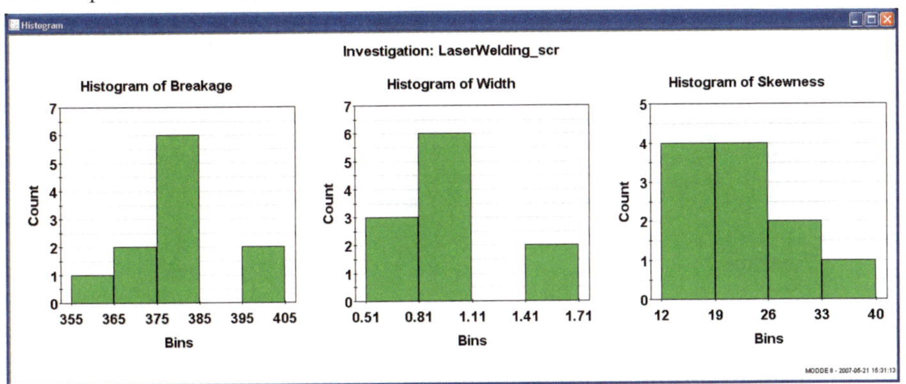

Figure 5.34: (left) Histogram of Breakage. (middle) Histogram of Width. (right) Histogram of Skewness.

Other aspects of the raw data that might be worthwhile to consider are the condition number of the fractional factorial design, and the inter-relatedness among the three responses. We notice from Figure 5.35 that the condition number is approximately 1.2, which lies within the interval of 1 – 3 in which standard screening designs typically are found. The correlation matrix listed in Figure 5.36 tells us that the three responses are strongly correlated, because their correlation coefficients range from 0.88 to 0.96. Hence, fitting three MLR models with identical composition of model terms appears reasonable.

	1	2
1		**Evaluation of MLR model**
2		All factors are orthogonally scaled
3	Condition number	1.1726
4	Worksheet runs	11
5	Model terms	5
6	DF residual	6
7	DF lack of fit	4
8	DF pure error (repl. runs)	2

	1	2	3	4	5	6	7	8
1		Po	Sp	No	Ro	Br	Wi	Sk
2	Po	1	0	0	0	0.319698	0.354061	0.321904
3	Sp	0	1	0	0	-0.832702	-0.811837	-0.923558
4	No	0	0	1	0	-0.104088	-0.132326	-0.0268254
5	Ro	0	0	0	1	0.289958	0.11802	0.130295
6	Br	0.319698	-0.832702	-0.104088	0.289958	1	0.957684	0.894094
7	Wi	0.354061	-0.811837	-0.132326	0.11802	0.957684	1	0.884856
8	Sk	0.321904	-0.923558	-0.0268254	0.130295	0.894094	0.884856	1

Threshold 0.3

Figure 5.35: *(left) Condition number evaluation of the laser welding design.*
Figure 5.36: *(right) Plot of correlation matrix of the laser welding design. Note the high inter-relatedness among the three responses.*

5.5.2 Regression analysis – I

We will now fit a linear model with five terms, the constant and four linear terms, to each response. The overall results of the model fitting are summarized in Figure 5.37. Evidently, the predictive power, as evidenced by Q^2, is acceptable only for Skewness. The prediction ability is poor for Breakage and Width. In trying to understand why this might be the case, it is appropriate to continue with the Model Validity diagnostic tool. Remembering that this diagnostic should be larger than 0.25, we see that all three models pass this test. Further, the Reproducibility diagnostic is good for all three responses.

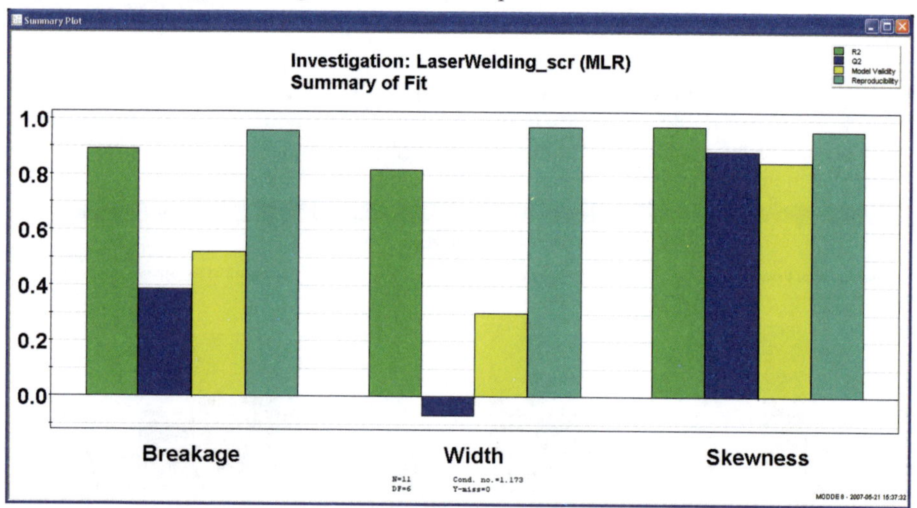

Figure 5.37: *Summary of fit plot of the laser welding model.*

The outcome of the next diagnostic tool, the normal probability plot of residuals, is shown in Figure 5.38. This plot indicates that the residuals for experiments 3 and 6 are unusually large, but we should not delete any experiment.

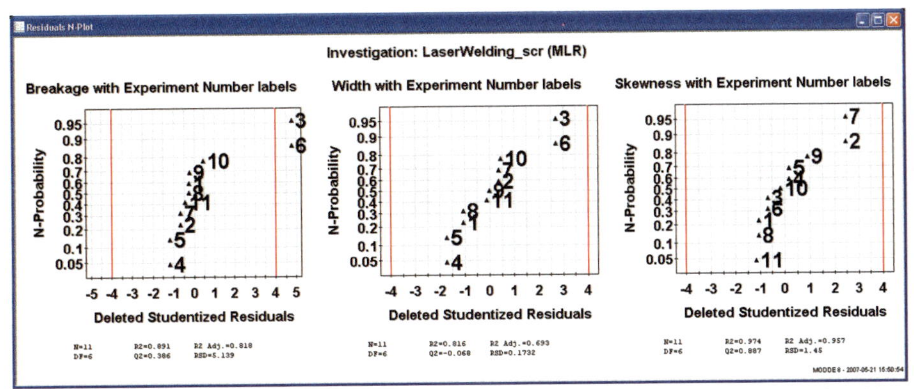

Figure 5.38: *(left) Normal probability plot of Breakage. (middle) Normal probability plot of Width. (right) Normal probability plot of Skewness.*

Since we have not yet obtained any tangible information as to why Q^2 is low for Breakage and Width, it is appropriate to consult plots of regression coefficients. Such regression coefficients are shown in Figure 5.39. Apparently, two factors dominate, that is, Power and Speed of laser. The third factor, NozzleGas, does not influence any response. The fourth factor, RootGas, exerts a minor influence with regards to Breakage. Based on this information, it appears reasonable to try to modify the original model by adding the two-factor interaction Po*Sp. However, in so doing, it must be remembered that this model term is confounded with another two-factor interaction, the one denoted No*Ro.

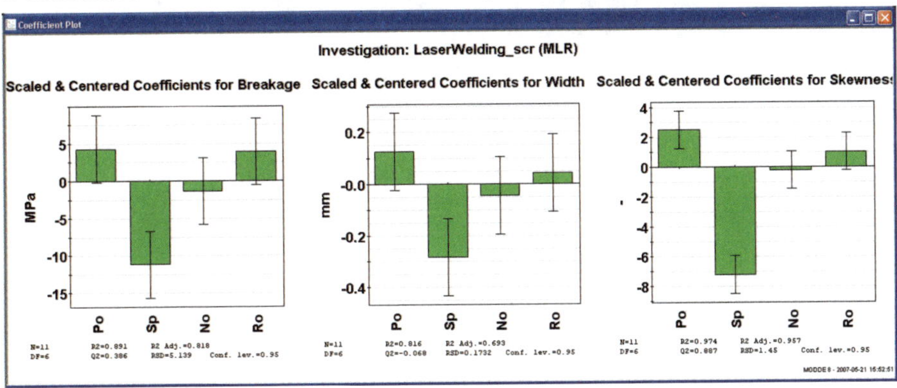

Figure 5.39: *Regression coefficients for Breakage, Width and Skewness.*

5.5.3 Model refinement – I

We will now carry out a refinement of the linear model. For reference purposes we re-display in Figures 5.40 and 5.41 the summary of fit plot and the coefficient overview plot of the original model. The progress of the model pruning may easily be overviewed using similar pairs of graphs.

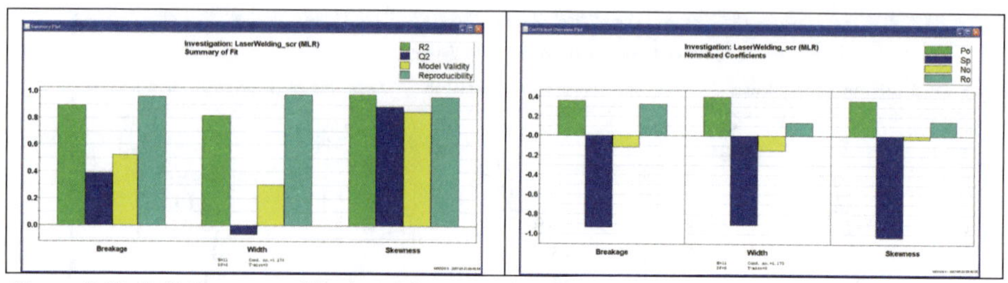

Figure 5.40: *(left) Summary of fit plot of the original model.*
Figure 5.41: *(right) Coefficient overview of the original model.*

In the first step, the Po*Sp two-factor interaction was incorporated, and the model refitted. As is evident from Figures 5.42 and 5.43, the inclusion of the interaction term is mainly beneficial for the second response, but its presence reduces the predictive power regarding the first response. We also observe that the linear term of NozzleGas is not influential in the model. Hence, it is relevant to remove this term and evaluate the consequences.

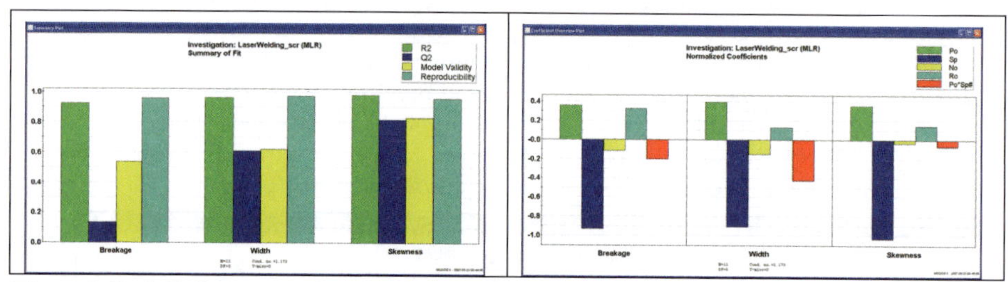

Figure 5.42: *(left) Summary of fit plot after inclusion of Po*Sp*
Figure 5.43: *(right) Coefficient overview after inclusion of Po*Sp.*

Figures 5.44 and 5.45 show that all three regression models are enhanced as a result of the exclusion of NozzleGas. The model summary shown in Figure 5.44 represents the "best" modelling situation that we have obtained. In general, a Q^2 above 0.5 is good and above 0.9 is excellent. Thus, Figure 5.44 demonstrates that we are doing a good job with regards to Skewness, but are doing less well with respect to Breakage. The gap between R^2 and Q^2 concerning Breakage is unsatisfactory. The modelling of Width is acceptable for screening purposes.

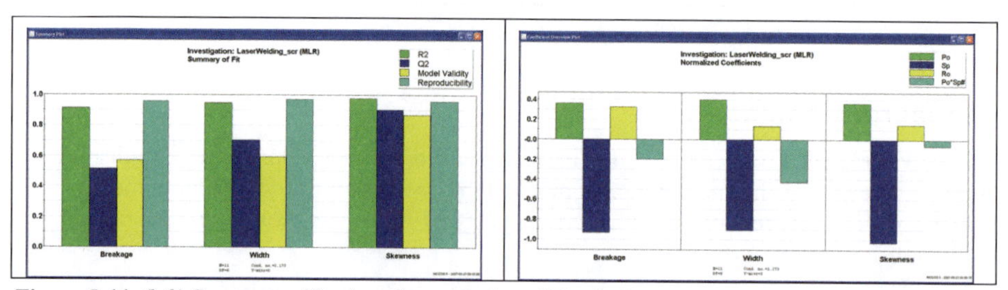

Figure 5.44: *(left) Summary of fit plot after exclusion of NozzleGas.*
Figure 5.45: *(right) Coefficient overview after exclusion of NozzleGas.*

We may now proceed by reviewing some other diagnostic tools. The Model Validity and Reproducibility diagnostics are good in all cases (Figure 5.44). The N-plots of residuals (Figure 5.46) indicate that one plausible reason for the low Q^2's of Breakage and Width might be the deviating behaviour of experiments 3 and 7. It may well be the case that some additional two-factor interaction is needed to better account for this deviation.

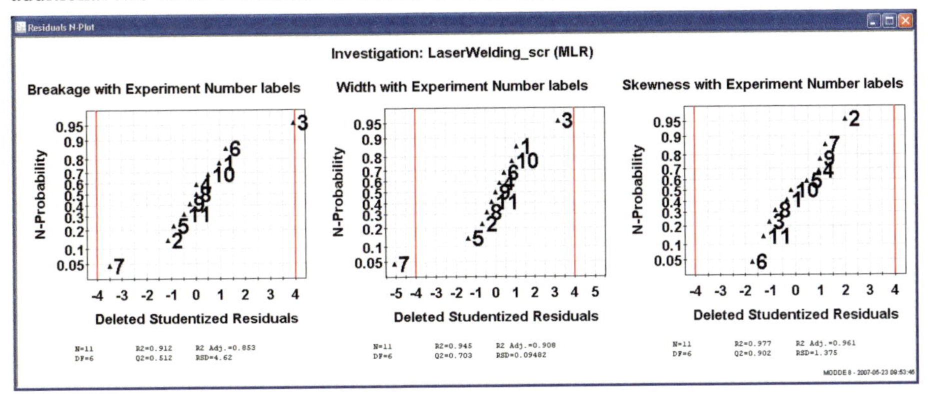

Figure 5.46: N-plots of residuals – after model refinement.

Because of the comparatively poor modelling of the first response, and the suspected need for more two-factor interactions, the investigator decided to carry out more experiments in order to upgrade the underlying design and, thereby, also the regression model. To perform this upgrade the investigator used a technique called fold-over. Fold-over will be explained in a moment (see Section 5.6.7).

5.5.4 Use of model – I

Although it has already been decided to do more experiments by using the fold-over technique, it is useful to interpret and use the existing model. This we do primarily to determine whether the desired response profile is likely to be realized within the investigated experimental region. Since the "working" regression model contains the Po*Sp two-factor interaction, it is logical to construct response contour plots focusing on this term. Such a triple-contour plot is shown in Figure 5.47.

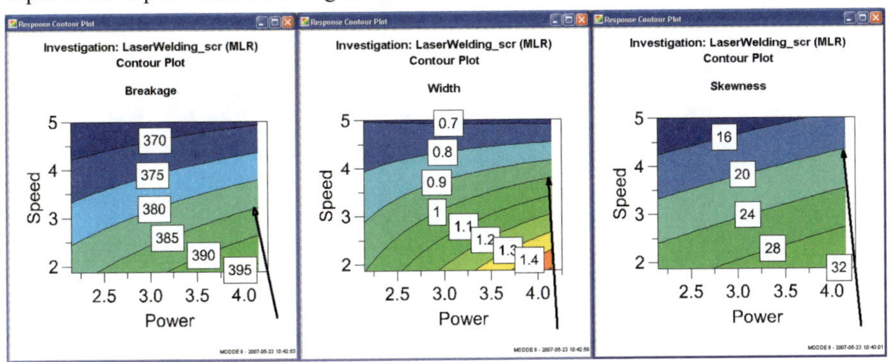

Figure 5.47: Triple-contour plot of Breakage, Width, and Skewness.

We can see how the three responses change as a result of changing the factors Speed and Power, while keeping RootGas fixed at its center level. Observe that NozzleGas is not included in the model, and hence does not affect the shape of these plots. The arrows indicate where the stated goals are fulfilled with regard to each individual response. It is evident that no unique point exists in these contours where all desirabilities are fulfilled. However, staying somewhere along the right-hand edge seems appropriate. Perhaps after the inclusion of the eleven additional experiments it will be possible to pinpoint an optimum. This remains to be seen. Later on when all 22 runs are analyzed together, we will use an automatic search functionality, an optimizer, to locate more precisely a possible optimum.

This concludes the first stage in the analysis of the laser welding data. We will now introduce some more theory relating to what to do after screening, as this matches our current situation. Then we will revert to the data analysis.

5.6 What to do after screening

5.6.1 Introduction

Exactly what to do after a screening investigation depends on a number of things. By necessity, this decision will be influenced by the quality of the regression model obtained, whether it is possible and/or necessary to modify the factor ranges, and whether some of the experiments already conducted are close to fulfilling the goals stated in the problem formulation.

If the obtained model is reliable it can be used for prediction. This is also termed either interpolation or extrapolation. Interpolation is based on using the derived regression model for predictions *inside* the explored experimental space (Figure 5.48). This is of interest when experimenting outside the investigated region is either impossible or undesired. Extrapolation, i.e., making predictions *outside* the investigated region, is relevant when it is possible to change factor settings (Figure 5.49).

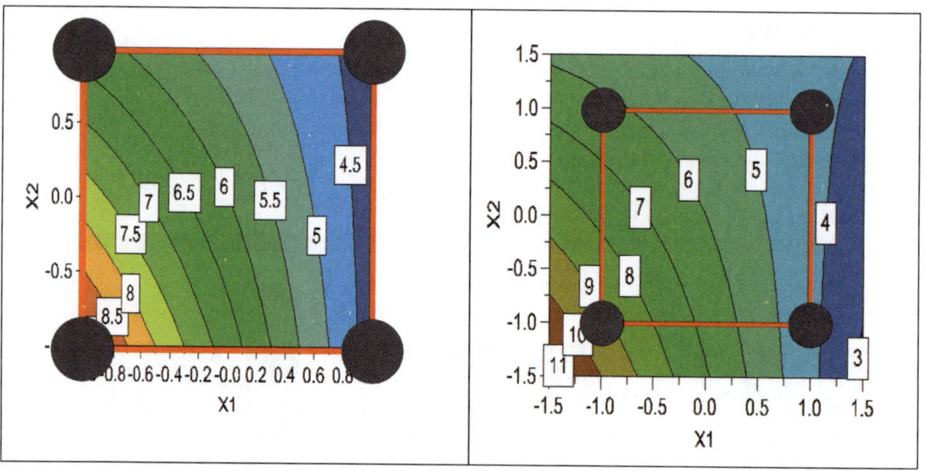

Figure 5.48: *(left) Principle for interpolation.*
Figure 5.49: *(right) Principle for extrapolation.*

5.6.2 Main outcomes of a screening campaign

After a completed screening design we can envision three main outcomes (Figure 5.50). The ideal case is of course when one of the performed experiments fulfills the experimental goals. In such a case, all we have to do is to make a limited set of new trials to verify the factor combination of this "golden run". A second and more common situation is when we have to rely on the derived regression model and compute predictions inside the investigated region (interpolation) to identify an interesting point or small area. Sometimes, however, it will turn out difficult or even impossible to locate an interesting area inside the experimental domain. Then, a third outcome is likely occur, i.e., when predictions outside the searched region have to be computed in order to lead forward to an interesting point or small region for further experimentation.

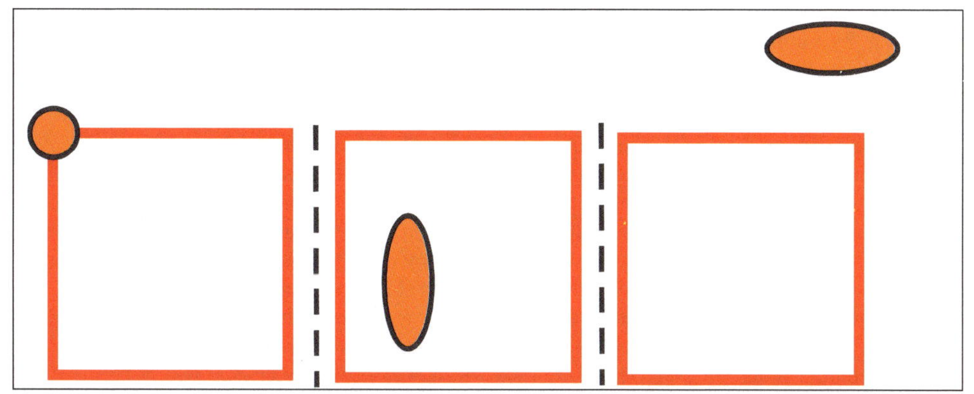

Figure 5.50: After the screening application three main outcomes are conceivable. In the ideal case, one of the performed experiments meets the demands set up in the problem formulation (illustrated in the left-hand part). We may call this experiment the "golden run". A very common result is when the obatined regression model must be used for interpolation to find an interesting point or small region for further experimentation (shown in the middle part). If such an interesting point or small region is not found inside the experimental domain, the obtained regression model may have to be used for extrapolation (shown in the right-hand part).

5.6.3 How do we find an interesting point or area?

As discussed above, if there is no golden run we are in the situation where we must use the derived regression model. And we want to use it to predict where to do our next experiment, which hopefully will correspond to, or be in the vicinity of, the optimal point. This means that we are just about to confront one of the minor experimental objectives, the one entitled *finding the optimal region*. Finding the optimal region is an objective which is often used to bridge the gap between screening and optimization. As seen in Figure 5.51, we want to understand how to alter the factor settings so that we enter into an area which does include the optimal point.

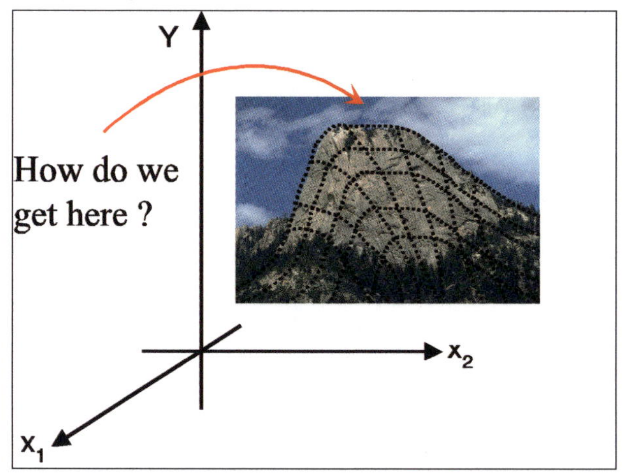

Figure 5.51: Graphical illustration of the experimental objective finding the optimal region.

At this stage, our intention is to describe two techniques for doing this. One technique is a graphically oriented gradient technique, and the other an automatic procedure based on running multiple simplexes in parallel. These techniques are used predominantly for extrapolation purposes, that is, for predicting outside the investigated experimental region. However, the automatic optimization procedure may be used for interpolation, as well.

5.6.4 Gradient techniques

Gradient techniques are useful when it is necessary to re-adjust factor settings and move the experimental design. Basically, one may envision two gradient techniques, *steepest ascent* and *steepest descent*. These are carried out using graphical tools for moving along the direction in which the best response values are expected to be encountered. Steepest ascent means that the movement is aiming for higher response values, that is, one wants to climb the mountain and reach its highest peak. Steepest descent means that the migration is directed towards lower response values, that is, one wishes to arrive at the bottom of the valley. Both gradient techniques are conveniently carried out with the help of graphical tools, e.g., response contour plots.

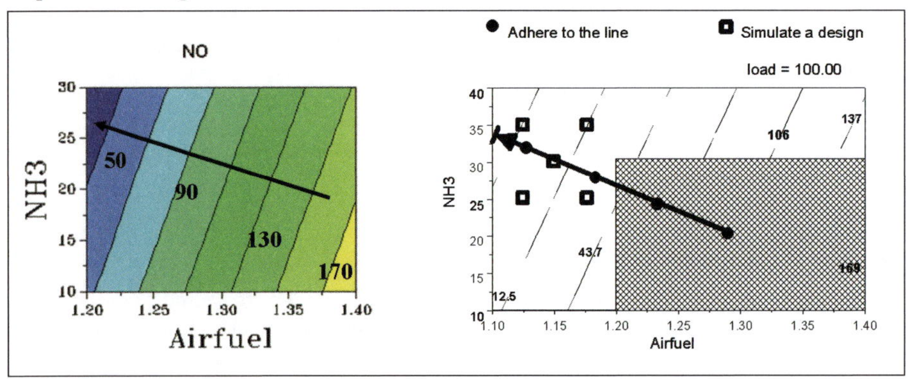

Figure 5.52: *(left) Example of steepest descent search for a lower response value.*
Figure 5.53: *(right) Steepest descent search extended outside the investigated area.*

Figure 5.52 shows an example of a steepest descent application. For clarity, we skip the details of the example, and focus on the procedural steps involved. Steepest descent is implemented by first identifying an interesting direction. This direction should be perpendicular to the parallel lines of the response contour plot. This is illustrated in Figure 5.53. Here, the factor axes have been stretched out in comparison with Figure 5.52, and the domain originally mapped is depicted by the hatched area. Then one moves along the reference direction, using the existing regression model to make response predictions at equidistant points. This is symbolized by the solid circles. In fact, with this approach we use the model both for inter- and extrapolation at the same time. Alternatively, one may simulate a design, as shown by the open squares in Figure 5.53, and make predictions in a square arrangement. Regardless of the method selected, when a promising predicted point has been found, it should be tested by performing verifying experiments.

Gradient techniques work best with few responses, and when two-factor interactions are fairly small. With large two-factor interactions, response contours tend to be rather twisted, making the identification of an interesting direction difficult and ambiguous.

5.6.5 Automatic search for an optimal point

MODDE contains an optimization routine, a so called optimizer. For the optimizer to work properly, the user must specify certain initial conditions. Firstly, the roles of the factors must be set. A factor can be allowed to change its value freely in the optimization, or it may be fixed at a constant value. In addition, for the free factors, low and high variation limits must be set. Secondly, certain criteria concerning the responses must be given. A response variable may either be maximized, minimized, directed towards an interval, or excluded in the optimization. Subsequent to these definitions, the optimizer will use the obtained regression model and the performed experiments to compute eight starting points for simplexes. In principle, the optimizer then uses these simplexes together with the fitted model, to optimize desirability functions, that is, mathematical functions representing the desirabilities, or goals, of the individual responses. These goals are taken from the criteria specified for the responses.

Figure 5.54 shows one simplex in action. For simplicity, this simplex is laid out in a two-dimensional factor space. The first simplex consists of the three experiments enumerated 1 – 3. Now, the idea is to mirror the worst experiment, here #1, through the line connecting the two best experiments, here #2 and 3, and perform a new experiment, here #4, at the position where the mirrored experiment "hit" the response contour plot. This means that the new simplex consists of runs 2, 3, and 4. In the next cycle, the worst experiment is mirrored in the line connecting the two best experiments, and so on. This is repeated until the peak of the elliptical mountain is reached by one simplex. We emphasize that this is only a simplified picture of how the simplex methodology works. In reality, when a new simplex is formed through a reflection, it may be reduced or enlarged in size in comparison with the precursor simplex.

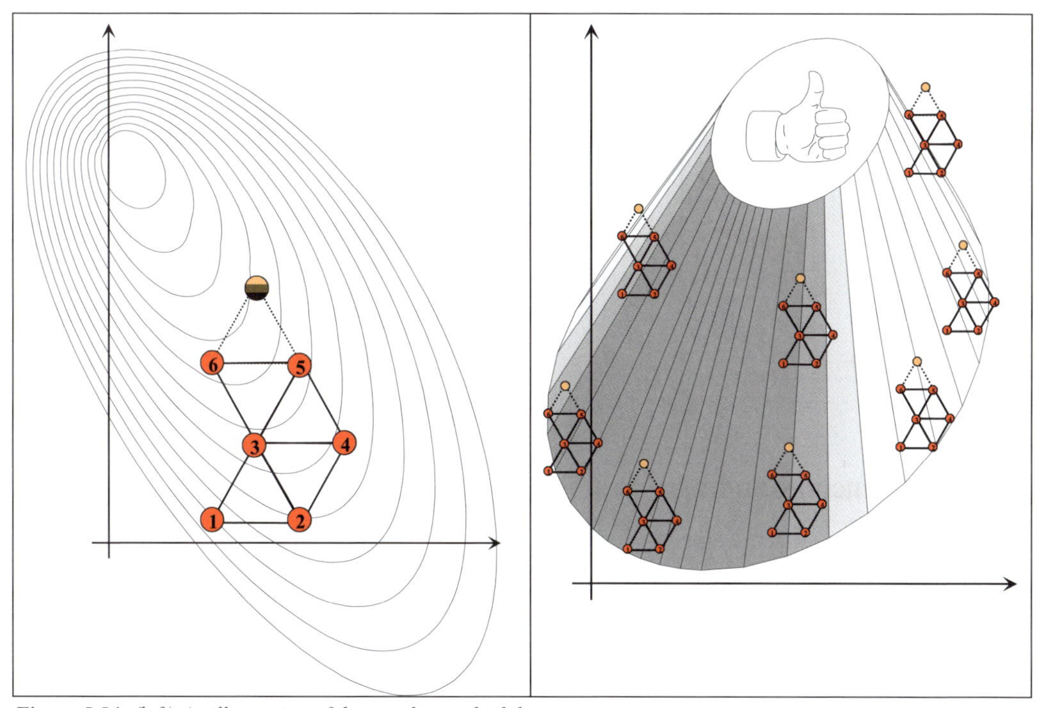

Figure 5.54: *(left) An illustration of the simplex methodology.*
Figure 5.55: *(right) Different simplexes get different starting points in the factor space.*

As mentioned, the optimizer will simultaneously start up to eight simplexes, from different locations in the factor space. Eight simplexes are initiated in order to avoid being trapped at a local minimum or maximum. This is illustrated in Figure 5.55. The co-ordinates for the simplex starting points are taken from the factors with the largest regression coefficients in the regression model. At convergence, each simplex will display the obtained factor settings, and it is then possible to compute predicted response values at each point. Observe that the drawing in Figure 5.55 is a schematic valid for chasing higher response values. In reality, one could also go for a minimum, and then the relevant response function would correspond to a hole or a valley.

5.6.6 Adding complementary runs

There are two primary motives for adding complementary runs to a screening design, namely complementing for unconfounding and complementing for curvature (Figure 5.56). Fractional factorial designs of resolution III and IV have confoundings among main effects and two-factor interactions. This means that these effects can not be computed clear of one another. It is possible to resolve confounded effects, that is, to unconfound, by adding extra experiments.

When *complementing for unconfounding* purposes, one may either (i) upgrade the whole design and the accompanying model, or (ii) selectively update the design with a minimum number of runs for unconfounding some specific effects that are of interest. Upgrading the entire design is usually accomplished with a technique called *fold-over*. However, this technique is unselective (see Section 5.6.7). It enables all terms of a certain complexity to be better resolved, and is therefore quite costly in terms of additional experiments. If a clear picture exists that, for example, two two-factor interactions need to be estimated free of each other, it is sufficient to selectively update the design with two additional experiments, rather than a bunch of new experiments. Such a selective design supplement is usually accomplished with a D-optimal design, but this an advanced topic.

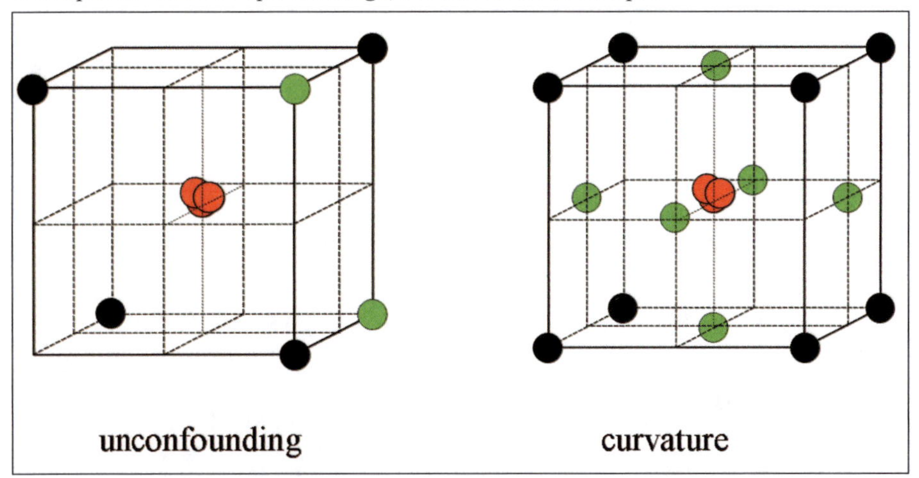

Figure 5.56: *When adding complementary runs to a screening design this can be done for two primary reasons. A few extra experiments can be added in order to unconfound two-factor interactions that are mixed-up and hence allow a better resolution of them. Alternatively, if there is a need to capture curvature, the entire screening design can be augmented to a central composite design.*

The second main reason for adding complementary experiments occurs when there is a need to model curvature, that is, non-linear relationships among factors and responses. Fractional and full factorial designs may be complemented by (i) upgrading the entire design to a central composite design, or by (ii) selective updating, as outlined above. In both instances, full or partial quadratic models result, which are capable of handling curved relationships. Central composite designs are discussed in detail in connection with General Example 2 in Chapter 6.

Thus, in summary, in the broadest sense, the measures to be taken after screening can be divided in two major approaches. One approach is based on changing the factor ranges and thus moving the experimental region. This is visualized in Figure 5.57. The other approach is based on *not* altering the factor settings.

We have now completed our initial discussion of what to do after screening. The next section will amplify on the use of the fold-over technique, as this was the complementarity principle selected in the laser welding study.

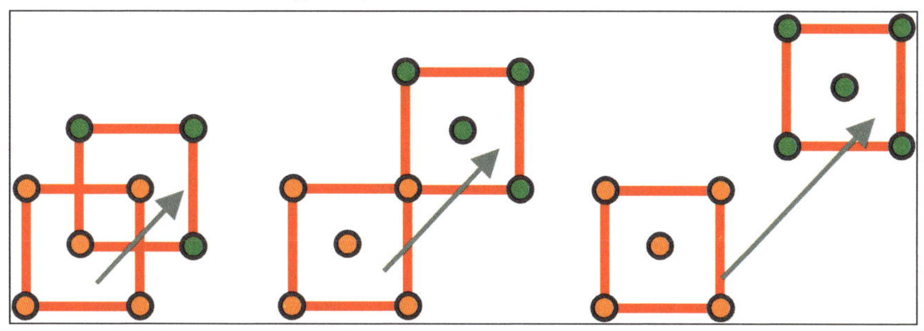

Figure 5.57: *It is important to realize that adding complementary runs to a design is not restricted to the situation when the experimental region is not moved. On the contrary, new experiments can be appended to the worksheet even when the factor settings are changed and the experimental region is permitted to move.*

5.6.7 Fold-over

The fold-over design is a complementary design that is usually added to achieve unconfounding in resolution III and resolution IV designs. It is a collection of experiments selected from the other existing, but unused, fractions. This is exemplified in Figure 5.58. The upper eight experiments originate from the 2^{5-2} fractional factorial design. The fold-over complement of this resolution III design is given by the lower eight experiments. When comparing the sign pattern of the upper quarter-fraction with that of the lower quarter-fraction, we can see that they have identical sequences except for a sign inversion. The fold-over complement was selected from all possible quarter-fractions, by switching the sign of the two columns used to introduce the fourth and the fifth factors.

			x4 =	x5 =
x1	x2	x3	x1x2	x1x3
-	-	-	+	+
+	-	-	-	-
-	+	-	-	+
+	+	-	+	-
-	-	+	+	-
+	-	+	-	+
-	+	+	-	-
+	+	+	+	+
+	+	+	-	-
-	+	+	+	+
+	-	+	+	-
-	-	+	-	+
+	+	-	-	+
-	+	-	+	-
+	-	-	+	+
-	-	-	-	-
			x4 =	x5 =
			-x1x2	-x1x3

Figure 5.58: *The 2^{5-2} design and its fold-over.*

When the initial design is of resolution III, as is the case with the 2^{5-2} fractional factorial design, the initial design together with its fold-over will be of resolution IV. This means that all main effects will be resolved from the two-factor interactions. Note, however, that this design is not as efficient as the 2^{5-1} fractional factorial design with resolution V. Moreover, when the original design is of resolution IV, the original design and its fold-over only sometimes become a resolution V design. MODDE automatically takes care of folding over a fractional factorial design, when this command is launched.

5.6.8 Creating the fold-over of the laser welding screening design

Because the starting design in the laser welding application was a resolution IV 2^{4-1} fractional factorial design, this design combined with its fold-over will be the complete 2^4 factorial design. The upgraded design is shown in Figure 5.59 in terms of the new worksheet. We can see that the worksheet has been appended with eleven new experiments, found in rows 12 through 22. Eight new corner experiments have been conducted, thus completing the factorial part of the design, as well as three additional center-points. Thus, we now have six replicated center-points.

Worksheet

	1	2	3	4	5	6	7	8	9	10	11	12
1	Exp No	Exp Name	Run Order	Incl/Excl	Power	Speed	NozzleGas	RootGas	$Block	Breakage	Width	Skewness
2	1	N1	5	Incl	2.15	1.875	27	27	-1	382.2	1.02	24.96
3	2	N2	4	Incl	4.15	1.875	27	42	-1	397.2	1.51	34.68
4	3	N3	2	Incl	2.15	5	27	42	-1	375.8	0.86	12.84
5	4	N4	7	Incl	4.15	5	27	27	-1	365.6	0.65	16.68
6	5	N5	10	Incl	2.15	1.875	36	42	-1	384.4	0.96	27.72
7	6	N6	6	Incl	4.15	1.875	36	27	-1	396.2	1.5	29.88
8	7	N7	3	Incl	2.15	5	36	27	-1	355.6	0.52	12.72
9	8	N8	9	Incl	4.15	5	36	42	-1	373.4	0.69	17.16
10	9	N9	1	Incl	3.15	3.4375	31.5	34.5	-1	377.6	0.97	23.31
11	10	N10	11	Incl	3.15	3.4375	31.5	34.5	-1	381.2	1.05	21.78
12	11	N11	8	Incl	3.15	3.4375	31.5	34.5	-1	376.5	0.95	20.45
13	12	C12	22	Incl	2.15	1.875	27	42	1	383.2	1.01	22.2
14	13	C13	13	Incl	4.15	1.875	27	27	1	397.2	1.41	65.4
15	14	C14	21	Incl	2.15	5	27	27	1	362.2	0.67	12.12
16	15	C15	14	Incl	4.15	5	27	42	1	369.6	0.66	21.48
17	16	C16	12	Incl	2.15	1.875	36	27	1	383.8	0.94	23.88
18	17	C17	17	Incl	4.15	1.875	36	42	1	402.4	1.53	32.04
19	18	C18	19	Incl	2.15	5	36	42	1	365.8	0.61	14.76
20	19	C19	15	Incl	4.15	5	36	27	1	368	0.73	14.4
21	20	C20	16	Incl	3.15	3.4375	31.5	34.5	1	374.6	1.06	24.3
22	21	C21	18	Incl	3.15	3.4375	31.5	34.5	1	382.9	0.89	22.1
23	22	C22	20	Incl	3.15	3.4375	31.5	34.5	1	378	0.93	26.8

Figure 5.59: *The new laser welding worksheet after fold-over.*

In addition, the software has added a block factor. This is a precautionary measure that is useful for probing whether significant changes over time have occurred in the response data. If small or near-zero, this block factor may be removed from the model, and hence the resulting design can be regarded as the 2^4 full factorial design. If the block factor is significant, the parent design combined with its fold-over corresponds to a fractional factorial design of resolution V+.

5.7 Laser welding application – Part II

5.7.1 Evaluation of raw data – II

We see from the three replicate plots in Figure 5.60, that the replicate error is small for every response. Because of the presence of the block factor, the two triplets of replicated center-points are recognized as different and therefore rendered on separate bars. We will evaluate in the data analysis whether the block term is essential or not. If it is small and insignificant, we have, in principle, no drift over time and in that case we have six identically replicated center-points in the design.

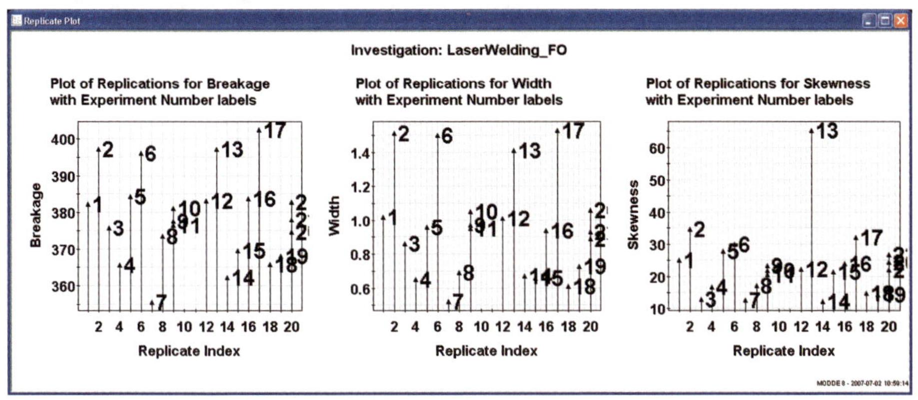

Figure 5.60: *(left) Replicate plot of Breakage. (middle) Replicate plot of Width. (right) Replicate plot of Skewness.*

Another noteworthy observation is the extreme value of trial #13 in the Skewness response. It indicates that this response is not normally distributed. Indeed, the histogram plot of Skewness in Figure 5.61 indicates that this is the case. Hence, it was decided to log-transform Skewness. Finally, the condition number evaluation shows that the condition number of the design is nearly 1.2. No plot is shown to illustrate this.

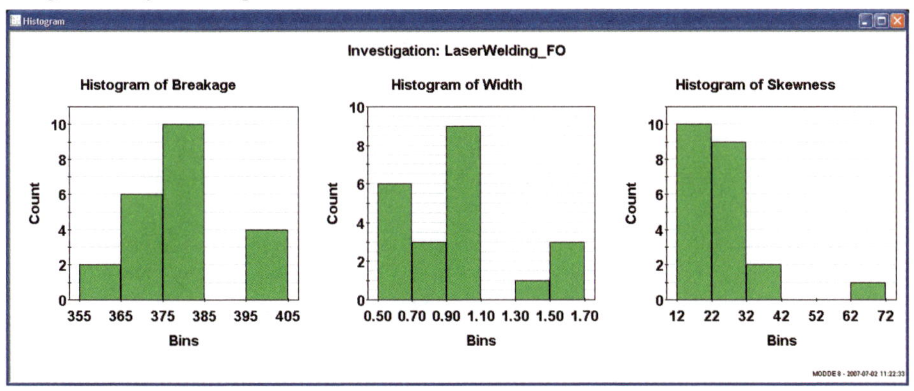

Figure 5.61: *(left) Histogram of Breakage. (middle) Histogram of Width. (right) Histogram of Skewness.*

5.7.2 Regression analysis – II

The result of fitting an interaction model with 15 terms to the three responses is shown in Figure 5.62. Evidently, we have obtained good models for Breakage and Width, but not for Skewness. No lack of fit is detected and the reproducibility is high.

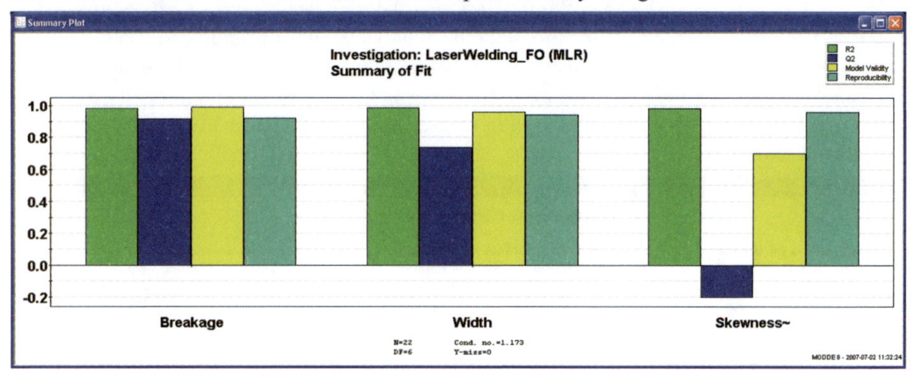

Figure 5.62: Summary of fit plot of the interaction models.

However, the N-plots of residuals, in Figure 5.63, display strange residual patterns. Although all points are located well within ± 4 standard deviations, the layered appearance of the points is conspicuous and should be more closely inspected.

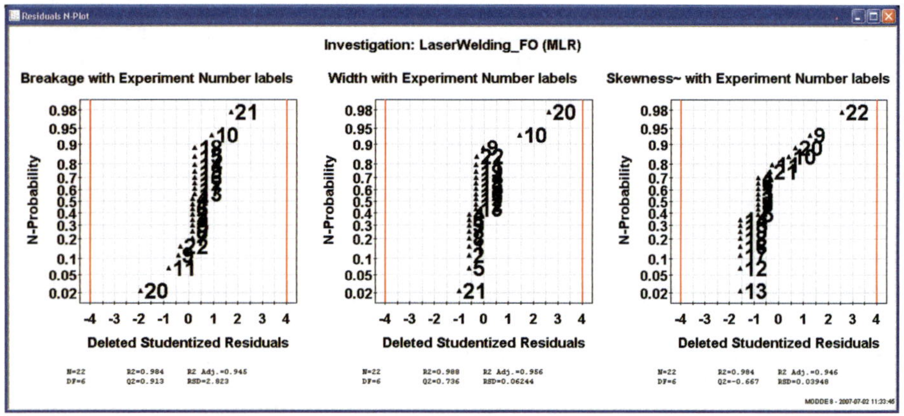

Figure 5.63: (left) N-plot of residuals of Breakage (interaction model). (middle) N-plot of residuals of Width (interaction model). (right) N-plot of residuals of Skewness (interaction model).

Next, we proceed to the plots of regression coefficients provided in Figure 5.64. After some scrutiny of these plots, it is obvious that the block factor is insignificant and should be deleted from the modelling. This indicates that there is no systematic drift between the two points in time at which the parent design and its fold-over were conducted. But it is not unusual that, for instance, variations between batches of raw materials, changes in ambient temperature and moisture, or other uncontrollable factors, may induce significant shifts in the response data. Furthermore, from an overall model interpretation viewpoint, all two-factor interactions except Po*Sp and Po*No are small and may be removed. Thus, it is sensible to delete nine model terms: the block factor and eight two-factor interactions. This is discussed further in the next section.

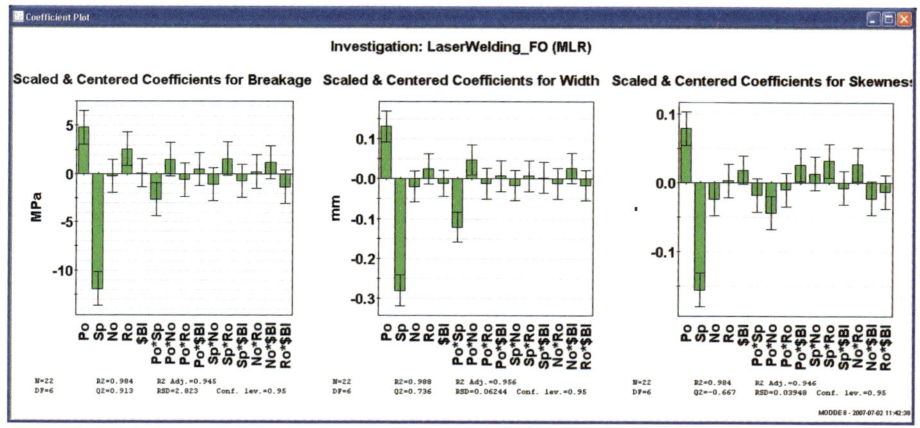

Figure 5.64: *(left) Regression coefficients of Breakage (interaction model). (middle) Regression coefficients of Width (interaction model). (right) Regression coefficients of Skewness (interaction model).*

5.7.3 Model refinement – II

Upon removal of the nine model terms and refitting the three regression models, the summary statistics given in Figure 5.65 were obtained. We now have Q^2's of 0.86, 0.94, and 0.68 for Breakage, Width, and Skewness, respectively. These values range from good to excellent. Hence, the pruning of the model has been beneficial. The Model Validity of each model is high and no lack of fit is detected.

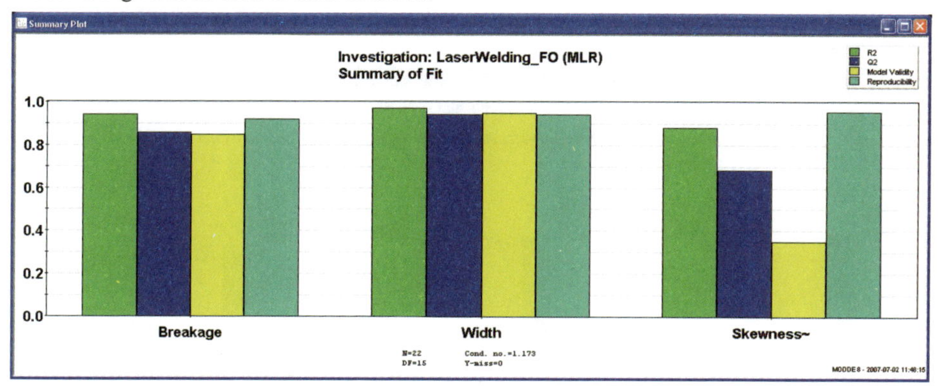

Figure 5.65: *Summary of fit plot of the refined interaction model.*

Moreover, the N-plots of residuals in Figure 5.66 are now much smoother. However, we have an indication that experiment #13 is deviating with respect to Skewness. This is the extreme value of this response, but it is a weak outlier. We define it as a weak outlier because Q^2 is relatively high at 0.68, and no lack of fit is discernible. However, if this experiment is removed and the Skewness model again refitted, the Q^2 for Skewness raises from 0.68 to 0.84. This constitutes some evidence that the presence of #13 slightly degrades Q^2 for Skewness. However, because #13 is well accounted for in the modelling of Breakage and Width, we decided to keep it in our model for Skewness as well.

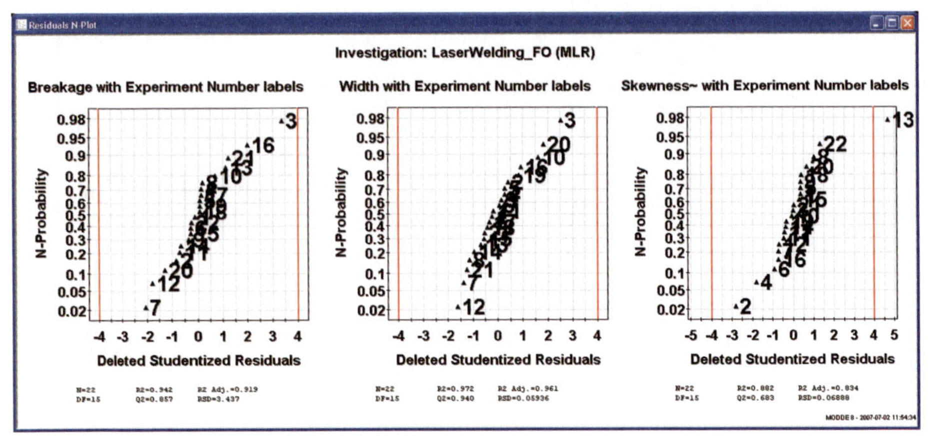

Figure 5.66: (left) N-plot of the refined interaction model of Breakage. (middle) N-plot of the refined interaction model of Width. (right) N-plot of the refined interaction model of Skewness.

The regression coefficients of the three models are given in Figure 5.67. Because the coefficient profiles are rather similar for all three responses, we may infer that the responses are correlated. A quick glimpse into the correlation matrix (no plot shown) corroborates this, as the correlation coefficients between the responses vary between 0.82 and 0.95. The two most important factors are the Power and Speed of the laser. NozzleGas is only of limited relevance, but it is kept in the final model since it participates in the Po*No two-factor interaction. The last factor, RootGas, is most meaningful for Breakage. Given that all three responses need to be well modelled at the same time, these three models represent the best compromise we can obtain. We will now use these models in trying to predict an optimal factor setting.

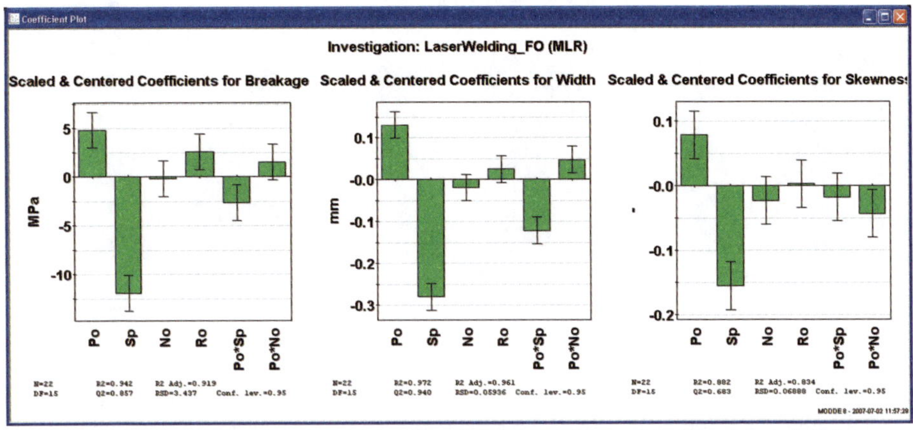

Figure 5.67: (left) Regression coefficients of the refined interaction model of Breakage. (middle) Regression coefficients of the refined interaction model of Width. (right) Regression coefficients of the refined interaction model of Skewness.

5.7.4 Use of model – II

Having acquired technically sound regression models of Breakage, Width, and Skewness, the next important step in the DOE process corresponds to identifying a factor combination at which all goals are likely to be achieved. This process is greatly facilitated by response contour plots, or response surface plots. An interesting triplet of response contour plots pertaining to the plane defined by Speed and Power, and keeping NozzleGas and RootGas fixed at their center-levels, is displayed in Figure 5.68. Alongside these plots the goals of the three responses are listed.

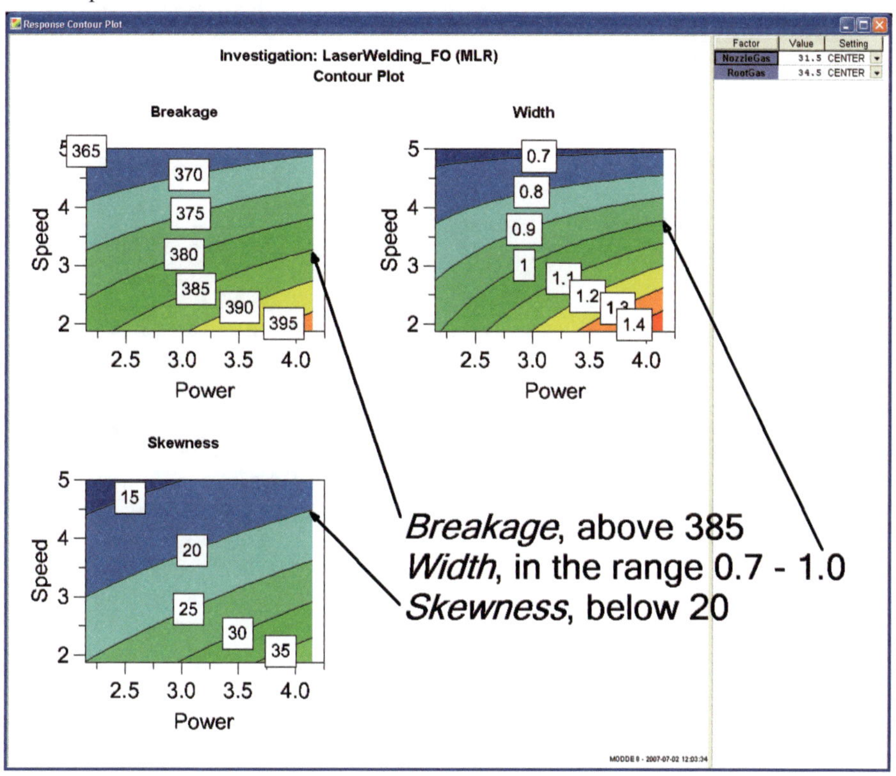

Figure 5.68: *Triple-contour plot of Breakage, Width, and Skewness, refined interaction model.*

The three arrows suggest that we should stick to the right-hand edge, but it is evident that there is no unique point at which all the goals are met simultaneously. In this situation, it is necessary to browse through many other such sets of contour triplets to identify an optimal point. However, this is both laborious and time-consuming, and there is really no guarantee that an optimal point will be found inside the investigated domain. Hence, it might be necessary to explore new territories, outside the mapped domain. Fortunately, two helpful techniques are available for this, the Gradient technique (introduced in Section 5.6.4) and the software optimizer (introduced in Section 5.6.5).

5.7.5 Gradient techniques applied to the laser welding application

Since the laser welding application comprises four factors and three responses, a large number of response contour plots are conceivable. This is an obstacle when carrying out gradient techniques. Fortunately, MODDE includes an informative 4D-response contour plot option, which permits the influence of four factors on a response to be explored at the same time. Such a plot regarding Breakage is displayed in Figure 5.69.

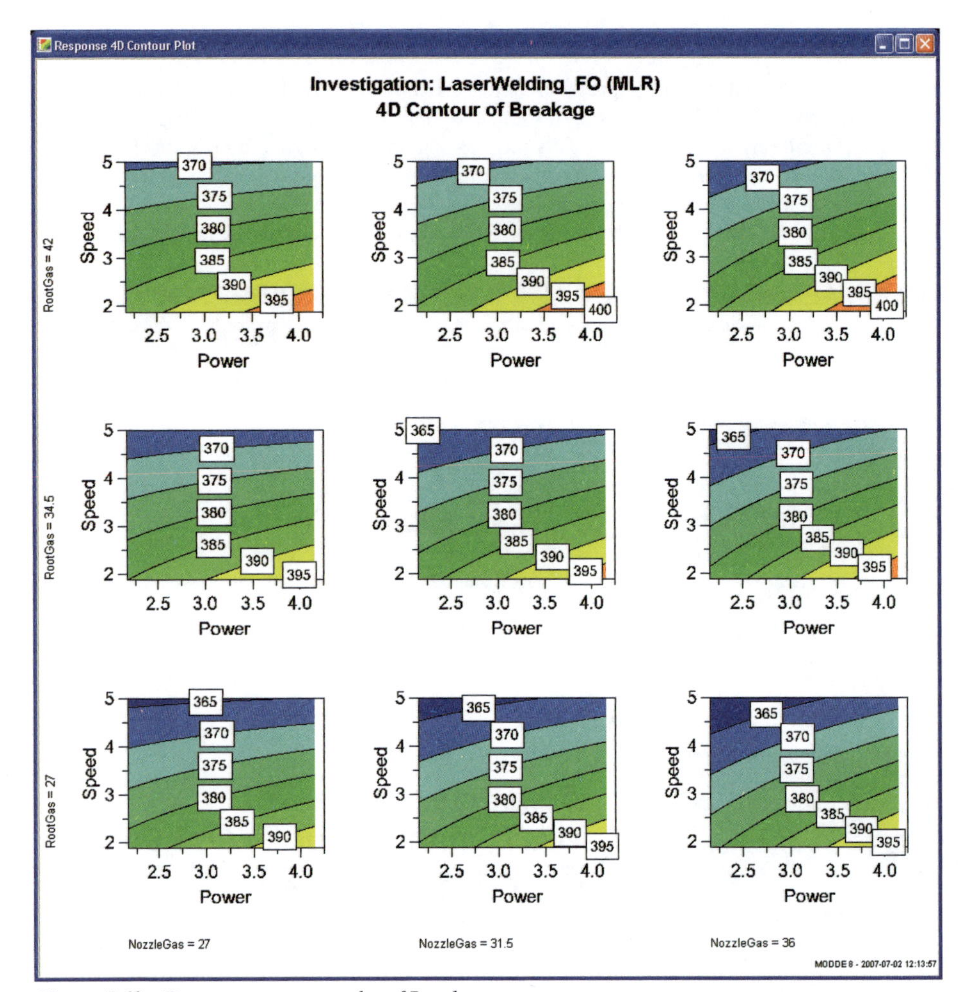

Figure 5.69: *4D-response contour plot of Breakage.*

The three by three grid of response contour plots was created by varying the four factors at three levels, low level, center level and high level. The inner factor array is made up of Power and Speed, and the outer array of NozzleGas and RootGas. Because the color coding, which goes from Breakage 360 to Breakage 410 in ten steps, is consistent over all plots, it is possible to see where the highest values of Breakage are predicted to be found. Recall that the goal of Breakage was to maximize this response, and preferably to get above 385. The predicted best point lies in the lower right-hand corner of the upper right-hand contour plot, that is, at low Speed, high Power, high NozzleGas and high RootGas.

Similar grids of contour plots can be constructed for Width and Skewness, but for reasons of brevity we do not present these. When weighing together the 4D-response contour plots of Breakage, Width and Skewness, the conclusion is that no optimal point is located within the scrutinized experimental region. We are rather close to the stipulated goals, but not sufficiently close. As a consequence, it is necessary to extrapolate. Such extrapolations can be carried out graphically with pictures such as the one shown Figure 5.69. However, a more efficient way of extrapolating is by using the automatic search routine built into MODDE. This will be discussed in the next section.

5.7.6 MODDE optimizer applied to the laser welding data

We will now relate what happened when applying the optimizer to the laser welding application. It is a good practice to conduct the optimization in stages. We recommend that the first phase is made according to the principles of interpolation, even though one may have obtained signals that extrapolation may be better. In interpolation, the low limits and high limits of the factors are taken from the factor definition. This is shown in Figure 5.70. We may disregard the block factor. It is listed, but as it has been excluded from the final model it will *not* influence the optimization.

Factor	Role	Value	Low Limit	High Limit
Power	Free		2.15	4.15
Speed	Free		1.875	5
NozzleGas	Free		27	36
RootGas	Free		27	42
$Block	Free		-1	1

Figure 5.70: *Factor settings used for interpolation.*

Next, it is necessary to give the desirability, or goal, of each response, that is, to specify a desired response profile. We can see in Figure 5.71 the criterion and goal set for each response. This table is understood as follows. Breakage is to maximized, and we want to obtain a value as high as 400, but will accept 385. Skewness is to be minimized, and we want to attain a value of 10, but will accept 20. Concerning Width the goal is neither maximization nor minimization, but it is desirable to direct it into an interval from 0.7 to 1.0. Technically, this is specified by setting the target to 0.85 and the limits to 0.7 and 1.0. Furthermore, it is also possible to assign weighting coefficients to each response. Such weights may range from 0.1 to 1.0. Here, we assign the same priority to all three responses, by stating that each weight should be 1.0. This weighting is an advanced topic that is not treated further in this course.

Response	Criteria	Weight	Min	Target	Max
Breakage	Maximize	1	385	400	
Width	Target	1	0.7	0.85	1
Skewness	Minimize	1		10	20

Figure 5.71: *Criteria of the responses used for interpolation.*

5.7.7 First optimization - Interpolation

Now we are ready to initiate the optimization. In doing so, the optimizer will first compute pertinent starting co-ordinates, factor combinations, for launching the eight simplexes. Five starting points are derived by laying out a 2^{3-1} fractional factorial design in the three most prominent model factors, and thereafter augmenting this design with one center-point. We see in Figure 5.72 that these three factors are Power, Speed and RootGas. The fractional factorial design corresponds to the five first rows in the table. In addition, the optimizer scans through the existing worksheet and picks out those three experiments which are closest to the specified response profile. These are the three runs listed in rows 6, 7 and 8.

	1	2	3	4	5	6	7	8	9	10
	Power	Speed	NozzleGas	RootGas	$Block	Breakage	Width	Skewness	iter	log(D)
1	2.15	1.875	31.5	42						
2	2.15	5	31.5	27						
3	4.15	1.875	31.5	27						
4	4.15	5	31.5	42						
5	3.15	3.4375	31.5	34.5						
6	2.15	5	27	42						
7	4.15	5	36	42						
8	2.15	5	27	27						

Figure 5.72: Factor settings for interpolation.

When pressing the arrow button (not shown), the eight simplexes will start their search for a factor combination, at which the predicted response values are as similar to the desired response profile as possible. We see the outcome of the eight simplexes in Figure 5.73. All proposed factor combinations are different, but some are rather near each other. However, none of these points is predicted to fulfill our requirements. The success of each simplex may be deduced from the numerical values of the rightmost column in Figure 5.73. This logD value represents a weighted average of the individual response desirabilities. It may be used to evaluate how the optimization proceeds. A positive logD is bad and undesirable. When logD is zero, all responses are predicted to be between their assigned target and limit values. This is a good situation. Better still is when logD is negative, and –10 is the lower, optimal, limit. Hence, in our case we are doing rather well. We are close, but not sufficiently close, to an optimal point. This means that we must try extrapolation.

	1	2	3	4	5	6	7	8	9	10
	Power	Speed	NozzleGas	RootGas	$Block	Breakage	Width	Skewness	iter	log(D)
1	2.15	2.0437	31.5246	41.9988		384.83	0.995	24.4017	340	0.0808
2	2.1502	3.032	28.234	41.9891		380.222	0.943	19.3219	283	0.0038
3	3.8788	3.431	35.9999	41.9222		385.763	1.0956	22.1675	464	0.2132
4	4.0684	4.0663	36	41.9991		381.158	0.97	19.1584	251	0.014
5	3.5097	3.638	36	41.9999		381.757	0.9872	20.481	253	0.0525
6	2.15	3.0079	27.0006	41.3269		380.611	0.9615	19.163	339	0.0147
7	4.0683	4.014	35.9998	41.9995		381.637	0.9832	19.4143	174	0.0284
8	2.15	2.9673	27.0003	31.7101		377.567	0.9343	19.164	249	0.0586

Figure 5.73: Summary of interpolation simplex searching.

5.7.8 Second optimization - Extrapolation

In the extrapolation phase, it is necessary to relax the factor limits. This is usually done in several steps. One may use the results of the interpolation to get some idea of how this should be done. For instance, seven of the eight simplexes (in Figure 5.73) have RootGas close or equal to 42. This is the high setting of this factor. Hence, it is reasonable to first relax this limit. It is seen in Figure 5.74 that we decided to raise this limit to 50.

Factor	Role	Value	Low Limit	High Limit
Power	Free		2.15	4.15
Speed	Free		1.875	5
NozzleGas	Free		27	36
RootGas	Free		27	50
$Block	Free		-1	1

	1	2	3	4	5	6	7	8	9	10
	Power	Speed	NozzleGas	RootGas	$Block	Breakage	Width	Skewness	iter	log(D)
1	2.15	2.6762	28.9805	49.9924		384.782	0.9942	21.0647	238	0.0154
2	2.1501	3.3256	27.0505	49.9987		381.669	0.9566	18.1126	317	-0.0404
3	2.9102	2.0369	35.4858	29.6427		385.585	1.1111	27.774	309	0.31
4	4.1479	4.1645	36	48.8941		383.016	0.9774	18.8883	257	-0.023
5	3.0639	3.4636	34.166	49.9999		383.276	0.9807	21.1365	189	0.0237
6	2.2503	3.1463	27.0055	49.9999		383.129	0.9862	19.3231	232	-0.0011
7	4.1321	4.1454	35.9999	50		383.491	0.984	18.9804	351	-0.0202
8	2.15	2.9869	27.0003	33.9089		378.201	0.9395	19.1238	208	0.047

Figure 5.74: *(left) Factor settings used in first extrapolation.*
Figure 5.75: *(right) Simplex results after first extrapolation.*

Figure 5.75 shows the results of the eight new simplexes. Simplex runs #2, 4 and 7 are the best. With their factor combination, the goals for Width and Skewness are met, and those for Breakage are almost met. We can also see that runs 4 and 7 have NozzleGas at its high limit. This made us consider relaxing the high limit of NozzleGas prior to the next round of simplex launches. As shown in Figure 5.76, the high limit of NozzleGas was changed to 45.

Factor	Role	Value	Low Limit	High Limit
Power	Free		2.15	4.15
Speed	Free		1.875	5
NozzleGas	Free		27	45
RootGas	Free		27	50
$Block	Free		-1	1

	1	2	3	4	5	6	7	8	9	10
	Power	Speed	NozzleGas	RootGas	$Block	Breakage	Width	Skewness	iter	log(D)
1	2.15	2.4849	30.5907	49.9996		385.302	0.9899	22.2573	237	0.0229
2	2.1584	3.2436	27.0008	45.3952		380.632	0.9516	18.378	391	-0.0153
3	4.05	3.5862	42.7656	35.5096		384.961	1.1008	17.3012	393	0.1689
4	4.0966	4.3122	44.9865	49.9994		384.206	0.9872	13.4821	257	-0.1484
5	3.332	3.5141	44.9999	49.9999		384.539	0.9868	18.1198	201	-0.0573
6	2.2758	3.2928	27.0158	49.9995		382.32	0.9724	18.8886	315	-0.0151
7	3.2453	3.3906	44.8502	48.3684		384.148	0.9821	18.8573	439	-0.041
8	2.3197	3.4032	27.2557	49.9742		381.701	0.9608	18.7221	328	-0.022

Figure 5.76: *(left) Factor settings used in second extrapolation.*
Figure 5.77: *(right) Simplex results after second extrapolation.*

Figure 5.77 reveals a somewhat better predicted point, that is, run #4. Interestingly, this fourth run has Power at its high level. In fact, this was also discernible for runs #4 and 7 in the first extrapolation round (Figure 5.75). Hence, we decided to increase the high level of Power from 4.15 to 5, and re-initiate the eight simplexes, the results of which are shown in Figures 5.78 and 5.79.

Factor	Role	Value	Low Limit	High Limit
Power	Free		2.15	5
Speed	Free		1.875	5
NozzleGas	Free		27	45
RootGas	Free		27	50
$Block	Free		-1	1

	1	2	3	4	5	6	7	8	9	10
	Power	Speed	NozzleGas	RootGas	$Block	Breakage	Width	Skewness	iter	log(D)
1	2.1501	2.461	30.6881	50		385.407	0.9909	22.3875	241	0.0254
2	2.1646	3.0681	27.0554	36.0173		378.475	0.9389	18.9356	374	0.036
3	4.9998	4.1201	42.0466	29.4399		384.642	1.1219	14.3665	229	0.1864
4	4.9988	4.8869	44.998	49.9459		385.096	0.9842	10.0933	175	-0.2248
5	3.3585	3.5408	43.6022	49.9982		384.536	0.9915	18.3722	214	-0.0421
6	2.2927	3.1474	27.8685	49.9958		382.991	0.9792	19.6743	254	-0.0029
7	3.2293	3.3881	44.992	49.9959		384.57	0.9829	18.892	313	-0.048
8	2.15	3.0901	27.0003	48.525		382.582	0.9765	18.9624	389	-0.013

Figure 5.78: *(left) Factor settings used in third extrapolation.*
Figure 5.79: *(right) Simplex results after third extrapolation.*

Again the fourth simplex is predicted as the most successful, and now we are close to a factor combination at which our goals are predicted to be accomplished. Of course, we may now proceed by extrapolating even further. However, one must remember that the further the extrapolation the greater the imprecision in the predictions. Hence, at some point, the extrapolation must be terminated and the usefulness of the predicted point be verified experimentally.

5.7.9 Local maximum or minimum

One problem with the simplex optimization approach is the risk of being trapped at a local response phenomenon, and thereby missing the global optimal point. In the optimization of a quadratic model (see Chapter 6) such a local response phenomenon might correspond to a minimum or maximum. This problem may be circumvented by taking the simplex predicted to be the best, and from its factor combination generate starting points for new simplexes. We decided to exploit this option. The best predicted point is shown as the fifth row in the table of Figure 5.80.

	1	2	3	4	5	6	7	8	9	10
	Power	Speed	NozzleGas	RootGas	$Block	Breakage	Width	Skewness	iter	log(D)
1	4.7138	4.5744	44.998	50						
2	4.7138	5	44.998	47.6459						
3	5	4.5744	44.998	47.6459						
4	5	5	44.998	50						
5	4.9988	4.8869	44.998	49.9459						

Figure 5.80: *The best predicted point is used as a starting point when generating a new series of simplexes.*

Around this point, four new starting points, rows 1-4, were defined. The co-ordinates of these new starting points were identified by moving away from the mother simplex to a distance corresponding to 20% of the factor ranges. By executing the four new simplexes and the old one, the results displayed in Figure 5.81 were obtained. Evidently, all five points are predicted to meet our experimental goals. Actually, by disregarding small variations among the decimal digits, we find that the five simplexes have converged to the same point. This means that we have not been trapped by a local response phenomenon. The identified point, with the factor settings Power = 5, Speed = 4.9, NozzleGas = 45 and RootGas = 50, is one that might be verified experimentally.

	1	2	3	4	5	6	7	8	9	10
	Power	Speed	NozzleGas	RootGas	$Block	Breakage	Width	Skewness	iter	log(D)
1	4.9619	4.8535	44.9989	49.9982		385.218	0.9893	10.2471	286	-0.2136
2	4.9998	4.8469	44.1062	49.9992		385.039	0.9842	10.7008	178	-0.2208
3	4.9869	4.8622	44.9997	49.9786		385.292	0.9904	10.1815	338	-0.2126
4	4.9995	4.8772	44.9994	50		385.225	0.9877	10.1193	193	-0.2188
5	5	4.8697	44.9703	49.9978		385.291	0.9897	10.1552	141	-0.2147

Figure 5.81: *Final results of optimization.*

5.7.10 Bringing optimization results into response contour plots

Prior to conducting the verifying experiment with the settings Power = 5, Speed = 4.9, NozzleGas = 45 and RootGas = 50, it is recommended that the relevance of this predicted optimal point be considered. This may be done by transferring the simplex optimization results into a graphical representation. Consider Figure 5.82. It provides a triplet of response contour plots, centered around the predicted optimal point. In the center of these plots, it is predicted that Breakage = 385, Width = 0.98 and Skewness = 10.1. The uncertainties in these predicted values are deduced from the associated 95% confidence intervals. These 95 % CIs indicate that at the optimal point Breakage is likely to vary between 372-398, Width between 0.76-1.20, and Skewness between 5.5-18.6. The unsymmetrical confidence intervals of the latter response have to do with the fact that Skewness is modelled in its log-transformed metric, and when the predicted values are re-expressed in the original metric their associated confidence intervals become skewed.

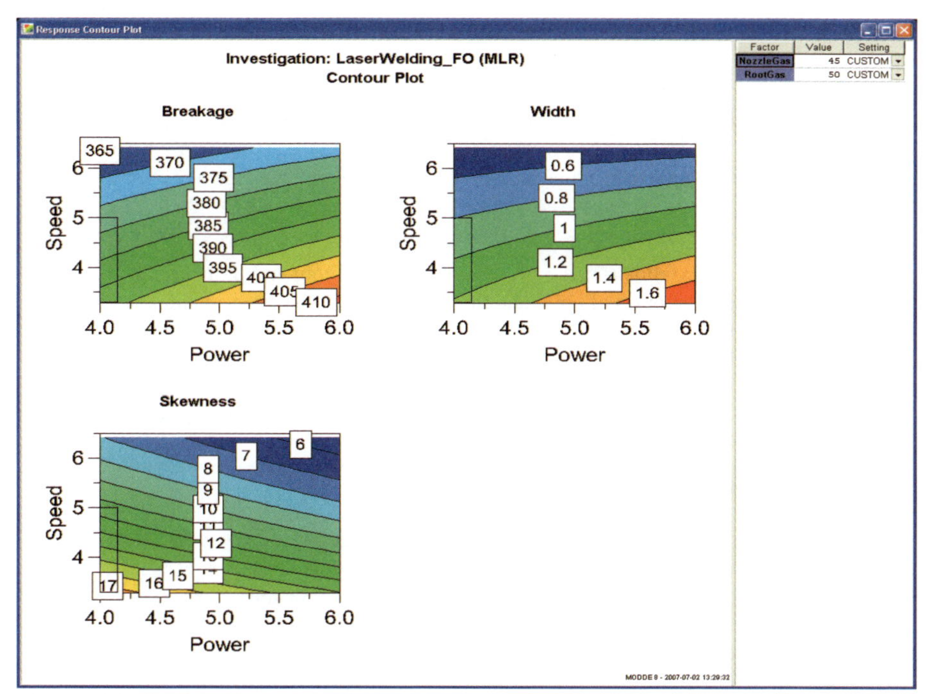

Figure 5.82: *Graphical summary of the optimization results. Plots are centered around the predicted optimal point.*

5.8 Summary of the laser welding application

The application of DOE in the production of gasket-free all-welded plate heat exchangers was very successful. With this approach it was possible not only to understand the mechanisms involved in the welding, but also to identify good operating conditions for the welding. We are, however, prevented from disclosing any detailed information related to the final choice of factor settings.

From an instructive point of view, the laser welding application is good for several reasons. For instance, it shows that applying DOE to technical problems often involves proceeding in stages. Initially, a screening design of only eleven experiments was laid out to probe the relevance of the defined experimental region. Once it was verified that this region was interesting, another set of eleven experiments was selected, using the fold-over technique, and used to supplement the initial experiments. With this combined set of 22 experiments it was possible to derive very good regression models for the three modelled responses Breakage, Width, and Skewness.

Furthermore, this data set also allowed us to study how tools for evaluation of raw data, tools for regression analysis, and tools for using models, were put into practice in a real situation. It should be clear from this application that the process of deriving the best possible regression model is often of an iterative nature, and frequently many repetitive modelling cycles are needed before a final model can be established.

The modelling step is of crucial importance for the future outcome of a DOE project. If an investigator overlooks checking the quality of a regression model, this may have devastating consequences at later stages when the model is used for predictive purposes. Hence, strong emphasis is placed on finding the most reliable and useful model.

We also used the laser welding study for reviewing what options are available after a screening investigation has been completed. In particular, we examined the graphically-oriented gradient techniques and the more automatic search technique based on executing multiple simplexes. With the MODDE optimizer we were able to identify the factor combination Power = 5, Speed = 4.9, NozzleGas = 45 and RootGas = 50 as the predicted best point. In principle, it is then up to the experimenter to determine whether it is relevant to proceed and verify this point, or whether the optimizer should be re-used to find an alternative point at some other location in the factor space.

Our intention has been only to illustrate the procedure for identifying an optimal point. One may also discuss what such an optimal point should be used for. If the experimenter is sure that the optimum will be found in the vicinity of a predicted point, he or she may proceed directly to optimization by anchoring a central composite design around the predicted optimal point. However, if he or she is not totally convinced that the optimum is within reach, it might be better to center a second screening design around the predicted best point.

5.9 Summary of fractional factorial designs

Fractional factorial designs are useful for screening large numbers of factors in few experiments. These designs are constructed by selecting fractions of corner experiments, drawn from the underlying full factorial designs. Because all possible corner experiments are not used with fractional factorial designs, these designs give rise to problems with the confounding of effects. Confounding of effects means that all effects cannot be estimated completely resolved from one another. A rapid insight into the confounding complexity of a fractional factorial design is given by its resolution. For general screening purposes, we recommend designs of resolution IV. Such designs have main effects unconfounded with two-factor interactions, which is a desirable situation. In summary, with fractional factorial designs it is feasible to answer questions, such as, which are the dominant factors? And what are their optimal ranges?

5.10 Questions for Chapter 5

- Which important questions regarding the factors are usually asked in screening?
- What is an Ishikawa diagram? In which respect is it helpful?
- Which types of regression model are normally utilized in screening?
- What is a fractional factorial design?
- How are fractional factorial designs constructed (describe in simple terms)?
- How many corner experiments are encoded by a 2^{4-1} fractional factorial design? By a 2^{5-1} design? By a 2^{5-2} design?
- What is a generator?
- What is a defining relation?
- What is meant by confounding of effects?
- What does the resolution of a design signify?
- Which resolution is recommended for screening?
- Which resolution is recommended for robustness testing?
- Which tools are useful for evaluating raw data?
- Which tools are meaningful for diagnostic checking of a model?
- What does interpolation mean? Extrapolation?

- What options are available after screening?
- What are the two primary reasons for adding complementary runs?
- What is fold-over?
- How are gradient techniques implemented?
- What does simplex optimization imply?

5.11 Summary and discussion

In this chapter, we have discussed the experimental objective called screening. For this purpose, we have used the laser welding application, consisting of four factors, three responses, and two series of eleven experiments. The problem formulation steps of this application were reviewed at the beginning of the chapter. The next part of the chapter was devoted to the introduction of fractional factorial designs, and a discussion relating to their advantages and limitations was given. Much emphasis was given to a geometrical interpretation of fractional factorial designs. After the introduction of the family of fractional factorial designs, the laser welding application was outlined in detail. Both the analysis of this data-set, leading towards useful models, and what to do after screening, were addressed. At the end of the chapter, procedures for converting the screening information into concrete actions and decisions of where to carry out new experiments were considered.

In principle, all the important features of fractional factorial designs are controlled by the selected generator(s). The features we have concentrated on are: the selected fraction of experiments, the confounding pattern, the defining relation and the resolution. Fortunately, with MODDE, the user only has to make sure that a resolution IV design is selected, and the rest will be automatically taken care of.

An important part of screening experimentation concerns what to do afterwards. Post-screening actions depend on (i) the quality of the obtained regression model, (ii) whether it is possible and necessary to modify the factor ranges, and (iii) whether some of the already conducted experiments are close to fulfilling the goals stated in the problem formulation. In order to further specify the exact path to follow the experimentalist has to consider if (a) experiments outside the investigated region are impossible and/or undesired, or if (b) experiments outside the investigated region are possible and/or desired.

6 Experimental objective: Optimization

6.1 Objective

The objective of this chapter is to use the DOE framework presented in Chapters 1-4 and discuss an optimization study. We will describe the truck engine application, in which the goal is to adjust three controllable factors so that a favorable response profile with regard to fuel consumption and exhaust gases is obtained. The experimental protocol underlying the truck engine application originates from the central composite design family. We will provide a detailed account of two commonly used members of the central composite family, the designs entitled CCC and CCF. These abbreviations refer to central composite circumscribed and central composite face-centered. Further, the analysis of the engine data involves the repeated use of a number of tools for raw data evaluation and regression analysis, which were introduced in earlier chapters. In addition, much emphasis will be given to how to solve an optimization problem using simplex searching and graphical representation of results.

6.2 Background to General Example 2

At Volvo Car Corporation and Volvo Truck Corporation, DOE is commonly applied in the early stages of new engine developments (Volvo Technology Report, #2, 1997). In this context, common goals are to reduce fuel consumption and reduce levels of unwanted species in exhaust gases. One way of decreasing the emission of nitrous oxides (NO_x) from a combustion engine is to decrease the combustion temperature. By lowering this temperature, smaller amounts of NO_x are formed by reaction between atmospheric nitrogen and oxygen. It is possible to lower this temperature by supplying an inert gas, which does not react but only absorbs heat. This kind of gas may originate from the car exhaust, and hence the term exhaust gas recirculation (EGR) is often used. A negative consequence of the EGR principle is that less air is available for combustion, since the EGR mass does not allow as much air to enter the cylinder. This may, in turn, cause elevated levels of Soot in the exhaust gases. By manipulating, for instance, inlet cylinder pressure, inlet temperature, and timing of fuel injection with regard to piston movement ("needle lift"), it is possible to influence the levels of Soot and NOx in the exhausts. A schematic of a typical testing environment is given in Figure 6.1.

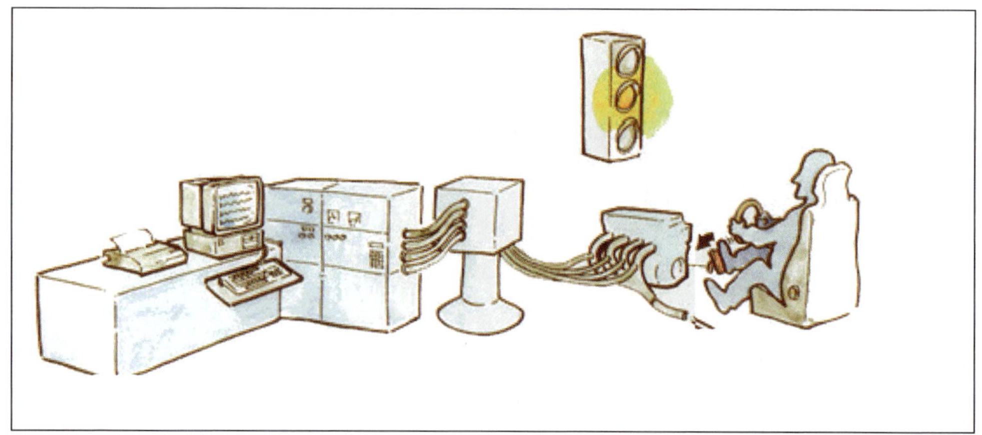

Figure 6.1: A schematic of a testing environment used in optimization of combustion engines.

The example that we will study was kindly contributed by Sven Ahlinder at Volvo Technical Development (VTD). It deals with the optimization of a truck piston engine. Sven Ahlinder studied how three factors, that is, *air mass used in combustion*, *exhaust gas re-circulation*, and *timing of needle-lift*, affected three responses, that is, *fuel consumption*, and levels of *NOx* and *Soot* in the exhaust gases. The units and the high and low settings of these factors are given in Figure 6.2. Analogously, Figure 6.3 discloses the units of the three responses.

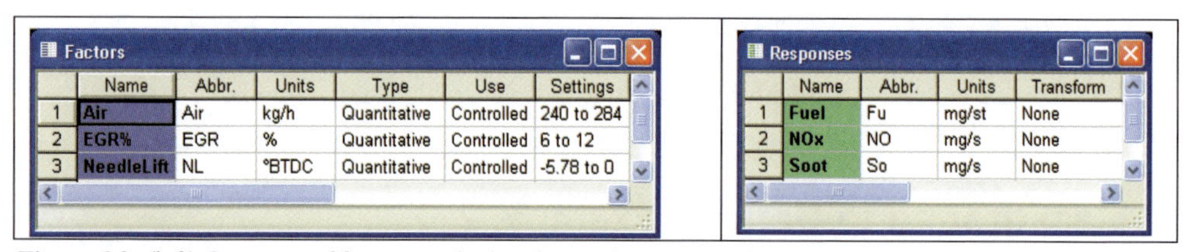

Factors

	Name	Abbr.	Units	Type	Use	Settings
1	Air	Air	kg/h	Quantitative	Controlled	240 to 284
2	EGR%	EGR	%	Quantitative	Controlled	6 to 12
3	NeedleLift	NL	°BTDC	Quantitative	Controlled	-5.78 to 0

Responses

	Name	Abbr.	Units	Transform
1	Fuel	Fu	mg/st	None
2	NOx	NO	mg/s	None
3	Soot	So	mg/s	None

Figure 6.2: *(left) Overview of factors studied in the truck engine application.*
Figure 6.3: *(right) Overview of measured responses in the truck engine application.*

As far as the responses are concerned, the desired response profile was fuel consumption as low as possible with a target of 230 mg/stroke (of piston), and simultaneously low levels of NOx and Soot in the exhaust emissions with targets of 25 mg/s and 0.5 mg/s, respectively.

This application is typical in that only a limited number of factors are studied. In optimization, usually between two and five factors are explored. This means that one can afford to make comparatively many experiments per varied factor and thus acquire a detailed understanding of the relationships between factors and responses. Sven Ahlinder used a CCF design in 17 experiments to determine whether the stipulated response profile was attainable. This design corresponds to a classical *response surface modelling*, RSM, design. It is extensively used in optimization as it supports quadratic polynomial models.

6.3 Problem Formulation (PF)

6.3.1 Selection of experimental objective

As discussed in Chapter 2, the problem formulation (PF) phase is important in DOE. In the truck engine application, the selected experimental objective was *optimization*. In optimization, the important factors, usually between two and five, have already been identified, and we now want to extract detailed information about them. It is of interest to reveal the nature of the relationships between these few factors and the measured responses. For some factors and responses the relationships might be linear, for others non-linear, that is, curved. For some factors and responses there exists a positive correlation, for others a negative correlation. These relationships are conveniently investigated by fitting a quadratic regression model.

A quadratic model may be used to predict response values for any factor combination in the region of interest. It may also be used for optimization, that is, for identifying the factor setting(s) fulfilling the desired response profile. Sometimes only *one* distinct point is predicted, sometimes several geometrically close factor combinations might be plausible. The latter case then corresponds to a so called *region of operability* or *Sweet Spot*. It is also possible that points clearly separated in factor space may be predicted to share the same response profile. We then have a situation of multiple optima.

6.3.2 Specification of factors

The second step of the problem formulation concerns the specification of factors. This is quite simple in optimization, because at this stage the important factors have already been identified. Since these are usually limited in number, it is not necessary to use the Ishikawa diagram for system overview. This diagram, which was introduced in Chapter 5, is most useful in screening. In the truck engine application, three factors were varied. Figures 6.4 – 6.6 show their details. The first factor, air mass used in combustion, is varied between 240 and 284 kg/h. The second factor, exhaust gas-recirculation, is varied between 6 and 12%. The last factor, NeedleLift, which has to do with the *timing* of the ignition procedure, is expressed in degrees before top dead-center (°BTDC) and varies between –5.78 and 0. The appropriateness of these values were determined during a previous screening phase. Notice, that all three factors are controlled and quantitative.

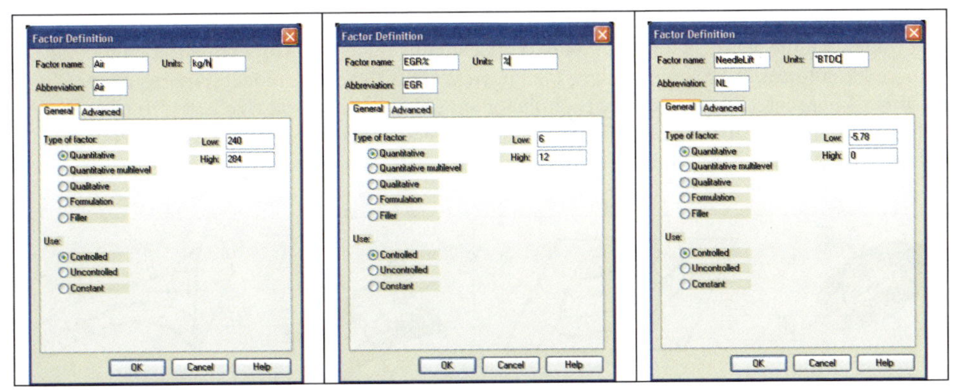

Figure 6.4: *(left) Factor definition of air mass used in combustion.*
Figure 6.5: *(middle) Factor definition of exhaust gas-recirculation.*
Figure 6.6: *(right) Factor definition of timing of needle lift.*

6.3.3 Specification of responses

The next step in the problem formulation deals with the specification of the responses. It is important to select responses that are relevant to the experimental goals. In the truck engine application, low fuel consumption and low levels of certain species in the exhaust emissions were the important properties. The investigator Sven Ahlinder registered four responses, but for simplicity we will study only three of them, namely the fuel consumption, and the amounts of NOx and Soot in the exhaust emissions. All details regarding these responses are given in Figures 6.7 – 6.9.

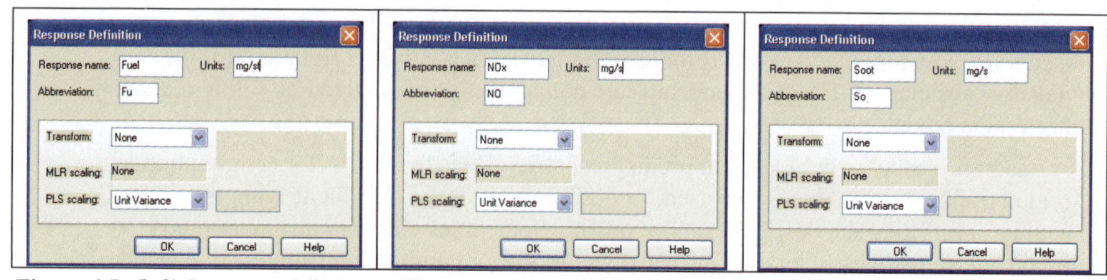

Figure 6.7: *(left) Response definition of fuel consumption.*
Figure 6.8: *(middle) Response definition of level of NOx.*
Figure 6.9: *(right) Response definition of level of Soot.*

The three responses are quantitative and untransformed. The first response is expressed in mg/st, the second in mg/s, and the third in mg/s. The overall goal is a response profile with a low value of fuel with a target of 230 mg/stroke, a low value of NOx with a target of 25 mg/s, and a low value of Soot with 0.5 mg/s specified as the target.

6.3.4 Selection of regression model

The selection of an appropriate regression model is the fourth step of the problem formulation. Remember that we may select between three different types of regression model (linear, interaction, quadratic). A quadratic model in all factors is a sound choice for the optimization objective. The reason for this is that the final goal is optimization and this often involves modelling curved response functions. Indeed, a quadratic model was selected in the truck engine application.

Quadratic models are flexible and can mimic many different types of response function. Figures 6.10 – 6.13 illustrate that quadratic models may take the shapes of stationary ridges, rising ridges, saddle surfaces, or mountains. This explains the popularity of quadratic models. Compared to a linear or an interaction model, however, the quadratic model requires many experiments per varied factor. Ideally each factor should be explored at five levels to allow a reliable quadratic model to be postulated. But, there are also good optimization designs available, which utilize only three levels per explored factor.

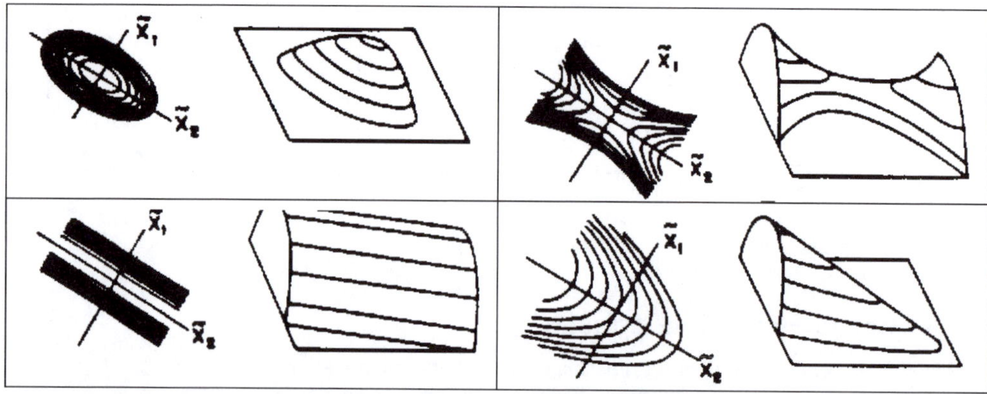

Figure 6.10: *(upper left) Surfaces mimicked by quadratic models: Top of mountain.*
Figure 6.11: *(upper right) Surfaces mimicked by quadratic models: Saddle surface.*
Figure 6.12: *(lower left) Surfaces mimicked by quadratic models: Stationary ridge.*
Figure 6.13: *(lower right) Surfaces mimicked by quadratic models: Rising ridge.*

6.3.5 Generation of design and creation of worksheet

The last two stages of the problem formulation deal with the generation of the statistical experimental design and the creation of the corresponding worksheet. For the truck engine application we will treat these two steps together. The experimenter selected a *central composite face-centered*, CCF, design in 17 runs. This is a standard design, which supports a quadratic model. Its worksheet is given in Figure 6.14.

Worksheet

	1	2	3	4	5	6	7	8	9	10
1	Exp No	Exp Name	Run Order	Incl/Excl	Air	EGR%	NeedleLift	Fuel	NOx	Soot
2	1	N1	2	Incl	241	6.1	-5.78	236.637	25.3414	0.14
3	2	N2	4	Incl	284	6	-5.78	254.722	26.2859	0.06
4	3	N3	10	Incl	241	12	-5.78	242.841	11.7348	1.15
5	4	N4	5	Incl	283	12.1	-5.78	262.626	12.6292	0.67
6	5	N5	14	Incl	241	6.2	0	217.069	29.6733	0.45
7	6	N6	11	Incl	283	6.1	0	230.527	32.7326	0.22
8	7	N7	6	Incl	242	12.1	0	221.274	13.6371	3.3
9	8	N8	13	Incl	284	11.8	0	236.25	15.2	1.94
10	9	N9	15	Incl	241	9	-3.16	231.77	17.48	1.24
11	10	N10	16	Incl	284	9.2	-3.16	248.064	19.3567	0.47
12	11	N11	3	Incl	261	6	-3.16	242.616	28.0458	0.19
13	12	N12	12	Incl	261	12	-3.16	245.303	12.7836	1.29
14	13	N13	7	Incl	262	9.3	-5.78	256.6	17.6	0.26
15	14	N14	9	Incl	262	9	0	227.266	22.0432	1.61
16	15	N15	17	Incl	261	9.2	-3.16	247.479	19.0623	0.44
17	16	N16	1	Incl	261	9	-3.16	243.15	19.125	0.35
18	17	N17	8	Incl	261	9.1	-3.16	242.982	19.1812	0.47

Figure 6.14: *The worksheet of the truck engine application.*

The three columns of the worksheet numbered 5, 6, and 7 represent the identified design. In the three right-most columns of the worksheet, we can see the measured response values. The first eight rows are the corner experiments of the basis cube, the next six rows the so-called axial points, and the last three rows the replicated center-points. Noteworthy is the fact that this design is slightly distorted. Because of minor experimental obstacles, it was difficult to adjust the three factors so that they exactly matched the specifications of the CCF design. For instance, the levels 240 and 284 were set for Air in the factor specification, but in reality small deviations around these targets were observed. Such small deviations from the factor specification are found also for the second factor, EGR%, and for the third factor, NeedleLift. However, for NeedleLift the deviation is found only at the center-level. Interestingly, we can use the condition number for evaluating the magnitude of the design distortion. The condition number for the conventional CCF in three factors is 4.438, but for the distorted truck engine design it amounts to 4.508, that is, an increase of 0.07. Changes in the condition number smaller than 0.5 are regarded as insignificant. We can therefore infer that the design distortion is almost non-existent, and will not compromise the validity or usefulness of the work.

6.4 Introduction to response surface methodology (RSM) designs

We have now completed the problem formulation phase of the truck engine application, and it is appropriate to take a closer look at the properties of the CCF design we have employed. The CCF design is often classified as an RSM design. Traditionally, RSM has been the acronym for *response surface methodology*, reflecting the predominant view that extensive use of response surface plots is advantageous for finding an optimal point. In more recent years, however, the re-interpretation *response surface modelling* has become more prevalent. This name emphasizes the significance of the modelling and understanding part, that is, the importance of verifying the reliability of a regression model prior to its conversion to response contour plots or response surface plots.

In RSM it is important to get good regression models and in general higher demands are placed on these models than on screening models. Consequently, there are certain fundamental demands which must be met by an RSM design. Good RSM designs have to allow the estimation of the parameters of the model with low uncertainty, which means that we want the confidence intervals of the regression coefficients to be as narrow as possible. Good RSM designs should also give rise to a model with small prediction error, and permit a judgement of the adequacy of this model. This latter aspect means that the design must contain replicated experiments enabling the performance of a lack of fit test. In addition, good RSM designs should encode as few experiments as possible. There are several classical RSM design families which meet these demands, and we will outline three of these. These are the central composite (Chapter 6), Box-Behnken (Chapter 9), and three-level full factorial (Chapter 9) design families. Computer generated D-optimal designs for irregular experimental regions may be tailored to fulfill such requirements, as well.

6.4.1 The CCC design in two factors

In the current section our intention is to describe the family of central composite designs. Box-Behnken, three-level full factorial and D-optimal designs are discussed later. The central composite designs are natural extensions of the two-level full and fractional factorial designs. We commence with the central composite circumscribed, CCC, design in two factors. Figure 6.15 displays this type of design. The CCC design consists of three building blocks, (i) regularly arranged corner (factorial) experiments of a two-level factorial design, (ii) symmetrically arrayed star points located on the factor axes, and (iii) replicated center-points. Hence, it is a *composite* design.

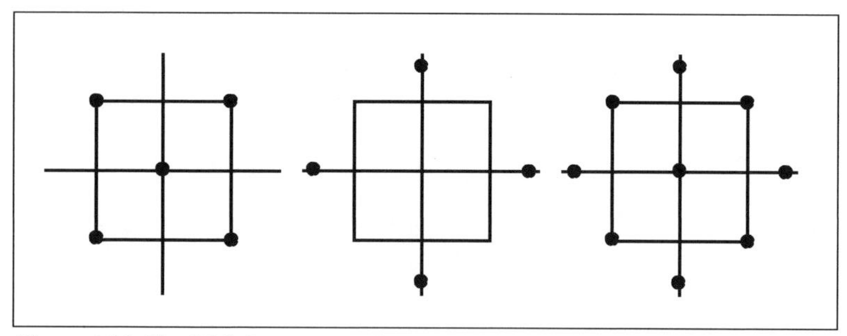

Figure 6.15: *The CCC design in two factors.*

Figure 6.16 gives the worksheet of an eleven run CCC design in two factors. Here, the first four rows represent the corner experiments, the next four rows the star points, and the last three rows the replicated center-points. The star points, also denoted axial points, represent the main difference between factorial and central composite designs. Thanks to them, all factors are investigated at five levels with the CCC design. This makes it possible to estimate quadratic terms with great rigor. We can see in Figures 6.15 and 6.16 that the corner experiments and the axial experiments are all situated on the circumference of a circle with radius 1.41, and therefore the experimental region is symmetrical.

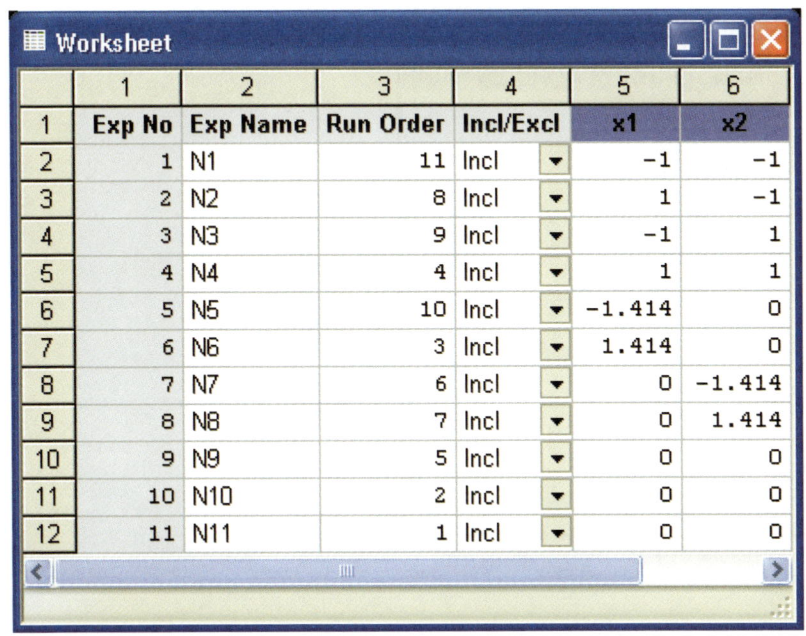

Worksheet

	1	2	3	4	5	6
1	Exp No	Exp Name	Run Order	Incl/Excl	x1	x2
2	1	N1	11	Incl	-1	-1
3	2	N2	8	Incl	1	-1
4	3	N3	9	Incl	-1	1
5	4	N4	4	Incl	1	1
6	5	N5	10	Incl	-1.414	0
7	6	N6	3	Incl	1.414	0
8	7	N7	6	Incl	0	-1.414
9	8	N8	7	Incl	0	1.414
10	9	N9	5	Incl	0	0
11	10	N10	2	Incl	0	0
12	11	N11	1	Incl	0	0

Figure 6.16: *Worksheet of the CCC design in two factors.*

With two, three and four factors, the factorial part of the CCC design corresponds to a full two-level factorial design. With five or more factors a fractional factorial design of resolution V is utilized. This means that with any CCC design, all linear, two-factor interaction, and quadratic model terms are estimable.

6.4.2 The CCC design in three factors

The CCC design in three factors is constructed in a fashion similar to that of the two factor analogue. Figure 6.17 displays its three components, which are (i) eight corner experiments, (ii) six axial experiments, (iii) and, at least, three replicated center-points.

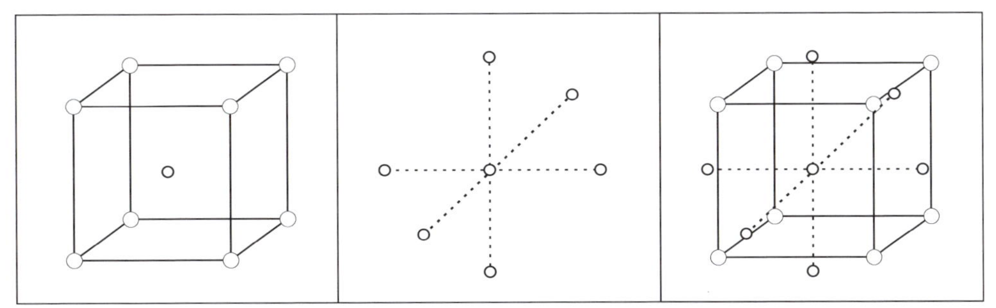

Figure 6.17: *The CCC design in three factors.*

The worksheet of such a 17 run design is displayed in Figure 6.18. Evidently, with three factors, the corner and axial experiments approximate the surface of a sphere and hence the experimental region is symmetrical. In a similar manner, four-factor and five-factor CCC designs define hyper-spherical experimental arrangements.

Worksheet

	1	2	3	4	5	6	7
1	Exp No	Exp Name	Run Order	Incl/Excl	x1	x2	x3
2	1	N1	1	Incl	-1	-1	-1
3	2	N2	16	Incl	1	-1	-1
4	3	N3	9	Incl	-1	1	-1
5	4	N4	8	Incl	1	1	-1
6	5	N5	7	Incl	-1	-1	1
7	6	N6	17	Incl	1	-1	1
8	7	N7	6	Incl	-1	1	1
9	8	N8	14	Incl	1	1	1
10	9	N9	15	Incl	-1.682	0	0
11	10	N10	12	Incl	1.682	0	0
12	11	N11	3	Incl	0	-1.682	0
13	12	N12	2	Incl	0	1.682	0
14	13	N13	11	Incl	0	0	-1.682
15	14	N14	13	Incl	0	0	1.682
16	15	N15	4	Incl	0	0	0
17	16	N16	10	Incl	0	0	0
18	17	N17	5	Incl	0	0	0

Figure 6.18: *Worksheet of the CCC design in three factors.*

6.4.3 The CCF design in three factors

The CCC design prescribes experiments, the axial points, whose factor values are located outside the low and high settings of the factor definition. Sometimes it is not possible to carry out this kind of testing. When it is desirable to maintain the low and high factor levels, and still perform an RSM design, the central composite face-centered (CCF) design is a viable alternative. This design in three factors is shown in Figure 6.19 and for comparison its CCC counterpart is plotted in Figure 6.20.

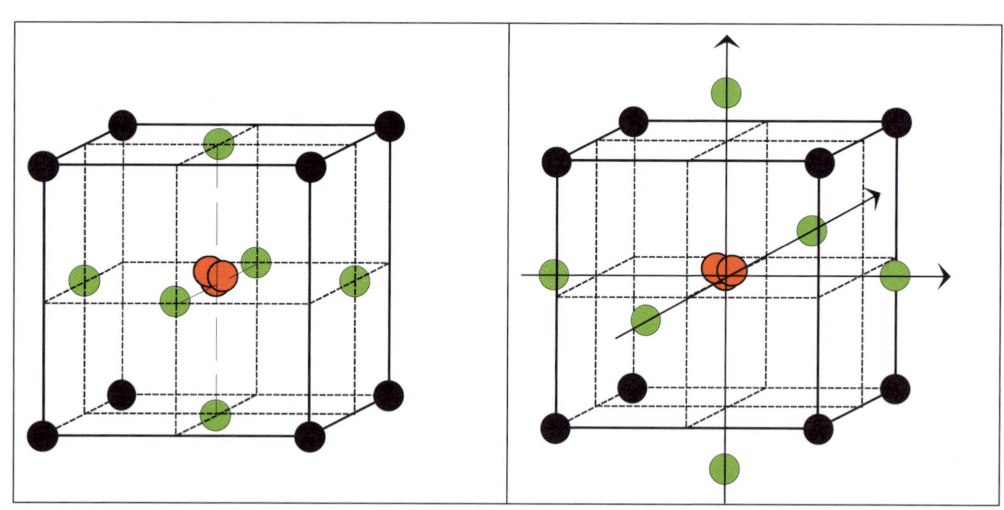

Figure 6.19: *(left) The CCF design in three factors.*
Figure 6.20: *(right) The CCC design in three factors.*

In the CCF design, the axial points are centered on the faces of the cube. This implies that all factors have three levels, rather than five, and that the experimental region is a cube, and not a sphere. CCF is the recommended design choice for pilot plant and full scale investigations.

Although it corresponds to only three levels of each factor, it still supports a quadratic model, because it contains an abundance of experiments. Theoretically, the CCF design is slightly inferior to the CCC design. Given the same settings of low and high levels of the factors, the CCC design spans a larger volume than does the CCF design. Five levels of each factor also means that the CCC design is better able to capture strong curvature; even a cubic response behaviour may be modelled. Figures 6.21 and 6.22, which provide the correlation matrices of the CCF and CCC designs, illustrate another argument in favor of the CCC design. It is seen that the quadratic model terms are less correlated in the CCC than CCF case, which is beneficial when it comes to regression modelling.

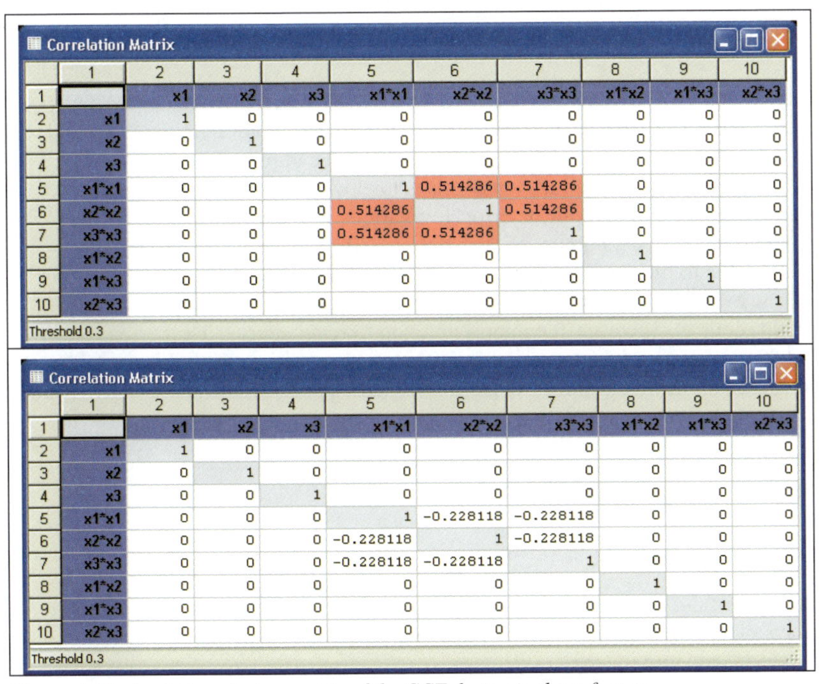

Correlation Matrix

	x1	x2	x3	x1*x1	x2*x2	x3*x3	x1*x2	x1*x3	x2*x3
x1	1	0	0	0	0	0	0	0	0
x2	0	1	0	0	0	0	0	0	0
x3	0	0	1	0	0	0	0	0	0
x1*x1	0	0	0	1	0.514286	0.514286	0	0	0
x2*x2	0	0	0	0.514286	1	0.514286	0	0	0
x3*x3	0	0	0	0.514286	0.514286	1	0	0	0
x1*x2	0	0	0	0	0	0	1	0	0
x1*x3	0	0	0	0	0	0	0	1	0
x2*x3	0	0	0	0	0	0	0	0	1

Threshold 0.3

Correlation Matrix

	x1	x2	x3	x1*x1	x2*x2	x3*x3	x1*x2	x1*x3	x2*x3
x1	1	0	0	0	0	0	0	0	0
x2	0	1	0	0	0	0	0	0	0
x3	0	0	1	0	0	0	0	0	0
x1*x1	0	0	0	1	-0.228118	-0.228118	0	0	0
x2*x2	0	0	0	-0.228118	1	-0.228118	0	0	0
x3*x3	0	0	0	-0.228118	-0.228118	1	0	0	0
x1*x2	0	0	0	0	0	0	1	0	0
x1*x3	0	0	0	0	0	0	0	1	0
x2*x3	0	0	0	0	0	0	0	0	1

Threshold 0.3

Figure 6.21: *(top) Correlation matrix of the CCF design in three factors.*
Figure 6.22: *(bottom) Correlation matrix of the CCC design in three factors.*

In summary, the CCF design is very good, and some of its properties are very tractable in pilot plant and large-scale operations. Interestingly, since the CCF and CCC designs require the same number of experiments, it is possible to tailor-make a design by adjusting the distance of the axial points from the design center. Each such CCF/CCC-hybrid is a very good RSM design, but one must remember to check the correlation matrix.

6.5 Overview of central composite designs

Usually, RSM designs, like CCC and CCF, are used in conjunction with two to five factors. Figure 6.23 provides a tabulated overview of how many experiments are necessary in these cases. For comparative purposes, the table is also extended to cover the situation with six and seven factors. We see that it is possible to explore as many as five factors in as few as 29 experiments. These 29 experiments consist of 16 corner experiments, 10 axial experiments, and three replicated center-points. However, when moving up to six factors, there is a huge increase in the number of experiments, which effectively means that RSM should be avoided in this case.

Number of factors	Number of experiments
2	8 + 3
3	14 + 3
4	24 + 3
5	26 + 3
6	44 + 3
7	78 + 3

Figure 6.23: *Overview of composite designs in 2 to 7 factors.*

Observe that the CCC and CCF designs in two, three, and four factors are based on two-level full factorial designs, whereas in the case of five or more factors two-level fractional factorial designs are employed as the design foundation. The recommended number of replicated center-points has been set to three for all designs. This is the lowest number needed to provide a reasonable estimate of the replicate error. For various reasons, however, it is sometimes desirable to increase the number of center-points. By adjusting the number of center-points it is possible to manipulate certain properties of the design, such as, its orthogonality and the overall precision of its predictions. This is an advanced topic that will not be pursued in this course. However, the important thing to remember is that for all practical purposes, three replicated center-points is an appropriate choice. With this discussion in mind, we will now continue with the truck engine application and the analysis of this data-set.

6.6 Truck engine application

6.6.1 Evaluation of raw data

We start the evaluation of the raw data by inspecting the replicate error of each response. Figure 6.24 displays plots of replications for the three responses Fuel, NOx, and Soot.

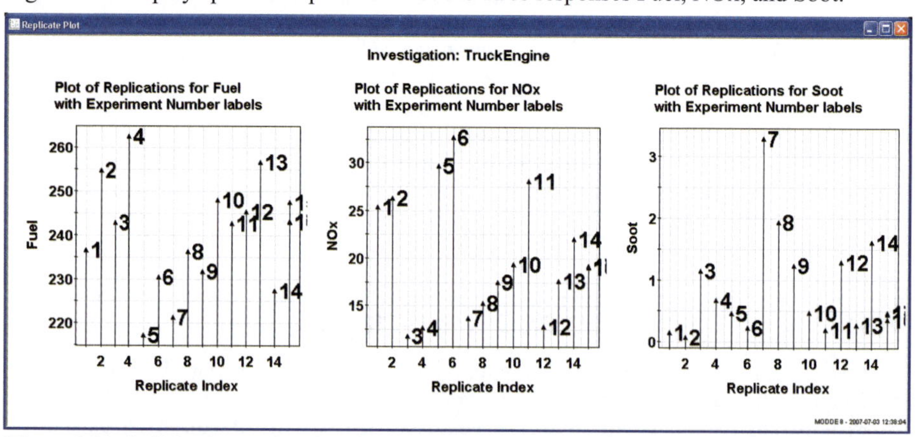

Figure 6.24: *(left) Replicate plot of Fuel. (middle) Replicate plot of NOx. (right) Replicate plot of Soot.*

We can see that the replicate errors are small in all cases, which is good for the forthcoming regression analysis. However, we can also see that for Soot we have one measurement, #7, which is much larger than the others. This means that we must expect the distribution of Soot to be positively skewed. That this is indeed the case is seen in the triplet of histogram plots in Figure 6.25.

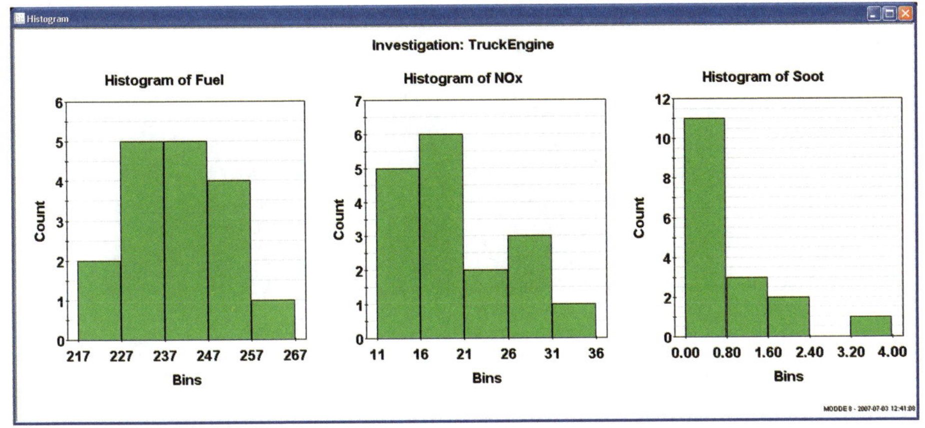

Figure 6.25: *(left) Histogram of Fuel. (middle) Histogram of NOx. (right) Histogram of Soot.*

Figure 6.25 reveals the need for log-transforming Soot. The appearance of the histogram and the replicate plot after log-transformation of Soot is given by Figures 6.26 and 6.27. It is clear that the log-transformation is warranted. A final check of the condition number in Figure 6.28 indicates that we have a pertinent design. Recall that classical RSM designs in two-to-five factors have condition number in the range 3-8.

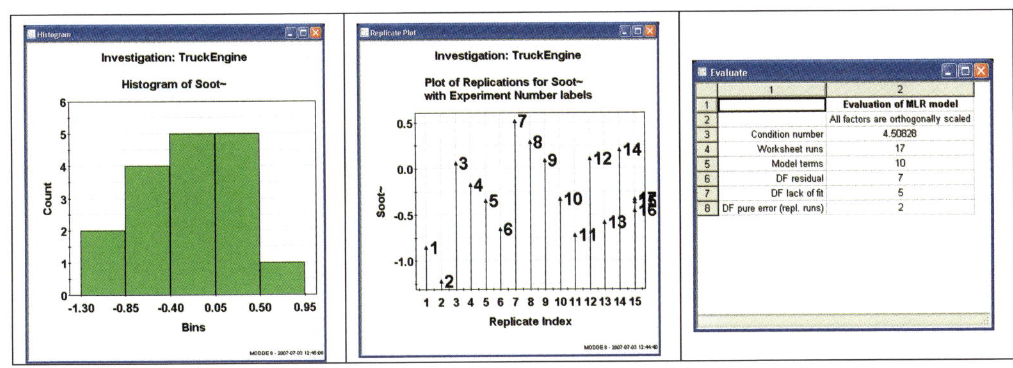

	1	2
1		Evaluation of MLR model
2		All factors are orthogonally scaled
3	Condition number	4.50828
4	Worksheet runs	17
5	Model terms	10
6	DF residual	7
7	DF lack of fit	5
8	DF pure error (repl. runs)	2

Figure 6.26: *(left) Histogram of Soot – after log-transformation.*
Figure 6.27: *(middle) Replicate plot of Soot – after log-transformation.*
Figure 6.28: *(right) Condition number evaluation.*

6.6.2 Regression analysis

We will now fit a quadratic model with 10 terms to each response. Each model will have one constant, three linear, three quadratic, and three two-factor interaction terms. The overall result of the model fitting is displayed in Figure 6.29. It is seen that the predictive power ranges from good to excellent. The Q^2 values are 0.93, 0.97, and 0.75, for Fuel, NOx, and Soot, respectively. All models also have high model validity indicating no lack of fit.

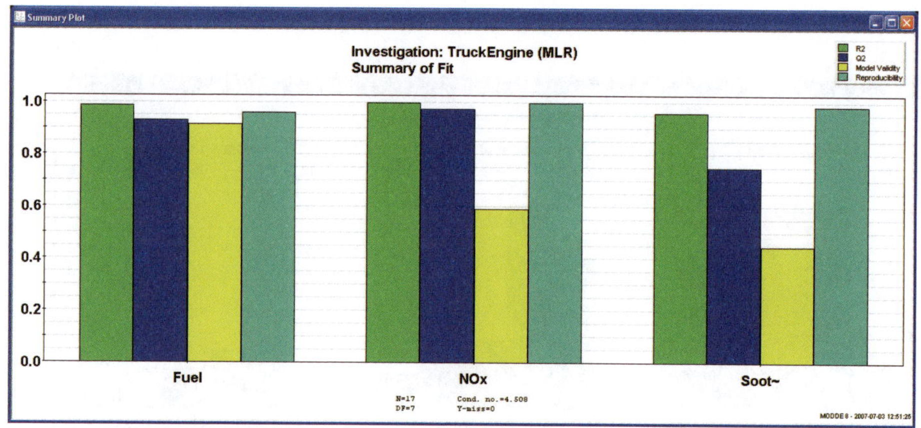

Figure 6.29: *Summary of fit of the initial modelling.*

Since this is RSM and the final goal is optimization, we would prefer a higher Q^2 of Soot. In order to see what may be done to achieve this, we proceed with the other diagnostic tools. The outcome of the normal probability plot of residuals is shown in Figure 6.30.

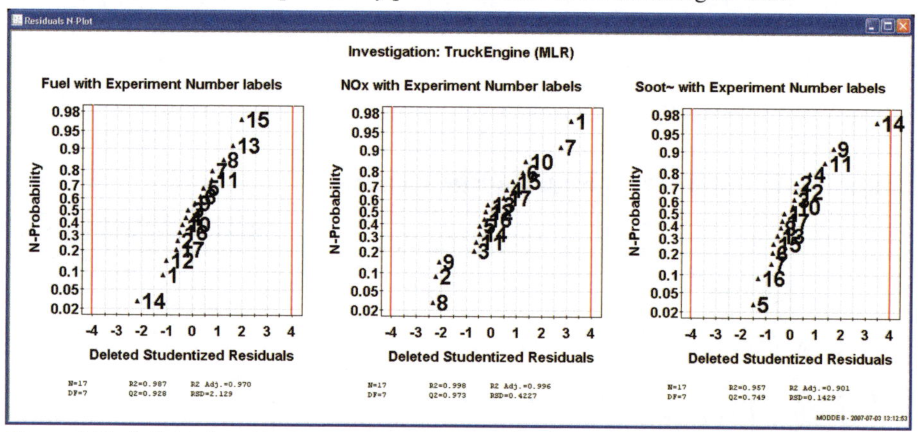

Figure 6.30: *(left) N-plot of Fuel residuals. (middle) N-plot of NOx residuals. (right) N-plot of Soot residuals.*

Primarily, the three N-plots of residuals suggest that attention should be given to experiment #14, as it deviates a little with respect to Soot. This deviation might be the cause of the comparatively low Q^2 for Soot. However, it is important to underscore that experiment 14 is *not* a strong outlier. This is inferred from the high R^2 value of the response.

Another possible reason as to why Q^2 is low for Soot might be that the regression model contains irrelevant terms. This may be checked through a bar chart of its regression coefficients. Figure 6.31 provides plots of the regression coefficients for each model. After some scrutiny of these plots, we can detect two model terms which are not significant for any one of the three responses. These are the Air*EGR and Air*NL two-factor interactions. Thus, in the model refinement step, it is logical to remove these and explore the modelling consequences.

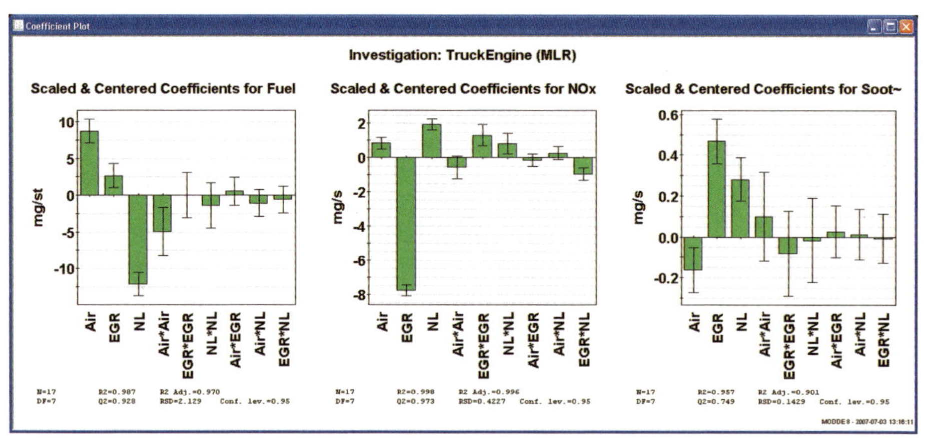

Figure 6.31: (left) Regression coefficients of the Fuel model. (middle) Regression coefficients of the NOx model. (right) Regression coefficients of the Soot model.

6.6.3 Model refinement

Upon deletion of the two two-factor interactions and refitting of the models, we obtained even better results. The summary of fit plot shown in Figure 6.32 shows that all three Q^2 values have increased, and now amount to 0.94, 0.99, and 0.84 for Fuel, NOx, and Soot, respectively. The models still display high model validity and no lack of fit. Thus the model refinement seems rewarding. However, before we can be sure that this model pruning is justifiable, we must consider the remaining diagnostic tools.

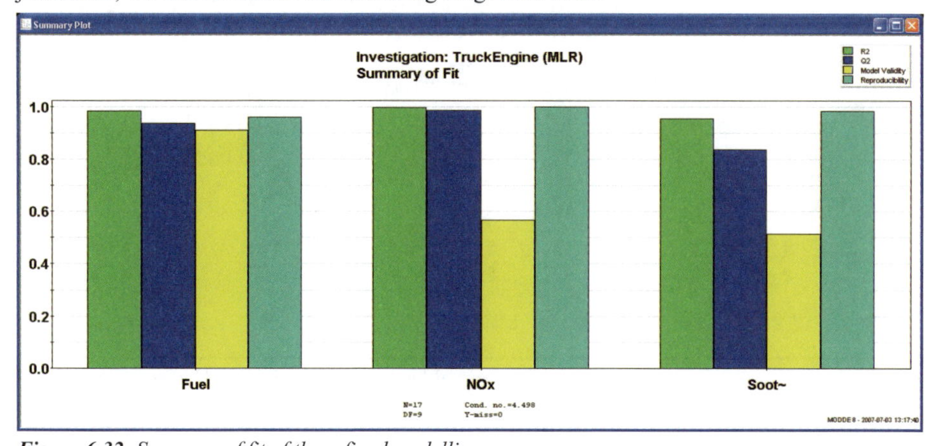

***Figure 6.32:** Summary of fit of the refined modelling.*

The relevant N-plots of residuals are shown in Figure 6.33. Everything looks good in these plots, except for the deviating behaviour of experiment 14. However, because of the very good R^2 and Q^2 of this model, we can conclude that #14 is a weak outlier, which does not influence the model decisively. We may regard these models as the best we can accomplish with the current data.

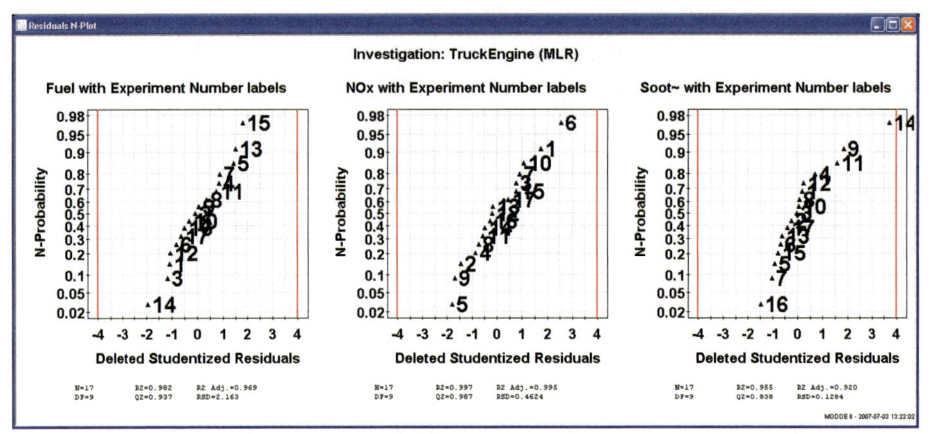

Figure 6.33: (left) N-plot of Fuel residuals – after model refinement. (middle) N-plot of NOx residuals – after model refinement. (right) N-plot of Soot residuals – after model refinement.

The models may be interpreted with the help of the regression coefficient plots shown in Figure 6.34. Because the coefficient patterns are quite variable among the models, the interpretation in this case is not straightforward. The overall picture is that linear terms dominate over terms of higher order. The main effects of Air and NeedleLift are influential for regulating Fuel, the main effect of EGR most meaningful for NOx, and the main effects of EGR and NeedleLift most important for Soot.

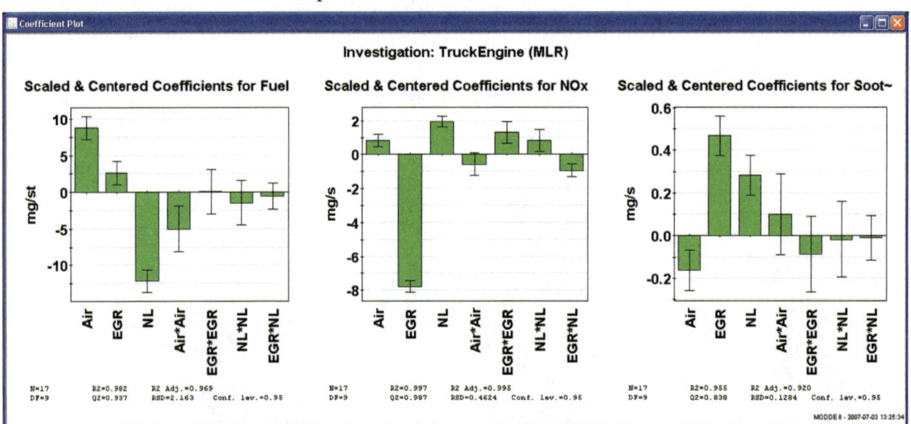

Figure 6.34: (left) Regression coefficients of the Fuel model. (middle) Regression coefficients of the NOx model. (right) Regression coefficients of the Soot model.

6.6.4 Use of model

When trying to extract concrete information from our models for deciding what to do next, it is convenient to use the regression coefficients plotted in Figure 6.34. It is not immediately clear which two factors should be put on the axes in the contour plots, and which factor should be held constant. One way to attack this problem could be as follows: We can see that Air, in principle, is only influential for Fuel. Since we want to get fuel consumption as low as possible, we should perhaps fix Air at its low level, and then let the other two factors vary. Such response contour plots are plotted in Figure 6.35. The solid arrows indicate where the desirabilities of the three responses are met.

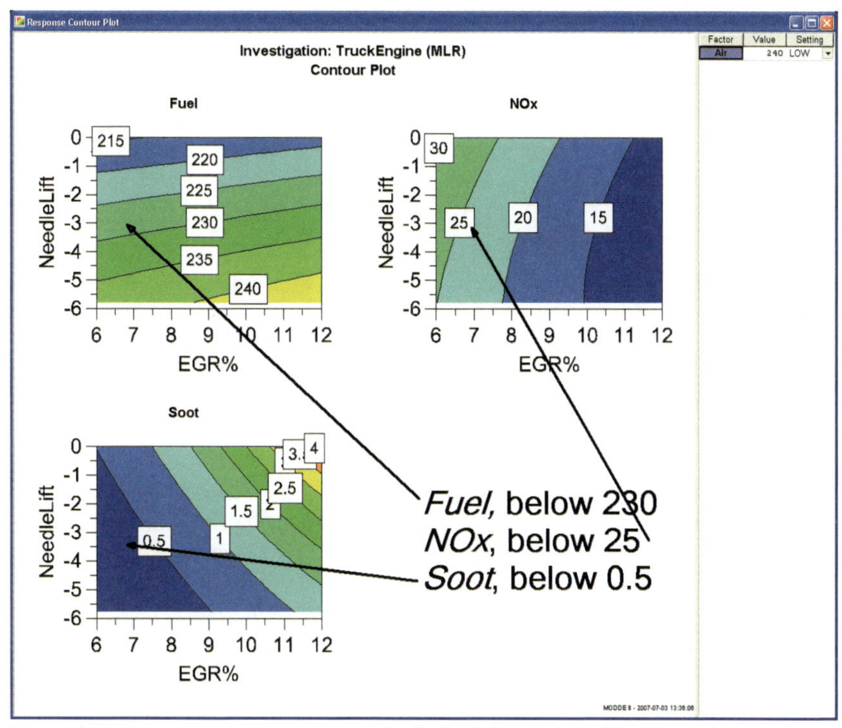

Figure 6.35: Response contour triplets of the final truck engine models.

Obviously, we should strive to stay somewhere in the mid left-hand section, that is, use low to medium levels of all three factors. The strong message conveyed by these contour plots is that the specified response profile is reasonable and will realistically be met within the mapped region. In principle, it is possible to carry on this procedure by using other triplets of response contour plots, in order to find the optimal point. However, this is both tedious and time-consuming. Fortunately, it is possible to automate this procedure with the optimizer in MODDE. We will investigate this option in a moment, but first provide some more theory relating to what to do after response surface modelling. Subsequently, we shall return to the analysis of the truck engine data.

6.7 What to do after RSM

6.7.1 Introduction

What to do after RSM depends on the results obtained. There are three ways in which we can proceed. The first way is appropriate when one of the experiments fulfills the goals stated in the problem formulation. This corresponds to the ideal case, and in principle only a couple of immediate verification experiments are needed to establish the usefulness of this factor combination. A second possibility involves regression modelling and use of the model for predicting the location of a new promising experimental point. This is the most common of the three options, and is deployed when none of the performed design runs are really close to fulfilling the experimental goals.

The third, and least frequently used, option is taken when a few extra experiments are needed to supplement the existing RSM design. For instance, it may be the case that a rather good quadratic model has been acquired. But, during the course of the data analysis, certain weaknesses of the quadratic model have been exposed, and it appears that the only remedy is the fitting of a cubic model term in one of the dominant factors. In this instance, it is possible to add to the parent design a couple of supplementary design runs, laid out D-optimally, which would enable the estimation of the lurking cubic term.

This was a short overview of the three primary measures that might be undertaken after RSM. In the following, we shall focus solely on the second situation, as it represents the most common action undertaken after RSM.

6.7.2 Interpolation & Extrapolation

When entering the stage of optimization, the user has a great deal of insight into the investigated system compared with the screening situation. Now, the investigator *knows* that the optimum must be close to the investigated region. The optimum is not far away, as may be the case in screening. This relative proximity to the optimal point implies that it is possible to use the RSM model for predicting the probable location of such an optimum. It is not – as in screening – necessary to employ any gradient technique for finding the optimal experimental region.

When we use the model for making predictions of the responses, and the predictions are made inside the explored experimental domain, we say that we conduct *interpolation*. Analogously, when we predict outside the region, we say that we perform *extrapolation*. Regardless of whether interpolation or extrapolation is taking place, it is important that the RSM model is predictively relevant. Earlier we presented a number of raw data evaluation tools and regression modelling diagnostic tools, which are useful for verifying that the model is reliable. When our model has been found relevant, the question is *in which direction of the experimental region do we make the predictions*? Without any rational strategy for this we could go on for a long time, because there are infinitely many factor combinations at which we could calculate response values. Fortunately, MODDE contains an optimization routine, which will automatically browse through a large number of factor combinations, and in the end select only a few at which fruitful predictions have been made.

6.7.3 Automatic search for an optimal point

The MODDE optimizer was introduced in Chapter 5 and we will now give a short resume of what was written there. Firstly, the roles of the factors must be given. A factor can be allowed to change its value freely in the optimization, or it may be fixed at a constant value. In addition, for the variable factors, low and high variation limits must be set. Secondly, certain criteria concerning the responses must be given. A response variable may either be maximized, minimized, directed towards an interval, or excluded from the optimization.

Subsequent to these definitions, the optimizer will use the fitted regression model and the experiments already performed to compute eight starting points for simplexes. In principle, the optimizer then uses these simplexes, together with the fitted model, to optimize the desirabilities, or goals, of the individual responses. These goals are taken from the criteria specified for the responses. Figure 6.36 shows one simplex in action. For simplicity this simplex is laid out in a two-dimensional factor space.

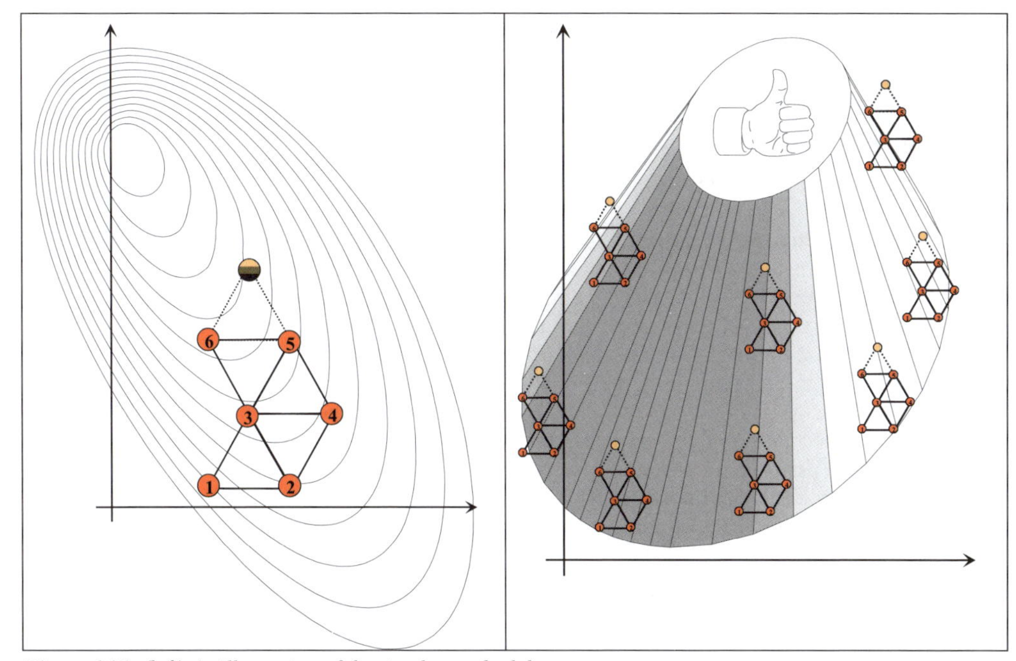

Figure 6.36: *(left) An illustration of the simplex methodology.*
Figure 6.37: *(right) Different simplexes get different starting points in factor space.*

The first simplex consists of the three experiments numbered 1, 2, and 3. Now, the idea is to mirror the worst experiment, here #1, through the line connecting the two best experiments, here # 2 and 3, and do a new experiment, here #4, at the position where the mirrored experiment "hits" the response contour plot. This means that the new simplex will consist of runs 2, 3, and 4. In the next cycle, the worst experiment is mirrored in the line connecting the two best experiments, and so on. This is repeated until the peak of the elliptical mountain is reached by one simplex. We emphasize that this is only a simplified picture of how the simplex methodology works. In reality, when a new simplex is formed through a reflection, it may be reduced or enlarged in size in comparison with the precursor simplex.

The optimizer will simultaneously start up to eight simplexes, from different locations in the factor space. Eight simplexes are initiated in order to avoid being trapped at a local minimum or maximum. This is illustrated in Figure 6.37. The co-ordinates for the simplex starting points are taken from the factors with the largest coefficients in the regression model. At convergence, each simplex will display the best factor settings, and it is then possible to compute response values at each point. Observe that the drawing in Figure 6.37 is a schematic valid when striving for higher response values. In reality, one could also look for a minimum, and the relevant response function would then correspond to a hole or a valley.

6.8 The MODDE optimizer applied to the truck engine data

We will now relate what happened when applying the optimizer to the truck engine application. It is a good practice to conduct the optimization in stages. We recommend that the first phase is conducted according to the principles of interpolation, even though extrapolation might sometimes (in screening) be the final result. In interpolation, the low and high limits of the factors are taken from the factor definition. This is shown in Figure 6.38.

Factor	Role	Value	Low Limit	High Limit
Air	Free		240	284
EGR%	Free		6	12
NeedleLift	Free		-5.78	0

Figure 6.38: *Factor settings in the optimizer study of the truck engine application.*

Next, it is necessary to give the desirability, or goal, of each response, that is, to specify a desired response profile. We see the criterion and goal set for each response in Figure 6.39. This table is understood as follows. All three responses, Fuel, NOx, and Soot are to be minimized. We want Fuel to be lower than 230, NOx lower than 25, and Soot below 0.5. Technically, these minimization requirements are specified by assigning two reference values for each response. One value is the target value, which represents the numerical value of each response towards which the optimization is driven. The other value is the maximum value which we can accept should the corresponding target value be unattainable. Furthermore, it is also possible to assign weighting coefficients to each response. Such weights may range from 0.1 to 1.0. Here, we assign the same priority to all three responses, by stating that each weight should be 1.0. This weighting procedure is an advanced topic that is not treated further in this course.

Response	Criteria	Weight	Min	Target	Max
Fuel	Minimize	1		225	230
NOx	Minimize	1		10	25
Soot	Minimize	1		0.05	0.5

Figure 6.39: *Response desirabilities in the optimizer study of the truck engine application.*

6.8.1 First optimization - Interpolation

Now we are ready to begin the optimization. In doing so, the optimizer will first compute appropriate starting co-ordinates, factor combinations, for launching the eight simplexes. Five starting points are derived by laying out a 2^{3-1} fractional factorial design with one center-point in the three model factors. This design corresponds to the five first rows in the spreadsheet of Figure 6.40. In addition, the optimizer browses through the existing worksheet and identifies those three experiments which are closest to the specified response profile. These are the three runs listed in rows 6, 7 and 8.

	1	2	3	4	5	6	7	8
	Air	EGR%	NeedleLift	Fuel	NOx	Soot	iter	log(D)
1	240	12	-5.78					
2	240	6	0					
3	284	6	-5.78					
4	284	12	0					
5	262	9	-2.89					
6	241	6.2	0					
7	242	6.1	0					
8	262	9	0					

Figure 6.40: *Starting co-ordinates of the simplexes in the first optimization round.*

On pressing the single-headed arrow (not displayed), the eight simplexes will start their search for a factor combination, at which the predicted response values are as similar to the desired response profile as possible. The outcome of the eight simplexes are seen in Figure 6.41. All proposed factor combinations are different, but some are rather similar. The success of each simplex may be deduced from the numerical values of the rightmost column. This logD value represents a weighted average of the individual response desirabilities. It may be used to evaluate how the optimization proceeds. A positive logD is bad. When logD is zero, all responses are predicted to be between their assigned target and maximum value. This is a good situation. Better still is when logD is negative, and –10 is the lower, optimal, limit. Hence, in our case we are doing rather well, and it seems possible to find a factor combination where all three responses will be below the critical limit. Therefore, we are close to detecting an optimal point *within* the investigated experimental region. This means that extrapolation is unnecessary.

	1	2	3	4	5	6	7	8
	Air	EGR%	NeedleLift	Fuel	NOx	Soot	iter	log(D)
1	240	6.9254	-3.1894	229.103	23.8056	0.3915	129	-0.1118
2	240.003	7.0058	-3.1932	229.191	23.5496	0.4055	103	-0.107
3	283.212	7.9535	-1.404	239.831	23.9633	0.4264	67	0.5454
4	240.002	7.1131	-2.7136	227.336	23.6293	0.4751	185	-0.1761
5	240.004	6.9491	-3.157	228.997	23.7582	0.3987	126	-0.1166
6	240	6.9956	-3.1948	229.185	23.5799	0.4036	148	-0.1074
7	240.001	7.029	-3.1945	229.214	23.4756	0.4097	117	-0.1057
8	240.001	6.9379	-3.2118	229.203	23.7472	0.3917	209	-0.1068

Figure 6.41: *Optimizer results of the first optimization round.*

6.8.2 Second optimization – New starting points around selected run

One problem which might compromise the validity of the simplex optimization approach is the risk of being trapped by a local minimum or maximum, and thereby missing the global optimal point. This may be circumvented by taking the simplex predicted to be the best, and from its factor combination generate new starting points for simplexes. We decided to exploit this option.

	1	2	3	4	5	6	7	8
	Air	EGR%	NeedleLift	Fuel	NOx	Soot	iter	log(D)
1	240	7.7131	-3.2916					
2	240	6.5131	-2.1356					
3	244.402	6.5131	-3.2916					
4	244.402	7.7131	-2.1356					
5	240.002	7.1131	-2.7136					

Figure 6.42: *Starting co-ordinates of the simplexes in the second optimization round.*

The initially predicted best point is shown as the fifth row in the spreadsheet of Figure 6.42. Around this point, four new starting points, rows 1-4, were defined. These co-ordinates were identified by moving away from the mother simplex, row 5, to a distance corresponding to 20% of the factor ranges. By executing the four new simplexes and the old one, the results displayed in Figure 6.43 were obtained.

	1	2	3	4	5	6	7	8
	Air	EGR%	NeedleLift	Fuel	NOx	Soot	iter	log(D)
1	240.013	6.9514	-3.1363	228.925	23.7693	0.4009	97	-0.1199
2	240.001	7.0371	-3.1915	229.208	23.4529	0.4114	102	-0.1059
3	240	7.0402	-3.1901	229.205	23.4443	0.4122	112	-0.1061
4	240.002	6.9283	-3.221	229.232	23.7701	0.3891	113	-0.1055
5	240.005	6.9948	-3.2024	229.219	23.5766	0.4026	84	-0.1057

Figure 6.43: *Optimizer results of the second optimization round.*

Evidently, all five points are predicted to meet our experimental goals. In fact, by neglecting some small fluctuations among the decimal digits, we find that the five simplexes have converged to approximately the same point, that is, Air = 240, EGR% ≈ 7.0, and NeedleLift ≈ -3.0. First of all, this means that these simplexes were not trapped by a local response phenomenon. Secondly, it indicates that the response surfaces do not display any dramatic changes in the vicinity of the identified optimal point.

6.8.3 Graphical evaluation of the optimization results

It is possible to investigate graphically the situation around the predicted optimal point, just to see if the results are sensitive to small changes in the factor values. Figure 6.44 provides the Sweet Spot plot indicating that an optimal area is located around EGR% = 7.0 and NeedleLift = -3.0, and holding Air = 240 fixed.

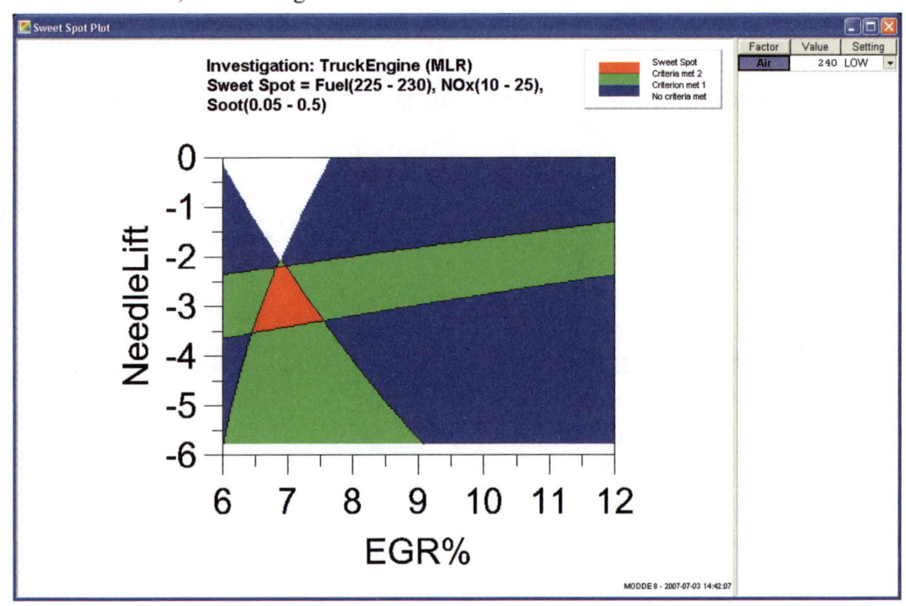

Figure 6.44: *Graphical exploration of the situation around the predicted optimal point.*

Evidently, there are no dramatic changes occurring in the predicted response functions. Rather, there are slow and controlled response changes. Hence, it appears relevant to conduct a few verifying experiments with the setting Air = 240, EGR% = 7.0, and NeedleLift = -3.0. The predicted response values at this point are shown in Figure 6.45 together with the 95% confidence intervals. It is seen that each point estimate fulfils the experimental goals. The investigator Sven Ahlinder carried out verifying experiments and found that the proposed optimum was well-founded.

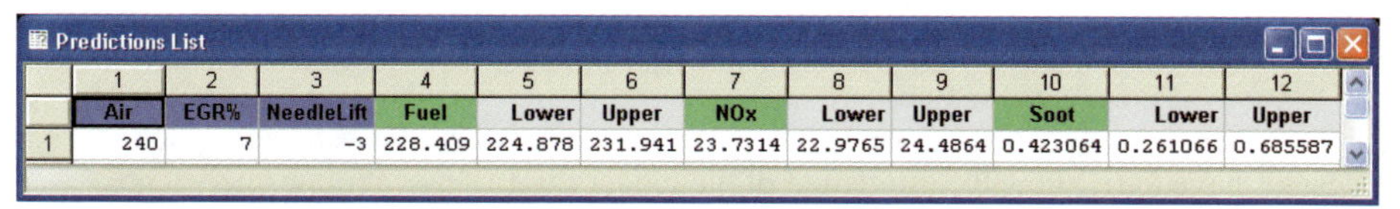

Predictions List

	1	2	3	4	5	6	7	8	9	10	11	12
	Air	EGR%	NeedleLift	Fuel	Lower	Upper	NOx	Lower	Upper	Soot	Lower	Upper
1	240	7	-3	228.409	224.878	231.941	23.7314	22.9765	24.4864	0.423064	0.261066	0.685587

Figure 6.45: *Confidence intervals (95%) of predicted response values at the optimal point.*

6.9 Summary of the truck engine application

The application of DOE in the construction of a well-behaved truck engine was successful. With this approach it was possible not only to understand the mechanisms involved in the combustion, but also to identify an optimal factor combination at which low fuel consumption and minimal pollution prevailed. From an educational point of view, the truck engine application is good for several reasons. For instance, it clearly shows that it is possible to lay out an optimization design and find an optimal point within the investigated experimental region. In this case, this was possible thanks to carefully conducted screening work.

Furthermore, this data set also allowed us to study how tools for evaluation of raw data, tools for regression analysis, and tools for model use, were put into practice in a real situation. It should be clear from this application that the process of deriving the best possible regression model is often iterative in nature, and many repetitive modelling cycles are often needed before a final model can be established. This modelling step is of crucial importance for the future relevance of a DOE project. If an investigator fails to ensure the quality of a regression model, this may have negative consequences at later stages when the model is used for predictions. Hence, strong emphasis is placed on finding the most reliable and useful model.

We also used the truck engine study for reviewing what options are available after RSM. The most common action after RSM is to use the developed model for optimization. With the software optimizer we rapidly identified an optimal point with the factor combination Air = 240, EGR% = 7.0, and NeedleLift = -3.0, at which the fuel consumption was predicted to be below 230, and the levels of NOx and Soot in the car exhaust gases modelled to be below 25 and 0.5, respectively. The principal experimenter Sven Ahlinder later verified this point experimentally.

6.10 Summary of central composite designs

The central composite designs CCC and CCF are natural extensions of the two-level full and fractional factorial designs, which explains their popularity in optimization studies. In fact, it is possible, by rather simple measures, to extend a screening design to become an RSM design. A central composite design consists of three building blocks, (i) regularly arranged corner experiments of a two-level factorial design, (ii) symmetrically arrayed star points located on the factor axes, and (iii) replicated center-points.

The CCC and CCF designs differ in how the star points, or axis points, are positioned. With the former design type, the axis points are located outside the factorial part of the composite design. This gives five levels for each investigated factor, which is advantageous in the estimation of a quadratic polynomial model. However, when it is desirable to keep the low and high factor levels unchanged, and still perform an RSM design, the CCF design is a viable alternative. In the CCF design, the axial points are centered on the faces of the cube or hypercube. This implies that all factors have three levels, rather than five, and that the experimental region is a cube or hypercube, and not a sphere. CCF is the recommended design choice for pilot plant and full scale investigations. Normally, RSM designs like CCC and CCF are used in conjunction with two to five factors. It is possible to explore as many as five factors in as few as 29 experiments. However, when moving up to six or more factors, there is a huge increase in the number of experiments, which means that RSM is less feasible in these cases.

We shall now examine some requirements on, and properties of, these RSM designs. Good RSM designs have to (i) allow the estimation of the model parameters with low uncertainty, (ii) give small prediction errors, (iii) provide prediction errors independent of direction, (iv) allow a judgement of model adequacy, and (v) limit the number of runs.

The first consideration means that the confidence intervals of the model's regression coefficients must be as narrow as possible. The second and third criteria combined imply that a model should predict well, and that the error in the predictions should not be affected by the direction in factor space in which predictions are made. The prediction error should only be influenced by the distance to the design center. This property of equal prediction precision in all directions is normally referred to as *rotatability*. Some of the designs studied here are rotatable, and some are not. Rotatability may be controlled in composite designs by regulating the distance of the star points to the design center, but this is an advanced topic.

Further, the fourth consideration is that replicated experiments must be incorporated in the design, so that a lack of fit test can be carried out in the ANOVA procedure. This is ensured with a suitable number of replicated center-point experiments. The number of center-points is also used for controlling the *orthogonality*, or *near-orthogonality*, of a design, which is another important property. Finally, we have the fifth consideration, which concerns the number of experiments. It is always desirable to reduce the number of experiments, but not by sacrificing the relevance or quality of the work. There is a trade-off between the sharpness of the information conveyed by a design and the number of runs it requires.

6.11 Questions for Chapter 6

- Which questions are typically asked in optimization?
- Which kind of regression model is usually fitted in optimization?
- What is RSM?
- What is a central composite design?
- What are the differences between the CCC and CCF designs?
- Which design is recommended in full scale applications?
- How many factors should, at maximum, be explored with an RSM design?
- How many experiments are needed to perform an RSM study of two factors?
- Which tools are useful for evaluation of raw data?
- Which tools are meaningful for diagnostic checking of a model?
- When and why is it necessary to log-transform responses?
- Which are the three actions that may occur after RSM?
- What does simplex optimization mean?
- How does the optimizer make use of the existing regression model for starting simplex optimization?

6.12 Summary and discussion

In this chapter, we have discussed the experimental objective called optimization. For this purpose, we have reviewed the truck engine application, consisting of three factors, three responses, and a CCF design comprising 17 experiments. The problem formulation steps of this application were reviewed at the beginning of this chapter. The next part was devoted to the introduction of central composite designs, and much emphasis was given to a geometrical interpretation of their properties. Following the introduction of the family of central composite designs, the truck engine application was outlined in detail. The analysis of this data-set resulted in three excellent models, which were used for optimization. The optimization, in turn, lead on to the identification of an optimal factor combination at which the desired response profile was satisfied. The detected optimal point was later verified experimentally.

7 Experimental objective: Robustness Testing

7.1 Objective

The objective of this chapter is to describe robustness testing, and this is done by adhering to the DOE framework outlined in Chapters 1-4. Our example regards the robustness assessment of a high-performance liquid chromatography (HPLC) analysis system, the kind of equipment routinely used in the analytical chemistry laboratory. This application involves the variation of five factors, four quantitative and one qualitative, and the measurement of three responses. Of these three responses, one is more important than the others and has to be robust towards changes in the five factors. The other two responses should also not change too much. Thus, the experimental objective is to verify the robustness of the important response. We will also highlight some statistical experimental designs which are well suited for the robustness testing experimental objective. In addition, the analysis of the HPLC data will show how a number of raw data evaluation tools, and regression modelling diagnostic tools, work on a real, complicated example. The chapter ends with a discussion of four limiting cases of robustness testing.

7.2 Introduction to robustness testing

The objective of robustness testing is to design a process, or a system, so that its performance remains satisfactory even when some influential factors are allowed to vary. In other words, what is wanted is to minimize our system's sensitivity to changes in certain critical factors. If this can be accomplished, it obviously provides many advantages, like easier process control, wider range of applicability of product, higher quality of product, and so on. A robustness test is usually carried out *before* the release of an almost finished product, or analytical system, as a last test to ensure quality. Such a design is usually centered around a factor combination which is currently used for running the analytical system, or the process. We call this point the *set point*. The set point might have been found through a screening design, an optimization design, or some other identification principle, such as, written quality documentation. The objective of robustness testing is, therefore, to explore robustness close to the set point.

7.3 Background to General Example 3

The example that we have chosen as an illustration originates from a pharmaceutical company. It represents a typical analytical chemistry problem within the pharmaceutical industry, and other industrial sectors. In analytical chemistry, the HPLC method is often used for routine analysis of complex mixtures. It is therefore important that such a system will work reliably for a long time, and be reasonably insensitive to varying chromatographic conditions. Some HPLC equipment is shown in Figure 7.1, and a typical chromatogram in Figure 7.2.

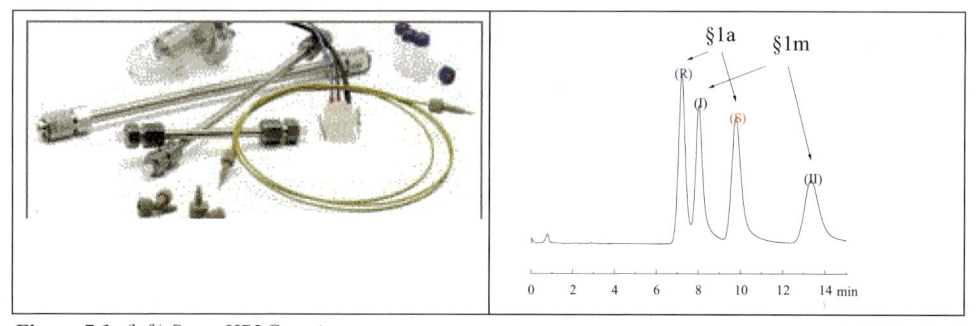

***Figure 7.1:** (left) Some HPLC equipment.*
***Figure 7.2:** (right) Schematic of an HPLC chromatogram.*

The investigators studied five factors, namely amount of acetonitrile in the mobile phase, pH of mobile phase, temperature, amount of the OSA counter-ion in the mobile phase, and batch of stationary phase (column), and mapped their influence on the chromatographic behaviour of two analyzed chemical substances. The low and high settings of these factors are found in Figure 7.3, together with the relevant measurement units. Observe that the last factor is of a qualitative nature. To study whether these factors had an influence on the chromatographic system, the researchers used a 12 run experimental design to encode 12 different chromatographic conditions. For each condition, three quantitative responses, reflecting the capacity factors of the two analytes (compounds) and the resolution between the analytes, were measured. These responses are summarized in Figure 7.4.

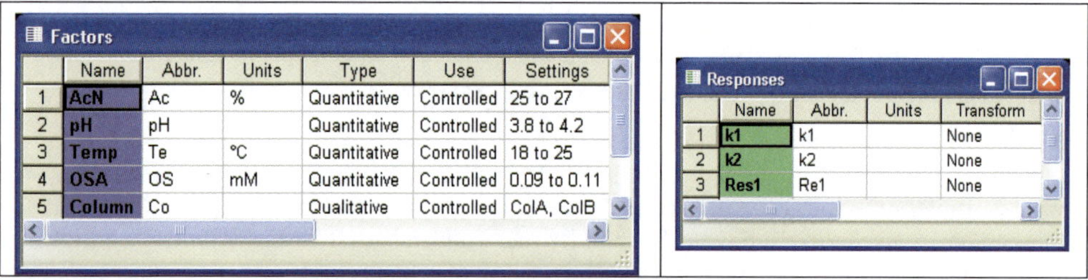

Factors

	Name	Abbr.	Units	Type	Use	Settings
1	AcN	Ac	%	Quantitative	Controlled	25 to 27
2	pH	pH		Quantitative	Controlled	3.8 to 4.2
3	Temp	Te	°C	Quantitative	Controlled	18 to 25
4	OSA	OS	mM	Quantitative	Controlled	0.09 to 0.11
5	Column	Co		Qualitative	Controlled	ColA, ColB

Responses

	Name	Abbr.	Units	Transform
1	k1	k1		None
2	k2	k2		None
3	Res1	Re1		None

***Figure 7.3:** (left) Overview of the factors studied in the HPLC application.*
***Figure 7:4:** (right) Overview of the responses measured in the HPLC application.*

In chromatography, the objective is separation of the analytes, or, rather, separation in a reasonable time. To get separation, one must first have retention. Thus, the retention of each analyte is important, and this response is given by the capacity factor, k. Furthermore, the degree of separation between two analytes is estimated as the resolution between two adjacent peaks in the chromatogram. A resolution of 1 is considered as the minimum value for separation between neighboring peaks, but for complete baseline separation a resolution of >1.5 is necessary. As the resolution value approaches zero, it becomes more difficult to discern separate peaks. The goal of this study was to constantly maintain a resolution of 1.5 or higher for all chromatographic conditions. No specific target values were given for the two capacity responses.

7.4 Problem Formulation (PF)

7.4.1 Selection of experimental objective

The problem formulation is important in DOE. The first step of the problem formulation is the selection of a relevant experimental objective. In the HPLC application, the chosen experimental objective was *robustness testing*. In robustness testing of an analytical equipment, the central idea is to lay out a condensed statistical design around the set point, and this design should have small changes in the factors. One then wants to find out how sensitive or insensitive the measured responses are to these small changes. Certain responses might be insensitive to such small changes in the factors and therefore claimed to be robust. Other responses might be sensitive to alterations in the levels of some factors, and these factors must then be better controlled. In the HPLC application, the set point was obtained from the written quality documentation of the analytical equipment. In the following, we will relate how the planning and execution of the robustness testing design was centered around this point.

7.4.2 Specification of factors

Robustness testing is useful for creating the final version of the quality documentation of an analytical system, in the sense that such a text should contain recommended factor settings between which robustness can be assured. Usually, the experimenter has a rough idea about such relevant factors settings for the system, and a robustness testing design is often used for critically corroborating these. In the HPLC case, five factors were selected for the robustness test. These factors are shown in Figures 7.5 – 7.9.

The factor levels were chosen by considering which changes may occur during a normal day in the laboratory, that is, mild but controlled shifts. The first factor is amount of acetonitrile in the mobile phase, for which 25 and 27% were set as low and high levels. Acetonitrile is used as an organic eluent modifier. The second factor is the pH of the mobile phase, and this factor was varied between 3.8 and 4.2. The third factor is the laboratory temperature, which was investigated between 18 and 25°C. The fourth factor is the amount of the counter-ion octanesulphonic acid, OSA, which is used for regulating a compound's retention. OSA was varied between 0.09 and 0.11 mM, and 0.01 is considered to be a large change. Finally, we have the fifth factor, the type of stationary phase (column), which display a qualitative, or discrete, nature. This factor represents the switch between two batches of stationary phase, for simplicity encoded as ColA and ColB.

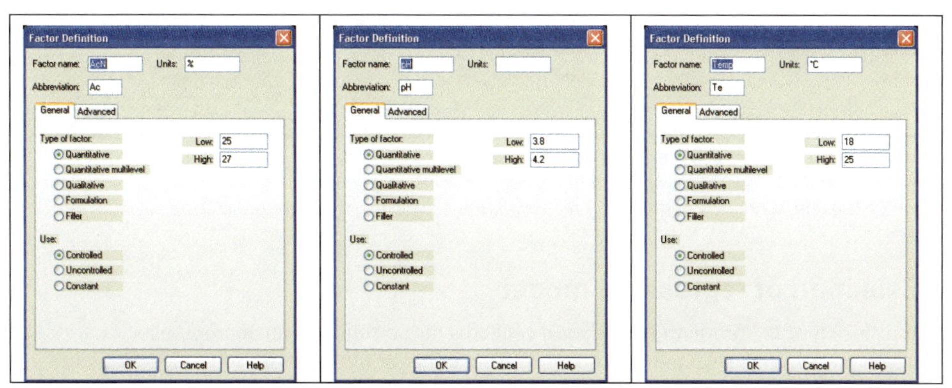

Figure 7.5: *(left) Factor definition of amount of acetonitrile in mobile phase.* ***Figure 7.6:*** *(middle) Factor definition of pH of mobile phase.* ***Figure 7.7:*** *(right) Factor definition of temperature.*

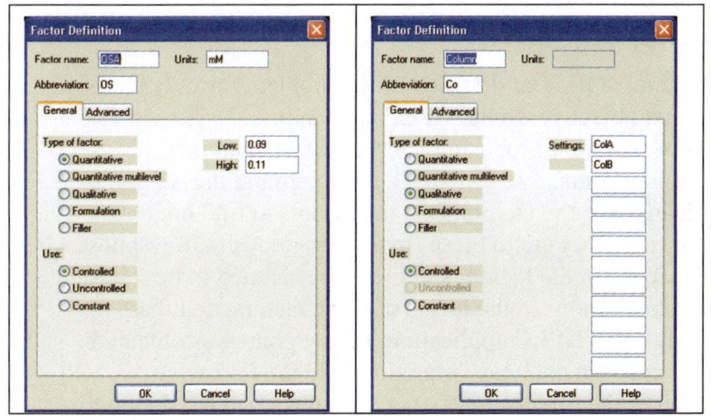

Figure 7.8: *(left) Factor definition of OSA counter-ion in mobile phase.*
Figure 7.9: *(right) Factor definition of batch of stationary phase.*

7.4.3 Specification of responses

The next step in the problem formulation concerns the specification of which responses to monitor. It is important that the selected responses reflect the important properties of the system studied. In HPLC, the resolution of the system, that is, the extent to which adjacent peaks in the resulting chromatogram can be separated from each other, is of paramount importance. A complete baseline separation corresponds to the ideal case, and this is inferred when the resolution exceeds 1.5. Therefore, the goal was to maintain a resolution of 1.5 or higher. The other type of response of interest is the retention, which corresponds to the time required for migration of a compound from the injector to the detector. A measure of the retention is given by the capacity factor. In this example, the goal was retention for a reasonable, short, time, but no specific targets were set. The three responses, denoted k_1 for capacity factor 1, k_2 for capacity factor 2, and Res1 for the resolution between compounds 1 and 2, are overviewed in Figures 7.10 – 7.12.

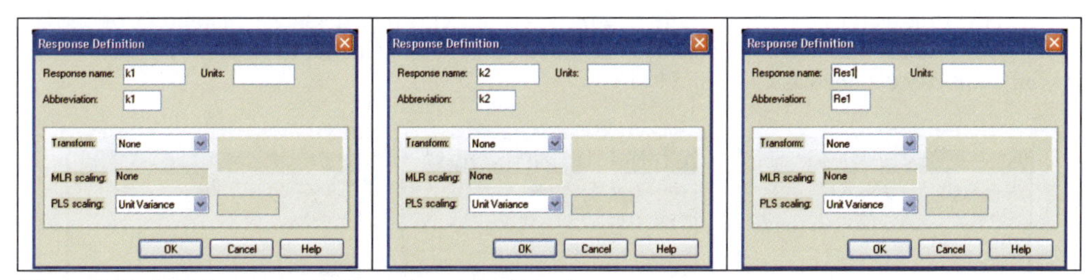

Figure 7.10: *(left) Response definition of capacity factor 1.*
Figure 7.11: *(middle) Response definition of capacity factor 2.*
Figure 7.12: *(right) Response definition of the resolution between compounds 1 and 2.*

7.4.4 Selection of regression model

The fourth step of the problem formulation concerns the selection of an appropriate regression model. This selection can be made from either of three classes of models, namely linear, interaction or quadratic regression models. A linear model is an appropriate choice for robustness testing. One reason for this is that we are usually exploring each factor within a narrow range, and hence strong departures from linearity are unlikely. Also, we are only interested in identifying which factors should be better controlled, and an adequate answer to this will normally be provided by a linear model. The investigators choose a linear model in all factors for the HPLC-data.

7.4.5 Generation of design and creation of worksheet

The last two stages of the problem formulation deal with the generation of the statistical experimental design and the creation of the corresponding worksheet. We will treat these two steps at the same. Because the ideal result in a robustness testing study is identical response values for each trial, the selected experimental protocol may well be a low-resolution screening design. In particular, we think here of resolution III fractional factorial designs, and designs drawn from the Plackett-Burman family. Fractional factorial designs were introduced in Chapter 5. Plackett-Burman designs are discussed in Chapter 8.

The experimenters behind the HPLC data selected a 2^{5-2} fractional factorial design consisting of eight runs. This design supports a linear polynomial model, and the worksheet is provided in Figure 7.13. The five columns with numbers 5 – 9 denote the factors, and the three columns with numbers 10 – 12 the responses. The first eight rows correspond to the eight corner-experiments. It is seen that these eight runs are supplemented with four experiments of a "center-point-nature". They have numerical settings close to the center-point values in the four quantitative factors, but not in the qualitative factor. No factor level located half-way between ColA and ColB is definable. This means that the worksheet can be said to contain two pairs of replicates, one pair being allotted to each level of the qualitative factor.

Worksheet

	1	2	3	4	5	6	7	8	9	10	11	12
1	Exp No	Exp Name	Run Order	Incl/Excl	AcN	pH	Temp	OSA	Column	k1	k2	Res1
2	1	N1	3	Incl	25	3.8	18	0.11	ColB	2.2906	3.3421	1.87
3	2	N2	8	Incl	27	3.8	18	0.09	ColA	1.7547	2.6802	1.75
4	3	N3	2	Incl	25	4.2	18	0.09	ColB	2.3933	3.4705	1.89
5	4	N4	1	Incl	27	4.2	18	0.11	ColA	1.823	2.8013	1.8
6	5	N5	6	Incl	25	3.8	25	0.11	ColA	2.1456	3.1599	1.83
7	6	N6	7	Incl	27	3.8	25	0.09	ColB	1.5031	2.4845	1.8
8	7	N7	5	Incl	25	4.2	25	0.09	ColA	2.2289	3.2715	1.86
9	8	N8	4	Incl	27	4.2	25	0.11	ColB	1.5994	2.6193	1.84
10	9	N12	9	Incl	26	4	22	0.1	ColA	2.0661	3.0592	1.81
11	10	N12	10	Incl	26	4	22	0.1	ColA	2.0253	3.0285	1.82
12	11	N12	11	Incl	26	4	22	0.1	ColB	2.0243	2.9903	1.79
13	12	N12	12	Incl	26	4	22	0.1	ColB	2.0131	3.0068	1.81

Figure 7.13: *Experimental data pertaining to the HPLC application.*

7.5 Common designs in robustness testing

Resolution III fractional factorial designs and Plackett-Burman designs are two design types often used for robustness testing. We will now briefly review their properties, and a detailed explanation is retrievable from Chapters 5 (FFDs) and 8 (PBs). A fractional factorial design is a design which consists of only a part, a fraction, of all the theoretically possible corner-experiments. As an example, consider Figure 7.14, which shows the 2^3 full factorial design of the CakeMix example. This 2^3 full factorial design may be split into two balanced half-fractions, one shown in Figure 7.15 and the other shown in Figure 7.16. These two half-fractions are encoded by two versions of the so called 2^{3-1} fractional factorial design, and imply that three factors can be investigated in four runs. From a design point of view, these two half-fractions are equivalent, but in reality one of them may be preferred to the other for practical reasons.

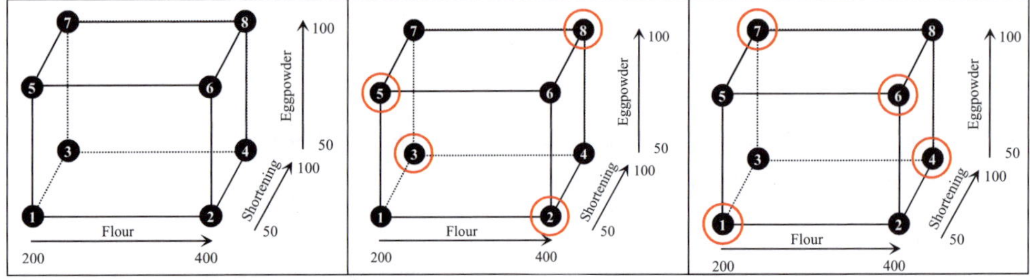

Figure 7.14: *(left) Geometry of the 2^3 design.*
Figure 7.15: *(middle) Geometry of one version of the 2^{3-1} design.*
Figure 7.16: *(right) Geometry of the second version of the 2^{3-1} design.*

In the HPLC application five factors are studied. With five factors in two levels, 32 corner-experiments are defined. These 32 corners may be divided into four quarter-fractions consisting of eight experiments each. The 2^{5-2} fractional factorial design used specifies the testing of one such quarter-fraction. This design has resolution III. The concept of design resolution is closely linked to the kind of regression model supported by the design. Resolution III designs are the most reduced, that is, contain fewest experiments, and are therefore useful for estimating linear models. This makes them well suited to robustness testing. Resolution IV and V fractional factorial designs include more experiments, and may be used to calculate more elaborate regression models. This was explained in Chapter 5.

We now turn to the other common design family in robustness testing, the Plackett-Burman class of designs. Plackett-Burman designs are orthogonal two-level experimental designs, which can be used to fit linear models. An example 12 run PB-design in 11 factors is given in Figure 7.17. Chapter 8 provides more details on the PB-design family.

Worksheet

	1	2	3	4	5	6	7	8	9	10	11	12	13	14	15
1	Exp No	Exp Name	Run Order	Incl/Excl	x1	x2	x3	x4	x5	x6	x7	x8	x9	x10	x11
2	1	N1	3	Incl	1	-1	1	-1	-1	-1	1	1	1	-1	1
3	2	N2	11	Incl	1	1	-1	1	-1	-1	-1	1	1	1	-1
4	3	N3	9	Incl	-1	1	1	-1	1	-1	-1	-1	1	1	1
5	4	N4	6	Incl	1	-1	1	1	-1	1	-1	-1	-1	1	1
6	5	N5	12	Incl	1	1	-1	1	1	-1	1	-1	-1	-1	1
7	6	N6	4	Incl	1	1	1	-1	1	1	-1	1	-1	-1	-1
8	7	N7	15	Incl	-1	1	1	1	-1	1	1	-1	1	-1	-1
9	8	N8	10	Incl	-1	-1	1	1	1	-1	1	1	-1	1	-1
10	9	N9	14	Incl	-1	-1	-1	1	1	1	-1	1	1	-1	1
11	10	N10	5	Incl	1	-1	-1	-1	1	1	1	-1	1	1	-1
12	11	N11	8	Incl	-1	1	-1	-1	-1	1	1	1	-1	1	1
13	12	N12	1	Incl	-1	-1	-1	-1	-1	-1	-1	-1	-1	-1	-1
14	13	N13	2	Incl	0	0	0	0	0	0	0	0	0	0	0
15	14	N14	13	Incl	0	0	0	0	0	0	0	0	0	0	0
16	15	N15	7	Incl	0	0	0	0	0	0	0	0	0	0	0

Figure 7.17: *Plackett-Burman design in 12 + 3 runs.*

One should observe that the number of runs in a PB-design is a multiple of four, which implies that the PB-designs nicely complement the fractional factorials family as their number of runs is a power of 2. In principle, this means that a fractional factorial design is selected whenever a robustness test is carried out in 8 or 16 runs, and a PB-design is selected when there is a need to use either 12 or 20 experimental trials. This concludes the introductory treatment of the robustness testing objective.

7.6 HPLC application

7.6.1 Evaluation of raw data

We start the evaluation of the raw data by inspecting the replicate error of each response. Three relevant plots of replications are displayed in Figure 7.18.

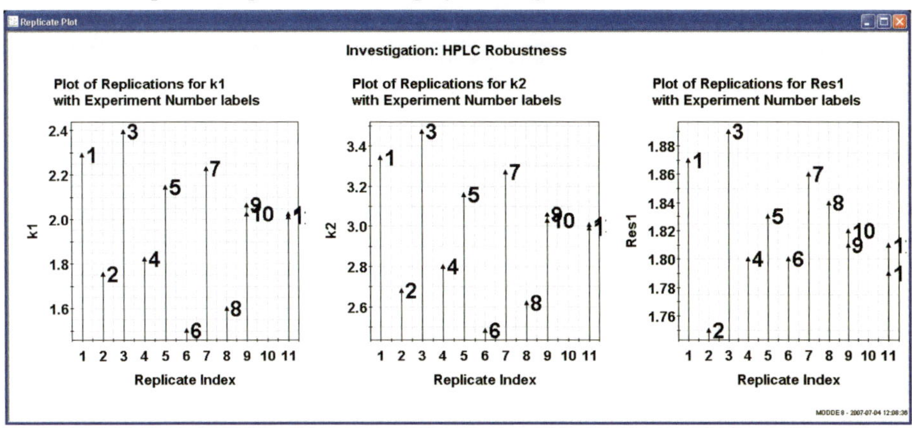

Figure 7.18: (left) Replication plot of capacity factor 1. (middle) Replication plot of capacity factor 2. (right) Replication plot of the resolution response.

As seen, the replicate error is small for each response, which is what was expected. We would not anticipate large drifts among the replicates, as we have deliberately set up a design where each run should ideally produce equivalent results. The numerical variation in the resolution response is small (Figure 7.18, right-hand part). The lowest measured resolution is 1.75 and the highest 1.89. Since the operative goal was to maintain the resolution above 1.5, we see even in the raw data that this goal is fulfilled, and this means that Res1 is robust.

Further, in the evaluation of the raw data, it is compulsory to check the data distribution of the responses, to reveal any need for response transformation. We may check this by plotting a histogram of each response. Such histograms are plotted in Figure 7.19, and they inform us that it is pertinent to work in the untransformed metric of each response.

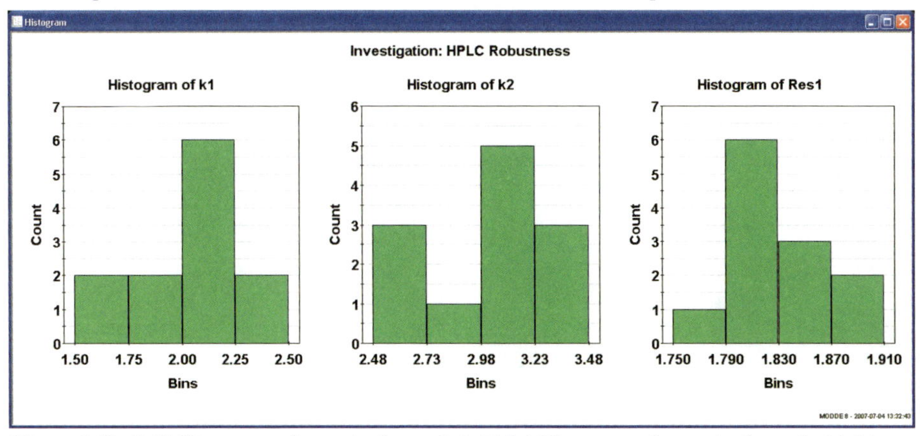

Figure 7.19: (left) Histogram of capacity factor 1. (middle) Histogram of capacity factor 2. (right) Histogram of the resolution response.

Another aspect of the raw data properties that might be worth considering is the condition number of the fractional factorial design. We observe in Figure 7.20 that the condition number is approximately 1.2, indicating a pertinent design. Furthermore, the correlation matrix, listed in Figure 7.21, indicates that the three responses are strongly correlated. Hence, the fitting of three regression (MLR) models with identical composition of model terms is reasonable.

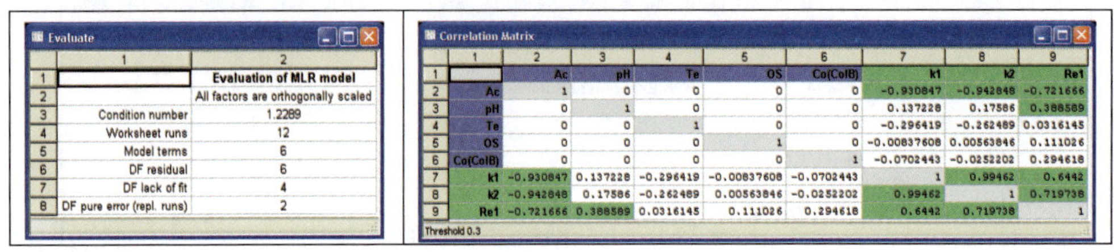

Evaluate

	1	2
1		Evaluation of MLR model
2		All factors are orthogonally scaled
3	Condition number	1.2289
4	Worksheet runs	12
5	Model terms	6
6	DF residual	6
7	DF lack of fit	4
8	DF pure error (repl. runs)	2

Correlation Matrix

	1	2	3	4	5	6	7	8	9
1		Ac	pH	Te	OS	Co(ColB)	k1	k2	Re1
2	Ac	1	0	0	0	0	-0.930847	-0.942848	-0.721666
3	pH	0	1	0	0	0	0.137228	0.17586	0.388589
4	Te	0	0	1	0	0	-0.296419	-0.262489	0.0316145
5	OS	0	0	0	1	0	-0.00837608	0.00563846	0.111026
6	Co(ColB)	0	0	0	0	1	-0.0702443	-0.0252202	0.294618
7	k1	-0.930847	0.137228	-0.296419	-0.00837608	-0.0702443	1	0.99462	0.6442
8	k2	-0.942848	0.17586	-0.262489	0.00563846	-0.0252202	0.99462	1	0.719738
9	Re1	-0.721666	0.388589	0.0316145	0.111026	0.294618	0.6442	0.719738	1

Threshold 0.3

Figure 7.20: *(left) Evaluation of the condition number.*
Figure 7.21: *(right) Plot of the correlation matrix revealing the strong inter-relatedness among the three responses.*

7.6.2 Regression analysis

The regression analysis phase in robustness testing is carried out in a manner similar to that of screening and optimization. However, here the focus is primarily on the R^2 and Q^2 parameters, and on the analysis of variance results, but not so much on residual plots and other graphical tools. The reason for this is that the interest in robustness testing lies in classifying the regression model as significant or not significant. With such information it is then possible to understand the robustness. Another modelling difference with respect to screening and optimization is that model refinement is not carried out.

We fitted a linear model with six terms to each response. The overall results of the model fitting is displayed in Figure 7.22. It is seen that the predictive power ranges from poor to excellent. The Q^2 values are 0.92, 0.96, and 0.12, for k_1, k_2, and Res1, respectively. In robustness testing the ideal result is a Q^2 near zero. Hence, the Q^2 of 0.12 for Res1 is an indication of an extremely weak relationship between the factors and the response, that is, it seems that the response is robust. The low Q^2 for Res1 might be explained by the fact that this response is almost constant across the entire design, and hence there is little response variation to account for. The high Q^2's for k_1 and k_2, on the other hand, indicate that these responses are sensitive to the small factor changes. However, for these latter responses, there is no point in making any robustness statement, as there are no specifications to be met.

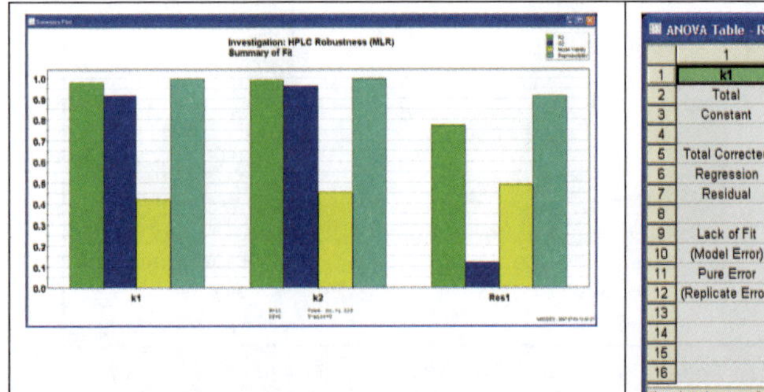

ANOVA Table - Response k1

	1	2	3	4	5	6	7
1	k1	DF	SS	MS (variance)	F	p	SD
2	Total	12	48.287	4.02392			
3	Constant	1	47.4711	47.4711			
4							
5	Total Corrected	11	0.815918	0.0741744			0.27235
6	Regression	5	0.798111	0.159622	53.785	**0.000**	0.39952
7	Residual	6	0.0178067	0.00296779			0.054477
8							
9	Lack of Fit	4	0.0169117	0.00422792	9.44746	**0.098**	0.065022
10	(Model Error)						
11	Pure Error	2	0.000895038	0.000447519			0.021154
12	(Replicate Error)						
13							
14		N = 12	Q2 =	0.915	Cond. no. =	1.229	
15		DF = 6	R2 =	0.978	Y-miss =	0	
16			R2 Adj. =	0.960	RSD =	0.05448	

The selected response 'k1'

Figure 7.22: *(left) Summary of fit plot for the HPLC data.*
Figure 7.23: *(right) ANOVA of capacity factor 1.*

ANOVA Table - Response k2

	1	2	3	4	5	6	7
1	k2	DF	SS	MS (variance)	F	p	SD
2	Total	12	108.479	9.03993			
3	Constant	1	107.485	107.485			
4							
5	Total Corrected	11	0.993942	0.0903584			0.300597
6	Regression	5	0.983463	0.196693	112.622	**0.000**	0.4435
7	Residual	6	0.0104789	0.00174649			0.041791
8							
9	Lack of Fit	4	0.00987157	0.00246789	8.12649	**0.113**	0.0496779
10	(Model Error)						
11	Pure Error	2	0.00060737	0.000303685			0.0174266
12	(Replicate Error)						
13							
14		N = 12	Q2 =	0.959	Cond. no. =	1.229	
15		DF = 6	R2 =	0.989	Y-miss =	0	
16			R2 Adj. =	0.981	RSD =	0.04179	

The selected response 'k2'

ANOVA Table - Response Res1

	1	2	3	4	5	6	7
1	Res1	DF	SS	MS (variance)	F	p	SD
2	Total	12	39.8743	3.32286			
3	Constant	1	39.8581	39.8581			
4							
5	Total Corrected	11	0.0162392	0.00147629			0.0384225
6	Regression	5	0.0125387	0.00250774	4.06608	0.059	0.0500773
7	Residual	6	0.00370047	0.000616746			0.0248344
8							
9	Lack of Fit	4	0.00345047	0.000862619	6.90096	**0.131**	0.0293704
10	(Model Error)						
11	Pure Error	2	0.00025	0.000125			0.0111803
12	(Replicate Error)						
13							
14		N = 12	Q2 =	0.121	Cond. no. =	1.229	
15		DF = 6	R2 =	0.772	Y-miss =	0	
16			R2 Adj. =	0.582	RSD =	0.02483	

The selected response 'Res1'

Figure 7.24: *(left) ANOVA of capacity factor 2.*
Figure 7.25: *(right) ANOVA of the resolution response.*

The results of the second diagnostic tool, the analysis of variance, are summarized in Figures 7.23 – 7.25. Remembering that the upper p-value should be smaller than 0.05 and the lower p-value larger than 0.05, we realize that the former test is a borderline case with respect to Res1, because the upper listed p-value is 0.059. This suggests that the model for Res1 is not significant, and therefore that Res1 is robust.

This concludes the regression analysis phase. The derived models will now be used in a general discussion concerning various outcomes of robustness testing.

7.6.3 First limiting case – Inside specification/Significant model

The first limiting case is *inside specification and significant model.* The HPLC application contains one example of this limiting case, the Res1 response. We know from the initial raw data assessment that this response is robust, because all the measured values are inside the specification, that is, above 1.5. Actually, as seen in Figure 7.26, the measured values are all above 1.75. The question of a significant model, however, is more debatable. It is possible to interpret the obtained regression model as a weakly significant regression equation. We will do so in this section for the sake of illustration. The classification of the model as significant is based on a joint assessment of the low, but positive, Q^2, seen in Figure 7.27, and the significant linear term of acetonitrile, seen in Figure 7.28. Hence, Res1 may be regarded as an illustration of the first limiting case.

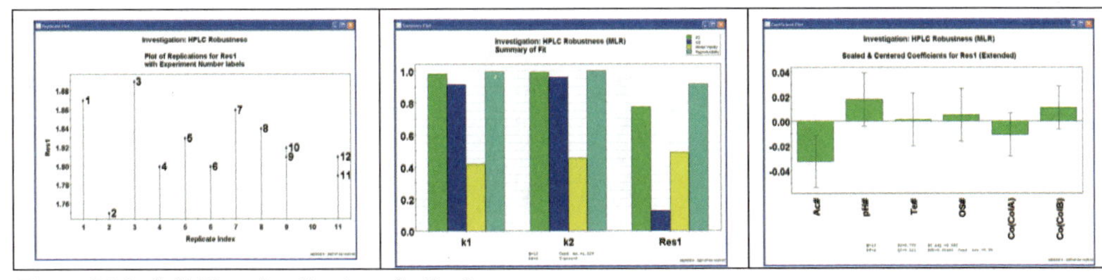

Figure 7.26: *(left) Plot of replications for Res1.*
Figure 7.27: *(middle) Summary of fit plot.*
Figure 7.28: *(right) Regression coefficients of the model for Res1.*

An interesting consequence of these modelling results is that it appears possible to relax the factor tolerances and still maintain a robust system. For instance, the model interpretation reveals that the amount of acetonitrile might be allowed to be as high as 28%, without compromising the goal of maintaining a resolution above 1.5.

Furthermore, in robustness testing it may also be of interest to estimate the response values of the most extreme experiments. Figure 7.28 gives guidance of how to obtain such estimates. We can see that one extreme experimental condition is given by the factor combination low AcN, high pH, high Temp, high OSA, and ColB, and the other extreme experiment by the reversed factor pattern. Figure 7.29 gives these Res1 predictions, and as seen they are both valid with regard to the given specification.

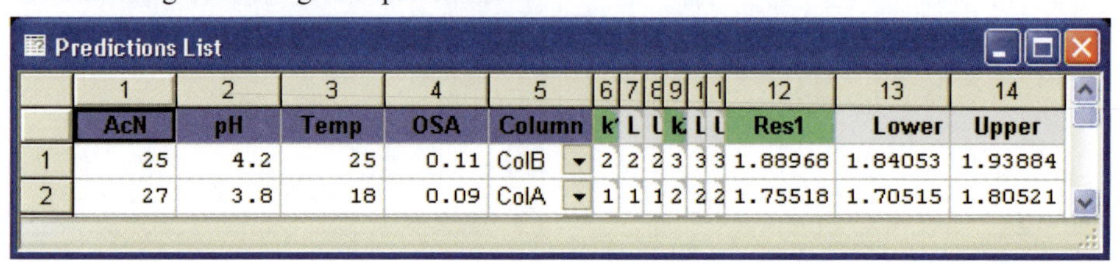

Predictions List

	1	2	3	4	5	6	7	8	9	1	1	12	13	14
	AcN	pH	Temp	OSA	Column	k'	L	L	k	L	L	Res1	Lower	Upper
1	25	4.2	25	0.11	ColB	2	2	2	3	3	3	1.88968	1.84053	1.93884
2	27	3.8	18	0.09	ColA	1	1	1	2	2	2	1.75518	1.70515	1.80521

Figure 7.29: *Extreme cases predictions for the Res1 response.*

7.6.4 Second limiting case – Inside specification/Non-significant model

The second limiting case is *inside specification and non-significant model.* This is the ideal outcome of a robustness test. Again, we may use the Res1 response as an illustration. We know that the measured values of this response are all inside specification, and it is also possible to interpret the regression model obtained as non-significant. This classification of the model as non-significant is contrary to the classification made in the previous section, but is still reasonable and is made for the purpose of illustrating the second limiting case.

In general, to assess model significance two diagnostic tools emerge as more appropriate than others. The first tool consists of the R^2 and Q^2 parameters. When these are simultaneously low or

near zero, as is the situation in Figure 7.30, we have the ideal case. This means that we are trying to model a system in which there is *no* relationship among the factors and the response in question. In reality, however, one has to expect slight deviations from this outcome. A typical result is the case when R^2 is rather large, in the range of 0.5-0.8, and Q^2 low or close to zero. As seen in Figure 7.31 this is the case for Res1, which points to an insignificant model.

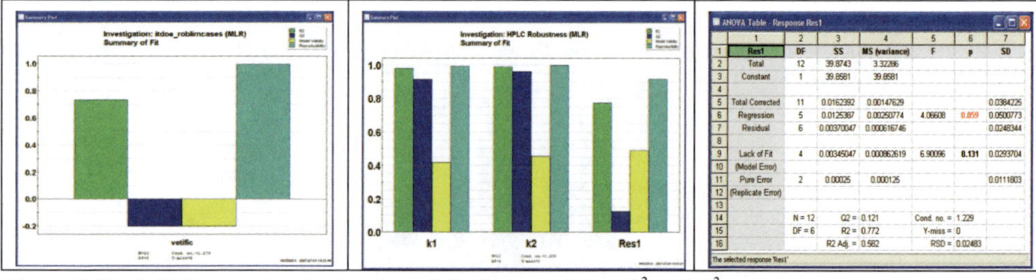

ANOVA Table - Response Res1

	1	2	3	4	5	6	7
1	Res1	DF	SS	MS (variance)	F	p	SD
2	Total	12	39.8743	3.32286			
3	Constant	1	39.8581	39.8581			
4							
5	Total Corrected	11	0.0162392	0.00147629			0.0384225
6	Regression	5	0.0125387	0.00250774	4.06608	0.059	0.0500773
7	Residual	6	0.00370047	0.000616746			0.0248344
8							
9	Lack of Fit	4	0.00345047	0.000862619	6.90096	0.131	0.0293704
10	(Model Error)						
11	Pure Error	2	0.00025	0.000125			0.0111803
12	(Replicate Error)						
13							
14		N = 12	Q2 =	0.121	Cond. no. =	1.229	
15		DF = 6	R2 =	0.772	Y-miss =	0	
16			R2 Adj. =	0.582	RSD =	0.02483	

The selected response 'Res1'

Figure 7.30: *(left) Schematic of a situation of simultaneously low R^2 and Q^2.*
Figure 7.31: *(middle) Summary of fit plot.*
Figure 7.32: *(right) ANOVA of Res1.*

The second important modelling tool relates to the analysis of variance, and particularly the upper F-test, which is a significance test of the regression model. We can see in Figure 7.32 that the Res1 model is weakly insignificant, because the p-value of 0.059 exceeds 0.05. Hence, we conclude that no useful model is obtainable. When no model is obtainable it is reasonable to anticipate that all the variation in the experiments can be interpreted as a variation around the mean. This variation can then be seen as the mean value +/- t-value * standard deviation.

7.6.5 Third limiting case – Outside specification/Significant model

The third limiting case is *outside specification and significant model.* This limiting case occurs whenever a significant regression model is calculated, and the raw response data themselves do not meet the goals of the problem formulation. We will use the second response, k_2, of the HPLC data to illustrate this limiting case. In order to accomplish a meaningful illustration, we will have to define a specification for k_2, for example that k_2 should be between 2.7 and 3.3. This kind of specification for a capacity factor is uncommon in the pharmaceutical industry, but one is set here for the sake of illustration.

We start by assessing the statistical behaviour of the k_2 regression model. This behaviour is evident from Figure 7.33, which indicates the sensitivity to small factor changes of k_2 (as well as k_1). In order to understand what is causing this susceptibility to changes in the factors, it is necessary to consult the regression coefficients displayed in Figure 7.34.

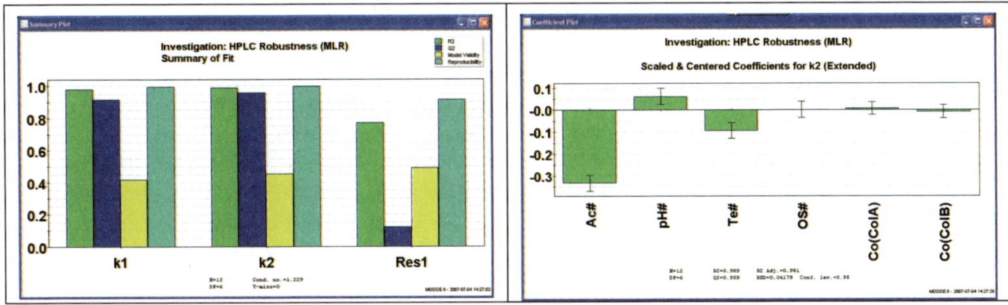

Figure 7.33: *(left) Summary of fit plot.*
Figure 7.34: *(right) Regression coefficients of the model for k_2.*

We can see that it is mainly acetonitrile, pH and temperature, which affect k_2. By using the procedure which was outlined in connection with the first limiting case, we can change the factor intervals to accomplish two things, namely (i) get k_2 inside specification and (ii) produce a non-significant model, that is, approach the second limiting case. First of all, it is possible to predict the most extreme experimental values (in the investigated area) of k_2. These are the predictions listed on the first two rows in Figure 7.35, and they amount to 2.50 and 3.49. Clearly, we are outside the 2.7 – 3.3 specification.

Predictions List

	1	2	3	4	5	6	7	8	9	10	11
	AcN	pH	Temp	OSA	Column	k	L	L	k2	Lower	Upper
1	25	4.2	18	0.11	ColA	2	2	2	3.49301	3.40882	3.57721
2	27	3.8	25	0.09	ColB	1	1	1	2.50145	2.41874	2.58417
3	25.5	4.1	19	0.11	ColA	2	2	2	3.2695	3.20274	3.33627
4	26.5	3.9	24	0.09	ColB	1	1	1	2.72496	2.65953	2.79039
5	25.9	4.1	20	0.11	ColA	2	2	2	3.11022	3.04953	3.17092
6	26.1	3.9	23	0.09	ColB	1	1	1	2.88424	2.82442	2.94406

Missing value.

Figure 7.35: *Predictions of k_2 corresponding to interesting factor combinations.*

In order to get within the specification, we must adjust the factor ranges of the three influential factors, and this is illustrated in rows three and four (Figure 7.35). To get a non-significant regression model even narrower factor intervals are needed. This is done as follows: The regression coefficient of acetonitrile is –0.33 and its 95% confidence interval ±0.036. These numbers mean that this coefficient must be decreased by a factor of ten, that is, be smaller than around –0.03, in order to make this factor non-influential for k_2. Since this coefficient corresponds to the response change when the amount of acetonitrile is increased by 1%, from 26 to 27%, we realize that the new high level must be lowered from 27 to 26.1%. A similar reasoning applies to the new lower factor level. Hence, the narrower, more robust, factor tolerances of acetonitrile ought to be between 25.9 and 26.1%. A similar reasoning for the temperature factor indicates that the factor interval should be decreased to one-third of the original size. Appropriate low and high levels thus appear to be 20 and 23 degrees centigrade. The obtained predictions are listed in rows five and six of Figure 7.35.

This concludes our treatment of the third limiting case, and our message has been that it is possible to use the modelling results for understanding how to reformulate the factor settings so that robustness can be obtained.

7.6.6 Fourth limiting case – Outside specification/Non-significant model

The fourth limiting case is *outside specification and non-significant model.* This limiting case may be the result when the derived regression model is poor, and there are anomalies in the data. It is important to uncover such anomalies, because their presence will influence the modelling. An informative graphical tool for understanding whether this is happening is the replicate plot. Figure 7.36 shows an example in which one strong outlier is present, which will invalidate the robustness. Figure 7.37 depicts a case where all the replicated center-points have much higher response values than the other runs. This pattern hints at curvature and implies non-robustness. A third common situation, which partly resembles the first case, is when one experiment deviates from the rest and also falls outside some predefined robustness limits. This is shown in Figure 7.38.

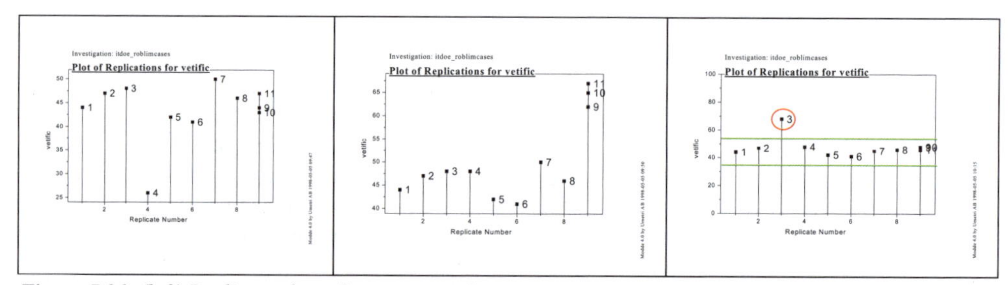

Figure 7.36: *(left) Replicate plot – One strong outlier.*
Figure 7.37: *(middle) Replicate plot – Center-points with higher values than other points.*
Figure 7.38: *(right) Replicate plot – One point outside specifications (horizontal lines).*

Evidently, there can be several underlying explanations for this limiting case, and we have just mentioned a few. Therefore, we consider this limiting case as the most complex one. In summary, we have now described four limiting cases of robustness testing, but it is important to realize that robustness testing results are not limited to these four extremes. In principle, there is a gradual transition from one limiting case to another, and hence an infinite number of outcomes are conceivable.

7.6.7 Summary of the HPLC application

The application of DOE in the robustness testing of the HPLC system was very successful. With this approach it was possible to infer the robustness of the Res1 response. From a tutorial point of view, the HPLC application is good for several reasons. It represents a realistic case in which all the necessary steps for verifying the robustness of an analytical system are illustrated. Furthermore, this data set also allowed us to study how tools for evaluation of raw data, tools for regression analysis, and tools for model use, were put into practice in a real situation. It should be clear from this application, that the modelling step is of crucial importance in robustness testing, as it is linked to an understanding of the nature of the robustness or non-robustness. We also used the HPLC study for discussing four limiting cases of robustness testing.

7.7 Questions for Chapter 7

- What is the typical objective in a robustness test?
- Which type of regression model is appropriate for robustness testing?
- Why is a low-resolution fractional factorial design appropriate?
- Which numbers of runs are most suitable for a Plackett-Burman design?
  A fractional factorial design?
- Which tools are useful for evaluation of raw data?
- Which tools are useful for diagnostic checking of a regression model?
- Which two evaluation criteria may be used for understanding the results of robustness testing?
- Which are the four limiting cases of robustness testing?

7.8 Summary and discussion

In this chapter, we have discussed the experimental objective called robustness testing. For this purpose, we have reviewed the HPLC application, consisting of five factors, three responses, and a 2^{5-2} fractional factorial design in 12 experiments. The problem formulation steps of this application were reviewed at the beginning of the chapter. The next part of the chapter was devoted to the discussion of common designs in robustness testing. Following this introduction, the HPLC application was outlined in detail. The analysis of this data-set resulted in excellent models for two responses and a weak model for one response. These models and the raw experimental data were used to establish the robustness of the Res1 response.

Part III

Chapters 8-18

Additional Topics

8 Additional screening designs for regular regions

8.1 Objective

When the design region is regular, two-level fractional factorial designs are frequently used for screening. This chapter outlines additional design families which can be used for screening regular experimental domains. To this end, we examine Plackett-Burman, Rechtschaffner and L-designs. Each design family is illustrated by one application.

8.2 Introduction

Screening is widely used to identify a few significant factors from a large number of potentially important ones. As discussed in Chapter 5, fractional factorial designs are the most commonly used design family for screening applications. They are useful for mapping many factors when the experimental domain is regular and is not deformed by any constraints. In this chapter, we describe alternative screening designs, which are applicable to regular experimental domains. The pros and cons of these design families are discussed.

First, we examine the Plackett-Burman group of designs. This type of design was introduced in 1946 by R.L. Plackett and J.P. Burman in their paper in Biometrika (vol. 33) entitled "The Design of Optimal Multifactorial Experiments". Plackett-Burman designs are very efficient for screening when only the main effects are of interest. These designs have run numbers that are multiples of four, rather than powers of two as is the case for fractional factorials. Thus, in a PB-design, the main effects are, in general, heavily confounded by two-factor interactions.

Another alternative to the fractional factorials, pioneered by Rechtschaffner in 1967, was described in the paper entitled "Saturated fractions of 2^n and 3^n factorial designs" published in Technometrics (vol. 9). Rechtschaffner designs are orthogonal, saturated fractions of resolution V of the 2^n and 3^n factorial designs. Hence, they can be used for both screening (2^n fractions) and optimization (3^n fractions). In this chapter, we focus on Rechtschaffner screening designs. These allow the estimation of all the main effects and all first order interactions without confounding effects. They are saturated designs, with no degrees of freedom remaining for the estimation of residuals and diagnostics thereof.

The final design type that we discuss is the family of three-level and mixed-level fractional factorial designs. These designs are fractional factorials drawn from the Graeco-Latin square family. They are sometimes referred to as L-designs. There are four designs available in MODDE, namely for 9, 18, 27 and 36 experiments.

The next three sections are devoted to illustrating and discussing the Plackett-Burman (Section 8.3), Rechtschaffner (Section 8.4) and L- (Section 8.5) designs. The chapter ends with a comparison of these designs and the two-level fractional factorials.

8.3 Application of Plackett-Burman designs

8.3.1 Introduction to Plackett-Burman designs

Plackett-Burman designs are orthogonal two-level experimental designs. They can only be used to fit linear models. Hence, they are useful in screening when only the main effects are of interest, and in robustness testing when, in principle, the absence of a model is the desirable outcome (see Chapter 7). The PB-designs are parsimonious, with the number of runs being a multiple of four. This implies that they nicely complement the two-level fractional factorials, since the number of runs in the latter is a power of two.

8.3.2 The Fluidized Bed Granulation application

The example that we have chosen to illustrate the use of Plackett-Burman design deals with the screening of process factors influencing a granulation process. The application, which involves studying the influence of six factors on the granule yield, is taken from the literature (Rambali, B., Baert, L., Thone, D., and MAssart, D.L., Using Experimental Design to Optimize the Process Parameters in Fluidized Bed Granulation, Drug Development and Industrial Pharmacy, 27, 2001, 47-55). The name, abbreviation and unit of each factor are presented in Figure 8.1.

Factors

	Name	Abbr.	Units	Type	Use	Settings
1	Air Flow Rate	Air	m3/hr	Quantitative	Controlled	140 to 286
2	Nozzle Diameter	Noz	mm	Quantitative	Controlled	1.2 to 2.2
3	Nozzle Height Position	No2		Quantitative	Controlled	1 to 3
4	Nozzle Air Pressure	No3	bar	Quantitative	Controlled	1.5 to 2.5
5	Inlet Air Temperature	Inl	°C	Quantitative	Controlled	50 to 70
6	Spray Rate	Spr	g/min	Quantitative	Controlled	77.5 to 135.6

Figure 8.1: *The six factors in the Fluidized Bed Granulation example.*

Briefly, it can be stated that granulation corresponds to a size-enlargement step in the production of tablets in the pharmaceutical industry. Many process and apparatus factors can influence the performance of a fluidized bed granulator and the objective of this study was to determine the most important ones. The aim was to identify the combination of factors that maximized the yield of granules between 75 and 500 µm (Figure 8.2).

Responses

	Name	Abbr.	Units	Transform	MLR Scale	PLS Scale	Type
1	Granule Yield	Gra	%	None	None	Unit Variance	Regular
				Double-click	here to	add a new	response

Figure 8.2: *The single response studied in the Fluidized Bed Granulation example.*

In order to screen the six factors, the investigators carried out a PB-design in twelve runs without center-points (Figure 8.3). With a linear model, this set-up corresponds to five degrees of freedom for the default linear model. It should be noted that there is an eight-run PB-analogue, but the investigators chose not to use this design because of its low number of degrees of freedom.

Exp No	Exp Name	Run Order	Incl/Excl	Air Flow Rate	Nozzle Diameter	Nozzle Height Position	Nozzle Air Pressure	Inlet Air Temperature	Spray Rate	Granule Yield
1	N1	6	Incl	286	1.2	3	1.5	50	77.5	84.7
2	N2	8	Incl	286	2.2	1	2.5	50	77.5	97.5
3	N3	10	Incl	140	2.2	3	1.5	70	77.5	59
4	N4	7	Incl	286	1.2	3	2.5	50	135.6	2.9
5	N5	9	Incl	286	2.2	1	2.5	70	77.5	90.3
6	N6	4	Incl	286	2.2	3	1.5	70	135.6	47.4
7	N7	3	Incl	140	2.2	3	2.5	50	135.6	0
8	N8	5	Incl	140	1.2	3	2.5	70	77.5	81.6
9	N9	12	Incl	140	1.2	1	2.5	70	135.6	0
10	N10	2	Incl	286	1.2	1	1.5	70	135.6	64.8
11	N11	1	Incl	140	2.2	1	1.5	50	135.6	0
12	N12	11	Incl	140	1.2	1	1.5	50	77.5	0

Figure 8.3: *The worksheet for the twelve-run Plackett-Burman design.*

8.3.3 Evaluation of raw data

The replicate plot in Figure 8.4 shows that in five of the experiments a zero or near-zero yield of granules was obtained. This phenomenon contributes to the bimodality seen in the histogram plot (Figure 8.4). Based on the histogram plot it was decided to analyze the data using the logit-transformed response variable.

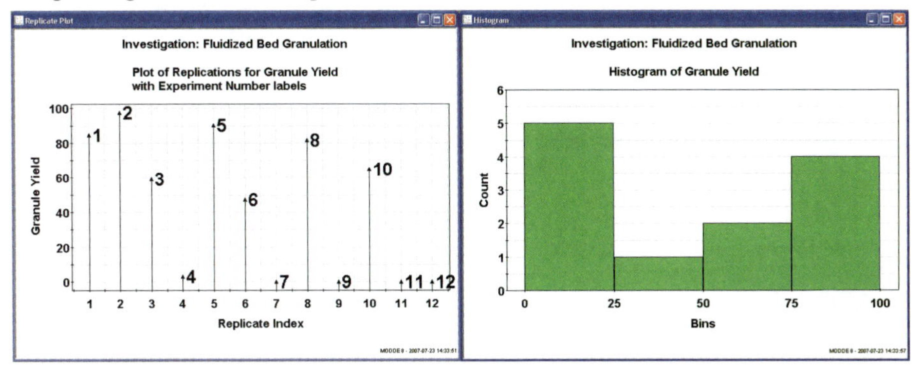

Figure 8.4: *Replicate plot (left) and histogram plot (right) of the Fluidized Bed Granulation example.*

8.3.4 Data analysis and model interpretation

When the linear model was fitted to the logit-transformed response variable, a poor model with $R^2 = 0.85$ and $Q^2 = 0.11$ was obtained (Figure 8.5). Such a large gap between R^2 and Q^2 is undesirable and indicates that something is not right with the model. As shown by the regression coefficient plot (Figure 8.5), the main reason that the model is poor is the presence of three small and unimportant main effects, i.e., the main effects of nozzle diameter (x_2), nozzle height position (x_3), and nozzle air pressure (x_4). Accordingly, these three main effects were discarded and the model was re-fitted to the response data.

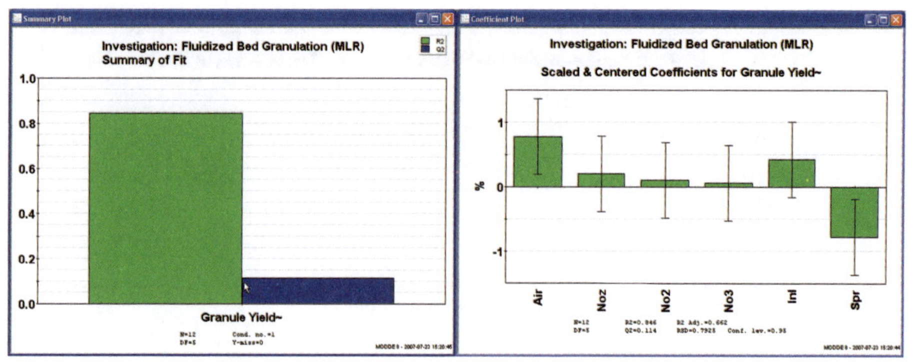

Figure 8.5: Summary of fit plot (left) and regression coefficient plot (right) of the Fluidized Bed Granulation example. Default linear model.

After revising the linear model to contain just the three important main effects (plus the constant term) a much better model with $R^2 = 0.81$ and $Q^2 = 0.58$ was obtained (Figure 8.6). Evidently, air flow rate (x_1) and spray rate (x_6) are the most influential factors (Figure 8.6). The factor inlet air temperature (x_2) is a borderline case, depending on the confidence interval, but was retained in the final model since it contributes to upholding the predictive power (Q^2 drops to 0.48 if x_2 is removed).

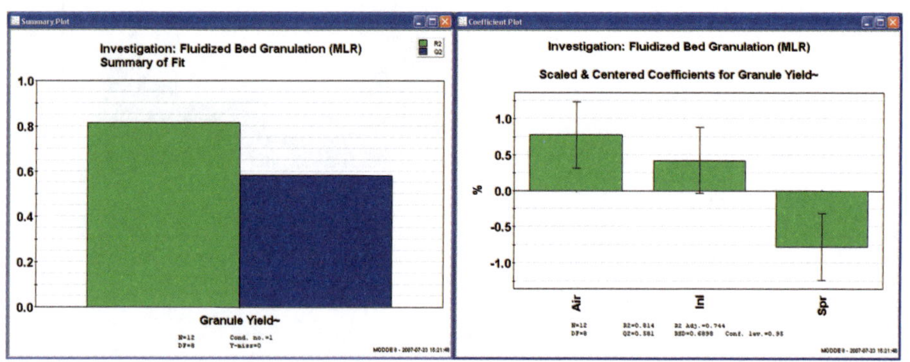

Figure 8.6: Summary of fit plot (left) and regression coefficient plot (right) of the Fluidized Bed Granulation data. Revised linear model.

8.3.5 Use of model

With three factors it is possible to create a trio of response contour plots to provide an overview of the system (Figure 8.7). From Figure 8.7 it can be deduced that the optimal factor combination, leading to the maximum granule yield, is high air flow rate (x_1), high inlet air temperature (x_5) and low spray rate (x_6).

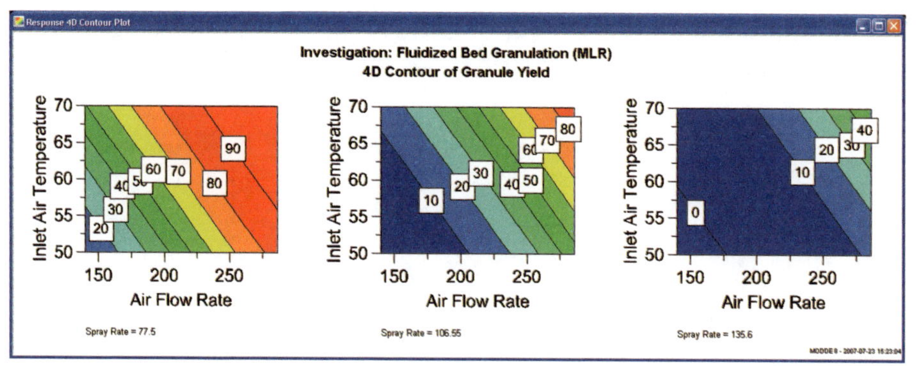

Figure 8.7: *Trio of response contour plots indicating that high x_1, high x_5 and low x_6 correspond to the most favorable process conditions for maximizing the granule yield.*

8.3.6 Follow-up

The three factors that were found to be most significant in the screening were set according to the results discussed above. Then, a second-phase screening design (a fractional factorial design) with five factors was constructed to further study the granulation process. With the new design even higher yields of granules were recorded, and, most importantly, low- or near-zero-yields were entirely avoided.

8.3.7 Summary of Plackett-Burman designs

Plackett-Burman protocols are very "compact" designs. A PB-design in twelve runs may, for example, be used for an investigation containing up to eleven factors. Such a worksheet, extended to include three center-points, is shown below (Figure 8.8). PB-designs also exist for 20-runs, 24-runs and 28-runs (and higher). In each case, the number of factors that the screening can examine is one less than the number of runs.

All PB-designs are at resolution III. They are sometimes referred to as *saturated main effect* designs because all degrees of freedom are utilized to examine the main effects. Moreover, PB-designs do not have a defining relation, since the interactions that may be involved in the study are not equal to the main effects. Hence it is not possible to identify confounding effects within MODDE when using a PB-design.

In summary, PB-designs are very useful for efficiently detecting large main effects, assuming all two-factor interactions are negligible in relation to the important main effects. In practice, a fractional factorial design tends to be used whenever a screening/robustness test is carried out in eight or 16 runs, and a PB-design is selected when either twelve or 20 experimental trials are required. The PB-designs with 24 and 28 runs are used less frequently.

Worksheet

	1	2	3	4	5	6	7	8	9	10	11	12	13	14	15
1	Exp No	Exp Name	Run Order	Incl/Excl	x1	x2	x3	x4	x5	x6	x7	x8	x9	x10	x11
2	1	N1	3	Incl	1	-1	1	-1	-1	-1	1	1	1	-1	1
3	2	N2	11	Incl	1	1	-1	1	-1	-1	-1	1	1	1	-1
4	3	N3	9	Incl	-1	1	1	-1	1	-1	-1	-1	1	1	1
5	4	N4	6	Incl	1	-1	1	1	-1	1	-1	-1	-1	1	1
6	5	N5	12	Incl	1	1	-1	1	1	-1	1	-1	-1	-1	1
7	6	N6	4	Incl	1	1	1	-1	1	1	-1	1	-1	-1	-1
8	7	N7	15	Incl	-1	1	1	1	-1	1	1	-1	1	-1	-1
9	8	N8	10	Incl	-1	-1	1	1	1	-1	1	1	-1	1	-1
10	9	N9	14	Incl	-1	-1	-1	1	1	1	-1	1	1	-1	1
11	10	N10	5	Incl	1	-1	-1	-1	1	1	1	-1	1	1	-1
12	11	N11	8	Incl	-1	1	-1	-1	-1	1	1	1	-1	1	1
13	12	N12	1	Incl	-1	-1	-1	-1	-1	-1	-1	-1	-1	-1	-1
14	13	N13	2	Incl	0	0	0	0	0	0	0	0	0	0	0
15	14	N14	13	Incl	0	0	0	0	0	0	0	0	0	0	0
16	15	N15	7	Incl	0	0	0	0	0	0	0	0	0	0	0

Figure 8.8: *Twelve-run Plackett-Burman design for eleven factors, augmented with three center-points.*

8.4 Application of Rechtschaffner designs

8.4.1 Introduction to Rechtschaffner designs

Rechtschaffner designs are applicable to screening and optimization. In this section we concentrate on screening protocols. Optimization protocols are dealt with in Chapter 9.

In the screening context, Rechtschaffner designs represent saturated two-level fractional factorials of resolution V, and they are used with interaction models. Saturation means that the number of terms in the model equals the number of factorial points in the design, e.g. 3 factors ↔ 7 runs, 4 factors ↔ 11 runs, 5 factors ↔ 16 runs, 6 factors ↔ 22 runs, 7 factors ↔ 29 runs, etc. Resolution V means that two-factor interactions are not confounded with other two-factor interactions. Hence, these designs are efficient in situations where interaction estimates are important.

8.4.2 The Cementation application

As an example of the applicability of Rechtschaffner designs, we discuss a cementation investigation. Cementation is a routine procedure to encapsulate radioactive waste generated at nuclear power plants. One difficulty that can arise is that some components of the waste may react chemically with the cement phases or the mixing water. Such reactions may accelerate, retard or even inhibit the cementation process. Some of these reactions may cause deleterious swelling and cracking of the cement material with concomitant loss of quality and performance. The data-set used in this example is taken from: C.C.D. Coumes and S. Courtois, Chemometrics and Intelligent Laboratory Systems, 80, 2006, 167-175.

The objective of this experiment was to identify the factors that influence variations in the length of cement blocks, and to select, from different types of cement, the best one to be used in the encapsulation of radioactive waste. The aim of the current data analysis exercise is to shed some light on the use of Rechtschaffner designs.

The dataset contains four factors, two qualitative and two quantitative (Figure 8.9):

- Cement Type (-), Type III and Type I.
- Oxide Density (-), High and Low, which indicates density of the oxide in the nuclear waste.
- Incorporation rate (%), which indicates waste incorporation rate as a percentage of the cement weight.
- Water/Cement ratio, this ratio influences plasticity and workability of the grout.

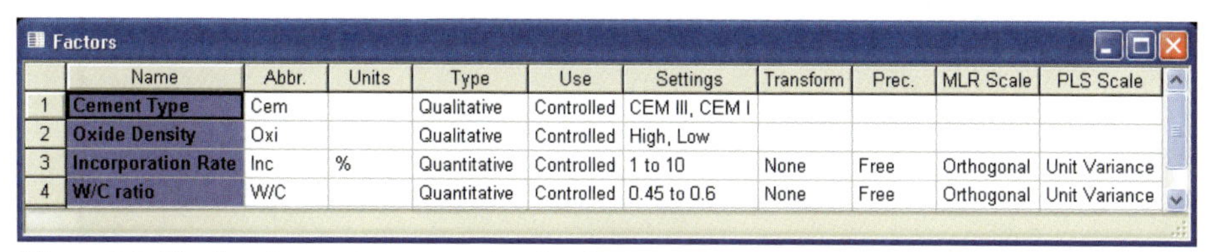

	Name	Abbr.	Units	Type	Use	Settings	Transform	Prec.	MLR Scale	PLS Scale
1	Cement Type	Cem		Qualitative	Controlled	CEM III, CEM I				
2	Oxide Density	Oxi		Qualitative	Controlled	High, Low				
3	Incorporation Rate	Inc	%	Quantitative	Controlled	1 to 10	None	Free	Orthogonal	Unit Variance
4	W/C ratio	W/C		Quantitative	Controlled	0.45 to 0.6	None	Free	Orthogonal	Unit Variance

Figure 8.9: *The four factors investigated in the Cementation example.*

The data-set contains a single response (Figure 8.10): Length variation (in μm/m) after 90 days (at room temperature). This value needs to be as low as possible, indicating little swelling.

	Name	Abbr.	Units	Transform	MLR Scale	PLS Scale	Type
1	Length Variation	Len	um/m	None	None	Unit Variance	Regular

Figure 8.10: *The response variable in the Cementation example.*

The design used was a Rechtschaffner design encoding 11+3 experiments (Figure 8.11). Note that the factor combination used in experiment 5 was replicated.

	1	2	3	4	5	6	7	8	9
1	Exp No	Exp Name	Run Order	Incl/Excl	Cement Type	Oxide Density	Incorporation Rate	W/C ratio	Length Variation
2	1	N1	13	Incl	CEM III	High	1	0.45	630
3	2	N2	2	Incl	CEM III	Low	10	0.6	3622
4	3	N3	10	Incl	CEM I	High	10	0.6	1342
5	4	N4	3	Incl	CEM I	Low	1	0.6	287
6	5	N5	1	Incl	CEM I	Low	10	0.45	1538
7	6	N6	12	Incl	CEM I	Low	1	0.45	250
8	7	N7	8	Incl	CEM I	High	10	0.45	1418
9	8	N8	4	Incl	CEM I	High	1	0.6	268
10	9	N9	5	Incl	CEM III	Low	10	0.45	5319
11	10	N10	7	Incl	CEM III	Low	1	0.6	700
12	11	N11	9	Incl	CEM III	High	10	0.6	2282
13	12	N12	6	Incl	CEM I	Low	10	0.45	1850
14	13	N13	11	Incl	CEM I	Low	10	0.45	1570
15	14	N14	14	Incl	CEM I	Low	10	0.45	1723

Figure 8.11: *The eleven-run Rechtschaffner design with three replicated points.*

8.4.3 Evaluation of raw data

The geometry of the design is shown below (Figure 8.12). As can be seen, eleven out of the 16 possible factorial combinations were examined. The arrow shows the combination of treatments selected for replication.

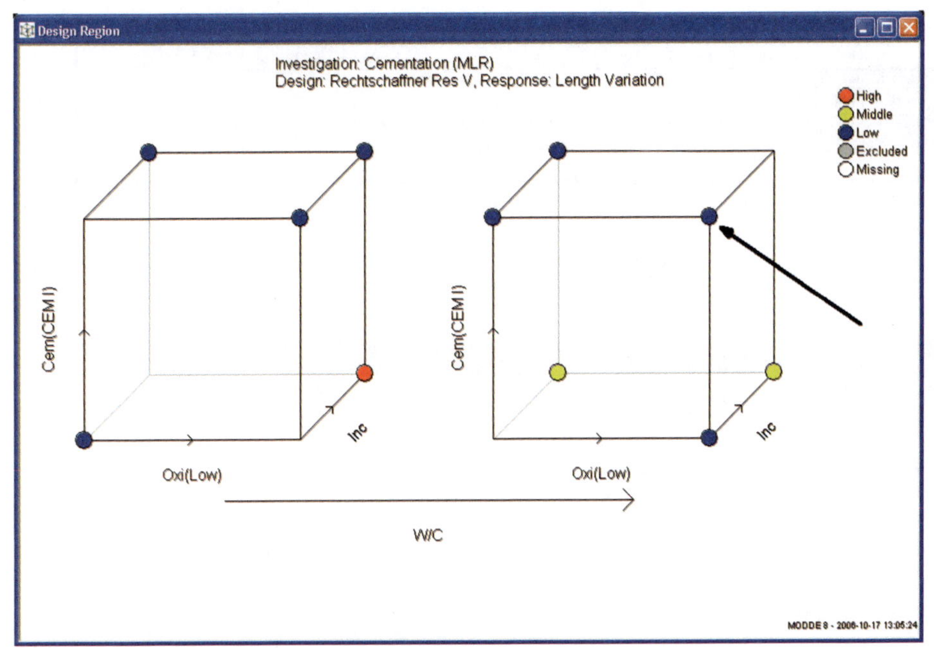

Figure 8.12: *The geometry of the 11-run Rechtschaffner design. The arrow indicates the point that was replicated.*

The replicate plot indicates low variability among the replicates (Figure 8.13). The distribution of the response is not even close to a normal distribution (Figure 8.13), so a log-transformation is required.

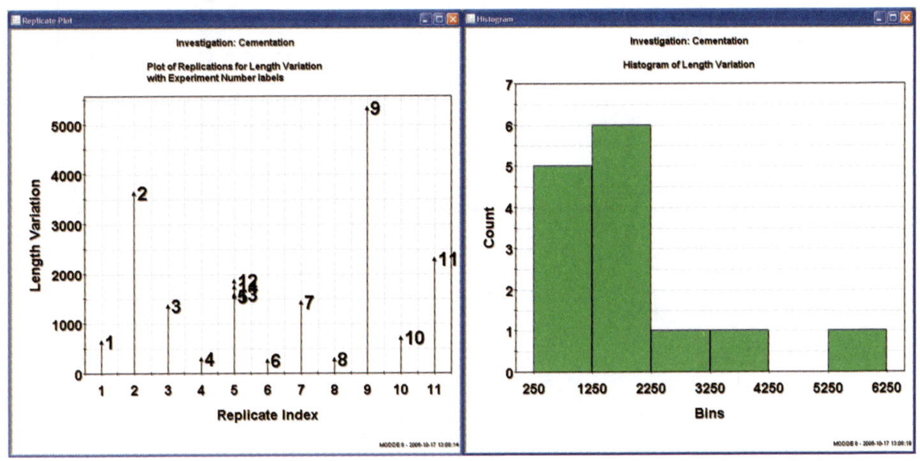

Figure 8.13: *Replicate plot (left) and histogram plot (right) of the Cementation data-set before transformation.*

After calculating the logarithm of the response values, the distribution is closer to being normal (Figure 8.14). With this improved input data, further analysis is possible.

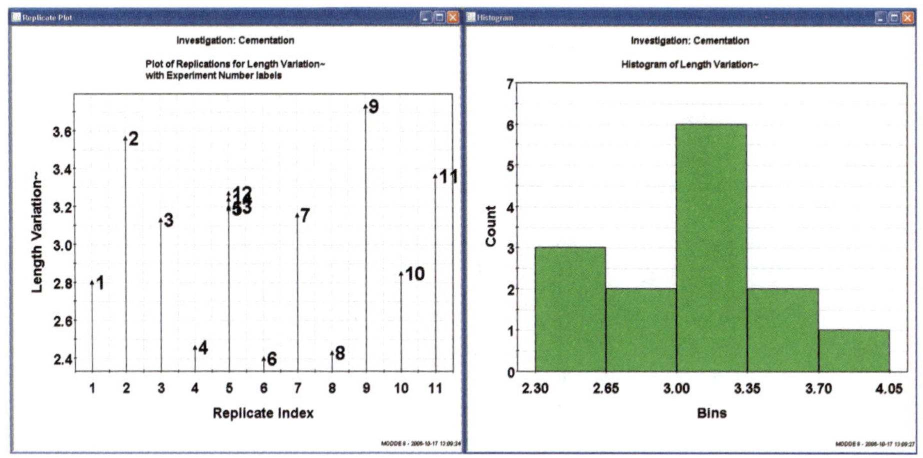

Figure 8.14: *Replicate plot (left) and histogram plot (right) of the Cementation data-set after transformation.*

8.4.4 Data analysis and model interpretation

The interaction model could now be calculated. Initially, however, the summary list presented below was obtained (Figure 8.15). The design is saturated and hence MODDE can not automatically display the traditional 'summary of fit' plot. This indicates that there are as many terms in the interaction model (eleven) as there are experiments (replicated points excluded) in the Rechtschaffner design.

Summary List

	1	2	3	4	5	6	7	8	9
1		R2	R2 Adj.	Q2	SDY	RSD	N	Model Validity	Reproducibility
2	Length Variation~	--	--	--	0.412902	0.0372072	14	--	0.99188
3									
4	N = 14	Cond. no. =	3.447						
5	DF = 3	Y-miss =	0						
6									

Figure 8.15: *Summary list acquired when fitting the interaction model to the data.*

The regression coefficients of the interaction model are shown below (Figure 8.16). By excluding some of the smallest model terms, the number of degrees of freedom (DF) increases. MODDE is then able to display the 'summary of fit' plot.

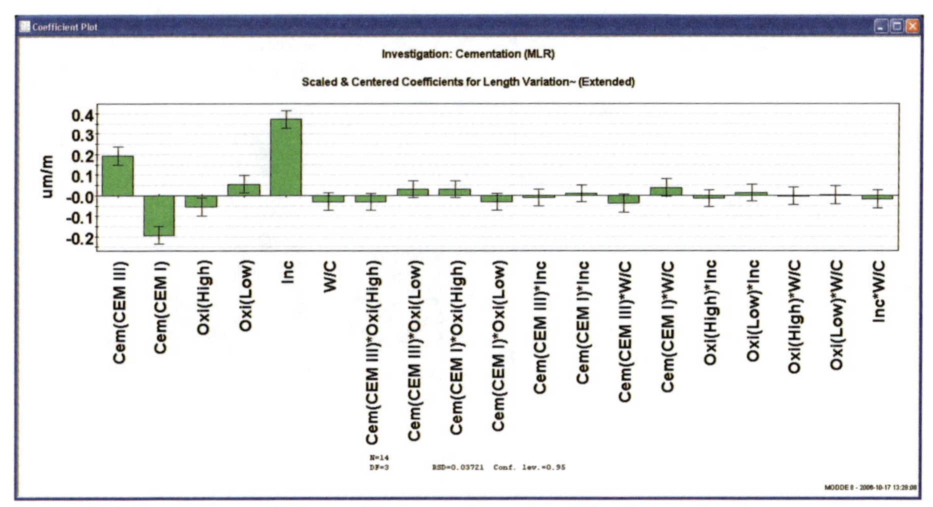

Figure 8.16: *Regression coefficients of the interaction model of the Cementation data-set.*

The interaction model was revised by omitting four two-factor interactions (Cem*Inc, Oxi*Inc, Oxi*W/C, and Inc*W/C). The resulting model has excellent performance indicators (Figure 8.17). The model shows that the incorporation rate should be kept low (Figure 8.17). From an industrial point of view, however, this rate should be kept high, since this corresponds to increased through-put. However, the higher the incorporation rate, the higher the risk of the solidified waste swelling. Other favorable settings are cement type I, high oxide density and low water/cement ratio.

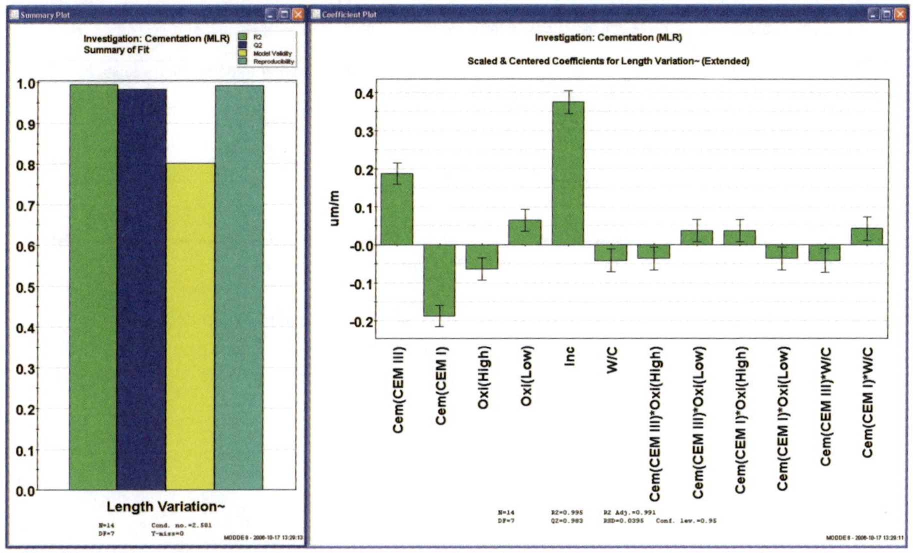

Figure 8.17: *Summary of fit plot (left) and regression coefficient plot (right) of the revised interaction model of the Cementation data-set.*

8.4.5 Use of model

In the prediction phase, there are two key issues to resolve:

- to identify the factor settings corresponding to the lowest degree of swelling, and
- to determine which type of cement is most robust (least prone to swelling).

The quartet of contour plots in Figure 8.18 indicates that the lowest risk of swelling is obtained using Cement type I, a low oxide density, and a low incorporation rate.

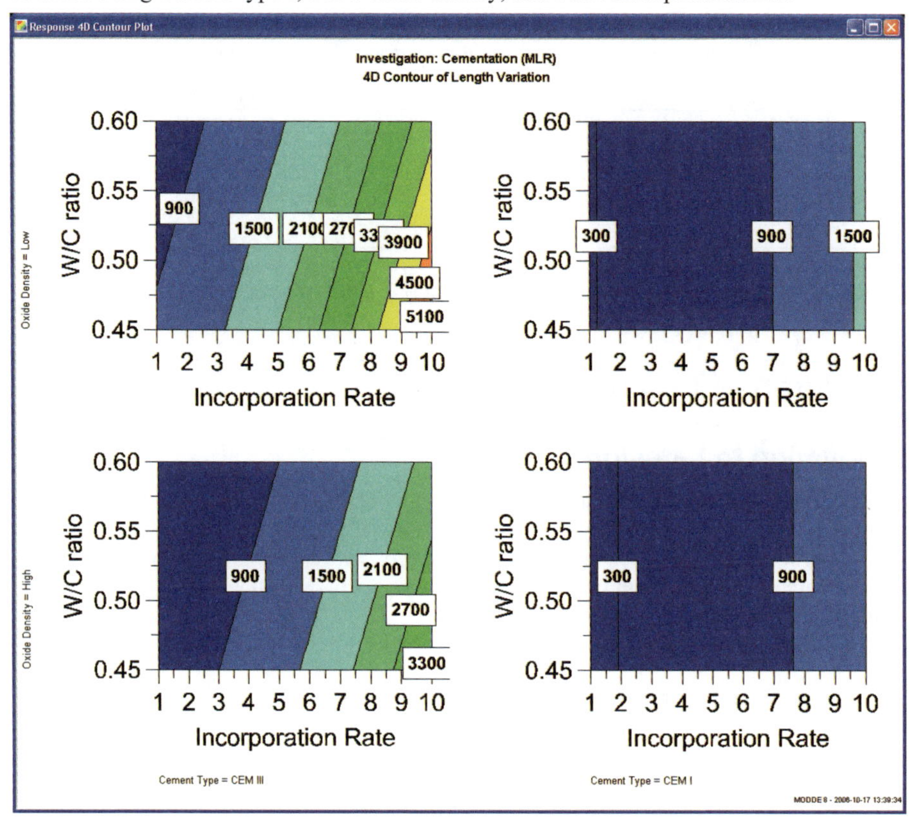

Figure 8.18: Response contour plots of the model of the Cementation data-set.

8.4.6 Conclusions of the study

Small length variation is achieved using Cement type I, a low oxide density, and a low incorporation rate. For this combination of factors, the choice of water/cement ratio is not critical. The water/cement ratio has a greater impact on the result if Cement type III is used.

8.4.7 Summary of Rechtschaffner designs

Rechtschaffner designs are saturated designs of resolution V in which the main effects and two-factor interactions can be estimated if three-factor and higher-order interactions are negligible. Studies of the statistical properties of such designs have shown they are also D-optimal.

An interesting feature of Rechtschaffner designs, in relation to the two-level fractional factorial designs, is their ability to reduce the number of runs required to an absolute minimum. They are, therefore, efficient in situations where interaction estimates are critical. The required number of runs (N) for Rechtschaffner designs with k factors is:

$N = 1 + k + k(k-1)/2$

In Figure 8.19 we present a table of the number of runs required to perform Rechtschaffner designs in relation to full and fractional factorial designs. The recommended design is indicated in bold. If experimentation is costly and time-consuming Rechtschaffner designs represent a viable alternative to full and fractional factorial designs, particularly when dealing with 4 or 6 factors.

# Factors	Linear model	Interaction model	Recht. Res V	Comment to MODDE proposal
3	N/A	**8**	7	Full factorial
4	8	**16**	11	Full factorial
5	8	**16**	16	Fractional factorial Res V
6	**16**	32	22	Fractional factorial Res IV
7	**16**	32	29	Fractional factorial Res IV
8	**16**	64	37	Fractional factorial Res IV
9	**32**	64	46	Fractional factorial Res IV
10	**32**	64	56	Fractional factorial Res IV

Figure 8.19: *Overview of the number of runs required in different Rechtschaffner screening designs. The recommended design is indicated in bold. The number of runs does not include center-points.*

8.5 Application of L-designs

8.5.1 Introduction to L-designs

The last group of designs we wish to discuss in this chapter is the L-group of orthogonal arrays, which have been promoted by Genichi Taguchi. These designs are three-level or mixed-level fractional factorial designs. In this context, *mixed-level* means a combination of two-level and three-level factors within the same design protocol. MODDE supports four L-designs, all of which support linear models:

- L9: Up to four factors at three levels.
- L18: Up to seven factors at three levels and one factor at two levels.
- L27: Up to 13 factors at three levels.
- L36: Up to 13 factors at three levels.

8.5.2 The BTX Detergent application

In order to exemplify the use of an L18-design, we discuss an example that concerns the production of detergents. At the MISKAR gas platform (in Tunisia) natural gas and condensate are produced and sent via a sub-sea pipeline for further processing at the Hannibal gas plant (Sfax, Tunisia). The gas fraction contains a relatively high concentration of aromatic compounds, such as benzene, toluene and xylenes (collectively designated BTX). These compounds need to be removed from the gas in order to avoid freezing problems during a low-temperature step in the process.

Once recovered, the BTX-fraction of a natural gas can be converted into the corresponding BTX-sulfonates by the addition of sulfuric acid. Such sulfonates are very useful in the manufacture of detergents. Typical applications of BTX-sulfonates include their use as anti-caking agents in powder detergents and as co-surfactants in rinsing aids for dishwashing machines.

The current example deals with a study of the factors that affect the preparation of BTX-sulfonates. The data-set used in this example is taken from: R. Baati, *et al.*, Chemometrics and Intelligent Laboratory Systems, 80, 2006, 198-208. The objective of this experiment was to identify the factors that influence the formation of a hydrotrope for liquid detergents. In this case, the hydrotrope corresponds to a mixture of sodium toluenesulfonate and sodium xylenesulfonates. The objective of the data analysis is to shed some light on the use of three-level and mixed-level fractional factorial designs, i.e., the so-called "L-family".

Briefly, sulfuric acid is used to sulfonate fractions of BTX mixtures. Factors x_2–x_4 and x_6 relate to the first sulfonation step. At the end of the sulfonation reaction, part of the residual organic phase is evaporated and some of the desulfation reagent is added to further react with the excess sulfuric acid. Factors x_1, x_5 and x_7 deal with this step. Thus, the data-set contains seven factors, of which the first varies at two levels and the rest at three (Figure 8.20).

- X_1: Desulfation Time (h), 1.5–3 h, the duration of the removal of excess sulfuric acid.
- X_2: Temperature (°C), 105 - 125, reaction temperature.
- X_3: Acid concentration (%), 96 – 104, concentration of sulfuric acid.
- X_4: Reaction Time (h), 6–9, duration of the sulfonation reaction.
- X_5: Desulfation agent (-), BTX, Xylene, Toulene, agent added to further react with any excess acid.
- X_6: Molar ratio (-), 0.8–1.2, defined as Acid/(Toluene + Xylenes) molar ratio.
- X_7: Amount of the desulfation agent (mol), 0.18–0.32, indicating the amount of desulfation agent.

Factors

	Name	Abbr.	Units	Type	Use	Settings	Transform	Prec.	MLR Scale	PLS Scale
1	Desulfation Time	DesT	h	Quantitative	Controlled	1.5 to 3	None	Free	Orthogonal	Unit Variance
2	Temperature	Temp	°C	Quantitative	Controlled	105 to 125	None	Free	Orthogonal	Unit Variance
3	Acid Concentration	Acid	%	Quantitative	Controlled	96 to 104	None	Free	Orthogonal	Unit Variance
4	Reaction Time	ReacT	h	Quantitative	Controlled	6 to 9	None	Free	Orthogonal	Unit Variance
5	Desulfation Agent	DesAg		Qualitative	Controlled	BTX, Xylene, Toluene				
6	Molar Ratio	Ratio	-	Quantitative	Controlled	0.8 to 1.2	None	Free	Orthogonal	Unit Variance
7	Amount DesAg	Amo	mol	Quantitative	Controlled	0.18 to 0.32	None	Free	Orthogonal	Unit Variance

Figure 8.20: *The seven factors explored in the BTX Detergent application.*

The data-set contains two responses (Figure 8.21); Residual Sulfate (%) and TX conversion (also %). The first response relates to the amount of residual sulfuric acid and needs to be minimized. A low value indicates an efficient manufacturing process. The second response measures the efficiency of the conversions of toluene and xylenes to the corresponding sulfonates. This value needs to be maximized.

Responses

	Name	Abbr.	Units	Transform	MLR Scale	PLS Scale	Type
1	Residual Sulfate	Res	%	None	None	Unit Variance	Regular
2	TX Conversion	TXC	%	None	None	Unit Variance	Regular

Figure 8.21: *The two responses in the BTX Detergent example.*

The design used was an L18-design encoding 18 experiments (Figure 8.22). Note that there were no replicate experiments. Ideally, experiment replication should always be incorporated into any experimental design.

Worksheet

Exp No	Exp Name	Run Order	Incl/Excl	Desulfation Time	Temperature	Acid Concentration	Reaction Time	Desulfation Agent	Molar Ratio	Amount DesAg	Residual Sulfate	TX Conversion
1	N1	1	Incl	1.5	105	96	6	BTX	0.8	0.18	22.5	38.4
2	N2	3	Incl	1.5	105	100	7.5	Xylene	1	0.25	14.4	48.6
3	N3	15	Incl	1.5	105	104	9	Toluene	1.2	0.32	11.2	52.2
4	N4	6	Incl	1.5	115	96	6	Xylene	1	0.32	16.7	47.3
5	N5	13	Incl	1.5	115	100	7.5	Toluene	1.2	0.18	20.4	56.8
6	N6	17	Incl	1.5	115	104	9	BTX	0.8	0.25	5.6	41.5
7	N7	9	Incl	1.5	125	96	7.5	BTX	1.2	0.25	21.3	54.9
8	N8	12	Incl	1.5	125	100	9	Xylene	0.8	0.32	2.5	45.1
9	N9	14	Incl	1.5	125	104	6	Toluene	1	0.18	8.7	48.7
10	N10	7	Incl	3	105	96	9	Toluene	1	0.25	21.6	45.1
11	N11	8	Incl	3	105	100	6	BTX	1.2	0.32	18.8	51.5
12	N12	5	Incl	3	105	104	7.5	Xylene	0.8	0.18	3.9	46.5
13	N13	16	Incl	3	115	96	7.5	Toluene	0.8	0.32	14.9	39.9
14	N14	2	Incl	3	115	100	9	BTX	1	0.18	11.1	52.9
15	N15	4	Incl	3	115	104	6	Xylene	1.2	0.25	6.3	60.8
16	N16	18	Incl	3	125	96	9	Xylene	1.2	0.18	13.7	64.5
17	N17	11	Incl	3	125	100	6	Toluene	0.8	0.25	7.2	43.5
18	N18	10	Incl	3	125	104	7.5	BTX	1	0.32	4.4	51

Figure 8.22: *The worksheet of the L18-design employed in the BTX Detergent example.*

8.5.3 Evaluation of raw data

The replicate plots indicate no extreme or outlying points (Figure 8.23). The residual sulfate response ranges between 2.5 and 22.5%. The TX conversion response ranges between 38.4 and 64.5%.

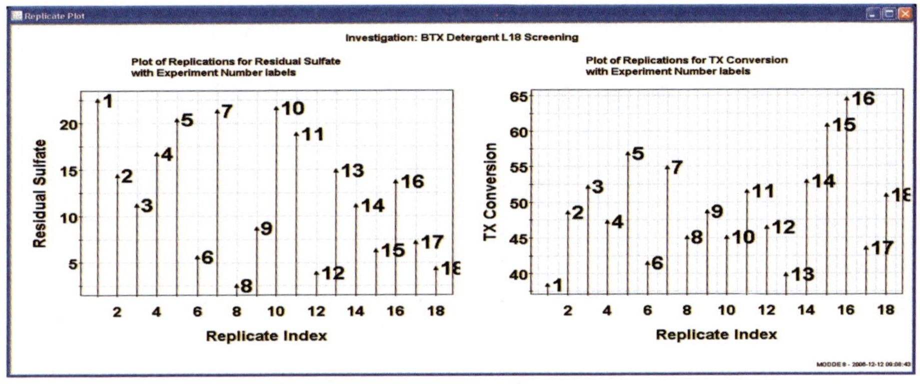

Figure 8.23: *Replicate plots of the BTX Detergent example.*

The distribution of each response is close to a normal distribution (Figure 8.24).

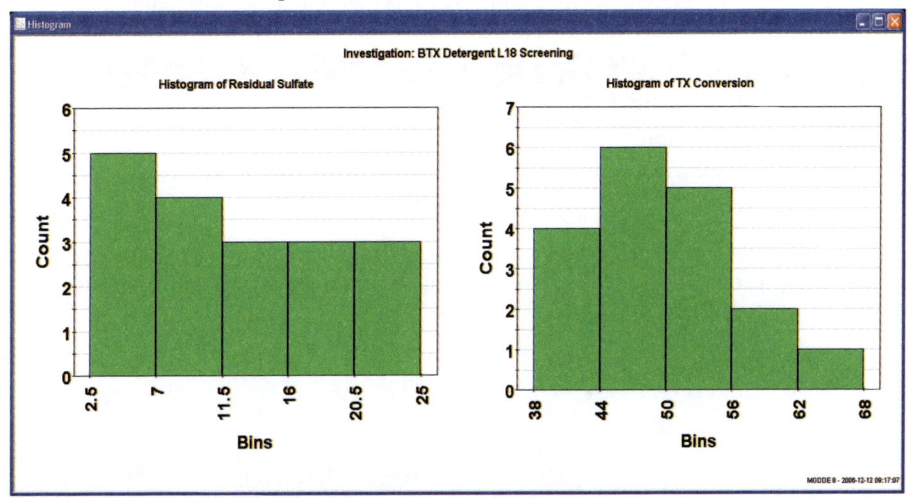

Figure 8.24: *Histogram plots of the BTX Detergent example.*

8.5.4 Data analysis and model interpretation

The default linear model is excellent in both cases (Figure 8.25). Since the design contains no replicates, model validity and reproducibility bars cannot be presented.

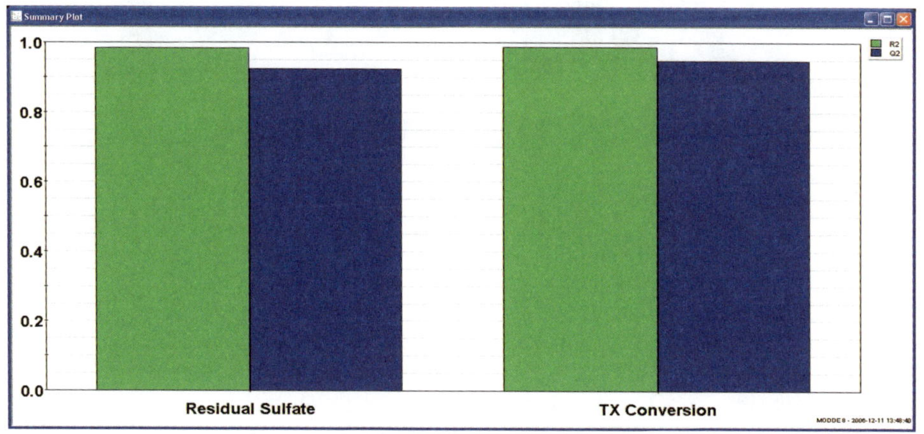

Figure 8.25: *Summary of fit plot of the BTX Detergent example.*

The relevance of each linear model is further emphasized by the two N-plots of residuals (Figure 8.26). These two plots are acceptable.

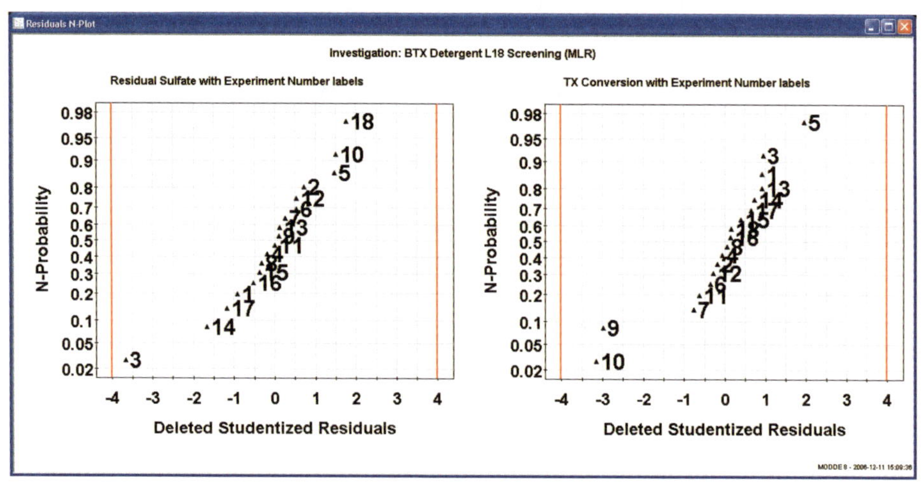

Figure 8.26: *N-plots of residuals of the BTX Detergent example.*

The regression coefficient plots reveal that x_3 (acid concentration) has the strongest effect on residual sulfate (Figure 8.27). A high concentration leads to low residual sulfate. With respect to TX conversion, factor x_6 (Molar ratio) has the highest influence. The higher the ratio the better the conversion efficiency.

A close inspection of the two linear models indicates that for five of the factors (x_1-x_5) there are no conflicts between the two response requirements. Factors x_1-x_4 should all be kept at the high level and xylene is the best setting for x_5 (desulfation agent). The last two factors (x_6 and x_7), however, conflict.

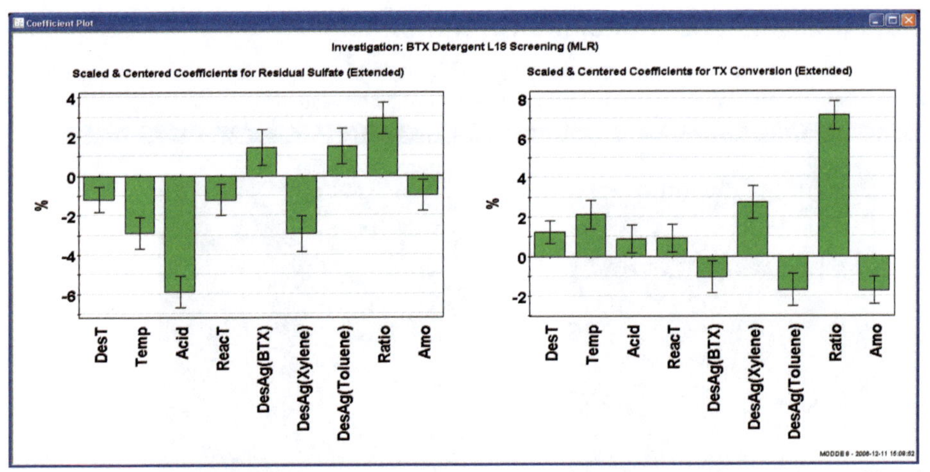

Figure 8.27: Plots of regression coefficients of the two linear models fitted to the response variables of the BTX Detergent data-set.

8.5.5 Use of model

One way to use response contour plots in this application is to place the two problematic factors (x_6 and x_7) as axes and lock the remaining five factors (x_1-x_5) at their optimum values (Figure 8.28). For the first five factors, this results in the profile: high, high, high, high and xylene.

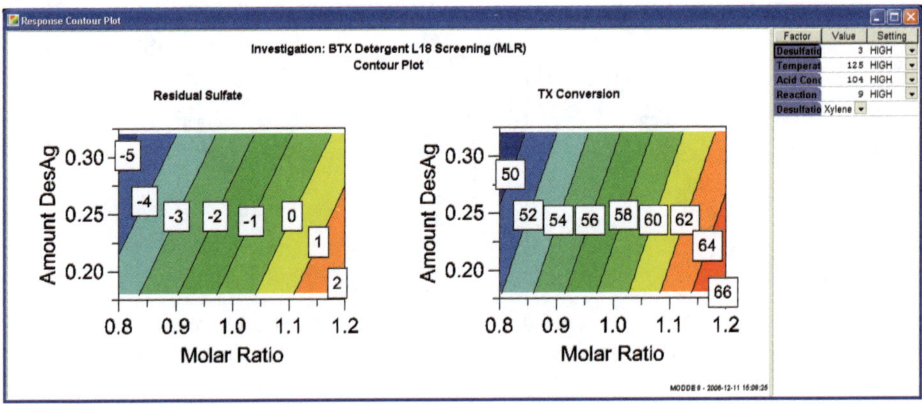

Figure 8.28: Response contour plots of the BTX Detergent example.

8.5.6 Conclusions of the study

Acid concentration has the greatest impact on Residual sulfate, and Molar ratio exerts the greatest influence on TX conversion. Fixing x_1–x_5 to the profile high, high, high, high and xylene would provide the best operating conditions. As seen from the contour plots above (Figure 8.28), this profile – supplemented with a low amount of the desulfation agent and a molar ratio of roughly 1.0 – is predicted to correspond to almost no residual sulfate and approximately 60% TX conversion (the latter value is at the high end of this response).

8.5.7 Summary of L-designs

The four L-designs implemented in MODDE represent experimental schemes encompassing 9, 18, 27 or 36 runs; these numbers exclude replicated center-points. With these designs MODDE (objective = screening) lets you select only the linear model from the outset, because these designs do not support interactions. However, in contrast to PB-designs (which support linear models) and Rechtschaffner designs (which support interaction models), L-designs allow you, in the Edit Model dialogue, to modify the default linear model and selectively include square terms. The latter is possible for all factors varied at three-levels. Hence, if quadratic behaviour is anticipated it may be appropriate to choose an orthogonal array, such as an L-design.

8.6 Questions for Chapter 8

- What is a Plackett-Burman design?
- Which type of regression model is supported by a Plackett-Burman design?
- What is the nature of the number of runs included in a Plackett-Burman design?
- What is a Rechtschaffner design?
- Which type of model is supported by a Rechtschaffner design?
- What is the nature of the number of runs included in a Rechtschaffner design?
- What is an L-design?
- What is a mixed-level design?
- Which type of regression model is supported by an L-design?
- What is the nature of the number of runs included in an L-design?

8.7 Summary and discussion

In this chapter we have considered three types of screening designs: Plackett-Burman designs, Rechtschaffner designs and L-designs. Together with the two-level full and fractional factorial designs discussed earlier (see Chapters 3–5), these design families provide solutions to almost any screening problem involving a regular experimental domain. The purpose of the discussion here is to compare and contrast these design families, using the summary presented in the table below (Figure 8.29) as its starting point.

Number of factors	Factorial/fractional factorial design	Plackett-Burman design	Rechtschaffner design	L-design
2	4_i	8_l	n/a	9_l(q)
3	8_i	8_l	7_i	9_l(q)
4	16_i	8_l	11_i	9_l(q)
5	16_i	8_l	16_i	18_l(q)
6	16_l or 32_i	8_l	22_i	18_l(q)
7	16_l or 32_i	8_l	29_i	18_l(q)
8	16_l or 64_i	12_l	37_i	18_l(q)
9	32_l or 64_i	12_l	46_i	27_l(q)
10	32_l or 64_i	12_l	56_i	27_l(q)

Figure 8.29: *Relationships among four families of screening designs. See text for further explanation.*

The summary provided in Figure 8.29 covers 2–10 factors and focuses on the number of runs required and the type of regression model supported by each design. It should be noted that the first three entries in the factorial design column are full factorial designs, while the rest are fractional factorial designs. A lower case 'l' designates a linear model and analogously a lower case 'i' an interaction model. For the L-design family, the lower case 'q' in parentheses indicates the ability of these designs to support selective quadratic terms, but not a full quadratic model with all of its two-factor interaction terms.

As noted in Chapter 5, the two-level fractional factorial designs of resolution IV (supporting an interaction model) are good starting points in screening. They allow investigation of the impact of many factors in relatively few runs. One advantage of these designs is that they often have sufficient degrees of freedom to allow a careful examination of residuals and other modelling diagnostics. Another advantage of the fractional factorials is that they can be easily modified to optimization central composite designs, except when there is a need to modify the low and high settings for some factors. Fractional factorial designs, therefore, remain our preferred choice for screening. They can be tailored to support either a linear or an interaction model.

Rechtschaffner designs are also very useful in screening, albeit suffering from a somewhat poor reputation in the DOE literature. They exclusively deal with interaction models and are available in MODDE for up to ten factors. The hallmark of the Rechtschaffner designs is their ability to reduce the number of runs required to an absolute minimum, meaning that the effective degrees of freedom (when center-points are ignored) is zero for the initial interaction model. As a consequence, small and non-significant terms must always be dropped in the data analysis to permit all modelling and diagnostic tools to be used.

Plackett-Burman designs are even more restrictive than the Rechtschaffner designs as far as the number of experimental trials is concerned. They only support linear regression models. In fact, the PB-designs represent the most parsimonious designs available in MODDE. Because of this property, Plackett-Burman designs are usually used as starting points for so-called super-saturated experimental designs, i.e., designs that allow exploration of more factors than there are experiments in the worksheet. The topic of super-saturated designs is, however, outside the scope of this course book.

Sometimes, weak non-linearity can be encountered in screening. If weak non-linearity is anticipated, the orthogonal L-arrays available in MODDE (L9, L18, L27 and L36) represent appropriate design choices. These designs encode three levels of each factor, except in the L18 case where one factor is a two-level factor. Hence, they can reliably estimate quadratic terms with little confounding from the linear terms. However, L-designs do not tolerate two-factor interactions, since such terms here are closely correlated with the main effects. Thus, although we should remember that a non-linear structure to the data is far better handled by optimization, it can be addressed reasonably well during screening using an L-design.

9 Additional optimization designs for regular regions

9.1 Objective

Chapter 6 was devoted to the CCC and CCF designs, which are widely used for optimization when the experimental region is regular. MODDE supports additional design families that can be used to achieve optimization. These designs are the three-level full factorial designs, Rechtschaffner designs, Box-Behnken designs and Doehlert designs. The aim of this chapter is to outline these additional design families. The Box-Behnken and Doehlert designs are also exemplified by data-sets from specific applications.

9.2 Introduction

The CCF and CCC designs introduced in Chapter 6 are the preferred choices for optimization. One reason for their popularity is that factorial and fractional factorial designs – often used in screening – with small measures can be readily converted into the corresponding CCF or CCC designs. The CCF and CCC designs permit efficient mapping of systems involving two to five factors. However, when the number of factors is six or more, CCF and CCC designs tend to involve a rather large number of experiments, and for that reason the user may wish to resort to other design alternatives.

Here, we discuss alternative design families that can be used for response surface modelling and, ultimately, for optimization. These are: (i) three-level full factorial designs; (ii) Rechtschaffner designs; (iii) Box-Behnken designs; and (iv) Doehlert designs. Our description of these families focuses on their geometric properties and their applicability. The latter two design types are also illustrated using pertinent application data-sets. A comparison with the central composite design family is also provided.

9.3 Three-level full factorial designs

Three-level full factorial designs are simple extensions of the two-level full factorials. Figure 9.1 shows the simplest design in this family, which is the 3^2 full factorial design. This notation represents two factors varied at three levels. We can see that the two factors are examined in a regular array consisting of nine experiments. The three-level full factorial design in three factors, denoted 3^3, is constructed in a similar manner. As shown in Figure 9.2, this design requires 27 experiments. The next design in this family is the 3^4 full factorial design, which requires as many as 81 experiments. Consequently, this design and other three-level factorials with more factors are rarely used, because the number of experiments becomes prohibitively large. In essence, this means that the 3^2 and 3^3 protocols are of most interest. Of course, both these designs benefit from the addition of replicated center-points. Also note that the 3^2 design is equivalent to the CCF design in two factors.

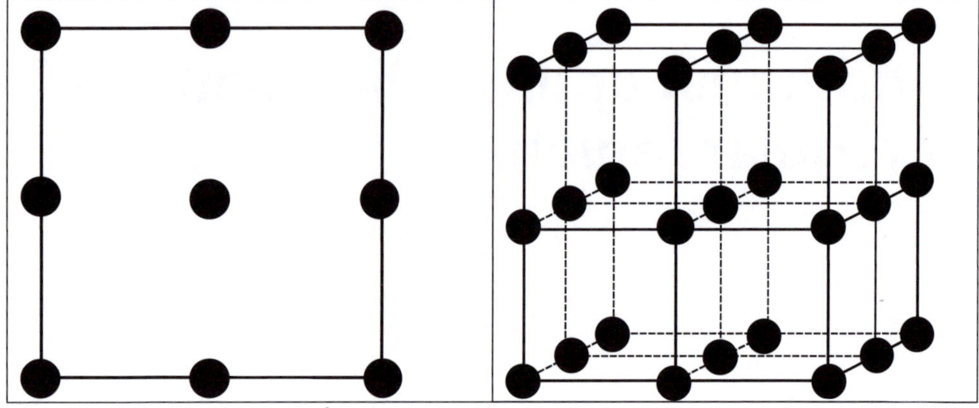

Figure 9.1: *(left) Geometry of the 3^2 full factorial design.*
Figure 9.2: *(right) Geometry of the 3^3 full factorial design.*

9.4 Rechtschaffner designs

Rechtschaffner designs can be used for both screening and optimization. The two-level Rechtschaffner designs used in screening were discussed in Chapter 8. In this section, we turn our attention to the three-level Rechtschaffner designs. The latter are well suited to the RSM objective with six or more factors, since they require fewer runs than the classical CCC or CCF designs. The aim of these designs is, after estimating all two-factor interactions, to eliminate the non-significant ones, and hence recover a number of degrees of freedom for diagnostics and residual analysis.

The required number of runs, N, for three-level Rechtschaffner designs with k factors is:

$N = 1 + k + k + k(k-1)/2$

It is recommended that three center-points should be added to these designs. Thus, the three-level Rechtschaffner design in three factors would require 10+3 runs. Interestingly, this means that four fewer experiments are required than with the corresponding CCC/CCF design. Figure 9.3 shows the orientation of the experiments in this Rechtschaffner design and Figure 9.4 provides the corresponding worksheet. It is clear that, compared to the closest CCC/CCF design, one corner experiment has been dropped plus three axial points.

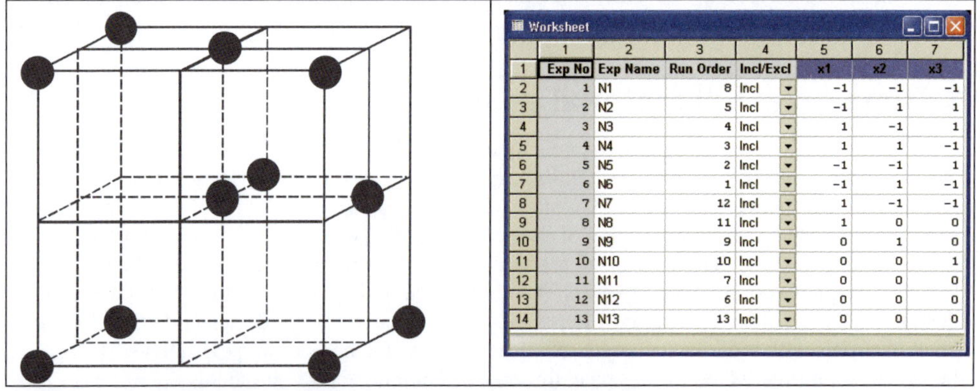

Worksheet

	1	2	3	4	5	6	7
1	Exp No	Exp Name	Run Order	Incl/Excl	x1	x2	x3
2	1	N1	8	Incl	-1	-1	-1
3	2	N2	5	Incl	-1	1	1
4	3	N3	4	Incl	1	-1	1
5	4	N4	3	Incl	1	1	-1
6	5	N5	2	Incl	-1	-1	1
7	6	N6	1	Incl	-1	1	-1
8	7	N7	12	Incl	1	-1	-1
9	8	N8	11	Incl	1	0	0
10	9	N9	9	Incl	0	1	0
11	10	N10	10	Incl	0	0	1
12	11	N11	7	Incl	0	0	0
13	12	N12	6	Incl	0	0	0
14	13	N13	13	Incl	0	0	0

Figure 9.3: *(left) The three-level Rechtschaffner design in three factors.*
Figure 9.4: *(right) Example worksheet of the design given in Figure 9.3.*

9.5 Box-Behnken designs

The Box-Behnken (BB) class of designs represents yet another family employing only three levels per varied factor. These designs are economical in the number of runs required, as exemplified by the three-factor version shown in Figure 9.5. An example worksheet for this design is presented in Figure 9.6. We can see that the experiments are located at the mid-point of each edge of the cube. An optional number of center-points are represented by the solid red dot in the interior of the design. BB-designs are useful when experiments at the corners are undesirable or impossible. Mostly, BB-designs are used when investigating three or four factors. Note that this design type is non-existent for the two-factor situation.

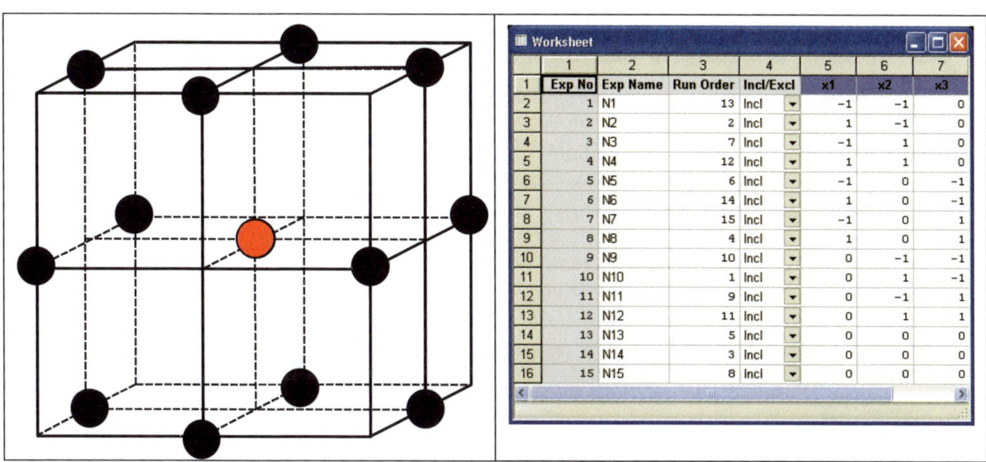

	1	2	3	4	5	6	7
1	Exp No	Exp Name	Run Order	Incl/Excl	x1	x2	x3
2	1	N1	13	Incl	-1	-1	0
3	2	N2	2	Incl	1	-1	0
4	3	N3	7	Incl	-1	1	0
5	4	N4	12	Incl	1	1	0
6	5	N5	6	Incl	-1	0	-1
7	6	N6	14	Incl	1	0	-1
8	7	N7	15	Incl	-1	0	1
9	8	N8	4	Incl	1	0	1
10	9	N9	10	Incl	0	-1	-1
11	10	N10	1	Incl	0	1	-1
12	11	N11	9	Incl	0	-1	1
13	12	N12	11	Incl	0	1	1
14	13	N13	5	Incl	0	0	0
15	14	N14	3	Incl	0	0	0
16	15	N15	8	Incl	0	0	0

Figure 9.5: *(left) The Box-Behnken design in three factors.*
Figure 9.6: *(right) Example worksheet for the design presented in Figure 9.5.*

9.5.1 An example using Box-Behnken design

To illustrate the use of a BB-design, we examine the RSM step of the BTX Detergent application introduced in Chapter 8. Recall that this example deals with the optimization of the factors associated with a detergent admixture preparation (see description in Chapter 8).

9.5.2 Recap of the main findings of the screening step

To accomplish the optimization step, the original investigators drew the following conclusions and made changes based on the outcome of the screening program (*italics indicate factors included in the RSM step*):

- X_1: Desulfation Time (h), original range 1.5–3. As this factor was among the weakest factors, it was fixed at the most economical setting, in this case 3h.
- *X_2: Temperature (°C), original range 105–125.* This factor had a positive effect on both responses. The modified range was 110–130 °C.
- *X_3: Acid concentration (%), original range 96–104*. This factor had a strong effect on residual sulfate, but economic reasons dictated that it be fixed at 98.5%.
- X_4: Reaction Time, (h), original range 6–9. As this factor had comparatively little influence, it was adjusted to a practical level, i.e., 6h.
- X_5: Desulfation agent (-), BTX, Xylene, Toulene. This factor was fixed as xylene. Hence, this qualitative factor disappeared from the RSM step.

- *X_6: Molar ratio (-), original range 0.8–1.2*. The investigators decided to study this conflicting factor more carefully. The range of low and high values was preserved.
- *X_7: Amount of desulfation agent (mol)*, original range 0.18–0.32. Because of conflicting results it was also decided to monitor this factor in more detail. Note, however, that there was a switch from expressing this factor in mol to %. The new range in % was 4.3–13%.

Thus three factors – x_2, x_6 and x_7 – were considered in the optimization phase.

9.5.3 Factors and responses and the corresponding desirabilities

An overview of the three factors examined in the RSM step is presented in Figure 9.7.

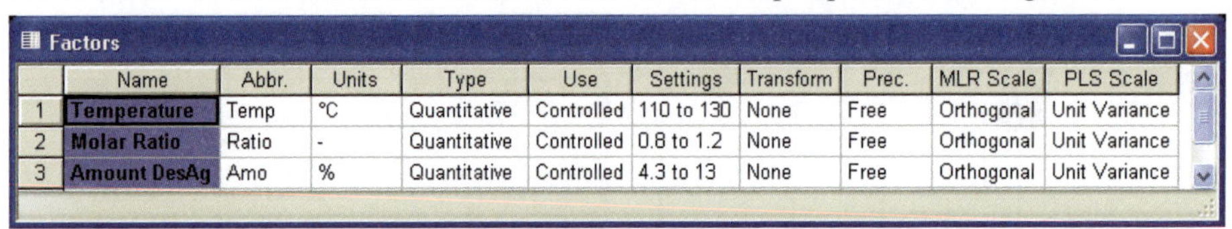
Factors

	Name	Abbr.	Units	Type	Use	Settings	Transform	Prec.	MLR Scale	PLS Scale
1	Temperature	Temp	°C	Quantitative	Controlled	110 to 130	None	Free	Orthogonal	Unit Variance
2	Molar Ratio	Ratio	-	Quantitative	Controlled	0.8 to 1.2	None	Free	Orthogonal	Unit Variance
3	Amount DesAg	Amo	%	Quantitative	Controlled	4.3 to 13	None	Free	Orthogonal	Unit Variance

Figure 9.7: *Factor definitions for the RSM step of the BTX Detergent application.*

To broaden the scope of the investigation, the data-set was expanded to contain five responses (Figure 9.8). Residual sulfate and % sodium benzenesulfonate (SBS) needed to be minimized and the other three responses maximized.

Responses

	Name	Abbr.	Units	Transform	MLR Scale	PLS Scale	Type
1	Residual Sulfate	Res	%	None	None	Unit Variance	Regular
2	SBS%	SBS	%	None	None	Unit Variance	Regular
3	STS%	STS	%	None	None	Unit Variance	Regular
4	SXS%	SXS	%	None	None	Unit Variance	Regular
5	TX Conversion	TXC	%	None	None	Unit Variance	Regular

Figure 9.8: *Response definitions for the RSM step of the BTX Detergent application.*

To facilitate the detection of a possible optimum combination the investigators specified individual desirabilities for the five responses (Figure 9.9).

	Response	Criteria	Weight	Min	Target	Max
1	Residual Sulfate	Minimize	1		1.5	6
2	SBS%	Minimize	1		4	6
3	STS%	Maximize	1	40	55	
4	SXS%	Maximize	1	35	50	
5	TX Conversion	Maximize	1	60	75	

Figure 9.9: *Desirabilities of the five responses.*

It was decided to vary the three factors according to a Box-Behnken design. The worksheet for this design is presented in Figure 9.10. Note that *four* replicated center-points were added to the twelve mid-point runs.

Worksheet

	1	2	3	4	5	6	7	8	9	10	11	12
1	Exp No	Exp Name	Run Order	Incl/Excl	Temperature	Molar Ratio	Amount DesAg	Residual Sulfate	SBS%	STS%	SXS%	TX Conversion
2	1	N1	15	Incl	110	0.8	8.65	7.6	5.1	42.2	44.9	54.7
3	2	N2	9	Incl	130	0.8	8.65	1.5	4.5	46.3	47.9	61.7
4	3	N3	16	Incl	110	1.2	8.65	14.6	4.8	44.6	36.9	76.6
5	4	N4	1	Incl	130	1.2	8.65	9	6.7	46.3	38	81.8
6	5	N5	12	Incl	110	1	4.3	13.3	5.5	45.9	35.2	70
7	6	N6	2	Incl	130	1	4.3	6.9	4.1	52.3	36.5	77.2
8	7	N7	14	Incl	110	1	13	10.4	4.5	42.6	42.6	62.2
9	8	N8	5	Incl	130	1	13	2.6	4.9	46.9	45.6	68
10	9	N9	7	Incl	120	0.8	4.3	5.9	5	47.7	41.3	63.3
11	10	N10	8	Incl	120	1.2	4.3	14.9	5.2	48.6	31.2	81.1
12	11	N11	11	Incl	120	0.8	13	3.2	3.8	43.4	49.7	55.9
13	12	N12	4	Incl	120	1.2	13	8.9	4.6	45.7	40.3	74.1
14	13	N13	6	Incl	120	1	8.65	8.9	4.2	46.1	40.6	69.1
15	14	N14	3	Incl	120	1	8.65	10	4.3	45.2	40.5	69.1
16	15	N15	10	Incl	120	1	8.65	8.7	4.3	44.9	41.6	62.8
17	16	N16	13	Incl	120	1	8.65	9.2	4.3	45	41.1	67

Figure 9.10: *Worksheet for the BB-design.*

9.5.4 Evaluation of raw data

The geometry of the Box-Behnken design is presented below (Figure 9.11). Each experiment is located at the mid-point of an edge. In reality, this means that all runs except the center-point experiment were situated on the surface of a sphere.

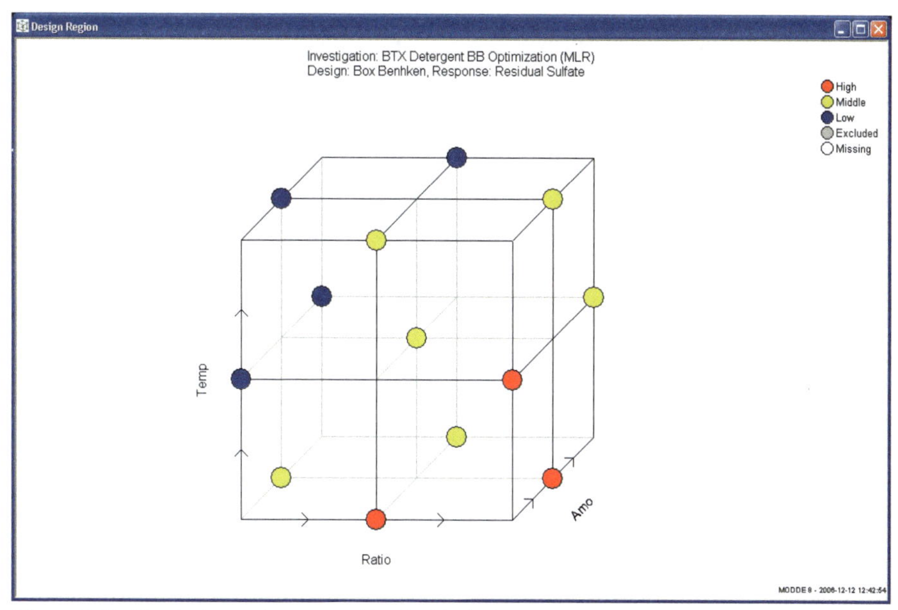

Figure 9.11: *The design region covered by the BB-design.*

Further inspection of the raw data indicates that there are small replicate errors for all responses (no replicate plots are given); two out of the five responses need to be transformed (see histogram plots in Figure 9.12).

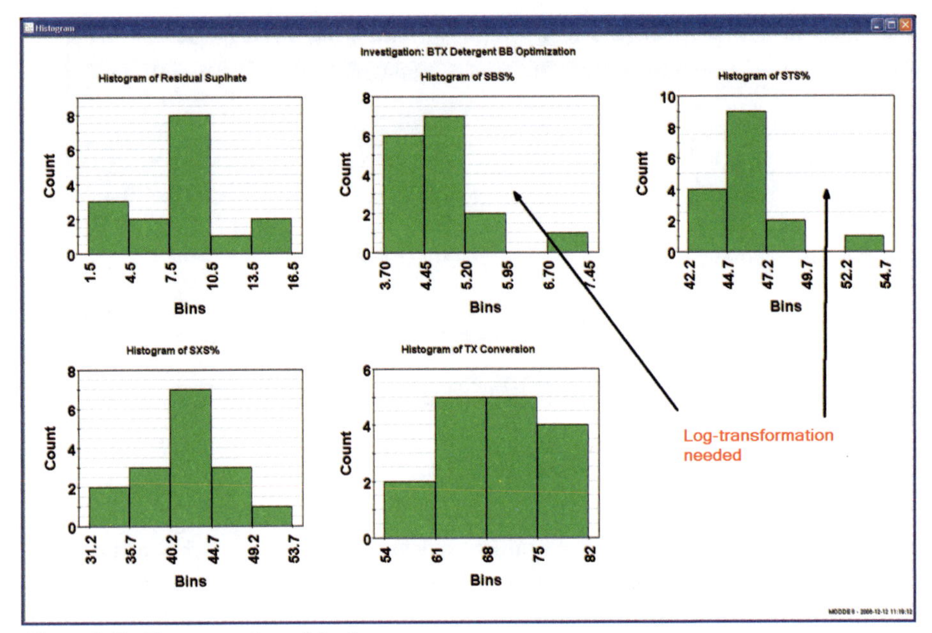

Figure 9.12: *Histogram plots of the five responses.*

The next two histogram plots (Figure 9.13) demonstrate the effect of log-transforming SBS% and STS%. Clearly, the log-transforms are appropriate.

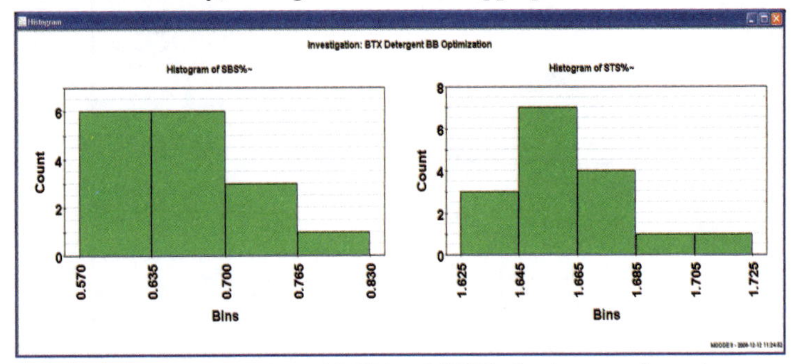

Figure 9.13: *Histogram plots of SBS% and STS% after log-transformation.*

9.5.5 Data analysis and model interpretation

The results of fitting the default quadratic model to the five responses are shown in Figure 9.14. Clearly, the models for Residual sulfate, SXS% and TX conversion are excellent, and the model for STS% is also acceptable. However, the outcome for SBS% is not satisfactory.

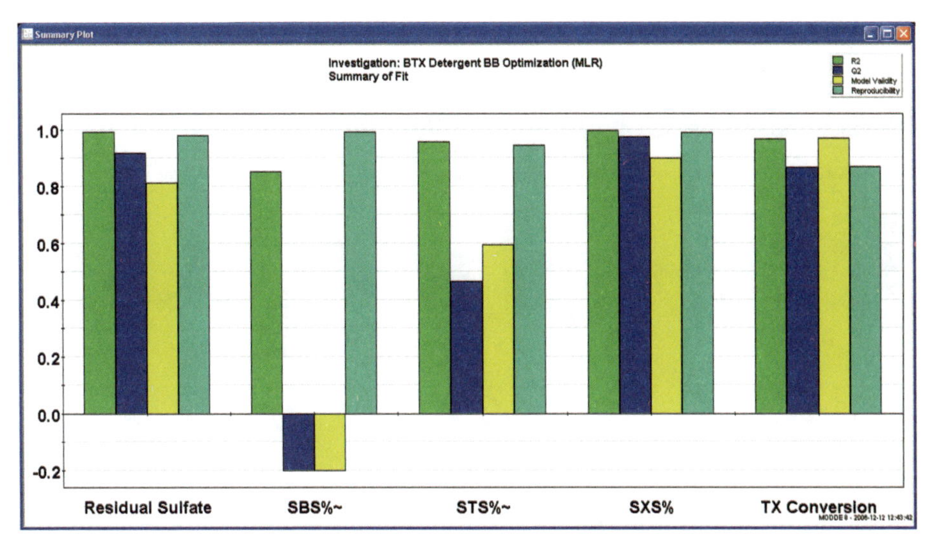

Figure 9.14: *Summary of fit plot of the RSM step of the BTX Detergent application.*

Further plotting of the model residuals – in terms of normal probability plots – suggests that the main reason for the negative Q^2 of SBS% is the existence of two outlying experiments (nos. 6 and 7; Figure 9.15). If either of these two experiments is deleted and the model re-computed, there is a dramatic improvement in the Q^2 of SBS%. However, since these two experiments fit well in the modelling of the other four response variables they were not omitted.

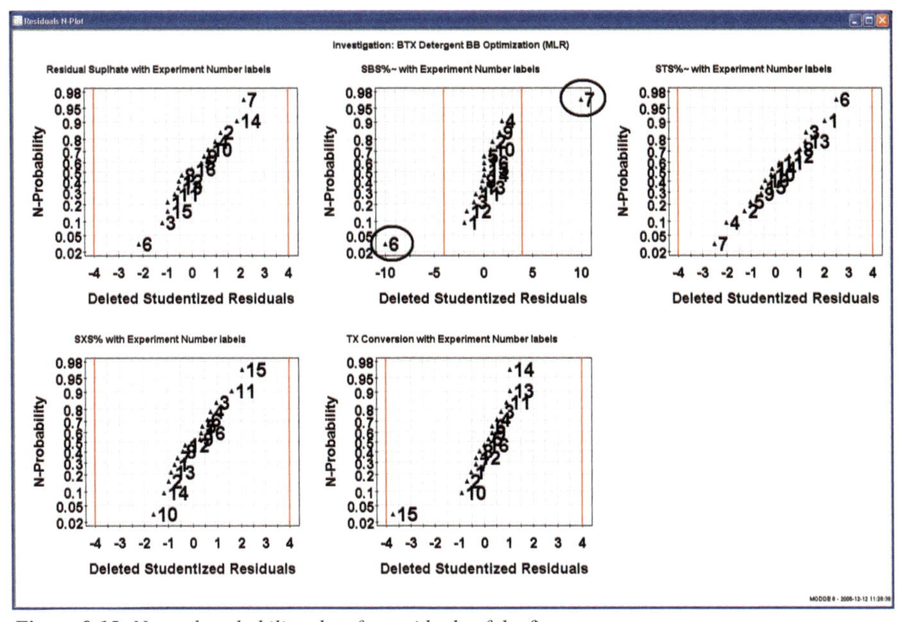

Figure 9.15: *Normal probability plots for residuals of the five responses.*

The regression coefficients show that, in general, linear terms dominate (Figure 9.16). However, Amount has a strong quadratic effect with respect to STS% and Temp with respect to SBS%.

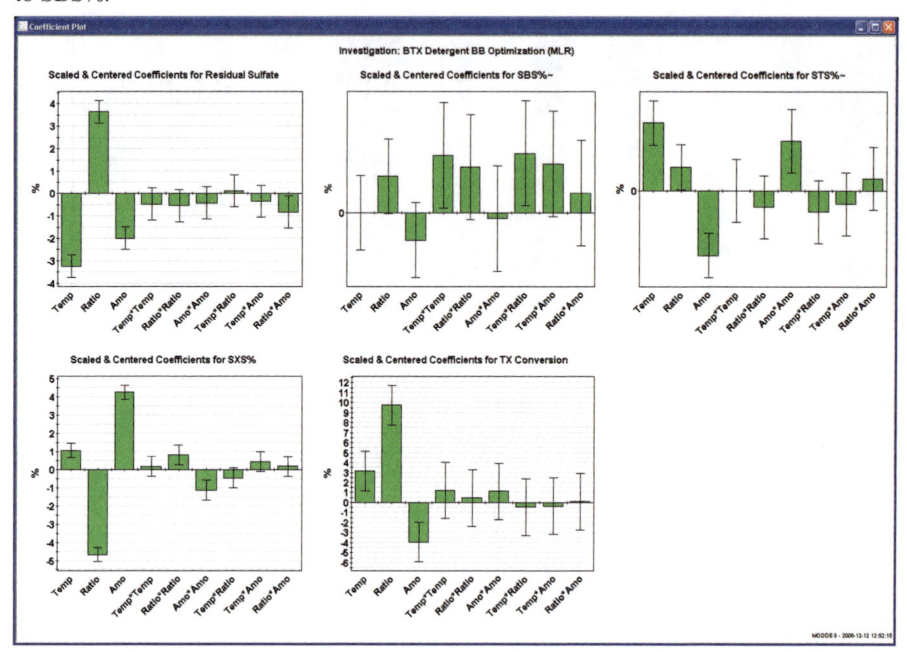

Figure 9.16: *Regression coefficients for the quadratic optimization models of the BTX Detergent application.*

9.5.6 Use of model

To optimize the system, the desirabilities for the responses were entered into the software optimizer. Figure 9.17 shows the optimization results. The best combination had high Temperature and low Molar Ratio and Amount of the Desulfation Agent. With this factor combination (= row 3) the responses are between the target and the concomitant min/max limits for all Y-variables.

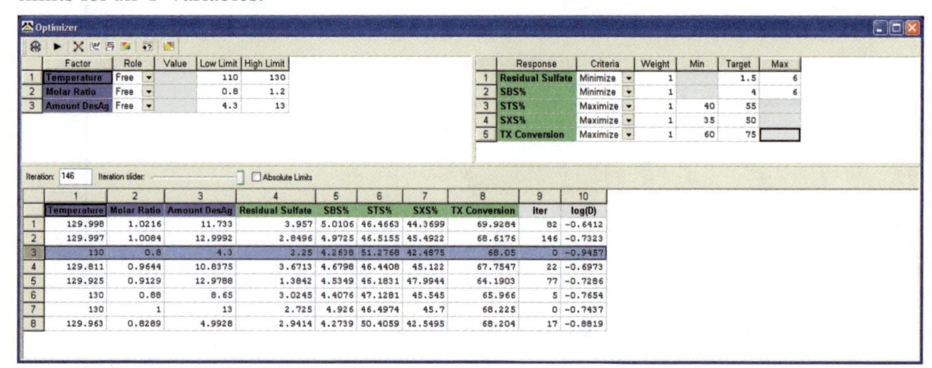

	Factor	Role	Value	Low Limit	High Limit
1	Temperature	Free		110	130
2	Molar Ratio	Free		0.8	1.2
3	Amount DesAg	Free		4.3	13

	Response	Criteria	Weight	Min	Target	Max
1	Residual Sulfate	Minimize	1		1.5	6
2	SBS%	Minimize	1		4	6
3	STS%	Maximize	1	40	55	
4	SXS%	Maximize	1	35	50	
5	TX Conversion	Maximize	1	60	75	

Iteration: 146 Iteration slider: Absolute Limits

	1 Temperature	2 Molar Ratio	3 Amount DesAg	4 Residual Sulfate	5 SBS%	6 STS%	7 SXS%	8 TX Conversion	9 iter	10 log(D)
1	129.998	1.0216	11.733	3.957	5.0106	46.4663	44.3699	69.9284	82	-0.6412
2	129.997	1.0084	12.9992	2.8496	4.9725	46.5155	45.4922	68.6176	146	-0.7323
3	130	0.8	4.3	2.25	4.2638	51.2768	42.4875	68.05	0	-0.9457
4	129.811	0.9644	10.8375	3.6713	4.6798	46.4408	45.122	67.7547	22	-0.6973
5	129.925	0.9129	12.9788	1.3842	4.5349	46.1831	47.9944	64.1903	77	-0.7286
6	130	0.88	8.65	3.0245	4.4076	47.1281	45.545	65.966	5	-0.7654
7	130	1	13	2.725	4.926	46.4974	45.7	68.225	0	-0.7437
8	129.963	0.8289	4.9928	2.9414	4.2739	50.4059	42.5495	68.204	17	-0.8819

Figure 9.17: *Optimization results in the BTX Detergent example.*

The strong message conveyed by the eight simplexes is that it is preferable for Temperature to be high. The Sweet Spot plot presented in Figure 9.18 was constructed around run 3 and by locating Temperature in the high region. This corresponds to visualizing the largest possible Sweet Spot area in any two-factor subspace in the relevant factor domain.

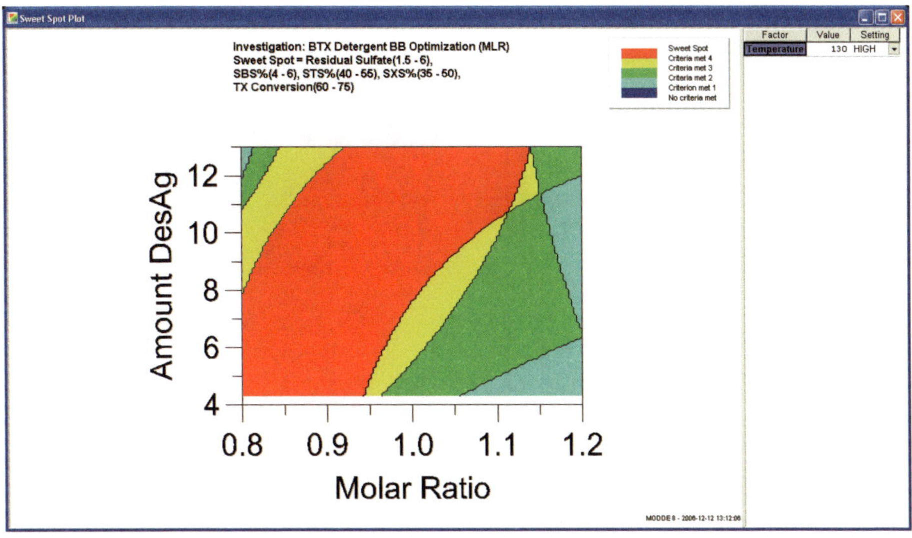

Figure 9.18: *Sweet Spot plot showing the optimal region for the BTX Detergent application.*

9.5.7 Conclusions of the study

By adjusting Temperature to a high level, a large region of satisfactory operability is created. The original authors identified one point in this area for a final verification of the experiment. At the factor combination Temp 130, Molar Ratio 0.82 and Amount 5% they were able to measure the following Y-values (values given on the second line are our model predictions, which are excellent):

	Res.Sulf	SBS%	STS%	SXS%	TX Conv.
Y_{obs}	1.5	3.8	51.1	43.9	67.9
Y_{pred}	2.8±1.3	4.3[-0.8/+1.0]	50.4[-2.0/+2.2]	42.8±1.1	67.6±5.5

The Box-Behnken design works well in this attempt at optimization. BB-designs are usually used when investigating three or four factors. Note that this design type is not possible for the two-factor situation.

9.6 Doehlert designs

The Doehlert design family is the fourth and final group of designs for RSM that will be discussed in this Chapter. Doehlert designs are quadratic RSM designs with some interesting properties, i.e., they can be built upon and extended to other factor intervals. These designs allow the estimation of all main effects, all first-order interactions, and all quadratic effects without any confounding effects.

Geometrically, the Doehlert designs are polyhedrons based on hyper-triangles (simplexes), with a hexagonal structure in the simplest case. This means they have uniform space-filling properties with an equally spaced distribution of points lying on concentric spherical shells. The geometries of the two- and three-factor analogues are presented in Figures 9.19 and 9.20.

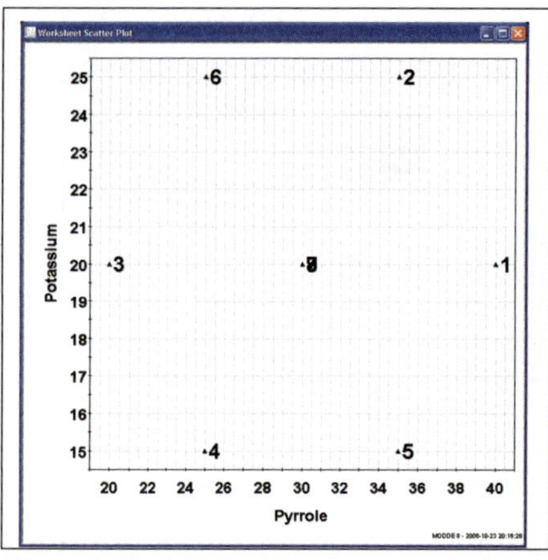

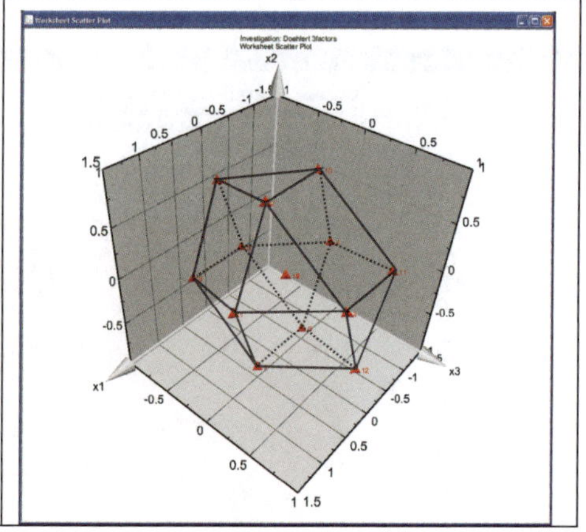

Figure 9.19: (left) Doehlert design in two factors.
Figure 9.20: (right) Doehlert design in three factors.

It is important to note that in a Doehlert design, the number of levels is not the same for all variables. In a two-factor problem, one factor is studied at five levels while the other is explored using only three levels. In fact, in the software, for any Doehlert design, the factor defined last will always be investigated using only three levels. Depending on the complexity of the design selected, the remaining factors will be investigated at five or sometimes even seven levels. This is further discussed in Section 9.6.5.

The Doehlert designs need at least one center-point, but preferably three, in order to include replication. Therefore, the number of runs, N, recommended for a Doehlert design in k factors is:

$$N = k + k^2 + 3$$

This is the number of runs that is considered in the final section (Section 9.8) of this chapter, in which the different design families are compared.

9.6.1 An example using Doehlert design

This example shows how DOE can be used to increase substantially the yield from an organic synthesis. In the manufacture of a compound to be used as an internal standard in gas chromatography, Torbjörn Lundstedt and co-workers needed to optimize the yield of an intermediary product. The intermediary had been synthesized previously, but the yield was poor (25%). Following successful screening, Lundstedt *et al.* were able to reduce the number of relevant factors from five to two. The levels of these two factors, amount of Pyrrole (the starting compound) and amount of Potassium (a catalyst), were fine-tuned using a Doehlert design. The two factors are summarized in Figure 9.21.

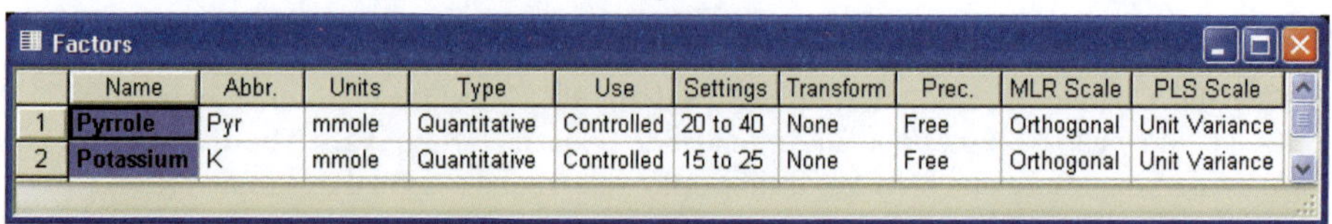

Factors

	Name	Abbr.	Units	Type	Use	Settings	Transform	Prec.	MLR Scale	PLS Scale
1	Pyrrole	Pyr	mmole	Quantitative	Controlled	20 to 40	None	Free	Orthogonal	Unit Variance
2	Potassium	K	mmole	Quantitative	Controlled	15 to 25	None	Free	Orthogonal	Unit Variance

Figure 9.21: Factor definitions for the Pyrrole example.

The data-set contains a single response, the yield of the intermediary product after 30 minutes (Figure 9.22).

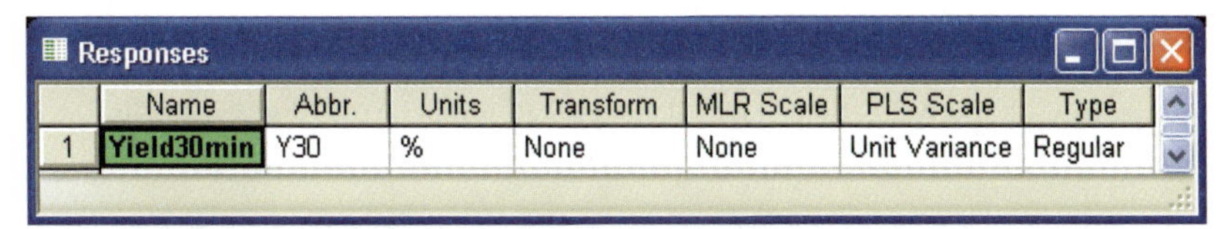

Responses

	Name	Abbr.	Units	Transform	MLR Scale	PLS Scale	Type
1	Yield30min	Y30	%	None	None	Unit Variance	Regular

Figure 9.22: Response definition for the Pyrrole example.

The design used was a Doehlert design in nine experiments. Its associated worksheet is presented in Figure 9.23.

Worksheet

	1	2	3	4	5	6	7
1	Exp No	Exp Name	Run Order	Incl/Excl	Pyrrole	Potassium	Yield30min
2	1	N1	9	Incl	40	20	45.7
3	2	N2	2	Incl	35	25	59.3
4	3	N3	7	Incl	20	20	52.4
5	4	N4	3	Incl	25	15	40.6
6	5	N5	1	Incl	35	15	36.1
7	6	N6	6	Incl	25	25	49.7
8	7	N7	8	Incl	30	20	66.2
9	8	N8	4	Incl	30	20	69.1
10	9	N9	5	Incl	30	20	67.4

Figure 9.23: The experimental worksheet for the Pyrrole application.

9.6.2 Evaluation of raw data

The replicate plot (Figure 9.24) indicates low variability among the replicates. However, the replicated center-points exhibit the highest response values (the highest yields). This phenomenon suggests that a quadratic regression model is needed. The distribution of the response variable (Figure 9.24) is not sufficiently close to a normal distribution. In spite of this, our initial attempt uses the untransformed data, because of the small size of the data set.

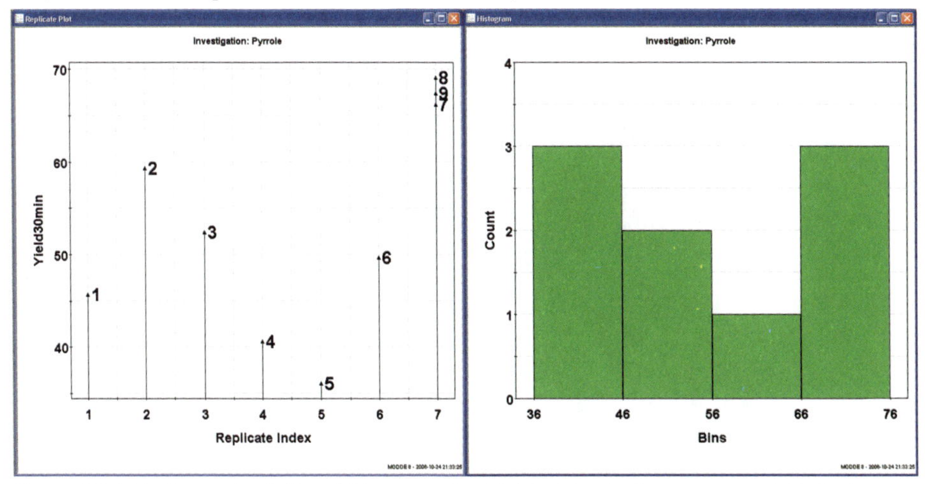

Figure 9.24: Replicate plot (left) and histogram plot (right) of the single response variable in the Pyrrole example.

9.6.3 Data analysis and model interpretation

When fitting the quadratic model to the data, the summary of fit and coefficient plots given below (Figure 9.25) were obtained. The quadratic dependence of the response on the factors is clear.

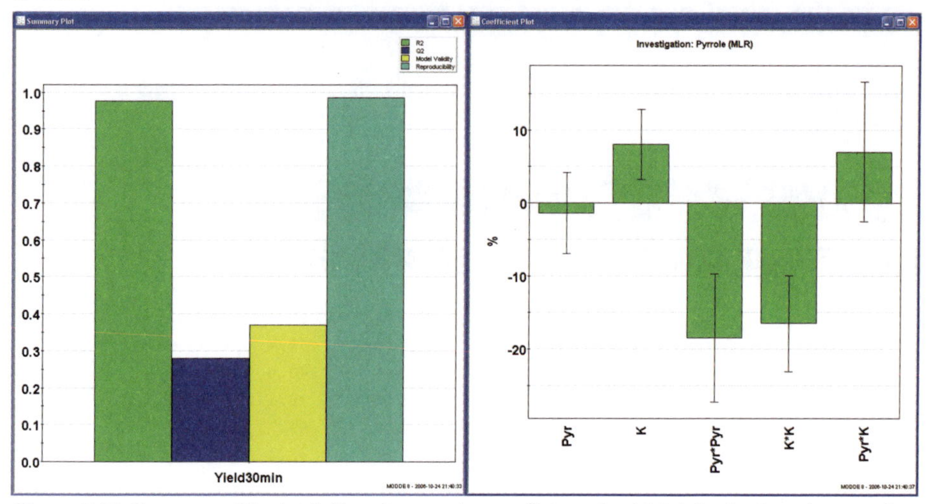

Figure 9.25: Summary of fit plot (left) and regression coefficient plot (right) for the Pyrrole example.

9.6.4 Use of model

In order to determine the factor combination with the predicted peak yield after 30 minutes, we constructed the response surface plot presented in Figure 9.26. It can be seen that an optimum does exist within the modelled factor space. Using the factor combination Pyrrole = 30 mmol and Potassium = 21 mmol, the yield of the product was estimated to be 68.5% ± 5.5%.

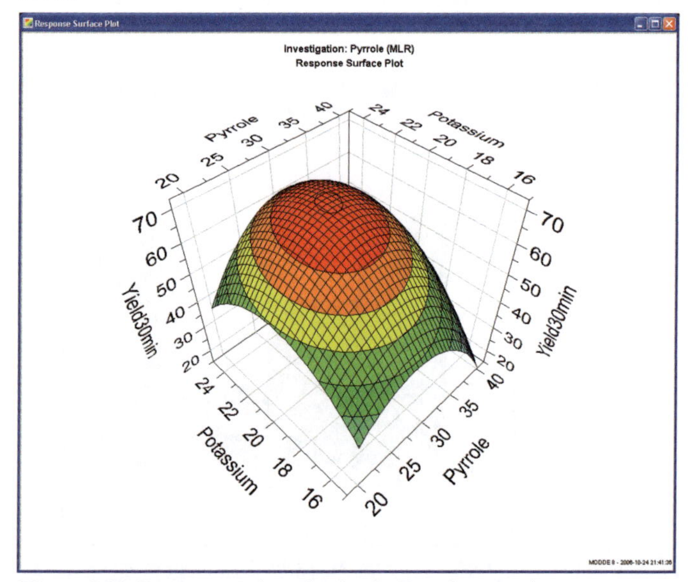

Figure 9.26: Response surface plot for the Pyrrole example.

9.6.5 Conclusions of the study

Using the RSM design described, it was possible to raise the yield substantially above the value originally reported in the literature. The modelled optimal conditions were subsequently verified experimentally.

This example shows that Doehlert designs are well suited to the aims of RSM. In fact, Doehlert designs apply well for up to five or six factors (see summary Table in Figure 9.27). The aim with such RSM designs is to obtain a precise model that can be used for optimization and to enhance our knowledge.

Factors	No levels of factors (first to last defined)	Number of runs	Condition number
2	5, 3	9	3.86
3	5, 5, 3	16	6.06
4	5, 7, 7, 3	23	7.41
5	5, 7, 7, 7, 3	33	9.08
6	5, 7, 7, 7, 7, 3	45	11.26

Figure 9.27: *Overview of Doehlert designs generated for two to six factors. The number of runs includes three center-points.*

The Doehlert designs can be readily expanded to produce upgraded designs, in which new designs are formed by re-using parts of the initial experiments. For instance, the Doehlert design in two factors with six runs + center-points, can be extended to a new design by adding three new experiments (Figure 9.28). Usually, one or two new center-points are added in the new design (i.e., in Figure 9.28 the point furthest to the right in the old design).

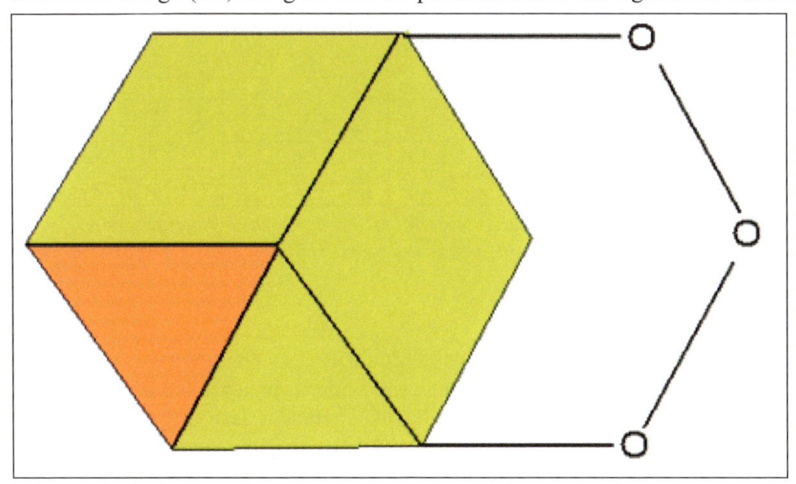

Figure 9.28: *Doehlert designs are extendable and buildable to form new designs.*

9.7 Questions for Chapter 9

- What is a three-level full factorial design?
- Why is a three-level full factorial design an unrealistic choice for four or more factors?
- What is a Rechtschaffner design?

- What is the purpose of a Rechtschaffner design?
- If complementing a three-level Rechtschaffner design, which design type does the complement converge towards?
- What is a Box-Behnken design?
- When is a BB-design appropriate?
- What is a Doehlert design?
- What are the characteristic features of a Doehlert design?
- How many levels are used for the different factors in a Doehlert design?

9.8 Summary and discussion

As discussed in Chapter 6 and in this chapter, there are a number of classical optimization designs that can be used in MODDE. These are useful when the experimental domain exhibits a regular geometry. Hence, it is appropriate to compare them with respect to their performance and the number of runs required. The table presented in Figure 9.29 provides an overview of the number of experiments incorporated into these designs – for tasks involving two to ten factors – together with the condition number of the resulting design and the largest term in the correlation matrix of the corresponding quadratic model.

Number of factors	Central composite designs			Three-level full factorial designs			Rechtschaffner designs			Box-Behnken designs			Doehlert designs		
	Runs	CondNo	Corr	Runs	CondNo	Corr	Runs	CondNo	Corr	Runs	CondNo	Corr	Runs	CondNo	Corr
2	11	3.62	-0.29	12	3.13	0.33	n/a	n/a	n/a	n/a	n/a	n/a	9	3.86	-0.20
3	17	4.99	-0.22	30	3.99	0.17	13	6.09	0.68	15	4.23	-0.07	15	6.06	-0.26
4	27	7.10	-0.20	84	5.39	0.07	18	8.35	0.75	27	5.84	-0.20	23	7.41	0.36
5	29	7.21	-0.13	246	6.73	0.02	24	10.33	0.80	43	7.52	-0.19	33	9.08	0.40
6	47	10.77	-0.14	732	7.92	0.008	31	13.09	0.83	51	8.25	0.37	45	11.26	0.44
7	81	15.19	-0.13	n/i	n/i	n/i	39	16.62	0.86	59	8.80	-0.12	59	13.60	0.47
8	83	16.44	-0.11	n/i	n/i	n/i	48	21.05	0.87	n/a	n/a	n/a	75	16.10	0.51
9	149	20.76	-0.10	n/i	n/i	n/i	58	26.45	0.89	123	12.73	-0.19	93	18.75	0.53
10	151	23.69	-0.09	n/i	n/i	n/i	69	32.87	0.90	163	15.05	-0.23	113	21.54	0.55

Figure 9.29: *A comparison of central composite, three-level full factorial, Rechtschaffner, Box-Behnken and Doehlert RSM designs. For the central composite design family, the statistics represent the CCC design.*
n/ i = not implemented. n/a = not available (non-existent).

The summary table above is valid for default settings in MODDE. As a consequence, in each case the number of runs listed includes three replicated center-points. In general, within each design family, the more factors and runs that are included in a particular design, the higher the condition number. One of the most interesting aspects of a design is the inter-relatedness of the terms in the regression model supported by the design. The largest correlation is listed for each design and its quadratic model. The largest correlation occurs between quadratic terms, with the exception of the Doehlert family, in which it occurs between two-factor interactions.

The performance summary shows that by far the fewest experiments are associated with the Rechtschaffner designs. The number is as low as possible while still enabling the estimation of quadratic models. The idea with the Rechtschaffner designs is to obtain an early indication of non-significant model terms, to remove these and hence recover degrees of freedom for the continued data analysis. The price we pay for the reduction in the number of experiments is somewhat weaker performance statistics. The Rechtschaffner designs have the highest condition numbers of all the designs considered, and the correlations between the quadratic terms are >0.75 for any design involving four or more factors.

An interesting aspect of the three-level Rechtschaffner designs is that they can be complemented by additional axial and corner experiments. If enough experiments are added the three-level Rechtschaffner designs eventually converge towards the closest member of the CCF design family.

Somewhat more experiments are required for the CCC/CCF and BB-designs, but the positive consequence is that this results in better (lower) correlations between model terms and much more manageable condition numbers. In fact, the CCF/CCC alternative is the first recommendation in MODDE for optimization studies involving up to 10 factors, with the exceptions of seven or nine factors, when the BB-design is the default.

One strong argument in favor of the central composite design family concerns information transfer from a screening design to a subsequent optimization design. If, during screening, a two-level fractional factorial design has been used, and one wishes to build on this design at a later stage, it is difficult to establish, for instance, a three-level full factorial or Box-Behnken design. Rather, the designs from the central composite family are more tractable in this respect. Hence, if screening is initiated with a two-level fractional factorial design, optimization is usually conducted using a central composite design.

The Doehlert designs exhibit uniform coverage properties, which can be a desirable feature in some applications. One distinct aspect which differentiates Doehlert designs from other designs is their ability to explore some of the investigated factors at as many as seven levels. They are also well adapted to sequential design development, where part of the old design is re-used to create the next generation design.

Finally, the three-level full factorial design family must be considered. The main drawback of these designs is the exponential increase in the number of experiments. They are only appropriate for two or three factors. Hence, in practice, this family is not really a serious contender when looking for a suitable optimization design.

10 D-optimal design

10.1 Objective

This chapter is devoted to an introduction of D-optimal design. Our aim is to describe the D-optimality approach and to point out when to use it. In the generation of a D-optimal design, the selection of experimental runs can be driven towards fulfilling different criteria. For this purpose, we will explain two common evaluation criteria, one called the G-efficiency and the other the condition number. We will describe a number of applications where D-optimal design is useful.

10.2 D-optimal design

10.2.1 When to use D-optimal design

D-optimal designs are used in screening and optimization, as soon as the researcher needs to create a non-standard design. The D-optimal approach can be used for the following types of problem:

- irregular experimental regions (Figures 10.1 – 10.3),
- multi-level qualitative factors in screening (Figure 10.4)
- optimization designs with qualitative factors (Figure 10.5)
- when the desired number of runs is smaller than required by a classical design (Figures 10.6 – 10.7)
- model updating (Figure 10.8 – 10.9)
- inclusions of already performed experiments (Figures 10.10 – 10.11)
- combined design with process and mixture factors in the same experimental plan (Figure 10.12)

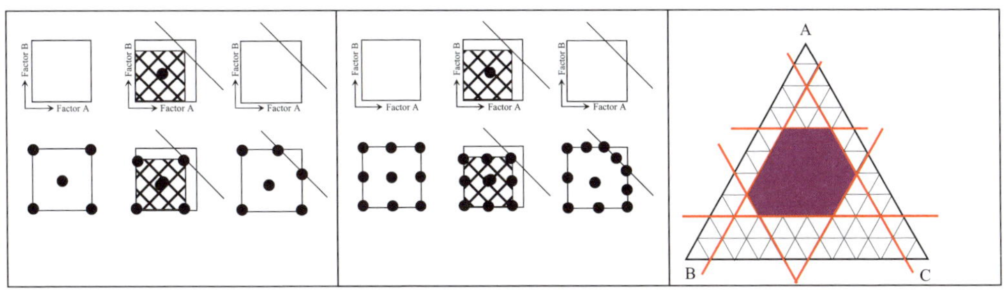

Figure 10.1: *(left) Irregular experimental region – screening.*
Figure 10.2: *(middle) Irregular experimental region – optimization.*
Figure 10.3: *(right) Irregular experimental region – mixture design.*

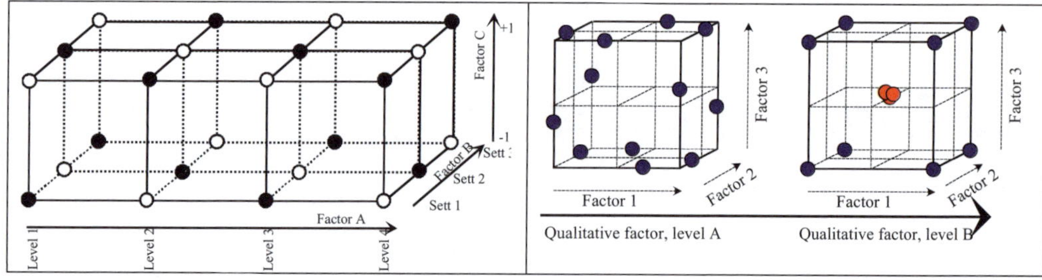

Figure 10.4: *(left) Multi-level qualitative factors in screening.*
Figure 10.5: *(right) Qualitative factors in optimization.*

# Runs	# Center-points	# Total runs	Design type
8	3	11	Frac Fac
11	3	14	D-optimal
12	3	15	PB
16	3	19	Frac Fac

#Factors	CCC/CCF	BB	D-opt
5	26 + 3	40 + 3	26 + 3
6	44 + 3	48 + 3	35 + 3
7	78 + 3	56 + 3	43 + 3

Figure 10.6: *(left) D-optimal design when number of runs is smaller than that for a classical design – screening.*
Figure 10.7: *(right) D-optimal design when number of runs is smaller than that for a classical design – optimization.*

$$y = b_0 + b_1x_1 + b_2x_2 + b_3x_3 + b_{12}x_1x_2 + b_{13}x_1x_3 + b_{23}x_2x_3 + b_{22}x_2^2 + e$$

$$\cancel{b_{11}x_1^2} \quad \cancel{b_{33}x_3^2}$$

$$y = b_0 + b_1x_1 + b_2x_2 + b_3x_3 + b_{12}x_1x_2 + b_{13}x_1x_3 + b_{23}x_2x_3 + b_{11}x_1^2 + b_{22}x_2^2 + b_{33}x_3^2 + b_{111}x_1^3 + e$$

Figure 10.8: *(left) Model updating – screening.*
Figure 10.9: *(right) Model updating – optimization.*

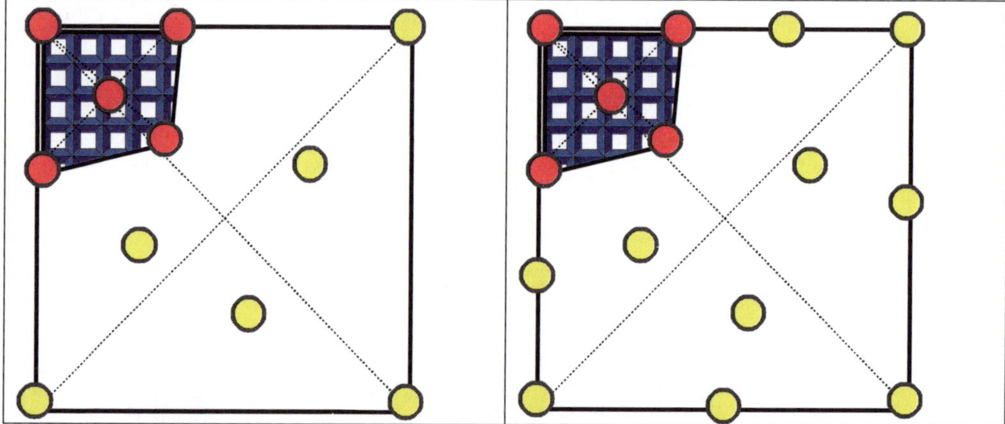

Figure 10.10: *(left) Inclusion of already performed experiments – screening.*
Figure 10.11: *(right) Inclusion of already performed experiments – optimization.*

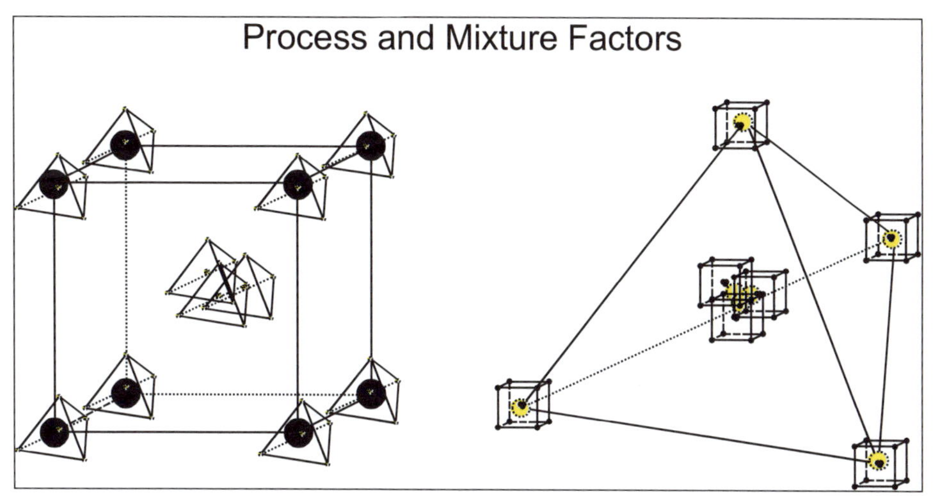

Figure 10.12: *Designs involving process and mixture factors at the same time.*

We will now give an explanation of D-optimal design, and thereafter use D-optimal design to illustrate three of these problem types, namely model updating, multi-level qualitative factors, and combined designs of process and mixture factors.

10.2.2 An introduction

A D-optimal design is a computer generated design, which consists of the *best* subset of experiments selected from a candidate set. The candidate set is the pool of theoretically possible and practically conceivable experiments. To understand what is the *best* design, one must evaluate the selection of experimental runs according to a given criterion. The criterion used here is that the selected design should maximize the determinant of the matrix X'X for a given regression model. This maximization criterion also explains how the letter "D" in D-optimal was derived, from D in determinant. Thus, the D-optimal approach means that N experimental runs are chosen from the candidate set, such that the N trials maximize the determinant of X'X. Equivalently, we can say that the N runs span the largest volume possible in the experimental region. This search for the best subset of experiments is carried out using an automatic search algorithm in MODDE.

In the following, we will present two ways of illustrating the principles underlying D-optimal design. The first example addresses D-optimality from a least squares analysis perspective. The second example is a geometric representation of what we hope to accomplish with a D-optimal selection of experiments.

10.2.3 An algebraic approach to D-optimality

We will now address D-optimality from an algebraic perspective, which is useful for the least squares analysis. Consider Figure 10.13, which shows the design matrix of the 2^2 full factorial design. We call the two factors x_1 and x_2. As seen in Figure 10.14, this design supports an interaction model in the two factors. This model, which is listed in equation form in Figure 10.14, may be rewritten in matrix form, as shown in Figure 10.15. This matrix expression hints at how to solve the problem algebraically, and estimate the model parameters, contained in the b-vector of regression coefficients. It can be shown (but not here) that the best set of coefficients according to least squares is given by the equation $b = (X'X)^{-1}X'y$.

run	x_1	x_2
1	-1	-1
2	1	-1
3	-1	1
4	1	1

Model

$y = b_0 + b_1x_1 + b_2x_2 +$

$+ b_{12}x_1x_2 + e$

Model in matrix form

$y = Xb + e$

➔ $b = (X'X)^{-1}X'y$

Figure 10.13: *(left) Design matrix of the 2^2 design.*
Figure 10.14: *(middle) Model supported by the design shown in Figure 10.13.*
Figure 10.15: *(right) Supported model in matrix notation.*

One crucial step in the estimation of the regression coefficients is the computation of the inverse of the X'X matrix, $(X'X)^{-1}$, and the properties of this matrix are intimately linked to the D-optimality approach. In order to compute the regression coefficients, we must extend the design matrix of Figure 10.13 to become the computational matrix, which is depicted in two versions in Figures 10.16 and 10.17. Figure 10.16 displays which column of the computational matrix is used for the estimation of which regression coefficient, and Figure 10.17 shows a condensed computational matrix devoid of the surrounding information. The matrix of Figure 10.17 is equivalent to the matrix X in Figure 10.15.

run	const.	x1	x2	x1x2
1	1	-1	-1	1
2	1	1	-1	-1
3	1	-1	1	-1
4	1	1	1	1
	b0?	*b1?*	*b2?*	*b12?*

$$X = \begin{pmatrix} 1 & -1 & -1 & 1 \\ 1 & 1 & -1 & -1 \\ 1 & -1 & 1 & -1 \\ 1 & 1 & 1 & 1 \end{pmatrix}$$

$$X' = \begin{pmatrix} 1 & 1 & 1 & 1 \\ -1 & 1 & -1 & 1 \\ -1 & -1 & 1 & 1 \\ 1 & -1 & -1 & 1 \end{pmatrix}$$

Figure 10.16: *(left) Alignment of columns and estimated coefficients.*
Figure 10.17: *(middle) The X-matrix.*
Figure 10.18: *(right) The X'-matrix.*

Figure 10.18 illustrates the transpose of X, that is X', which is needed to compute the X'X matrix displayed in Figure 10.19. Evidently, X'X is a symmetrical matrix with the element four in all entries of the diagonal. The value of four arises because the design has four corner-point experiments. When inverting this matrix, the matrix of Figure 10.20 is the result. The properties of this inverse matrix strongly influence the precision of the estimated regression coefficients. Figure 10.21 reveals that three terms are used in the calculation of confidence intervals for coefficients, namely, (i) the inverse of X'X, (ii) the residual standard deviation of the model, and (iii) the Student's t parameter. From this it follows that the smallest $(X'X)^{-1}$, or, alternatively, the largest X'X is beneficial for the precision of the regression coefficients.

$$(X'X) = \begin{pmatrix} 4 & 0 & 0 & 0 \\ 0 & 4 & 0 & 0 \\ 0 & 0 & 4 & 0 \\ 0 & 0 & 0 & 4 \end{pmatrix}$$

$$(X'X)^{-1} = \begin{pmatrix} 0.25 & 0 & 0 & 0 \\ 0 & 0.25 & 0 & 0 \\ 0 & 0 & 0.25 & 0 \\ 0 & 0 & 0 & 0.25 \end{pmatrix}$$

Precision in b from:

sqrt $[(X'X)^{-1}]$ * RSD * t

➔ smallest $(X'X)^{-1}$

➔ largest X'X

Figure 10.19: *(left) The X'X-matrix.*
Figure 10.20: *(middle) The $(X'X)^{-1}$-matrix.*
Figure 10.21: *(right) How the properties of $(X'X)^{-1}$ influence regression coefficient estimation.*

10.2.4 A geometric approach to D-optimality

We will now use another simple example to illustrate the D-optimal approach from a geometrical perspective. Imagine that our experimental objective is to uncover the effects of two factors, x_1 and x_2, on a response y. Furthermore, both x_1 and x_2 are investigated at three levels, denoted –1, 0 and +1. Then, as seen in Figure 10.22, we have a candidate set comprising nine experiments. Our proposed regression model is that $y = b_0 + b_1x_1 + b_2x_2 + e$. This model contains only three parameters and hence only three experiments are needed. For the sake of simplicity, let us also assume that we wish to do only three experiments. In such a case the question that we have to ask ourselves is *how are we going to select these N = 3 runs from the nine-run candidate set*? Observe that there are (9! / (3!*6!)) = 84 ways of selecting three trials out of nine.

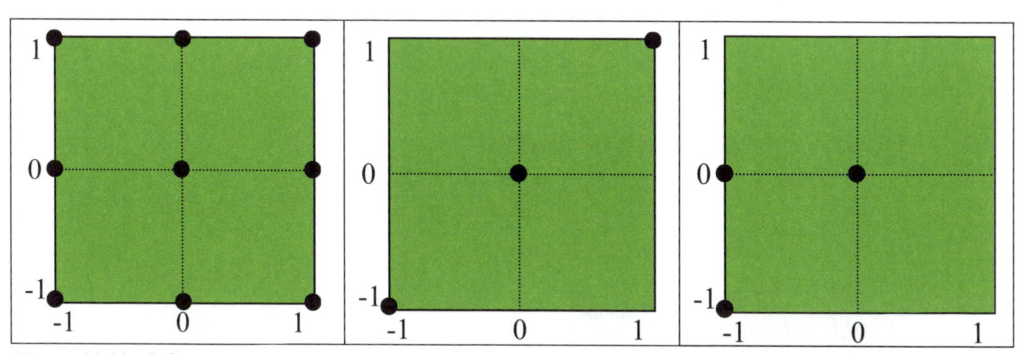

Figure 10.22: *(left) Nine-run candidate set.*
Figure 10.23: *(middle) Three-run selection with determinant 0*
Figure 10.24: *(right) Three-run selection with determinant 1.*

We will employ the D-optimal approach and choose the three runs that maximize the determinant of the relevant X'X matrix. And, for comparative purposes, we provide four supplementary choices of experiments, which are *not* D-optimal. Figures 10.23 – 10.26 show some alternative choices of three experiments. The three runs selected in Figure 10.23 encode an X'X matrix of determinant 0, that is, the experiments span no volume, or, as here, area, of the experimental region. Next, we have Figure 10.24, which illustrates a positioning of experiments corresponding to a determinant of 1. Similarly, Figure 10.25 corresponds to a determinant of 4, Figure 10.26 a determinant of 9, and Figure 10.27 a determinant of 16. We can see that the selection in Figure 10.27 has the best coverage of the experimental region. It also exhibits the largest determinant, and would therefore provide the most precise estimates of the coefficients of the proposed regression model $y = b_0 + b_1x_1 + b_2x_2 + e$. Observe that any selection of three corners out of four, would result in a determinant of 16.

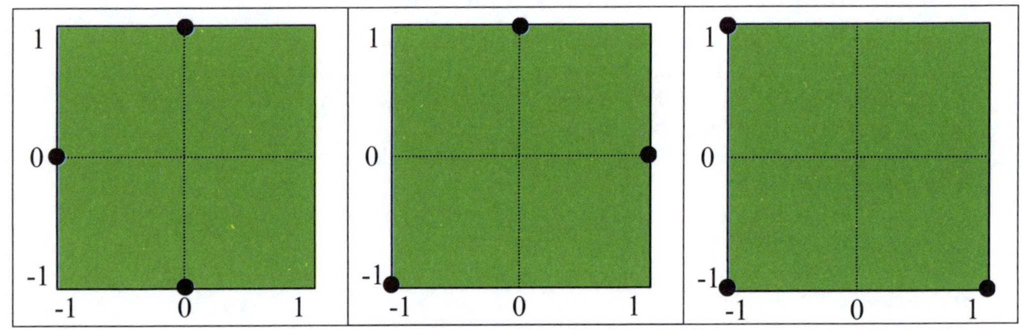

Figure 10.25: *(left) Three-run selection with determinant 4.*
Figure 10.26: *(middle) Three-run selection with determinant 9.*
Figure 10.27: *(right) Three-run selection with determinant 16.*

10.2.5 How to compute a determinant

We will now review how to compute a determinant. As an illustration, we shall consider the selection of three experiments corresponding to a determinant of 4. The distribution of these three experiments are portrayed in Figure 10.28.

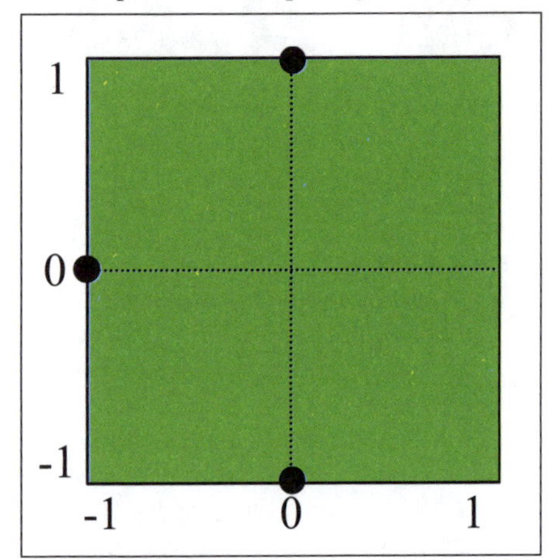

Figure 10.28: *Three-run selection with determinant 4.*

The three selected experiments represent the matrix X shown in Figure 10.29, which, in turn, may be transposed into the matrix given in Figure 10.30. The resulting matrix, X'X, is listed in Figure 10.31, and it is for this matrix that we wish to calculate the determinant.

1	**-1**	**0**	**1**	**1**	**1**	**3**	**-1**	**0**
1	**0**	**-1**	**-1**	**0**	**0**	**-1**	**1**	**0**
1	**0**	**1**	**0**	**-1**	**1**	**0**	**0**	**2**

Figure 10.29: *(left) Matrix X corresponding to the example shown in Figure 10.28.*
Figure 10.30: *(middle) Matrix X' corresponding to the example shown in Figure 10.28.*
Figure 10.31: *(right) Matrix X'X corresponding to the example shown in Figure 10.28.*

In order to compute the determinant of X'X one proceeds as is schematically illustrated in Figures 10.32 and 10.33, and arithmetically outlined in Figure 10.34. Figures 10.32 and

10.33 display a slightly elongated version of X'X. Firstly, the elements of three successive diagonals going from top left to bottom right are multiplied diagonal-wise, and then summed. Secondly, the elements of three successive diagonals running from top right to bottom left are multiplied diagonal-wise, and then summed. Finally, as shown in Figure 10.34, the second sum is then subtracted from the first. The result is the determinant of the X'X matrix.

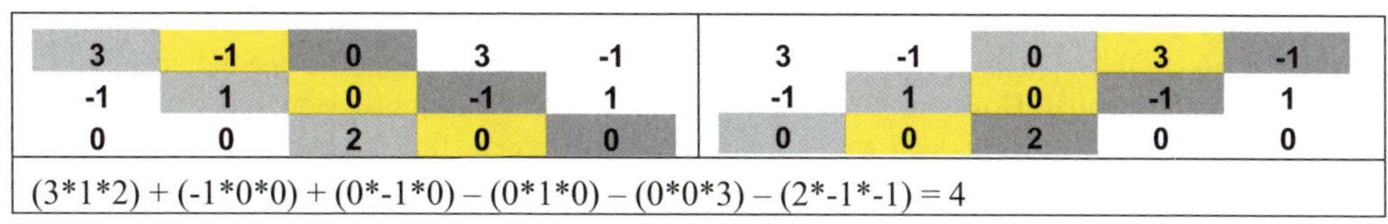

3	-1	0	3	-1	3	-1	0	3	-1
-1	1	0	-1	1	-1	1	0	-1	1
0	0	2	0	0	0	0	2	0	0

(3*1*2) + (-1*0*0) + (0*-1*0) – (0*1*0) – (0*0*3) – (2*-1*-1) = 4

Figure 10.32: *(upper left) Computation of determinant partial sum.*
Figure 10.33: *(upper right) Computation of determinant partial sum.*
Figure 10.34: *(lower part) Computation of determinant.*

10.2.6 Features of the D-optimal approach

The D-optimal approach assumes that the selected regression model is "correct" and "true". This means that experiments will be selected that are maximally suited for the identified model. Usually, such experiments are positioned on the boundary of the experimental region. As a consequence, the D-optimal approach is sensitive to the choice of model. The D-optimal algorithm requires the number of design points to equal the number of parameters of the model, but often the selected number of design runs is greater than the number of parameters. Further, from a user point of view, it is important to know that some of the weaknesses of the D-optimal approach are counterbalanced by the software. For instance, the concept of lack of fit is not recognized by the D-optimal criterion, and to compensate for this MODDE automatically appends three replicates at the overall center-point. Also, the software mitigates against the sensitivity to the selected model by considering *potential terms*. A potential term is a higher order term not included in the original model, but which may be used if necessary. The selected design contains experiments which will support the estimation of potential terms.

10.2.7 Evaluation criteria

The D-optimal selection of experiments may be evaluated by means of several criteria. Here, we will describe two criteria, termed the *condition number* and the *G-efficiency*. The condition number is a measure of the sphericity and symmetry of a D-optimal design. Recall that this diagnostic tool was introduced in Chapter 4. Formally, the condition number is the ratio of the largest and smallest singular values of the X-matrix. Informally, it is roughly equivalent to the ratio of the largest and smallest design diagonals. For an orthogonal design the condition number is 1, and the higher the number the less orthogonality. Thus, if the D-optimal approach is directed towards this evaluation criterion, designs are proposed in which the experiments are positioned as orthogonally as possible. The second evaluation criterion, the G-efficiency, compares the efficiency, or performance, of a D-optimal design to that of a fractional factorial design. G-efficiency is computed as Geff = 100*p/n*d, where p is the number of model terms, n is the number of runs in the design, and d the maximum relative prediction variance across the candidate set. The upper limit of Geff is 100%, which implies that the fractional factorial design was returned by the D-optimal search. We recommend a G-efficiency above 60-70%.

10.3 Model updating

We will now describe three applications of D-optimal designs and start with model updating. Model updating is common after screening, when it is necessary to unconfound two-factor interactions. Consider the laser welding application described in Chapter 5. The four factors varied in this application are listed in Figure 10.35. In the first instance, the researcher conducted a 2^{4-1} fractional factorial design with three center-points, that is, eleven experiments. In the analysis it was found that one two-factor interaction, the one between Po and Sp, was influential. However, because of the low resolution of this design, this two-factor interaction was confounded with another two-factor interaction, namely No*Ro. An escape-route out of this problem was to carry out the fold-over design, enabling resolution of Po*Sp from No*Ro, as well as resolution of the remaining four two-factor interactions.

Factors

	Name	Abbr.	Units	Type	Use	Settings	Transform	Prec.	MLR Scale	PLS Scale
1	Power	Po	kW	Quantitative	Controlled	2.15 to 4.15	None	Free	Orthogonal	Unit Variance
2	Speed	Sp	m/min	Quantitative	Controlled	1.875 to 5	None	Free	Orthogonal	Unit Variance
3	NozzleGas	No	l/min	Quantitative	Controlled	27 to 36	None	Free	Orthogonal	Unit Variance
4	RootGas	Ro	l/min	Quantitative	Controlled	27 to 42	None	Free	Orthogonal	Unit Variance

Figure 10.35: *The four factors of the laser welding application.*

The disadvantage of making the fold-over was a lot of extra experiments. Eleven additional runs were necessary. An alternative approach in this case, less costly in terms of experiments, would be to make a D-optimal design updating, adding only a limited number of extra runs. In theory, only two extra experiments are needed to resolve the Po*Sp and No*Ro two-factor interactions. In practice, however, four additional runs plus one or two center-points to test that the system is stable over time, is a better supplementation principle. Such an updated design is listed in Figure 10.36. Many other D-optimal proposals with similar performance measures exist, and it is up to the user to select a preferred version for the given application.

Worksheet

	1	2	3	4	5	6	7	8	9
1	Exp No	Exp Name	Run Order	Incl/Excl	Power	Speed	NozzleGas	RootGas	Width
2	1	N1	1	Incl	2.15	1.875	27	27	1.02
3	2	N2	2	Incl	4.15	1.875	27	42	1.51
4	3	N3	3	Incl	2.15	5	27	42	0.86
5	4	N4	4	Incl	4.15	5	27	27	0.65
6	5	N5	5	Incl	2.15	1.875	36	42	0.96
7	6	N6	6	Incl	4.15	1.875	36	27	1.5
8	7	N7	7	Incl	2.15	5	36	27	0.52
9	8	N8	8	Incl	4.15	5	36	42	0.69
10	9	N9	9	Incl	3.15	3.4375	31.5	34.5	0.97
11	10	N10	10	Incl	3.15	3.4375	31.5	34.5	1.05
12	11	N11	11	Incl	3.15	3.4375	31.5	34.5	0.95
13	12	C12	15	Incl	4.15	5	36	27	
14	13	C13	13	Incl	4.15	5	27	42	
15	14	C14	17	Incl	2.15	5	27	27	
16	15	C15	16	Incl	2.15	5	36	42	
17	16	C16	12	Incl	3.15	3.4375	31.5	34.5	
18	17	C17	14	Incl	3.15	3.4375	31.5	34.5	

Figure 10.36: *Worksheet from model updating by D-optimal principle.*

10.4 Multi-level qualitative factors

10.4.1 Degrees of freedom

In the second example of this chapter, our aim is to demonstrate that multi-level qualitative factors are well handled by D-optimal design. We shall here consider the Cotton study (more details are given in Chapter 12), where two multi-level qualitative factors, the cotton Variety and the cultivation Center, are explored at four and seven levels respectively. In screening, one way to address these factors is with the full 4*7 factorial design in 28 runs, but this is overly many experiments. Since in screening a linear model is usually adequate, a design with around 15 runs would suffice. An estimate of how many runs might be appropriate can be deduced from the number of degrees of freedom (DF). In this case, a linear model would at least require one DF for the constant, six DF for the linear term of Center, and three DF for the linear term of Variety. This adds up to ten DF. How to calculate these degrees of freedom is explained in Chapter 12. In addition, as a precautionary measure, we recommend that five extra DF be added to facilitate the regression modelling. This is how we ended up with the estimate of 15 DF, that is, that the design should contain around 15 runs. As part of the next step, the D-optimal algorithm is then instructed to search for the best possible design centered around the use of 15 runs.

10.4.2 Evaluation of alternative designs

The D-optimal algorithm was instructed to search for designs with 14, 15 and 16 runs, and for each number of runs five alternative proposals were formulated. Thus, in total, MODDE generated 15 D-optimal designs. It turned out that all five 14 run designs had the same G-efficiency, 57.1%. Similarly, among the 15 run designs all five had a G-efficiency of 54.5%, and among the 16 run designs the result was always 60.2%. Consequently, the ranking according to G-efficiency is that a 16 run design is the best choice, followed by the 14 and 15 run counterparts. The identical trend holds true when the condition number is used as the evaluation criterion. However, in addition to this technical judgement, it is often favorable to obtain a geometric appraisal of the situation. In particular, we think here of the concept of *balancing*. Balancing implies that each level of a qualitative factor is explored with the same number of runs. This is a feature which is not absolutely necessary, but is desirable since it makes the design easier to understand. Balancing may be checked manually.

In the case of varying number of levels among qualitative factors, balancing is typically sought for the qualitative factor with most levels. We will now explore the balancing in the designs generated with 14 and 16 runs. Figures 10.37 and 10.38 display the arrangement of experiments of two alternative 14 run proposals. With 14 runs in the design, each level of Center is tested twice, and each level of Variety either three of four times.

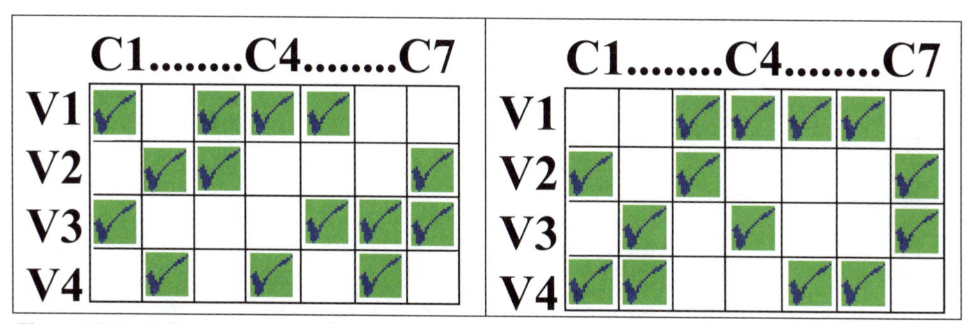

Figure 10.37: *(left) Arrangement of experiments of a 14 run D-optimal design.*
Figure 10.38: *(right) Alternative appearance of a 14 run design.*

In an analogous fashion, Figures 10.39 and 10.40 display two alternative 16 run designs. Here, the balancing takes place with the four-level qualitative factor in focus. Apparently, each level of Variety is investigated using four experiments and each level of Center in either two or three experiments.

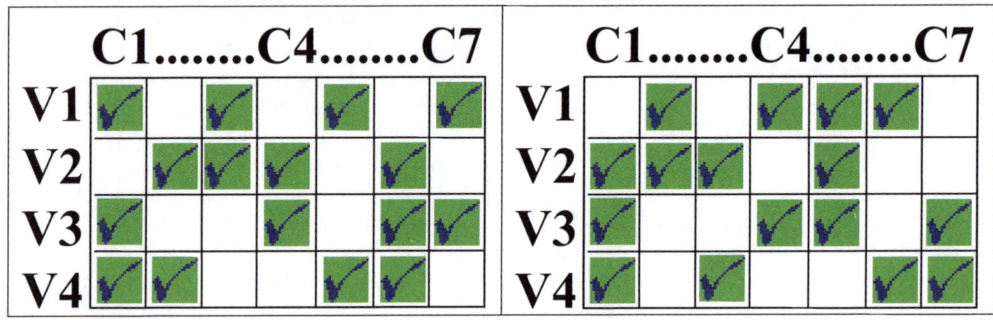

Figure 10.39: *(left) Arrangement of experiments of a 16 run D-optimal design.*
Figure 10.40: *(right) Alternative appearance of a 16 run design.*

In conclusion, in the Cotton application we would recommend the use of one of the 14 run designs, as such a design is balanced with respect to the seven-level factor Center. This kind of design is comparable to any one of the D-optimal designs comprising 16 experiments. It is also recommended that two replicated experiments be included, and these should be positioned in different rows and columns. According to Figure 10.37, a possible combination of replicates might be V1/C1 and V2/C2, but not the combination V1/C1 and V1/C3.

10.5 Combining process and mixture factors in the same design

Now, as a third example, our intention is to outline how process and mixture factors can be combined in one design. For this purpose, we will study an optimization of bubble formation. This application has little commercial value, but is instructive as it reflects one of the most complex types of problem to which D-optimal design is applied.

Children like to blow bubbles, but dislike bubbles which burst too rapidly. We decided to use the mixture design approach for investigating which factors may affect bubble formation and lifetime. We browsed the Internet looking for a suitable bubble mixture recipe, which we could use as a starting reference mixture. Then this recipe was modified using mixture design, and bubbles were blown for each new mixture. The four ingredients, mixture factors, are overviewed in Figure 10.41. These are dish-washing liquid 1 (Skona, ICA, Sweden), dish-washing liquid 2 (Neutral, ADACO, Sweden), tap water (Umeå, Sweden) and glycerol (Apotekets, Sweden).

Factors

	Name	Abbr.	Units	Type	Use	Settings	Transform	Prec.	MLR Scale	PLS Scale
1	Temp	Te		Quantitative	Controlled	7 to 21	None	Free	Orthogonal	Unit Variance
2	Time	Ti		Quantitative	Controlled	1 to 25	None	Free	Orthogonal	Unit Variance
3	DWL1	DW1	Fraction	Formulation	Controlled	0 to 0.4	None	Free	None	Unit Variance
4	DWL2	DW2	Fraction	Formulation	Controlled	0 to 0.4	None	Free	None	Unit Variance
5	Water	Wa	Fraction	Formulation	Controlled	0.4 to 0.9	None	Free	None	Unit Variance
6	Glycerol	Gly	Fraction	Formulation	Controlled	0 to 0.2	None	Free	None	Unit Variance

Figure 10.41: *The six factors of the bubble formation example.*

We can see that both dish-washing liquids are allowed to vary between 0 and 0.4. In addition to these restrictions, a relational constraint was specified stating that the sum of the dish-washing liquids ought to be between 0.2 and 0.5. This extra constraint is given in Figure 10.42.

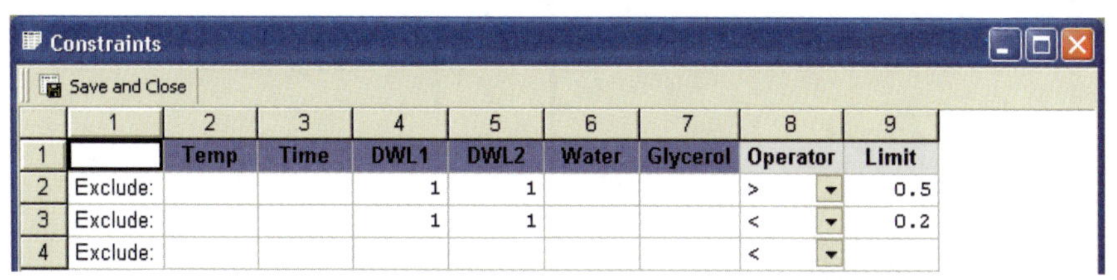

	1	2	3	4	5	6	7	8	9
1		Temp	Time	DWL1	DWL2	Water	Glycerol	Operator	Limit
2	Exclude:			1	1			>	0.5
3	Exclude:			1	1			<	0.2
4	Exclude:							<	

Figure 10.42: *The relational constraint imposed on the two dish-washing liquid factors.*

Besides the four mixture factors, the influences of two "process" factors, storage temperature and mixture settling time, were investigated. As seen in Figure 10.43, the measured response was the lifetime of bubbles produced with a child's bubble wand. The time until bursting was measured for bubbles of 4 – 5 cm diameter.

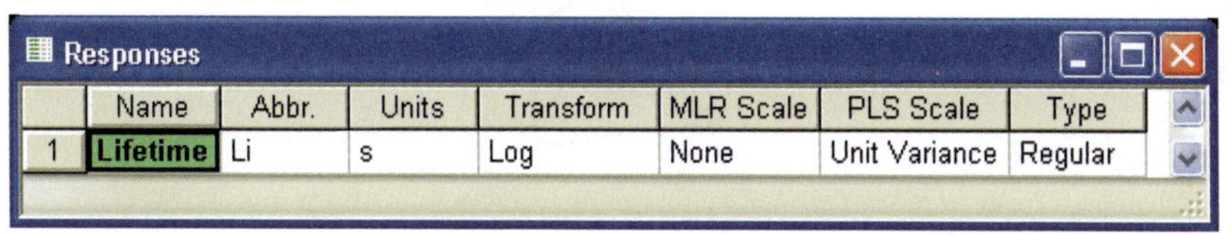

	Name	Abbr.	Units	Transform	MLR Scale	PLS Scale	Type
1	Lifetime	Li	s	Log	None	Unit Variance	Regular

Figure 10.43: *The measured response in the bubble formation example.*

10.6 Bubble application

10.6.1 How many runs are necessary?

The experimental objective selected was screening and the model was one with linear and interaction terms. Interactions were allowed among the process factors, between the process and mixture factors, but not among the mixture factors themselves. In order to understand the complexity of the problem, we may look at the schematic in Figure 10.44. We may view this problem as one based on a two-level factorial design, on which is superimposed an array of mixture tetrahedrons. Observe that in the drawing the tetrahedral structures are depicted as symmetrical, whereas in reality they are mutilated.

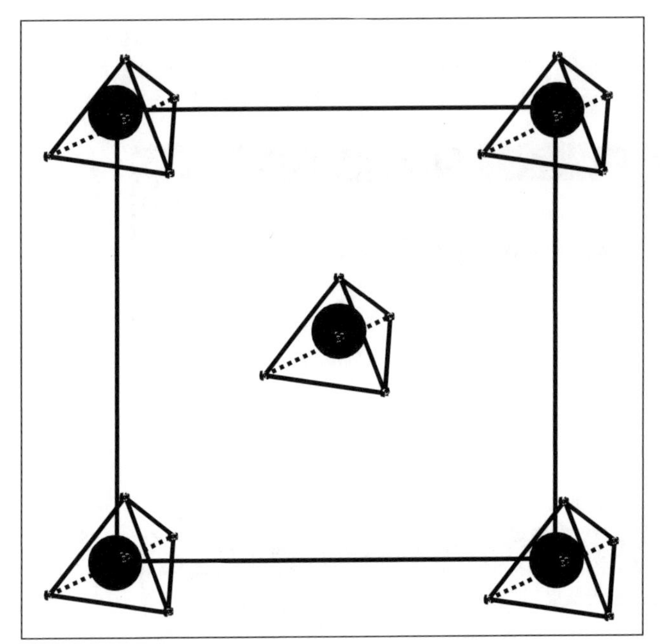

Figure 10.44: *Geometric representation of the bubble formation problem.*

In this case, the candidate set contains 169 experimental points, allocated as 48 extreme vertices, 48 edge points, 72 centroids of high-dimensional surfaces, and the overall centroid. This is a small candidate set and it was left intact by us. With six factors and a screening objective, the lead number of experiments N = 20 was suggested by MODDE. In the computation of this lead number, the number of degrees of freedom (DF) of the proposed model was taken into account. The necessary DF are calculated as follows: (a) 1 DF for the constant term, (b) 2 DF for the linear process terms, (c) 3 DF for the linear mixture terms, (d) 1 DF for the process*process interaction, and (e) 6 (2*3) DF for the process*mixture interactions. This makes 13 DF, and by adding 5 extra experiments to help the regression analysis we end up with N = 18. In addition, two supplementary runs were added to handle the additional complexity introduced by the linear constraint, thus giving N = 20 as the final lead number of experiments. Note that no replicates are included in this N = 20 estimate.

10.6.2 Generation of alternative D-optimal designs

In the generation of a D-optimal design an element of randomness is normally incorporated. In principle, this means that the D-optimal search is initiated from different starting points within the experimental region. Each selection may reach a local optimum, and the element of randomness means that one may get slightly different design proposals with the same number of runs.

Furthermore, the success of the D-optimality search also depends on the size of N. The recommended procedure would be to identify a lead number of design runs, N (here 20), selected by considering the number of model parameters plus some extra degrees of freedom, and then generate designs with N ± 4 runs and, say, five alternative versions for each level of N ± 4 runs. Adhering to this procedure, we generated 45 alternative D-optimal designs ranging from N = 16 to N = 24 and with five versions for each N. The G-efficiencies of these designs are given in Figure 10.45. The best resulting design was one with N = 16 showing a G-efficiency of 76.1% and a condition number of 2.7. Its summary statistics are given in Figure 10.46, and they indicate a very good design.

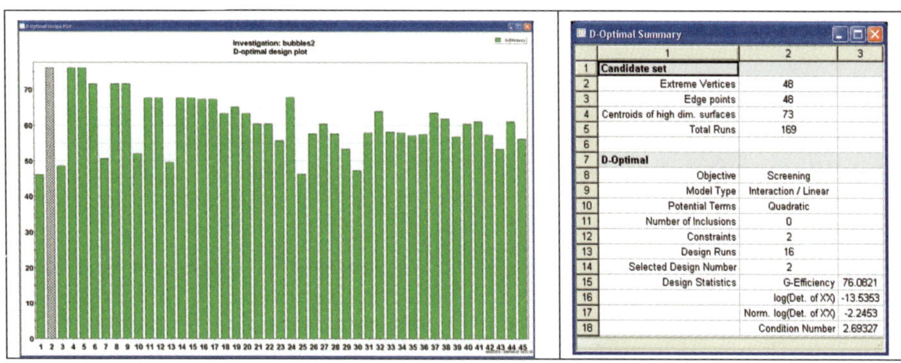

D-Optimal Summary

	1	2	3
1	**Candidate set**		
2	Extreme Vertices	48	
3	Edge points	48	
4	Centroids of high dim. surfaces	73	
5	Total Runs	169	
6			
7	**D-Optimal**		
8	Objective	Screening	
9	Model Type	Interaction / Linear	
10	Potential Terms	Quadratic	
11	Number of Inclusions	0	
12	Constraints	2	
13	Design Runs	16	
14	Selected Design Number	2	
15	Design Statistics	G-Efficiency	76.0821
16		log(Det. of X'X)	-13.5353
17		Norm. log(Det. of X'X)	-2.2453
18		Condition Number	2.69327

Figure 10.45: *(left) Evaluation of proposed D-optimal designs (G-efficiency).*
Figure 10.46: *(right) Summary statistics of the selected D-optimal design.*

10.6.3 The experimental results

To the 16 experiments, we added two series of four replicates, making the entire screening design comprise 16 + 8 = 24 experiments. The resulting worksheet is listed in Figure 10.47. In retrospect, it may seem that overly many replicates were carried out, but at the time of experimenting we were unsure of the experimental reproducibility obtainable in a non-hi-tech kitchen. Since we wanted to obtain good insight into this source of uncertainty, we conducted 2*4 replicates. In other situations resembling ours, we would recommend 2*2 replicates.

Worksheet

	1	2	3	4	5	6	7	8	9	10	11
1	Exp No	Exp Name	Run Order	Incl/Excl	Temp	Time	dwliq2	dwliq1	water	glycerol	resp1
2	1	N1	15	Incl	7	1	0.4	0	0.4	0.2	139
3	2	N2	3	Incl	7	1	0.1	0.4	0.5	0	19
4	3	N3	22	Incl	7	1	0.2	0	0.8	0	14
5	4	N4	7	Incl	7	1	0	0.2	0.6	0.2	60
6	5	N5	4	Incl	21	1	0	0.4	0.4	0.2	208
7	6	N6	18	Incl	21	1	0.4	0.1	0.5	0	15
8	7	N7	17	Incl	21	1	0	0.2	0.8	0	11
9	8	N8	16	Incl	21	1	0.2	0	0.6	0.2	35
10	9	N9	2	Incl	7	25	0	0.4	0.4	0.2	362
11	10	N10	19	Incl	7	25	0.4	0.1	0.5	0	25
12	11	N11	14	Incl	7	25	0	0.2	0.8	0	26
13	12	N12	12	Incl	7	25	0.2	0	0.6	0.2	52
14	13	N13	21	Incl	21	25	0.4	0	0.4	0.2	213
15	14	N14	20	Incl	21	25	0.1	0.4	0.5	0	40
16	15	N15	13	Incl	21	25	0.2	0	0.8	0	33
17	16	N16	5	Incl	21	25	0	0.2	0.6	0.2	74
18	17	N17	10	Incl	7	13	0.2	0.2	0.5	0.1	94
19	18	N18	6	Incl	7	13	0.2	0.2	0.5	0.1	78
20	19	N19	11	Incl	7	13	0.2	0.2	0.5	0.1	132
21	20	N20	23	Incl	7	13	0.2	0.2	0.5	0.1	117
22	21	N21	1	Incl	21	13	0.2	0.2	0.5	0.1	61
23	22	N22	8	Incl	21	13	0.2	0.2	0.5	0.1	54
24	23	N23	9	Incl	21	13	0.2	0.2	0.5	0.1	77
25	24	N24	24	Incl	21	13	0.2	0.2	0.5	0.1	43

Figure 10.47: *Experimental worksheet of the bubble formation example.*

When we consider the resulting worksheet, we should remember its very special feature of simultaneously encoding changes in both process and mixture factors. It can be seen that the measured lifetime values vary between 11 seconds and 362 seconds (6.02 min), that is, a span of 1.5 orders of magnitude. Undoubtedly, this large difference must be attributed to the use of a sound D-optimal design. When the bubble-data obtained were further analyzed with regression analysis, a direction in which the bubble lifetime would be improved was discovered. Subsequently, a new D-optimal *optimization* design was laid out, and with this design a bubble lifetime as high as 22.28 min was scored. The key to prolonging the lifetime was to substantially increase the amount of glycerol.

Based on these bubble experiments, it was possible to conclude that the sequential strategy of first carrying out a screening study, followed by an optimization study, was very fruitful. Also, the D-optimal approach worked excellently for combining mixture and process factors.

10.7 Questions for Chapter 10

- When is a D-optimal design applicable?
- What is the meaning of the "D" in the name D-optimal?
- Which property of the X'X matrix does the determinant reflect?
- What is the basic assumption of the D-optimal approach concerning the selected model?
- What is a potential term?
- What is the condition number?
- What is the G-efficiency?
- What is the recommended procedure for establishing a D-optimal design?

10.8 Summary and discussion

In this chapter, we have provided an introduction to D-optimal design. Initially, the algebraic and geometrical aspects of the D-optimality criterion were introduced. We described how to use two criteria, the condition number and the G-efficiency, for evaluating a D-optimal search. Our treatment of D-optimal designs provided an overview of when D-optimal design is applicable, and three of these situations were discussed more thoroughly. These cases were (i) model updating, (ii) multi-level qualitative factors, and (iii) combined design of process and mixture factors.

11 Mixture design

11.1 Objective

Many commercially available chemical products are manufactured and sold as mixtures. The word mixture in this context designates a blend in which all ingredients sum to 100%. Examples of such mixtures are pharmaceuticals, gasolines, plastics, paints, and many types of food and dairy products. In most mixtures, the number of major components, sometimes also called constituents or ingredients, can vary between three and eight. Because mixtures may contain large numbers of components, it is necessary to use mixture design for finding mixture compositions with improved properties. Mixture design is a special topic within DOE, and many of the basic DOE concepts and modelling principles discussed so far, are applicable for design of mixtures. However, there are also some fundamental differences. It is the objective of this chapter to provide a short introduction to mixture design.

11.2 A working strategy for mixture design

The characteristic feature of a mixture is that the sum of all its ingredients is 100%. Firstly, this means that these components, mixture factors, cannot be manipulated completely independently of one another. Secondly, it means that their proportions must all lie somewhere between 0 and 1. As an example, we shall consider a tablet manufacturing application. The data set is taken from P.J. Waaler, Acta Pharm Nord 4: 9-16, 1992. Three constituents, microcrystalline cellulose, lactose and dicalciumphosphate-dihydrate, were mixed according to a ten-run mixture design. These factors will be referred to as *cellulose*, *lactose*, and *phosphate*. For each tablet produced, the release rate of the active substance (the active pharmaceutical ingredient, the API) was monitored. In the sections to come, we shall use this tablet example as an illustration of a proposed working strategy for mixture design. This strategy is influenced by the DOE framework put forward in this course book.

11.2.1 Step 1: Definition of factors and responses

The first step of the mixture design strategy corresponds to the *definition of factors and responses*. For each factor low and high levels must be defined, as with regular process designs. In the context of mixture design, the term *bound* is used more frequently than *level*, and the lower and upper bounds of each mixture factor are given as proportions. As seen in Figure 11.1, all three mixture factors have the lower bound equal to 0 and the upper bound equal to 1. It is important to check the consistency of these bounds, because some combinations of bounds might be incompatible. For instance, proposing the alternative bounds cellulose (0/1), lactose (0/1) and phosphate (0.1/1) would not work in reality. The adjustments cellulose (0/0.9) and lactose (0/0.9) are necessary to make this mixture system realistic. The detection and fixing of inconsistent bounds is automatically taken care of by MODDE. Further, as part of the first step, the responses of interest have to be specified. We can see from Figure 11.2 that the response of interest is the release rate of the active substance, which is to be maximized.

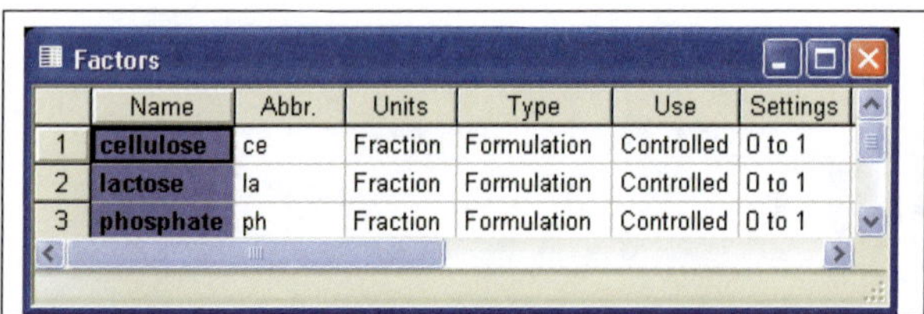

Factors

	Name	Abbr.	Units	Type	Use	Settings
1	cellulose	ce	Fraction	Formulation	Controlled	0 to 1
2	lactose	la	Fraction	Formulation	Controlled	0 to 1
3	phosphate	ph	Fraction	Formulation	Controlled	0 to 1

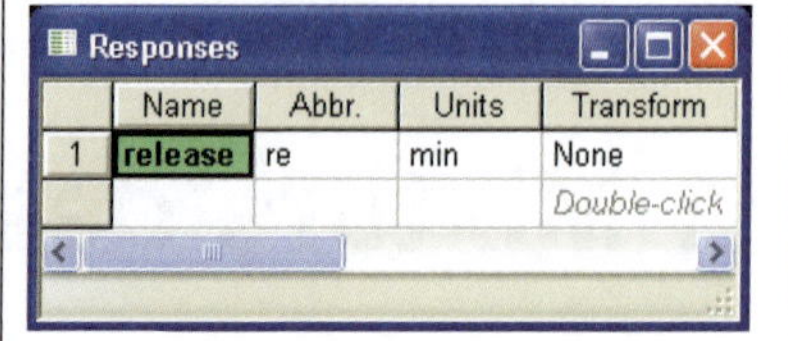

Responses

	Name	Abbr.	Units	Transform
1	release	re	min	None
				Double-click

Figure 11.1: *(left) Factors of the tablet manufacturing example.*
Figure 11.2: *(right) Response of the tablet manufacturing example.*

11.2.2 Step 2: Selection of experimental objective and mixture model

The second step is the selection of experimental objective and mixture model. We have previously learnt that the choice of experimental objective, that is, screening, optimization, or robustness testing, has a decisive impact on the number of experimental trials. The experimental objective selected here was optimization. This objective needs the quadratic regression model:

$$y = \beta_0 + \beta_1 x_1 + \beta_2 x_2 + \beta_3 x_3 + \beta_{11} x_1^2 + \beta_{22} x_2^2 + \beta_{33} x_3^2 + \beta_{12} x_1 x_2 + \beta_{13} x_1 x_3 + \beta_{23} x_2 x_3 + \varepsilon$$

with the overall external constraint $x_1 + x_2 + x_3 = 1$, and where x_k are the fractions of the three mixture ingredients, expressed in the 0 - 1 range.

The overall mixture constraint introduces a closure which produces some problems for the regression analysis, and some mathematical reparametrization is needed to alleviate the impact of this constraint. One approach is to calculate the regression coefficients with constraints imposed, linking certain groups of model terms to each other and hence making the individual terms inseparable. This is very similar to the situation occurring with expanded terms of multi-level qualitative factors (see Chapter 12). For instance, with the above model, it would be impossible to delete only the x_3^2 term. A removal of x_3^2 also requires that the two cross-terms x_1x_3 and x_2x_3 be removed from the model. Thus, these three terms should be treated as a unit, and be deleted from, or incorporated in, the model together.

11.2.3 Step 3: Formation of the candidate set

In the third step of the strategy, the candidate set is compiled. This is the pool of theoretically possible experiments from which is drawn a subset of experiments comprising the actual mixture design. Unwanted experiments may be deleted from the candidate set prior to the generation of the design. In the current example, the candidate set is small, because the mixture region is *regular* in geometry. This makes it easy to compute the extreme vertices, the centers of edges, the overall centroid, and so on, which build up the candidate set. Here, the extreme vertices are the extreme points of the experimental domain, and the overall centroid has properties reminiscent of a center-point experiment.

In the tablet formulation example, the candidate set comprises three extreme vertices, three centers of edges, three interior points and one overall centroid. However, in more complicated mixture applications, with several factors and elaborate constraints, the experimental region might be highly irregular. For such an application, the candidate set is usually larger in size and also difficult to compile. Typically, the candidate set for an irregular mixture region consists of (i) the extreme vertices, (ii) the centers of edges, (iii) the centers of high-dimensional surfaces, and (iv) the overall centroid. From the resulting candidate set several alternative designs may be proposed D-optimally, each one corresponding to a unique selection of experiments.

11.2.4 Step 4: Generation of the mixture design

In the generation of the experimental design, which is the fourth step of the working strategy, attention has to be paid to the specifications set in the three preceding steps. By considering these specifications, one finds that the classical *simplex centroid* design is applicable in the tablet example. This experimental protocol is shown in Figure 11.3.

Worksheet

	1	2	3	4	5	6	7	8
1	Exp No	Exp Name	Run Order	Incl/Excl	cellulose	lactose	phosphate	release
2	1	N1	10	Incl	1	0	0	197
3	2	N2	7	Incl	0	1	0	110
4	3	N3	4	Incl	0	0	1	324
5	4	N4	9	Incl	0.5	0.5	0	67
6	5	N5	2	Incl	0.5	0	0.5	362
7	6	N6	6	Incl	0	0.5	0.5	312
8	7	N7	1	Incl	0.666667	0.166667	0.166667	206
9	8	N8	3	Incl	0.166667	0.666667	0.166667	171
10	9	N9	8	Incl	0.166667	0.166667	0.666667	344
11	10	N10	5	Incl	0.333333	0.333333	0.333333	214

Figure 11.3: *Experimental protocol of the tablet manufacturing example.*

We will now examine the geometrical properties of this design. With three mixture factors ranging between 0 and 1, the experimental region is a regular simplex. Such a simplex is illustrated in Figure 11.4. In this simplex, the orientation is as follows: Each vertex corresponds to a pure component, that is, only cellulose, only lactose, or only phosphate. To each vertex a component axis is associated, and at the common intersection of these axes the overall centroid of composition 1/3,1/3,1/3 is located. In order to well map this experimental region it is essential that the experiments are spread as evenly as possible. Figure 11.5 indicates the distribution of experiments of the selected design.

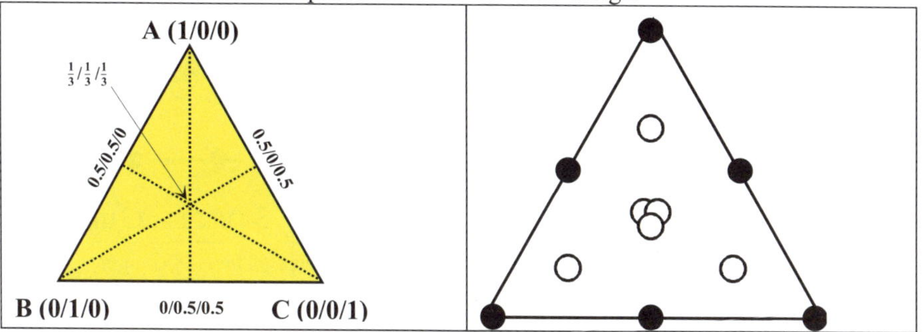

Figure 11.4: *(left) Simplex-shaped experimental region.* ***Figure 11.5:*** *(right) Experimental design used in the tablet manufacturing example. Solid points indicate mandatory experiments, open circles optional runs. Note that only one experiment was carried out at the overall centroid.*

11.2.5 Step 5: Evaluation of size and shape of the mixture region

In the fifth step of the strategy, the size and the shape of the mixture region is considered. This is particularly important when the mixture region is irregular. MODDE contains the Show/Design Region functionality, which facilitates an overview of the design region. When used on the Bubble formation example, introduced in Chapter 10, this functionality clearly shows the irregularity of the mixture region (Figure 11.6).

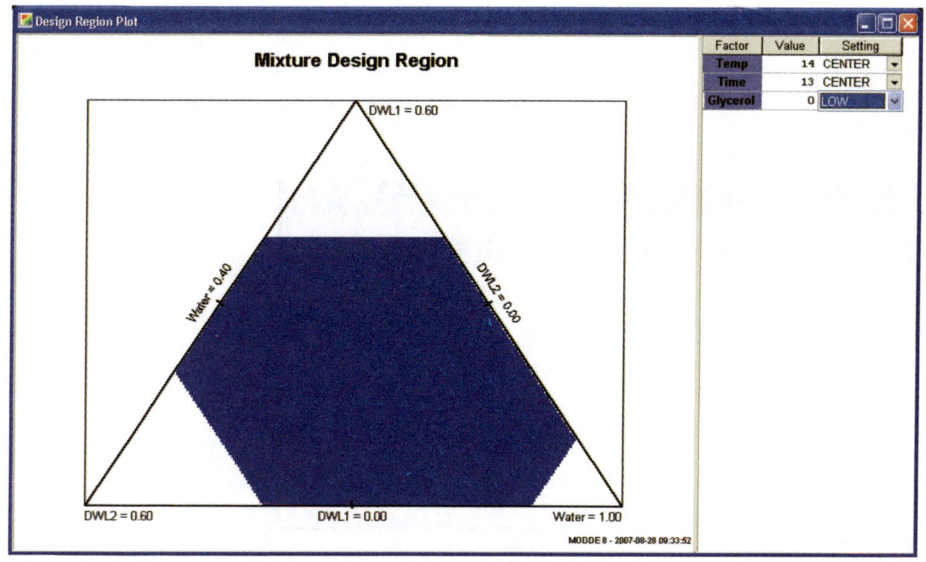

Figure 11.6: *Illustration of the irregular mixture design region in the Bubble formation example.*

In many cases, however, the mixture region can be a regular simplex. The design region plot is still useful. In Figure 11.7 alternative mixture designs for a regular mixture region are shown. As is easily realized, the choice of regression model will be important.

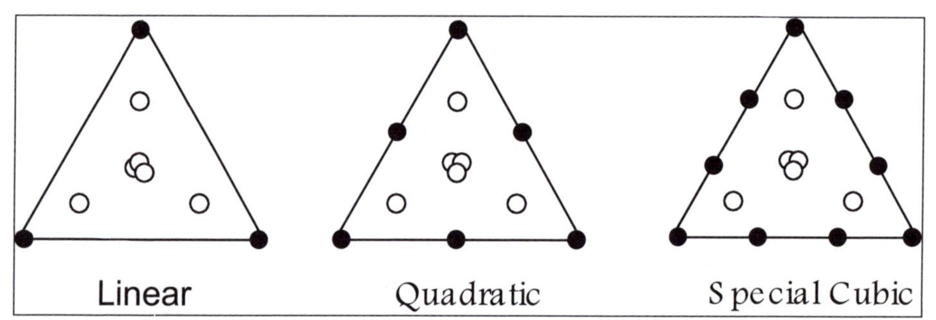

Figure 11.7: *Alternative mixture designs for a regular mixture region in three factors. Open circles denote optional runs.*

11.2.6 Step 6: Definition of the reference mixture

The sixth step of the strategy focuses on the definition of the reference mixture. This reference mixture plays an important role in the regression analysis, because the regression coefficients are expressed in relation to its co-ordinates. In a non-constrained simplex-shaped mixture region, the reference mixture corresponds to the overall centroid, that is, the 1/3,1/3,1/3 mixture. This is shown in Figure 11.8. For a regular mixture region, the identification of the reference mixture is facile. However, with irregular experimental regions the task is more taxing, and an efficient algorithmic approach is often needed. The co-ordinates of the reference mixture may well be used for replication. However, the problem is often that the reference mixture is quoted to 4-5 decimal places, which might be difficult to achieve in practice. Thus, it might be necessary to manually modify the calculated reference mixture and round off to a more manageable precision.

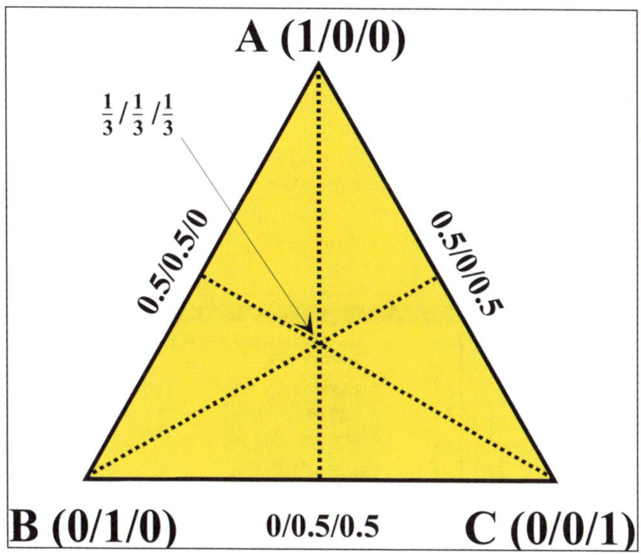

Figure 11.8: *Graphical illustration of the location of the overall centroid in the case of a non-constrained simplex-shaped mixture region.*

11.2.7 Step 7: Execution of the mixture design

Once the mixture design has been generated, evaluated, and found to encode a region of sufficient size and acceptable shape, and the reference mixture has been found and possibly adjusted, the next step is to carry out the experiments. It is important that these experiments, as far as is practically possible, are conducted in random order. This is in order to convert any systematic time trend which might occur into random unsystematic behaviour. Such unintentionally influencing trends might, for example, be an "improvement" in the laboratory skill of the experimenter, the degrading or aging of an HPLC-column, a constantly decreasing ambient temperature, and so on. Figure 11.9 displays the tablet manufacturing worksheet. It contains a column giving a random run order, which could be employed if these experiments were carried out anew.

Worksheet

	1	2	3	4	5	6	7	8
1	Exp No	Exp Name	Run Order	Incl/Excl	cellulose	lactose	phosphate	release
2	1	N1	10	Incl	1	0	0	197
3	2	N2	7	Incl	0	1	0	110
4	3	N3	4	Incl	0	0	1	324
5	4	N4	9	Incl	0.5	0.5	0	67
6	5	N5	2	Incl	0.5	0	0.5	362
7	6	N6	6	Incl	0	0.5	0.5	312
8	7	N7	1	Incl	0.666667	0.166667	0.166667	206
9	8	N8	3	Incl	0.166667	0.666667	0.166667	171
10	9	N9	8	Incl	0.166667	0.166667	0.666667	344
11	10	N10	5	Incl	0.333333	0.333333	0.333333	214

Figure 11.9: *The experimental worksheet of the tablet manufacturing application.*

11.2.8 Step 8: Analysis of data and evaluation of the model

Step 8 of the strategy is the analysis of the data and the evaluation of the model. This step is carried out according to the principles outlined in Chapters 1 – 4, the difference being that the PLS regression technique is used instead of MLR. PLS is detailed in Chapter 18 and in the Statistical Appendix. It is appropriate to commence with the evaluation of the raw data. The histogram shown in Figure 11.10 indicates that the response data can be used in the untransformed metric. Furthermore, since the design does not contain any replicated experiments, the replicate plot will not give any information about the replicate error.

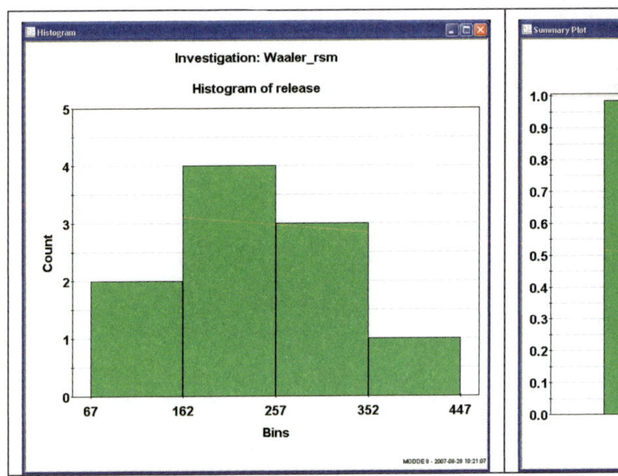

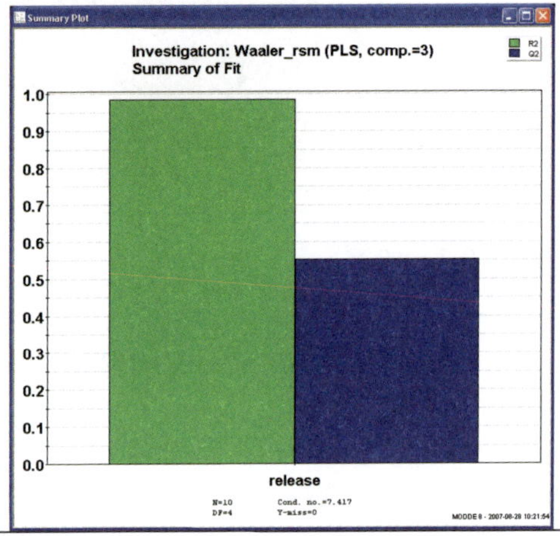

Figure 11.10: *(left) Histogram of the Release response.*
Figure 11.11: *(right) Summary of fit of the quadratic model fitted to Release.*

The PLS analysis of the tablet data gave a model with $R^2 = 0.98$ and $Q^2 = 0.55$ (Figure 11.11). These statistics point to an imperfect model, because R^2 substantially exceeds Q^2. Unfortunately, the second diagnostic tool (Figure 11.12), the ANOVA table, is incomplete because the lack of fit test could not be performed. However, a possible reason for the poor modelling is found when looking at the N-plot of the response residuals given in Figure 11.13. Experiment number 10 is an outlier and degrades the predictive ability of the model. If this experiment is omitted and the model refitted, Q^2 will increase from 0.55 to 0.69. We decided *not* to remove the outlier, however, primarily to conform with the modelling procedure of the original literature source.

ANOVA Table - Response release

	1	2	3	4	5	6	7
1	release	DF	SS	MS (variance)	F	p	SD
2	Total	10	624571	62457.1			
3	Constant	1	532225	532225			
4							
5	Total Corrected	9	92346.1	10260.7			101.295
6	Regression	5	90944.8	18189	51.9202	**0.001**	134.866
7	Residual	4	1401.3	350.325			18.717
8							
9	Lack of Fit	4	--	--	--	--	--
10	(Model Error)						
11	Pure Error	0	--	--			--
12	(Replicate Error)						
13							
14		N = 10	Q2 =	0.553	Cond. no. =	7.417	
15		DF = 4	R2 =	0.985	Y-miss =	0	
16		Comp. = 3	R2 Adj. =	0.966	RSD =	18.72	

The selected response 'release'

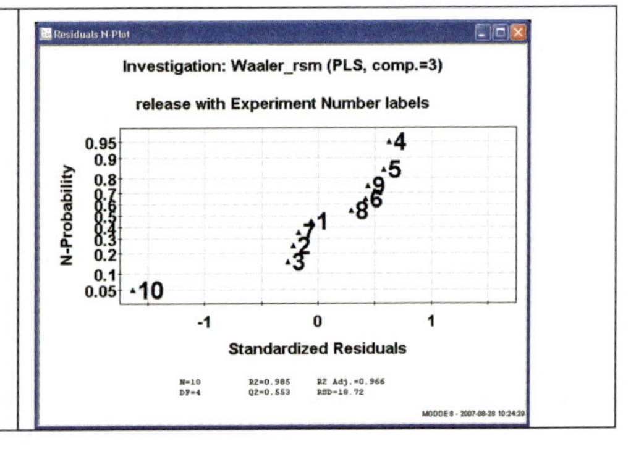

Figure 11.12: *(left) ANOVA of Release.*
Figure 11.13: *(right) N-plot of the residuals of Release.*

11.2.9 Step 9: Visualization of the modelling results

The ninth step of the strategy concerns the visualization of the modelling results. Scaled and centered regression coefficients of the computed model are plotted in Figure 11.14. This coefficient plot shows that in order to maximize the release rate, the amount of lactose in the recipe should be kept low and the amount of phosphate high. The presence of significant square and interaction terms indicate the existence of quadratic behaviour and non-linear blending effects. These effects are more easily understood by means of the trilinear mixture contour plot shown in Figure 11.15. This plot suggests that with the mixture composition 0.32/0/0.68 one may expect a response value above 350. This point should be tested in reality, thus functioning as an experimental verification of the model.

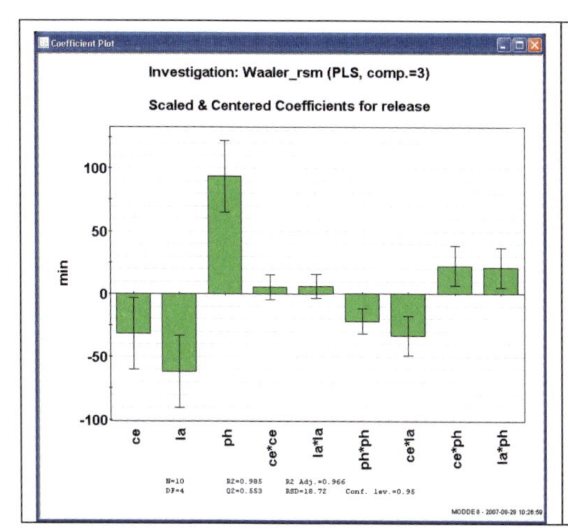

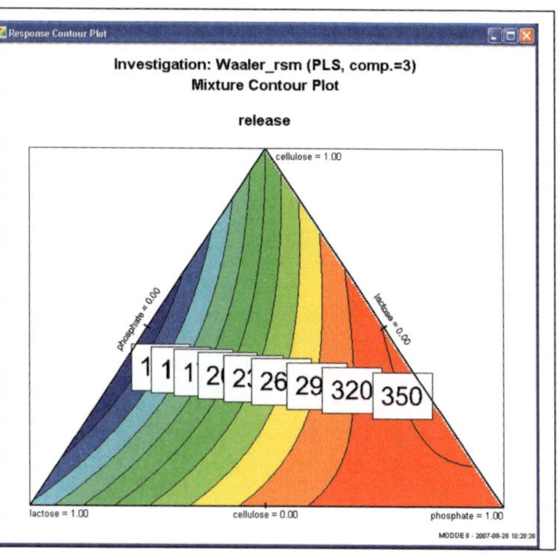

Figure 11.14: *(left) Regression coefficient plot of the model for Release.*
Figure 11.15: *(right) Trilinear mixture contour plot for Release.*

11.2.10 Step 10: Use of model

Transferring the modelling results into decisions, plans, and concrete action, is of utmost importance in mixture design. This is the essence of step 10 of the strategy, called use of model. With a screening design one may, for instance, use the resulting model to get an appreciation of where to carry out future experiments, that is, how to modify the factor ranges and thus "move" the experimental domain towards a more interesting region. This region may be mapped with another screening design or an RSM design. In order to accomplish this, the analyst will have to specify a profile of desired response values and then find out which settings of the mixture and/or process factors best correspond to the desired profile. The identification of the optimal factor settings is found with the software optimizer.

In the tablet example, the optimizer identified only one point, the mixture 0.32/0/0.68, where maximum release rate was predicted at 363 minutes. This point was not tested in the original work, but one close to it was. The experimenters performed three verifying experiments and these results together with model predictions are summarized in Figure 11.16. As seen, the model predicts well except for the mixture 0.5/0.125/0.375.

Pred No	cellulose	lactose	phosphate	release (obs)	release(pred)	Lower	Upper
1	0.32	0	0.68	---	363	322	404
2	0.5	0.125	0.375	370	293	262	324
3	0.333	0	0.667	340	363	322	405
4	0.667	0	0.333	345	320	278	361

Figure 11.16: *Model predictions of Release.*

11.3 Advanced mixture designs

Usually, it is not relevant to allow mixture factors to vary between 0 and 1, and other, narrower bounds are therefore employed. Figure 11.17 illustrates a more complex mixture region arising when three mixture factors are constrained. Here, factor A is allowed to vary between 0.2 and 0.6, factor B between 0.1 and 0.6, and factor C between 0.1 and 0.6. This experimental region is irregular and cannot be addressed by any classical mixture design. A D-optimal design is necessary. Another complicated problem which is becoming increasingly addressed is to vary both process factors and mixture factors within the same experimental protocol. An example of this latter case was outlined in Chapter 10.

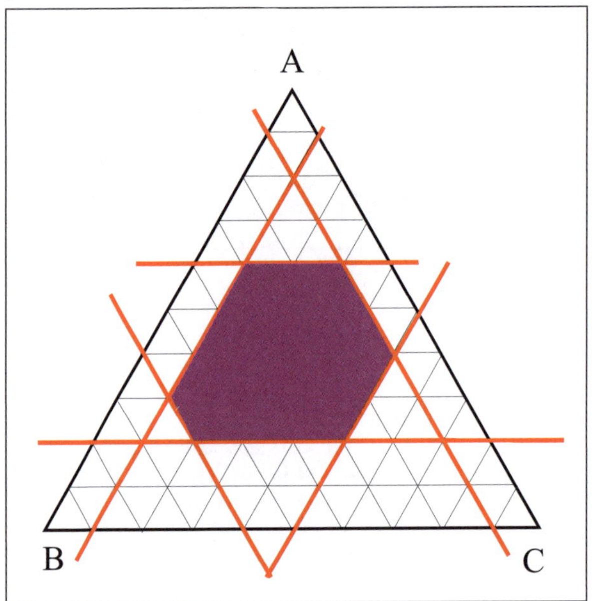

Figure 11.17: *An irregular mixture region in three mixture factors.*

11.4 Question for Chapter 11

- What is a mixture in the context of DOE?
- What is a factor bound?
- Which experimental objectives can be addressed in mixture design?
- What is an extreme vertex?
- What is the overall centroid?
- What is the reference mixture?

- Which three diagnostic tools may be used in the regression modelling?
- How can the modelling results be converted into concrete action?
- Which kind of design must be used when the experimental region is irregular?

11.5 Summary and discussion

In this chapter we have provided an introduction to mixture design. The characteristic feature of a mixture is that all ingredients sum to 100%. This imposes a restriction on the factors, so that they can no longer be varied independently of one another. Hence, orthogonal factor arrays no longer apply, and instead mixture designs must be used. We used a tablet manufacturing application to illustrate how mixture design and analysis of data may be carried out in the case when all factors vary between 0 and 1. We also outlined a ten-step strategy for mixture design. By way of example, it was demonstrated that the same kind of diagnostic and modelling tools, which were introduced for ordinary DOE designs, are useful also for mixture design data. Finally, this chapter ended with a brief look at more complicated, and perhaps more realistic, applications, where the mixture design region is irregular. For such situations there are no classical mixture designs available, and the problems must be addressed with D-optimal designs.

12 Multilevel qualitative factors

12.1 Objective

Qualitative factors for which more than two levels have been defined have certain special properties that must be considered in regression modelling. This chapter explains how to deal with designs containing such factors. For this purpose, we concentrate on the use of the regression coefficient plot and the interaction plot. In this case, the coefficient plot is a little trickier to interpret than when there are only quantitative factors or when there are qualitative factors at only two levels. The interaction plot is useful since it facilitates model interpretation. In order to illustrate these modelling principles, we examine a data-set related to cotton cultivation, which contains two qualitative factors at four and seven levels, respectively.

In addition, we outline a special case in which the temporary use of multilevel qualitative factors plays a key role. This concerns the situation when we wish to propose a diverse set of factor combinations and where the factors are (i) quantitative and (ii) explored at multiple levels. To accomplish this, D-optimal design is needed, together with a temporary re-coding of the quantitative factors into multilevel qualitative factors.

12.2 Introduction

Usually, a full factorial design constructed using qualitative factors at many levels is experimentally laborious. Imagine a situation where two qualitative factors and one quantitative factor are varied. Factor A is a qualitative factor with four levels, factor B a qualitative factor with three settings, and factor C a quantitative factor varying between –1 and +1. A full factorial design in this case would correspond to 4*3*2 = 24 experiments; this is depicted by all the filled and open circles in Figure 12.1.

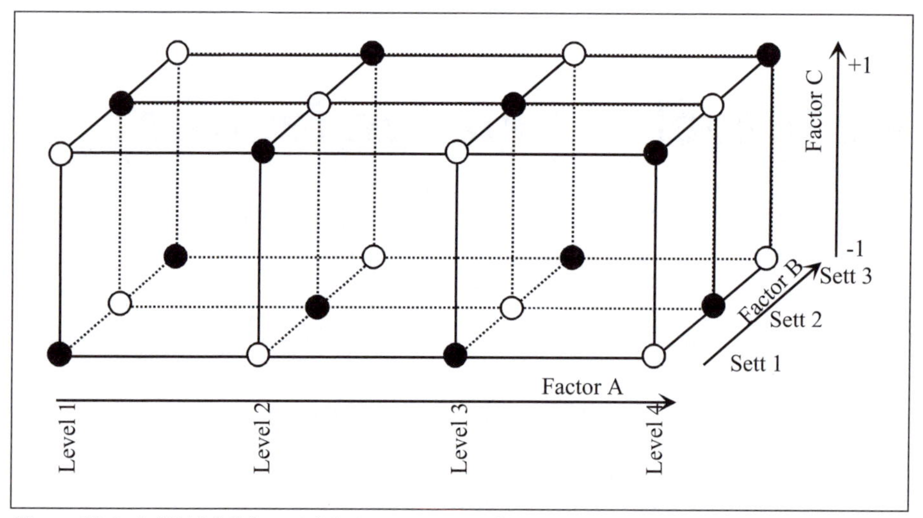

*Figure 12.1: A 4*3*2 full factorial design comprising 24 runs.*

However, for screening, and with a linear model, the full factorial design displayed in Figure 12.1 includes more runs than are necessary. An alternative design with fewer experiments might, therefore, be legitimate. It is possible to identify an appropriate reduced design using D-optimal techniques (see discussion in Chapter 10 and below).

We now attempt to illustrate how one may *analyze* data originating from a design constructed using only qualitative factors.

12.3 Example – Cotton cultivation

To demonstrate data analysis involving multilevel qualitative factors, we discuss an example relating to cotton cultivation. The Cotton example has two factors, which are summarized in Figure 12.2. One factor is related to the Variety, or species, of cotton, and is varied in four levels, denoted V1–V4. The other factor represents the Center, or location, of cotton cultivation. For this factor, as many as seven levels are defined, denoted C1–C7.

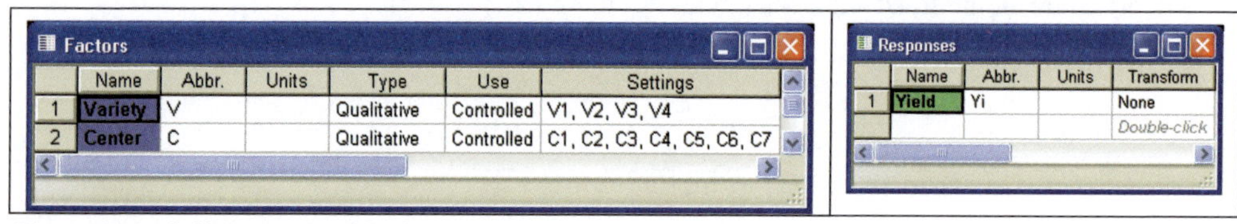

Factors

	Name	Abbr.	Units	Type	Use	Settings
1	Variety	V		Qualitative	Controlled	V1, V2, V3, V4
2	Center	C		Qualitative	Controlled	C1, C2, C3, C4, C5, C6, C7

Responses

	Name	Abbr.	Units	Transform
1	Yield	Yi		None
				Double-click

Figure 12.2: (left) Summary of the factors in the Cotton example.
Figure 12.3: (right) The Yield response measured in the Cotton example.

To describe the growth of cotton a response called Yield was measured (Figure 12.3). The Yield was recorded relative to a standard crop; a high numerical value indicates good growth. To investigate the impact of the two factors on the Yield, the experimenters conducted a full factorial design in 28 runs. The worksheet for this design is presented in Figure 12.4. Because of the abundance of experimental data points it is possible to fit an interaction model.

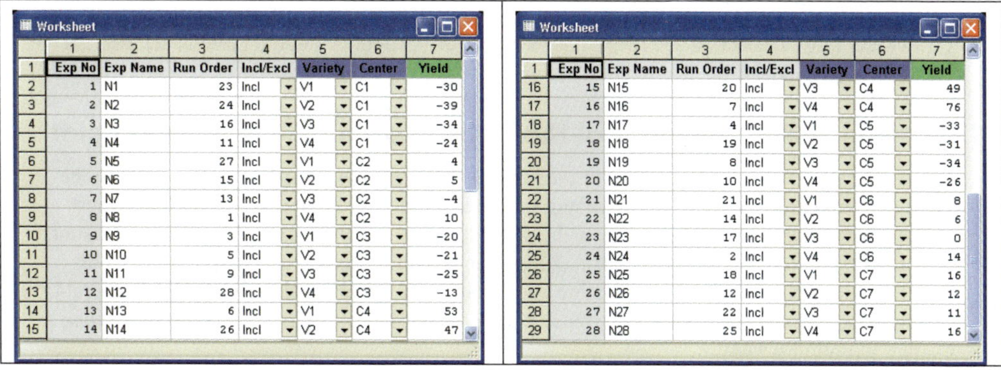

Exp No	Exp Name	Run Order	Incl/Excl	Variety	Center	Yield
1	N1	23	Incl	V1	C1	-30
2	N2	24	Incl	V2	C1	-39
3	N3	16	Incl	V3	C1	-34
4	N4	11	Incl	V4	C1	-24
5	N5	27	Incl	V1	C2	4
6	N6	15	Incl	V2	C2	5
7	N7	13	Incl	V3	C2	-4
8	N8	1	Incl	V4	C2	10
9	N9	3	Incl	V1	C3	-20
10	N10	5	Incl	V2	C3	-21
11	N11	9	Incl	V3	C3	-25
12	N12	28	Incl	V4	C3	-13
13	N13	6	Incl	V1	C4	53
14	N14	26	Incl	V2	C4	47

Exp No	Exp Name	Run Order	Incl/Excl	Variety	Center	Yield
15	N15	20	Incl	V3	C4	49
16	N16	7	Incl	V4	C4	76
17	N17	4	Incl	V1	C5	-33
18	N18	19	Incl	V2	C5	-31
19	N19	8	Incl	V3	C5	-34
20	N20	10	Incl	V4	C5	-26
21	N21	21	Incl	V1	C6	8
22	N22	14	Incl	V2	C6	6
23	N23	17	Incl	V3	C6	0
24	N24	2	Incl	V4	C6	14
25	N25	18	Incl	V1	C7	16
26	N26	12	Incl	V2	C7	12
27	N27	22	Incl	V3	C7	11
28	N28	25	Incl	V4	C7	16

Figure 12.4: *Experimental data from the Cotton example.*

12.3.1 Regression analysis of the Cotton example – Coefficient plot

We now discuss the process of fitting an interaction model to the Yield response. Our primary aim, here, is to highlight certain technical characteristics of qualitative factors, rather than providing a detailed account of the regression analysis. The interaction model in question had four parent terms: the constant term, the two main effects of Variety and Center, and their two-factor interaction. When fitting the interaction model to the data, the regression coefficient plot shown in Figure 12.5 was produced. This plot has 39 bars, each representing one coefficient. Four bars are associated with Variety, seven bars with Center, and 28 are derived from the two-factor interaction. Note that we do not give a summary of fit plot because the model is saturated and hence R^2 is 1.0.

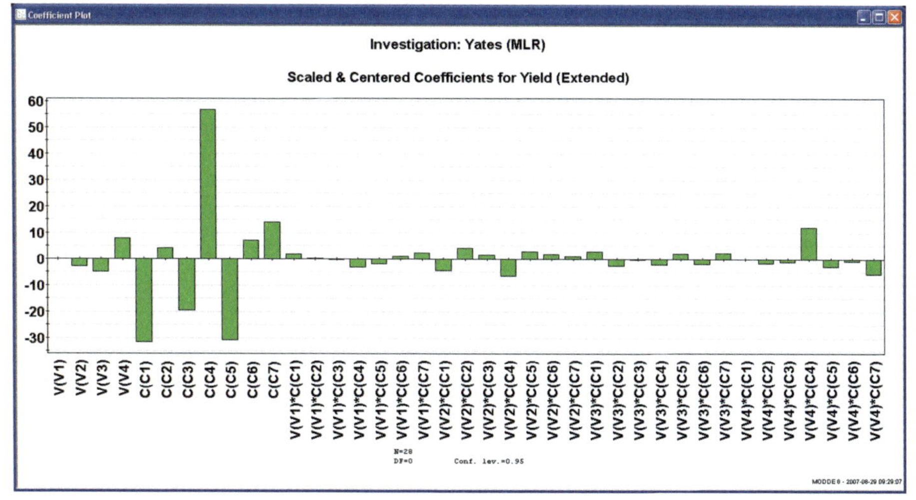

Figure 12.5: *Regression coefficient plot of the interaction model used in the Cotton example.*

An abundance of regression coefficients is one characteristic feature of designs containing multilevel qualitative factors. Furthermore, each block of regression coefficients, here those arising from Center and Variety, must each be treated as a single entity. For instance, it is not possible to delete only some of the 28 coefficients pertaining to the V*C two-factor interaction. On the contrary, these terms must be included in, or removed from, the model as a whole.

Now we interpret the model. Although the coefficient plot is cluttered, its message is that Center has more impact on the result than Variety. A particularly good Yield is achieved at Center #4.

12.3.2 Regression analysis of the Cotton example – Interaction plot

In the case of many multilevel qualitative factors, the regression coefficient plot tends to be cluttered and hard to interpret. Fortunately, there is another graphical tool available which facilitates model interpretation. This is the so-called interaction plot. Figure 12.6 shows such a plot for the model fitted to the cotton cultivation data. Looking at this graph, it does not take long to realize that the best possible combination of factors is Variety #4 and Center #4. In addition, it appears that a change in cultivation center has more impact on the result than a change in cotton variety.

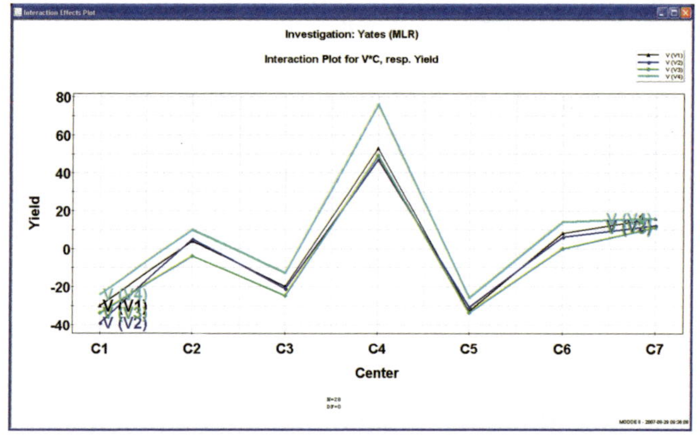

Figure 12.6: Interaction plot of the model used in the Cotton example.

This concludes our presentation of the modelling results. In the rest of this chapter, we describe the reasons why multilevel qualitative factors appear as they do in a coefficient plot.

12.4 Regression coding of qualitative variables

Qualitative variables require a special form of coding for regression analysis to work properly. More specifically, a qualitative factor with k levels, will have k-1 expanded terms in the model calculations. Let us consider the Variety factor in the Cotton example. This factor has four levels and requires three degrees of freedom, thus three experiments, to be estimable. In order to be able to estimate its impact, a minor mathematical re-expression is necessary prior to the data analysis, as outlined in Figure 12.7.

	Expanded term		
Level of factor	V(V2)	V(V3)	V(V4)
V1	-1	-1	-1
V2	1	0	0
V3	0	1	0
V4	0	0	1

Figure 12.7: Regression coding of the four-level qualitative factor Variety.

We can see that the Variety factor is expanded into three artificial categorical variables. In a similar manner, the Center factor of seven levels will require six categorical variables to code its information. In regression modeling, each such expanded term gives rise to one regression coefficient, and may therefore be inspected visually in a regression coefficient plot.

12.5 Regular and extended lists of coefficients

All expanded model terms that relate to a qualitative factor are presented either as a *regular* or an *extended coefficient* display. Figure 12.8 shows the Cotton example presented in this way. In the *regular* mode, the coefficients of the expanded terms of Variety are presented as the coefficients for level 2, V2, level 3, V3, and level 4, V4. In the *extended* mode the coefficient for level 1, V1, is computed as the negative sum of the coefficients of the other expanded terms, as shown in Figure 12.8. Since the four coefficients sum to zero, only three degrees of freedom are needed for the variable Variety.

Regular		Extended	
Yield	**Coeff.**	**Yield**	**Coeff.**
Constant	**-0.25**	**Constant**	**-0.25**
		V	**DF = 3**
		V(V1)	**-0.035**
V(V2)	**-2.75**	**V(V2)**	**-2.75**
V(V3)	**-5.036**	**V(V3)**	**-5.036**
V(V4)	**7.821**	**V(V4)**	**7.821**
Sum	***0.035***	***Sum***	***0***
		C	**DF = 6**
		C(C1)	**-31.5**
C(C2)	**4**	**C(C2)**	**4**
C(C3)	**-19.5**	**C(C3)**	**-19.5**
C(C4)	**56.5**	**C(C4)**	**56.5**
C(C5)	**-30.75**	**C(C5)**	**-30.75**
C(C6)	**7.25**	**C(C6)**	**7.25**
C(C7)	**14**	**C(C7)**	**14**
Sum	***31.5***	***Sum***	***0***

Figure 12.8: *Illustration of regular and extended lists of regression coefficients.*

In the lower part of Figure 12.8, the corresponding coefficients related to the expansion of the Center variable are presented. We can see that six degrees of freedom are sufficient for the seven-level factor. The V*C two-factor interaction is expanded and calculated in a similar manner. This means that 3*6 =18 or 4*7 = 28 coefficients would be displayed, depending on whether the regular or extended mode is used. In the analysis, it is the user's responsibility to specify whether the regular or extended mode is required. When the extended mode is used, a total of 39 coefficients are displayed for the Cotton application, namely four for variety, seven for Center, and 28 for their interaction (Figure 12.5).

12.6 Generation of designs with multilevel qualitative factors

Next, we wish to discuss design generation. When we are dealing with only two-level qualitative factors, the design generation is simple, since regular two-level full and fractional factorial designs are readily available. When a qualitative factor is explored at three or more levels, however, the task of creating a good experimental design with relatively few runs becomes more problematic.

Let us return to the example which we described at the beginning of this chapter; this had one four-level and one three-level qualitative factor, plus a quantitative factor. A full factorial design for screening would, in this case, correspond to 4*3*2 = 24 experiments, as depicted by all the filled and open circles in Figure 12.9. In addition to these mandatory "factorial" experiments, between three and five replicates ought to be included in the final design. Hence, between 27 and 29 experiments would be necessary; this is too many for a screening program. An alternative design with fewer experiments would be preferable. One such design, with only 12 experiments, is represented by the solid circles in Figure 12.9. The open circles represent an alternative design. Note that replicated experiments are not included in the selected dozen, but must be added.

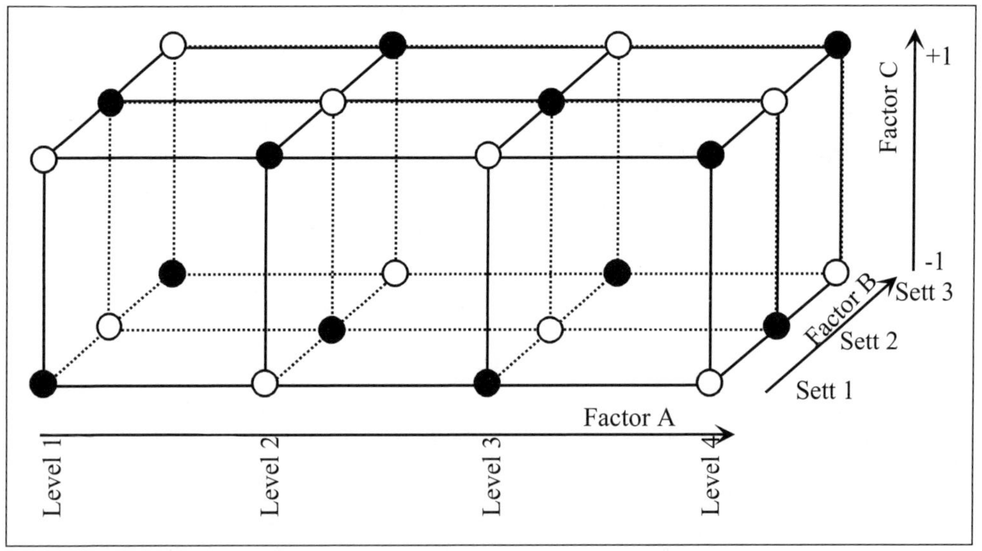

Figure 12.9: *A 4*3*2 full factorial design comprising 24 runs (filled and open points), or a related reduced design in 12 runs (filled or open points).*

These subsets, shown in Figure 12.9, were selected using a theoretical algorithm, a D-optimal algorithm. This identified those experiments with the best spread and best balanced distribution. A balanced design has the same number of runs for each level of a qualitative factor. This feature is not absolutely necessary, but is often convenient and makes the design easier to understand. D-optimal designs were discussed in Chapter 10.

12.7 Obtaining a diverse set of factor combinations

The last part of this chapter is devoted to a technique from which the end-result is *not* a design in multilevel qualitative factors, but a design in multilevel quantitative factors where all levels of the factors are explored in a balanced way. The reason for covering this subject here in Chapter 12 is that the required design features – uniform coverage of many factor combinations using few and diverse experiments – are not easy to achieve without making a little detour from quantitative factor space into the domain of qualitative factors.

12.7.1 Background to the problem

In some applications of DOE the experimenter may *not* want to investigate quantitative factors in just two levels, but rather the goal may be to study them at multiple levels using experimental designs comprising as few and diverse experiments as possible. This means that the problem has similarities to onion and space-filling designs, where the prime objective is to accomplish a uniform or near-uniform coverage of the accessible experimental domain.

When quantitative factors are specified at two levels, the design generation phase is straightforward. There are then several classical design families to choose from and the resulting designs often have a limited number of runs. Such designs for regular experimental regions were discussed and compared in Chapters 5 and 8. However, when quantitative factors at three, four, or even more levels are involved, a challenge arises in the sense that the number of factor combinations grows substantially, far beyond what is required by e.g. a two-level fractional factorial design. Consequently, a D-optimal design is usually considered to be the only realistic choice for problems involving several multilevel quantitative factors.

As discussed in Chapter 10, a D-optimal design maximizes the determinant of the X'X matrix (the variance–covariance matrix) of the selected design and regression model. This means that the selected experiments will be positioned such that they span the largest possible volume of the permissible experimental region. For multilevel quantitative factors, this space-spanning characteristic of D-optimal designs unavoidably means that only the lowest and highest levels, not the intermediate ones, will be populated by experiments in the final design. In particular, this is true for a screening application with an associated linear regression model. Quadratic models may occasionally produce a more uniform coverage of all the levels of multilevel quantitative factors, but still there is no guarantee that all levels will be tested.

Fortunately, however, it is still possible to devise a screening design involving several multilevel quantitative factors in which (i) all levels are tested and (ii) the number of experiments is kept low. As outlined below, the work-around is to re-code the quantitative factors into qualitative ones and then generate a D-optimal design. After the D-optimal design has been generated, a back-transformation into the original continuous factor space takes place. This process is illustrated below for a real application.

12.7.2 Example: Equipment Performance Testing

In performance testing of new equipment it is imperative to test many and diverse factor combinations. The goal is to stress the system under as many conditions as possible and thereby obtain a reliable understanding of its properties and behaviour. Our example dataset relates to a set of granulators and their effect on the properties of granules and tablets. The data-set was kindly provided by P. Hoppu, AZ Mölndal, Sweden and University of Helsinki, Finland.

The data-set has four factors: (i) type of granulator; (ii) water level; (iii) wet massing time; and (iv) impeller speed (Figure 12.10). As shown in Figure 12.10, the first and last factors are qualitative, each allocated three levels. The second factor is a four-level quantitative factor and the third a three-level quantitative factor. Thus, a full factorial mixed design would correspond to 108 runs (= 3 * 4 * 3 * 3), even without replicated experiments. The experimental objective was a screening design supporting an interaction model.

Factors

	Name	Abbr.	Units	Type	Use	Settings	Transform
1	Granulator	Gra		Qualitative	Controlled	MTR, MiPRO, Diosna	
2	Water level	Wat	ml/g	Multilevel	Controlled	0.35, 0.5, 0.65, 0.8	None
3	Wet massing time	Time	min	Multilevel	Controlled	1, 5, 10	None
4	Impeller speed	Imp		Qualitative	Controlled	low, medium, high	

Figure 12.10: *The four factors in the Equipment Performance Testing example.*

12.7.3 Set-up of the initial design

Although we have already hinted at the work-around, we begin our discussion by reviewing the initial design that was proposed. We do so because we believe it provides an insight into the problem. The basic premises to consider are screening and an interaction model. In order to get a preliminary indication of how many experiments might be necessary, we consider the degrees of freedom (DF) of the interaction model chosen:

- constant term 1 DF
- linear terms of the two quantitative factors 2 DF
- linear terms of the two qualitative factors (2*2) 4 DF
- interaction term between quantitative factors 1 DF
- interaction term between qualitative factors4 DF
- interaction terms between quant. and qual. factors (4*2) 8 DF

Hence, we realize that a minimum of 20 experiments will be necessary to support the interaction model. In addition, about five DF is recommended to stabilize residual analysis and modelling diagnostics. This gives N=25 experiments as the lead number of runs in the proposed design.

Another feature to pay attention to is balancing. As discussed in Chapter 10, balancing means that all levels of a qualitative factor are explored using the same number of experiments. Strictly speaking, however, balancing is not completely necessary in a design with qualitative factors, but if the design is balanced it is a lot easier to understand geometrically and to evaluate statistically.

If 24 experiments are chosen, excluding center-points, balancing is possible, and permitted by the software for all factors, but the number of experiments will be low. We could also opt for 27 experiments, which would permit balancing with respect to all factors except the third (water level). Alternatively, we could consider 36 experiments, which would result both in enough trials and balancing with respect to all factors. As shown in Figure 12.11, it was decided to opt for 36 design runs.

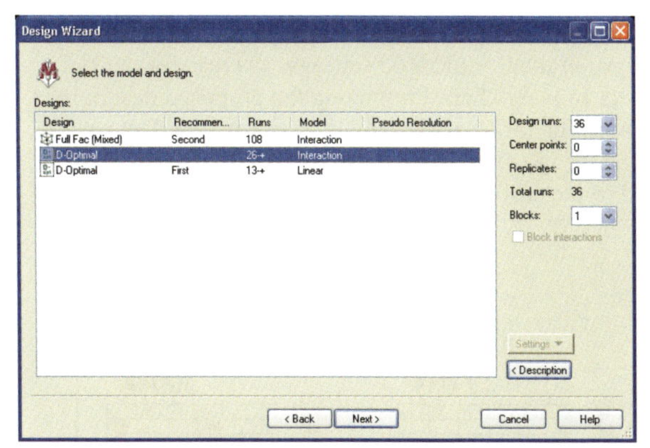

Figure 12.11: *In the initial D-optimal design the lead number of runs was 36.*

Subsequently, MODDE was instructed to consider balanced D-optimal designs only and 50 D-optimal designs with 36 runs were computed. When these candidate designs were evaluated, it turned out that they were all the same, with a G-efficiency of 100% and a condition number of 3. The worksheet for one such design is presented in Figure 12.12.

Worksheet

	1	2	3	4	5	6	7	8
1	Exp No	Exp Name	Run Order	Incl/Excl	Granulator	Water level	Wet massing time	Impeller speed
2	1	N1	14	Incl	MTR	0.35	1	low
3	2	N2	26	Incl	MiPRO	0.35	1	low
4	3	N3	1	Incl	Diosna	0.35	1	low
5	4	N4	19	Incl	MTR	0.8	1	low
6	5	N5	15	Incl	MiPRO	0.8	1	low
7	6	N6	17	Incl	Diosna	0.8	1	low
8	7	N7	12	Incl	MTR	0.35	10	low
9	8	N8	20	Incl	MiPRO	0.35	10	low
10	9	N9	16	Incl	Diosna	0.35	10	low
11	10	N10	18	Incl	MTR	0.8	10	low
12	11	N11	35	Incl	MiPRO	0.8	10	low
13	12	N12	28	Incl	Diosna	0.8	10	low
14	13	N13	6	Incl	MTR	0.35	1	medium
15	14	N14	5	Incl	MiPRO	0.35	1	medium
16	15	N15	24	Incl	Diosna	0.35	1	medium
17	16	N16	9	Incl	MTR	0.8	1	medium
18	17	N17	31	Incl	MiPRO	0.8	1	medium
19	18	N18	11	Incl	Diosna	0.8	1	medium
20	19	N19	10	Incl	MTR	0.35	10	medium
21	20	N20	23	Incl	MiPRO	0.35	10	medium
22	21	N21	25	Incl	Diosna	0.35	10	medium
23	22	N22	13	Incl	MTR	0.8	10	medium
24	23	N23	32	Incl	MiPRO	0.8	10	medium
25	24	N24	29	Incl	Diosna	0.8	10	medium
26	25	N25	30	Incl	MTR	0.35	1	high
27	26	N26	34	Incl	MiPRO	0.35	1	high
28	27	N27	7	Incl	Diosna	0.35	1	high
29	28	N28	22	Incl	MTR	0.8	1	high
30	29	N29	21	Incl	MiPRO	0.8	1	high
31	30	N30	36	Incl	Diosna	0.8	1	high
32	31	N31	8	Incl	MTR	0.35	10	high
33	32	N32	3	Incl	MiPRO	0.35	10	high
34	33	N33	33	Incl	Diosna	0.35	10	high
35	34	N34	2	Incl	MTR	0.8	10	high
36	35	N35	4	Incl	MiPRO	0.8	10	high
37	36	N36	27	Incl	Diosna	0.8	10	high

Figure 12.12: *The worksheet of the initial design considered in the Equipment Performance Testing example.*

The initial design – which is presented in Figure 12.12 – is balanced with respect to the two qualitative factors Granulator and Impeller speed. This can be readily seen in Figure 12.13, which provides a graphical representation of the spread of the experiments. It is also evident from Figure 12.13 that the two quantitative factors are not explored as required. For the water level factor, no experiments are proposed using the 0.5 and 0.65 settings, whereas for the wet massing time the 5 min level is not covered. In fact, when scrutinizing the proposed design in more detail, it appears we have obtained a full factorial design with the quantitative factors reduced to two levels, i.e., a $2^2 * 3^2$ factorial design. This design also represents the limiting case of the D-optimal approach as applied to the current problem.

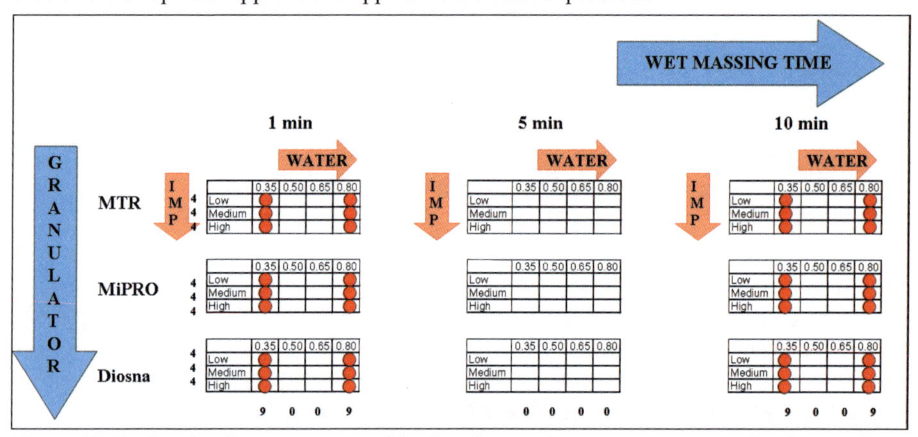

Figure 12.13: *Graphical representation of the distribution of experiments contained in the initial design proposed for the Equipment Performance Testing example.*

The geometric properties of the initial design make it an optimal design for supporting the specified interaction model. Thus, in that sense, the D-optimal algorithm has achieved what is was instructed to do. Unfortunately, however, supporting an interaction model is not our only goal. We are also interested in obtaining a diverse set of factor combinations, where all levels of the two quantitative factors are studied in a uniform manner. In order to achieve this latter objective, we must redefine our factors and generate a new design.

12.7.4 A second "work-around" design in multilevel qualitative factors

In the next stage, the two quantitative factors were redefined as two qualitative factors (Figure 12.14). Thus, in this modified design environment, we now have four multilevel qualitative factors.

Factors

	Name	Abbr.	Units	Type	Use	Settings
1	Granulator	Gra		Qualitative	Controlled	MTR, MiPRO, Diosna
2	Water level	Wat		Qualitative	Controlled	L0.35, L0.5, L0.65, L0.8
3	Wet massing time	Time		Qualitative	Controlled	T1, T5, T10
4	Impeller speed	Imp		Qualitative	Controlled	low, medium, high

Figure 12.14: *The four qualitative factors – after factor re-coding – of the Equipment Performance Testing example.*

The desired number of design runs was retained (Figure 12.15). However, the design problem is now more complex and 36 runs are, therefore, not sufficient to support a full interaction model. Hence, a linear model has to be used as the basis (Figure 12.15). This linear model will *not* consume all 36 degrees of freedom, and therefore it will be possible to supplement the linear model with some selected two-factor interactions. Using some loosely defined DOE jargon, such a model can be referred to as a "linear-plus" model.

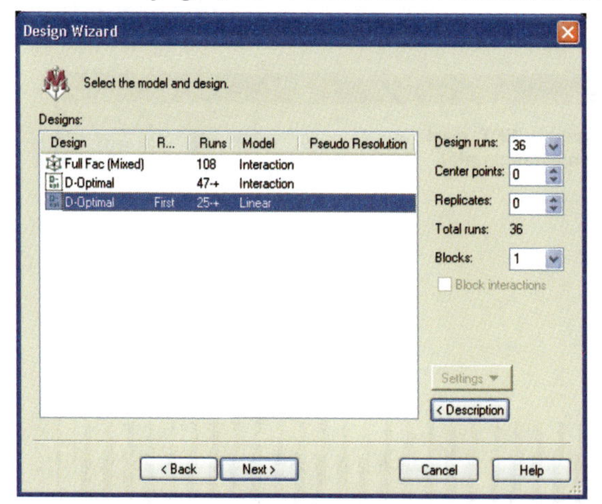

Figure 12.15: In the second design the desired number of experiments remained unchanged.

When it comes to adding a small number of selected two-factor interactions, different principles can be used for determining which terms are likely to be most appropriate and useful. For example, existing mechanistic information may suggest that certain two-factor interactions may be relevant. Or, as in our case, specific features of the problem definition may dictate which terms are added. As shown by Figure 12.16, it was decided to add cross-terms for the second factor (Water level) simply because it has the largest number of levels.

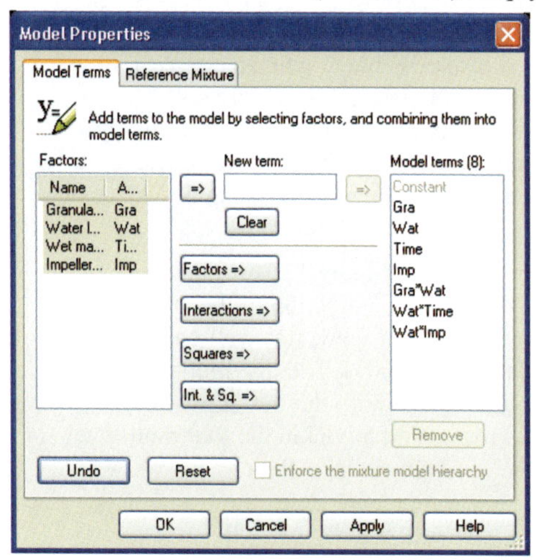

Figure 12.16: The default linear model was augmented by incorporating three cross-terms.

Thereafter, the D-optimal algorithm was launched. The outcome of this is plotted in Figure 12.17. As can be seen, there are many D-optimal selections which have a comparatively low G-efficiency, below 40, but interestingly there is also a handful which have an optimal G-efficiency of 100%.

In the following text we focus on this sub-set of high-performance D-optimal designs. From a mathematical point of view they are all equivalent since they have identical G-efficiencies (100 %). However, it is important to realize that mathematical equivalence does not correspond directly to geometric equivalence. In other words, when comparing two arbitrarily chosen D-optimal designs, it must remembered that the geometric configuration of their experiments is not necessarily the same, even though they exhibit identical mathematical performance criteria. This special feature of D-optimal design was highlighted in Chapter 10.

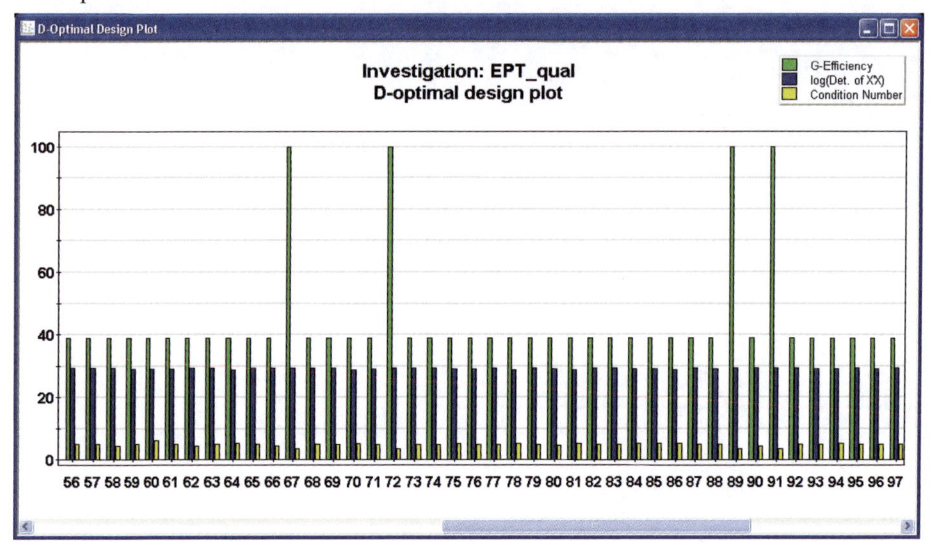

Figure 12.17: *Design overview plot indicating a few D-optimal designs that exhibit a G-efficiency of 100%.*

There are two principal ways of comparing two or more D-optimal proposals. The first and most obvious way consists of creating the worksheet to examine the different factor combinations (see e.g. Figure 12.12). In this way it is also possible to gain an appreciation of the proposed experiments in order to verify that they are all reasonable and are not disqualified by e.g. incompatibilities between the various experimental settings. Secondly, for D-optimal proposals that are considered to be really interesting, it is always warranted to examine the distribution of the experiments graphically. This can be done using design arrangements similar to the lay-out in Figure 12.13.

The worksheet of the design chosen in our case is presented in Figure 12.18. A detailed inspection of this worksheet shows that the design is an exact 1/3-fraction of the underlying full $4 * 3^3$ design in 108 experiments. From a geometric point of view, this well-balanced selection of experiments is the reason for the G-efficiency of 100%. The distribution of experiments in the second design is presented in Figure 12.19. With this set of experiments we have an optimal basis for estimating a linear-plus regression model in the four multilevel qualitative factors.

Worksheet

	1	2	3	4	5	6	7	8
1	Exp No	Exp Name	Run Order	Incl/Excl	Granulator	Water level	Wet massing time	Impeller speed
2	1	N1	29	Incl	Diosna	L0.35	T1	low
3	2	N2	27	Incl	MiPRO	L0.5	T1	low
4	3	N3	36	Incl	MiPRO	L0.65	T1	low
5	4	N4	7	Incl	Diosna	L0.8	T1	low
6	5	N5	21	Incl	MiPRO	L0.35	T5	low
7	6	N6	5	Incl	Diosna	L0.5	T5	low
8	7	N7	3	Incl	MTR	L0.65	T5	low
9	8	N8	35	Incl	MiPRO	L0.8	T5	low
10	9	N9	11	Incl	MTR	L0.35	T10	low
11	10	N10	16	Incl	MTR	L0.5	T10	low
12	11	N11	30	Incl	Diosna	L0.65	T10	low
13	12	N12	4	Incl	MTR	L0.8	T10	low
14	13	N13	24	Incl	MiPRO	L0.35	T1	medium
15	14	N14	10	Incl	MTR	L0.5	T1	medium
16	15	N15	22	Incl	MTR	L0.65	T1	medium
17	16	N16	12	Incl	MiPRO	L0.8	T1	medium
18	17	N17	20	Incl	MTR	L0.35	T5	medium
19	18	N18	23	Incl	MiPRO	L0.5	T5	medium
20	19	N19	8	Incl	Diosna	L0.65	T5	medium
21	20	N20	1	Incl	MTR	L0.8	T5	medium
22	21	N21	34	Incl	Diosna	L0.35	T10	medium
23	22	N22	32	Incl	Diosna	L0.5	T10	medium
24	23	N23	6	Incl	MiPRO	L0.65	T10	medium
25	24	N24	18	Incl	Diosna	L0.8	T10	medium
26	25	N25	19	Incl	MTR	L0.35	T1	high
27	26	N26	2	Incl	Diosna	L0.5	T1	high
28	27	N27	31	Incl	Diosna	L0.65	T1	high
29	28	N28	25	Incl	MTR	L0.8	T1	high
30	29	N29	26	Incl	Diosna	L0.35	T5	high
31	30	N30	13	Incl	MTR	L0.5	T5	high
32	31	N31	9	Incl	MiPRO	L0.65	T5	high
33	32	N32	33	Incl	Diosna	L0.8	T5	high
34	33	N33	15	Incl	MiPRO	L0.35	T10	high
35	34	N34	28	Incl	MiPRO	L0.5	T10	high
36	35	N35	17	Incl	MTR	L0.65	T10	high
37	36	N36	14	Incl	MiPRO	L0.8	T10	high

Figure 12.18: The worksheet for the second design created using the four multilevel qualitative factors.

Figure12.19 shows that one way of understanding this application is through a system of nine "boxes". There are four experiments within each box.

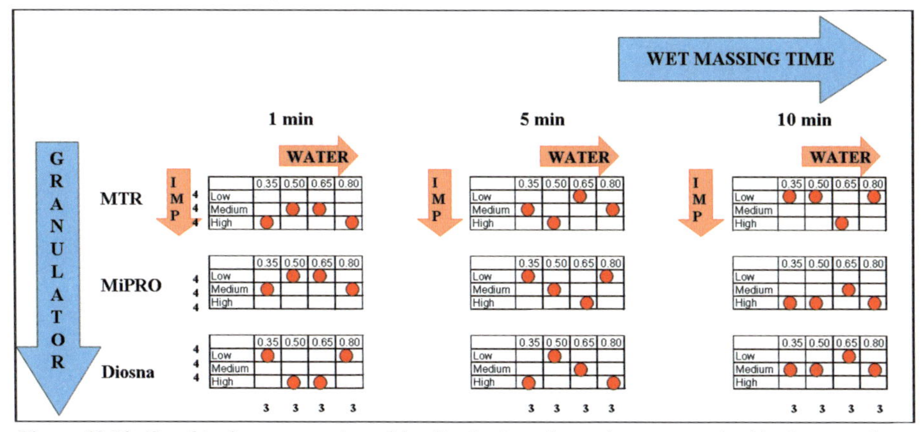

Figure 12.19: Graphical representation of the distribution of experiments contained in the second design proposed for the Equipment Performance Testing example.

The four experiments inside each box change according to three alternative patterns and each pattern is permuted across a vertical trio of boxes. In order to simplify the problem, the investigator (P. Hoppu) selected only one permutation pattern to be systematically applied to all boxes (Figure 12.20). This gives an optimal design for space-filling coverage. The worksheet of the re-arranged design is presented in Figure 12.21.

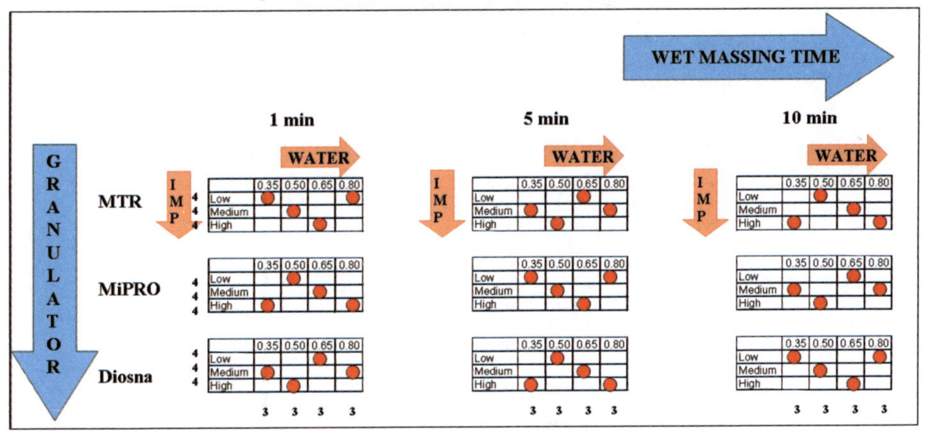

Figure 12.20: A simpler design based on just one permutation pattern.

Worksheet

	1	2	3	4	5	6	7	8
1	Exp No	Exp Name	Run Order	Incl/Excl	Granulators	Water levels	Wet massing times	Impeller speed
2	1	N1	22	Incl	MTR	L0.35	T1	low
3	2	N2	24	Incl	MTR	L0.5	T1	medium
4	3	N3	6	Incl	MTR	L0.65	T1	high
5	4	N4	17	Incl	MTR	L0.8	T1	low
6	5	N5	16	Incl	MTR	L0.35	T5	medium
7	6	N6	18	Incl	MTR	L0.5	T5	high
8	7	N7	11	Incl	MTR	L0.65	T5	low
9	8	N8	8	Incl	MTR	L0.8	T5	medium
10	9	N9	10	Incl	MTR	L0.35	T10	high
11	10	N10	36	Incl	MTR	L0.5	T10	low
12	11	N11	19	Incl	MTR	L0.65	T10	medium
13	12	N12	9	Incl	MTR	L0.8	T10	high
14	13	N13	13	Incl	MiPRO	L0.35	T1	high
15	14	N14	7	Incl	MiPRO	L0.5	T1	low
16	15	N15	34	Incl	MiPRO	L0.65	T1	medium
17	16	N16	30	Incl	MiPRO	L0.8	T1	high
18	17	N17	3	Incl	MiPRO	L0.35	T5	low
19	18	N18	31	Incl	MiPRO	L0.5	T5	medium
20	19	N19	33	Incl	MiPRO	L0.65	T5	high
21	20	N20	14	Incl	MiPRO	L0.8	T5	low
22	21	N21	26	Incl	MiPRO	L0.35	T10	medium
23	22	N22	28	Incl	MiPRO	L0.5	T10	high
24	23	N23	35	Incl	MiPRO	L0.65	T10	low
25	24	N24	1	Incl	MiPRO	L0.8	T10	medium
26	25	N25	15	Incl	Diosna	L0.35	T1	medium
27	26	N26	23	Incl	Diosna	L0.5	T1	high
28	27	N27	32	Incl	Diosna	L0.65	T1	low
29	28	N28	25	Incl	Diosna	L0.8	T1	medium
30	29	N29	21	Incl	Diosna	L0.35	T5	high
31	30	N30	12	Incl	Diosna	L0.5	T5	low
32	31	N31	20	Incl	Diosna	L0.65	T5	medium
33	32	N32	2	Incl	Diosna	L0.8	T5	high
34	33	N33	29	Incl	Diosna	L0.35	T10	low
35	34	N34	4	Incl	Diosna	L0.5	T10	medium
36	35	N35	5	Incl	Diosna	L0.65	T10	high
37	36	N36	27	Incl	Diosna	L0.8	T10	low

Figure 12.21: The worksheet for the modified second design.

The re-arranged version of the second design is mathematically equivalent to its predecessor and the two have identical performance statistics. The great advantage of the modified design is that its geometry is easy to understand. It accomplishes a spread of trials analogous to a space-filling or onion design. Hence, its worksheet (Figure 12.21) can be used as a template when generating the final design.

12.7.5 Obtaining the final design

Now, that we have obtained a good design template, the next step requires us to return to the continuous metrics of the two factors Water level and Wet massing time. This is much simpler than it may sound. All that is needed is to manually modify the original worksheet (Figure 12.12) according to the template (Figure 12.21). The resulting "real" worksheet is presented in Figure 12.22.

Worksheet

	1	2	3	4	5	6	7	8
1	Exp No	Exp Name	Run Order	Incl/Excl	Granulator	Water level	Wet massing time	Impeller speed
2	1	N1	37	Incl	MTR	0.35	1	low
3	2	N2	12	Incl	MTR	0.5	1	medium
4	3	N3	16	Incl	MTR	0.65	1	high
5	4	N4	2	Incl	MTR	0.8	1	low
6	5	N5	30	Incl	MTR	0.35	5	medium
7	6	N6	17	Incl	MTR	0.5	5	high
8	7	N7	38	Incl	MTR	0.65	5	low
9	8	N8	1	Incl	MTR	0.8	5	medium
10	9	N9	8	Incl	MTR	0.35	10	high
11	10	N10	24	Incl	MTR	0.5	10	low
12	11	N11	34	Incl	MTR	0.65	10	medium
13	12	N12	21	Incl	MTR	0.8	10	high
14	13	N13	25	Incl	MiPRO	0.35	1	high
15	14	N14	19	Incl	MiPRO	0.5	1	low
16	15	N15	35	Incl	MiPRO	0.65	1	medium
17	16	N16	11	Incl	MiPRO	0.8	1	high
18	17	N17	32	Incl	MiPRO	0.35	5	low
19	18	N18	27	Incl	MiPRO	0.5	5	medium
20	19	N19	5	Incl	MiPRO	0.65	5	high
21	20	N20	15	Incl	MiPRO	0.8	5	low
22	21	N21	13	Incl	MiPRO	0.35	10	medium
23	22	N22	39	Incl	MiPRO	0.5	10	high
24	23	N23	3	Incl	MiPRO	0.65	10	low
25	24	N24	6	Incl	MiPRO	0.8	10	medium
26	25	N25	36	Incl	Diosna	0.35	1	medium
27	26	N26	33	Incl	Diosna	0.5	1	high
28	27	N27	7	Incl	Diosna	0.65	1	low
29	28	N28	4	Incl	Diosna	0.8	1	medium
30	29	N29	14	Incl	Diosna	0.35	5	high
31	30	N30	10	Incl	Diosna	0.5	5	low
32	31	N31	18	Incl	Diosna	0.65	5	medium
33	32	N32	29	Incl	Diosna	0.8	5	high
34	33	N33	9	Incl	Diosna	0.35	10	low
35	34	N34	31	Incl	Diosna	0.5	10	medium
36	35	N35	26	Incl	Diosna	0.65	10	high
37	36	N36	20	Incl	Diosna	0.8	10	low

Figure 12.22: *The final worksheet obtained by manually editing the worksheet for the initial design on the basis of the structure of the second design template.*

Finally, we must evaluate the new design, which can and should be undertaken in the light of the performance statistics of the original design. The final design is capable of supporting the interaction model given in Figure 12.23. The condition number of the selected design and model is approximately 3.3 (Figure 12.23). Hence, in relation to the initial design, there is a slight increase in the condition number, from 3.0 to 3.3. This is mostly attributable to the non-linear time factor.

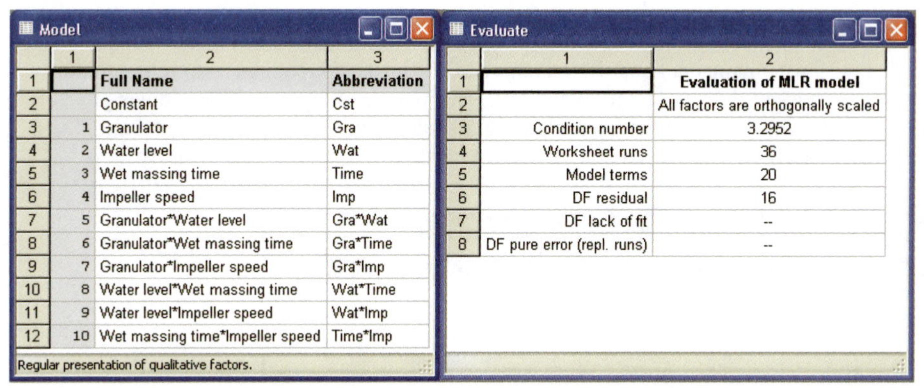

Model

	1	2	3
1		Full Name	Abbreviation
2		Constant	Cst
3	1	Granulator	Gra
4	2	Water level	Wat
5	3	Wet massing time	Time
6	4	Impeller speed	Imp
7	5	Granulator*Water level	Gra*Wat
8	6	Granulator*Wet massing time	Gra*Time
9	7	Granulator*Impeller speed	Gra*Imp
10	8	Water level*Wet massing time	Wat*Time
11	9	Water level*Impeller speed	Wat*Imp
12	10	Wet massing time*Impeller speed	Time*Imp

Regular presentation of qualitative factors.

Evaluate

	1	2
1		Evaluation of MLR model
2		All factors are orthogonally scaled
3	Condition number	3.2952
4	Worksheet runs	36
5	Model terms	20
6	DF residual	16
7	DF lack of fit	--
8	DF pure error (repl. runs)	--

Figure 12.23: *The interaction model supported by the final design (left) and the corresponding matrix condition number (right).*

12.7.6 Summary of the example

From the example, it is apparent that one problem associated with using multilevel quantitative factors is achieving a balanced number of experiments for each defined level of each factor. Going from a linear to an interaction model, or an interaction to a quadratic model, may sometimes alleviate part of this problem. However, a wide variation among a large set of factor combinations is possible by re-coding the multilevel quantitative factors as multilevel qualitative factors. The resultant design, with the qualitative factors, may then be used as a blueprint for creating the final design with the multilevel quantitative factors.

12.8 Questions for Chapter 12

- What is a multilevel qualitative factor?
- Which type of plot is particularly informative for evaluating multilevel qualitative factors?
- How many degrees of freedom are associated with the main term of a five-level qualitative factor?
- How many coefficients of the expanded terms of a five-level qualitative factor are encountered in the regular coefficient mode? In the extended mode?
- What type of design is useful for multilevel qualitative factors?

12.9 Summary and discussion

In this chapter, we described some interesting features pertaining to experimental design with multilevel qualitative factors. One characteristic property of multilevel qualitative factors is that their expanded terms are linked to one another. Hence, each block of associated terms must be treated as an aggregate, and cannot be subdivided in order to consider individual terms separately. In connection with this, we demonstrated that regression coefficient plots may be presented in two styles: regular or extended. Another feature of multilevel qualitative factors is that full factorial designs are usually not useful, because they require too many experimental trials; D-optimal designs are more appropriate.

13 Onion design

13.1 Objective

Onion design is a recent addition to the DOE toolbox. An onion design represents the experimental space as comprising a number of sub-spaces. The experimental domain can be defined by regular factors, such as concentration, temperature and time, but can equally well be constructed from score variables derived from a multivariate projection (PCA or PLS) model. Here we refer to such a sub-space as a "layer" or "shell". A local design is associated with each layer. These local designs can be classical or D-optimal. We discuss two main uses of onion design, namely (i) onion design in multivariate calibration and (ii) onion design in statistical molecular design.

13.2 Introduction

The basic idea with onion design is to select a diverse range of experiments, with uniform coverage across the experimental domain. In this respect, onion designs have similar scope to the better-known uniform coverage designs, such as space filling and grid-based designs. An additional key feature of onion designs is their objective to restrict the actual number of trials as much as possible. Onion designs based on the D-optimal approach combine the best properties of other design types, notably the ability of D-optimal designs to encode relatively few points along with the good inner coverage associated with grid-based and space-filling designs. It is also important to note that, unlike space-filling and grid-based designs, onion designs can be modified to correspond to different regression models, i.e., linear, interaction, quadratic, etc.

The selection of factor combinations encoded by an onion design can be used for two main purposes. The first is to begin from a single experiment and then propose *new experiments* around it. The most logical approach in this case is to use the center-point co-ordinate as the point of departure and then try to populate the experimental space around it with a well-balanced spread of experiments. The second is to start with an inventory of *existing experiments* or trials, a so-called candidate set, from which a sub-set of experiments is selected to represent the variation across the entire candidate set. We envision the first type of onion design to be applicable in multivariate calibration and the second type primarily in statistical molecular design. Examples of both types of application are presented below.

In general, onion design, like other design families discussed in this book, can be used in both screening and optimization. However, unlike other designs, an onion design is not just a single isolated design, but is actually an aggregate design consisting of several smaller building block designs. The interesting aspect is that each individual design is tailored to cover just a small part of the available experimental space. Once created, these individual designs are combined to produce the final onion design. This introduces a flexibility and adaptability not seen in the other design types.

One critical step in the generation of an onion design is the division of the design space into smaller parts. In MODDE, this partitioning is achieved in different ways depending on the type of factor involved, i.e. regular factors or score variables drawn from a projection (PCA/PLS) model. In both approaches, the sub-division always emanates from the defined center-point. When score variables are used, the sub-set definition is distance-based (see Section 13.5). This means that each sub-division of the experimental domain can be interpreted as a spherical "layer" or "shell" surrounding the center-point co-ordinate (Figure 13.1). Thus, the structure of these designs resembles an onion, hence their name.

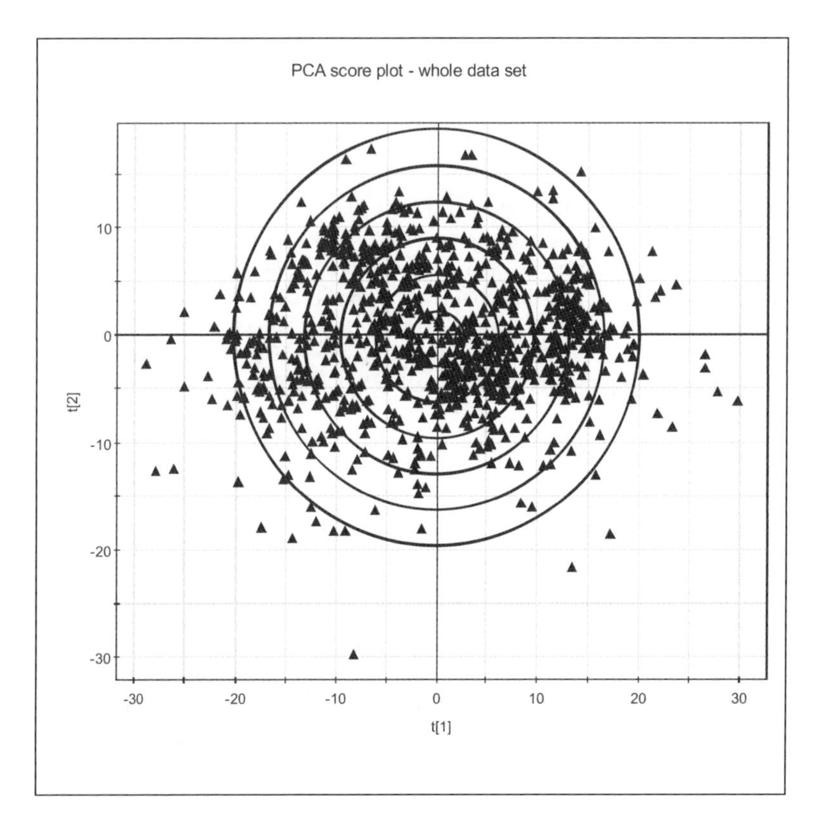

Figure 13.1: *Schematic of the layer-formation in an onion design laid out using score variables.*

When the experimental domain is constructed from regular factors, the division into smaller design regions is also layer-based. In contrast to the process for the score variables, however, the layer formation is driven by binning the original factor ranges into smaller intervals. The layering is therefore *not* distance-based. This means that each resulting layer will appear as a regular square in two dimensions (Figure 13.2), and a cube or hypercube when there are three or more factors. In any onion design, the number of layers is adjustable. The default number of layers in MODDE is three, regardless of the type of onion design.

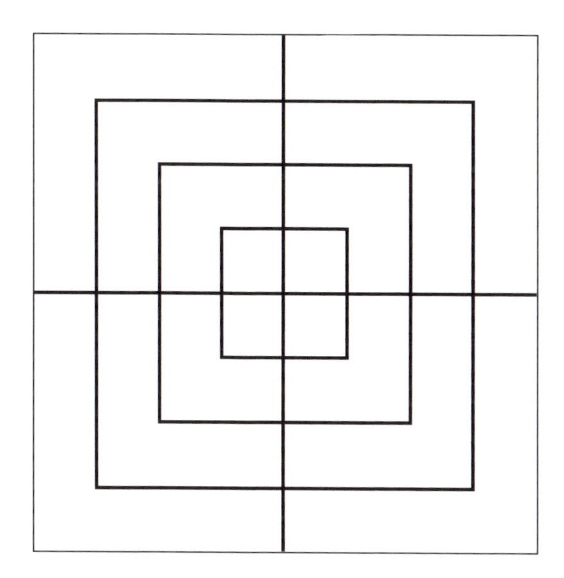

Figure 13.2: Schematic of the layer-formation in an onion design created in regular factors.

13.3 Onion design based on the D-optimal approach

In MODDE, onion design in regular factors is achieved using "Onion D-optimal" design. This design type is applicable to both screening and optimization. In order to discuss some key aspects of this type of design, we shall construct a fictitious example with four regular factors (Figure 13.3). It will be an RSM application. For clarity, the settings of all factors run from -1 to +1.

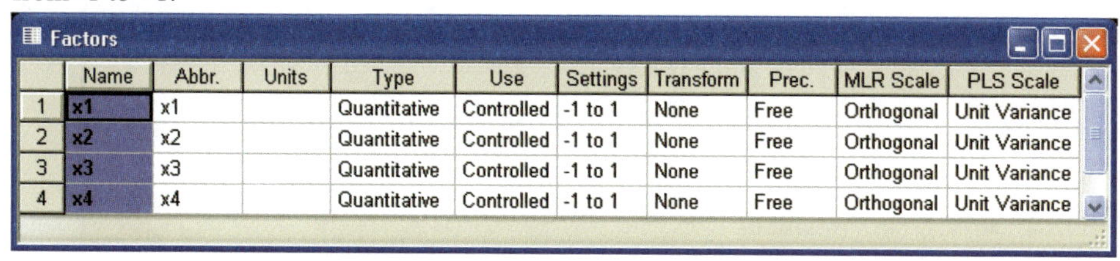

Factors

	Name	Abbr.	Units	Type	Use	Settings	Transform	Prec.	MLR Scale	PLS Scale
1	x1	x1		Quantitative	Controlled	-1 to 1	None	Free	Orthogonal	Unit Variance
2	x2	x2		Quantitative	Controlled	-1 to 1	None	Free	Orthogonal	Unit Variance
3	x3	x3		Quantitative	Controlled	-1 to 1	None	Free	Orthogonal	Unit Variance
4	x4	x4		Quantitative	Controlled	-1 to 1	None	Free	Orthogonal	Unit Variance

Figure 13.3: The four factors used in the illustration of an onion D-optimal design.

Compared with setting up a "normal" design, there are some additional key points to consider when configuring an onion D-optimal design. One such issue concerns the number of layers (Figure 13.4). The default setting is three layers and the higher the number of layers, the more experiments will appear in the final worksheet. Another issue is the type of regression model to be supported in each layer (Figure 13.5). Note that the model that is selected for the outermost layer will determine the model for the entire investigation in MODDE (Figure 13.4). The default setting in RSM is for a linear model in all layers, with the exception of the outermost one, which supports a quadratic model (Figure 13.5).

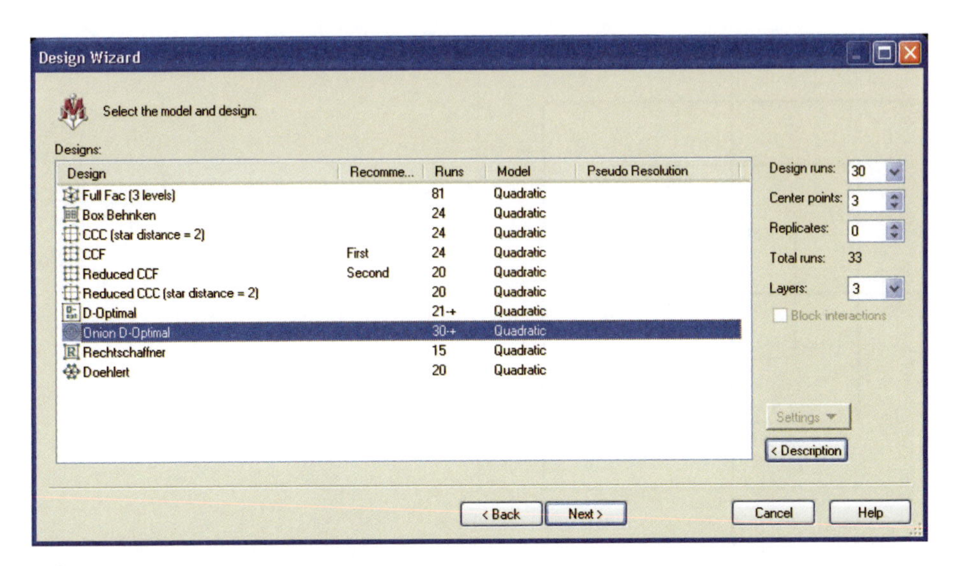

Figure 13.4: When choosing an onion D-optimal design the default model is a quadratic one and the default protocol a design in three layers.

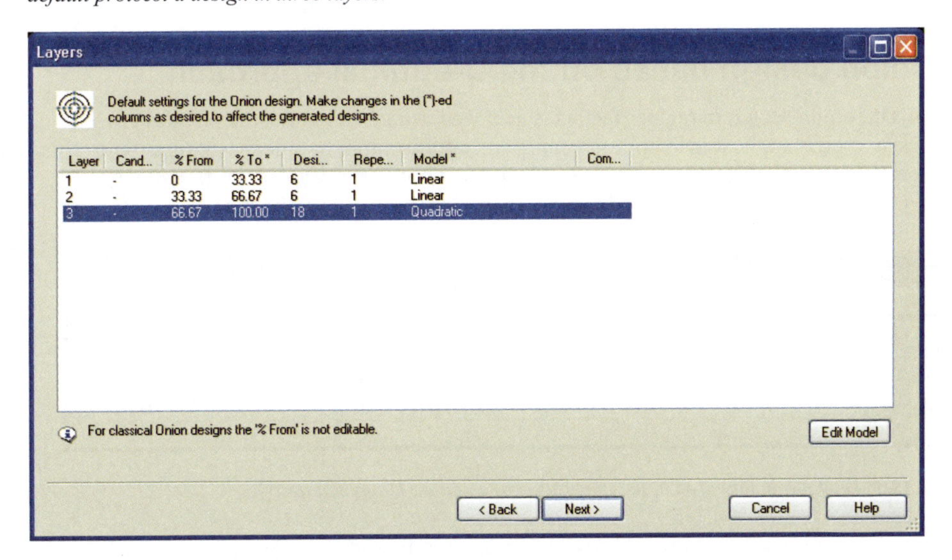

***Figure 13.5:** In an onion D-optimal design the selection of experiments inside each layer can be chosen to support different models, i.e., linear, interaction or quadratic. It is the model selected for the outermost layer that determines the model for the whole investigation in MODDE.*

Of course, it is possible to opt for a quadratic model in all layers of an RSM Onion D-optimal design. However, this will lead to a relatively large number of runs in the final design. The number of runs required can be drastically reduced if linear models are selected for all inner layers in the onion design, since a linear model needs fewer experiments than a quadratic model. We can see from Figure 13.5 that the lead number of runs in our example is six for a linear model and 18 for a quadratic model. The corresponding value for an interaction model would be 12.

After generating a number of tentative D-optimal designs, tailored to the various layers, MODDE displays the recommended number of runs for each layer in a separate plot (Figure

13.6). As seen in Figure 13.6, eight runs are optimal for the two innermost layers and 20 runs for the outer layer. The G-efficiencies for the two inner layers are 100%, indicating that fractional factorials (2^{4-1} designs) have been identified. The G-efficiency of the design covering the third layer is approximately 58%, suggesting that a good design has also been obtained for this design sub-space.

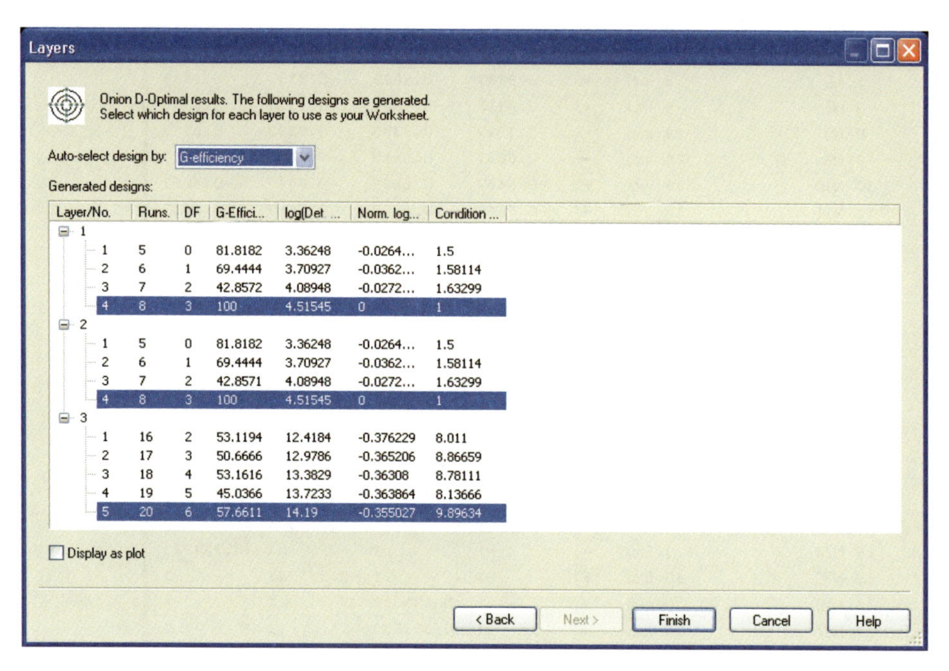

Figure 13.6: *When generating an onion D-optimal design one local D-optimal design is laid out in each layer. Because the performance properties of a D-optimal design may vary within a layer, sub-set selections with different numbers of runs must be tested. The design with the best performance is then chosen to represent the layer. Here, designs are auto-selected according to their G-efficiencies.*

If the recommendations put forward by the software are accepted, the final onion D-optimal design will have 8 + 8 + 20 + 3 (replicates) = 39 runs. The resulting worksheet is presented in Figure 13.7. In this case, the first eight rows are the runs covering the inner layer. Geometrically, they form a regular 2^{4-1} fractional factorial design. Recall from Chapter 10 that factorial and fractional factorial designs are also D-optimal, in fact, they represent the limiting situation for D-optimality. Similarly, the next eight runs, rows 9-16, form another 2^{4-1} fractional factorial design representing the middle layer. The lower part of the worksheet contains experiments covering the outer layer together with the replicated center-points.

Worksheet

	1	2	3	4	5	6	7	8
1	Exp No	Exp Name	Run Order	Incl/Excl	x1	x2	x3	x4
2	1	N1	14	Incl	0.3333	-0.3333	-0.3333	-0.3333
3	2	N2	36	Incl	-0.3333	0.3333	-0.3333	-0.3333
4	3	N3	28	Incl	-0.3333	-0.3333	0.3333	-0.3333
5	4	N4	8	Incl	0.3333	0.3333	0.3333	-0.3333
6	5	N5	21	Incl	-0.3333	-0.3333	-0.3333	0.3333
7	6	N6	5	Incl	0.3333	0.3333	-0.3333	0.3333
8	7	N7	4	Incl	0.3333	-0.3333	0.3333	0.3333
9	8	N8	34	Incl	-0.3333	0.3333	0.3333	0.3333
10	9	N9	24	Incl	0.6667	-0.6667	-0.6667	-0.6667
11	10	N10	31	Incl	-0.6667	0.6667	-0.6667	-0.6667
12	11	N11	18	Incl	-0.6667	-0.6667	0.6667	-0.6667
13	12	N12	13	Incl	0.6667	0.6667	0.6667	-0.6667
14	13	N13	35	Incl	-0.6667	-0.6667	-0.6667	0.6667
15	14	N14	9	Incl	0.6667	0.6667	-0.6667	0.6667
16	15	N15	6	Incl	0.6667	-0.6667	0.6667	0.6667
17	16	N16	26	Incl	-0.6667	0.6667	0.6667	0.6667
18	17	N17	16	Incl	-1	-1	-1	-1
19	18	N18	11	Incl	1	1	1	-1
20	19	N19	15	Incl	-1	1	-1	1
21	20	N20	29	Incl	1	1	-1	1
22	21	N21	33	Incl	1	-1	1	1
23	22	N22	1	Incl	-1	-1	1	0.333333
24	23	N23	27	Incl	-1	-1	0.333333	1
25	24	N24	3	Incl	-1	1	1	-0.333333
26	25	N25	10	Incl	-1	1	-0.333333	-1
27	26	N26	20	Incl	-1	0.333333	1	-1
28	27	N27	37	Incl	-1	-0.333333	1	1
29	28	N28	19	Incl	1	-1	-1	0.333333
30	29	N29	2	Incl	1	-1	-0.333333	-1
31	30	N30	38	Incl	1	1	0.333333	1
32	31	N31	22	Incl	1	0.333333	-1	-1
33	32	N32	12	Incl	0.333333	-1	-1	1
34	33	N33	17	Incl	-0.333333	-1	1	-1
35	34	N34	23	Incl	0.333333	1	-1	-1
36	35	N35	30	Incl	-0.333333	1	1	1
37	36	N36	39	Incl	0	0	0	0
38	37	N37	32	Incl	0	0	0	0
39	38	N38	7	Incl	0	0	0	0
40	39	N39	25	Incl	0	0	0	0

Figure 13.7: *The worksheet for the onion D-optimal design in four factors.*

We shall end the discussion of this fictitious onion D-optimal example by considering the relationship between the final design and the candidate set. For this purpose, we examine the two scatter plots in Figure 13.8. The left-hand plot shows the distribution of experiments in the sub-space defined by the two factors x1 and x2. (Similar scatter plots can be constructed for other combinations of factors, but such plots are not shown.) The design identified has a uniform coverage across the experimental domain. There are relatively few experiments in the inner part of the experimental domain, and more experiments around the periphery. This phenomenon is a consequence of the sequence of models chosen (linear-linear-quadratic).

To provide a balanced picture of the situation, we should also consider the right-hand part of Figure 13.8. Here, open symbols denote the points in the candidate set and the contours of the different layers are clear. The number of entries in the candidate set stemming from the different layers reflect the type of model that was selected for each layer. The sub-set of experiments drawn from the candidate set to form the final design is highlighted by small triangles positioned inside the open symbols.

A detailed inspection of Figure 13.8 reveals that, for the inner and middle layers only, the factorial points have been selected from the candidate set. The axial points have been ignored. This gives two levels per factor per layer. In the outer layer, four levels per factor are exploited, arising from a combination of factorial and non-factorial (but not axial) runs. Evidently, the design runs representing the outer layer have a near-uniform, but not perfectly symmetrical, coverage of "their" sub-space. The arrow in Figure 13.8 indicates the location of a factor combination that is not mapped. This point is not essential to support the outer quadratic model.

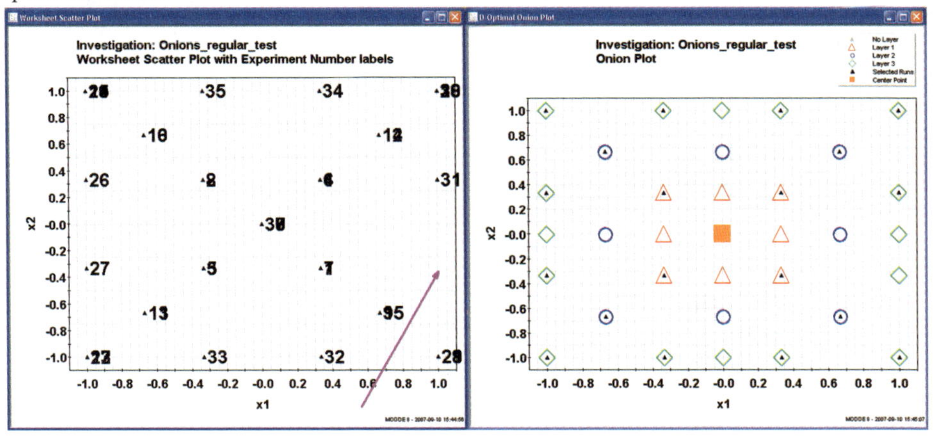

Figure 13.8: Scatter plot in factors x_1 and x_2 showing (left) the distribution of the selected experiments and (right) the size of the candidate set in each of the three layers. In the right-hand plot a little triangle located inside an open symbol denotes a factor combination that is included in the final worksheet.

In summary, by adhering to the default settings in MODDE (three layers with a linear-linear-quadratic chain of models) you will be able to produce an onion D-optimal design with uniform coverage of the accessible experimental domain and where each factor is assigned seven levels. Of course, more levels for the factors are possible if more layers are defined, but there is then a price to pay in terms of a rapidly increasing number of runs. However, regardless of the number of layers, some "obvious" experiments may be left out (cf. the arrow in Figure 13.8), especially in the outermost layer, simply because they are not needed to allow estimates to be made of the chosen model.

In the next section, we shall outline an alternative procedure for onion design in regular factors, which (a) results in a maximum of three layers, (b) provides nine levels of each factor, and (c) yields a symmetrical and balanced selection of experiments in each layer.

13.4 Onion design by overlaying classical designs

In multivariate calibration, DOE is used to define desirable concentrations or "levels" of the analytes in a sample. Thus, a set of reference samples is first mixed according to a selected design and then subjected to spectroscopic measurement. In the subsequent multivariate data analysis, the spectral data are used as the X-block and the known concentrations (from the underlying experimental design) as the Y-block. In that respect, DOE in multivariate calibration is a little different from regular DOE, since the experimental design is used to encode the Y-block and not the X-block.

Because it is desirable for analytical chemists to explore many different concentrations of the various analytes, the designs which are employed usually have a relatively high number of levels, generally in the range seven to nine. This implies that there is a risk that the size of the design – and thereby the set of reference samples – will grow to a considerable size, particularly when there is a large number of analytes, the concentrations of which need to be systematically varied. Onion designs, with their ability to limit the number of required trials, are therefore valuable in multivariate calibration.

Here we devise a procedure for setting up some straightforward onion designs for multivariate calibration. As outlined below, these designs are based on combining two or three classical experimental designs. The basic premise is the combination of one outer CCC with one or two CCFs or full factorials. This results in onion designs each with either seven or nine levels. The designs presented here encompass between two and five factors (i.e., analytes).

When dissecting the onion design structure in more detail it is apparent that we will have to deal with three types of experimental points, namely axial points, center-points and factorial points (Figure 13.9). Some of these points are more important than others, and, depending on the exact positioning of the experiments, some may even be unnecessary.

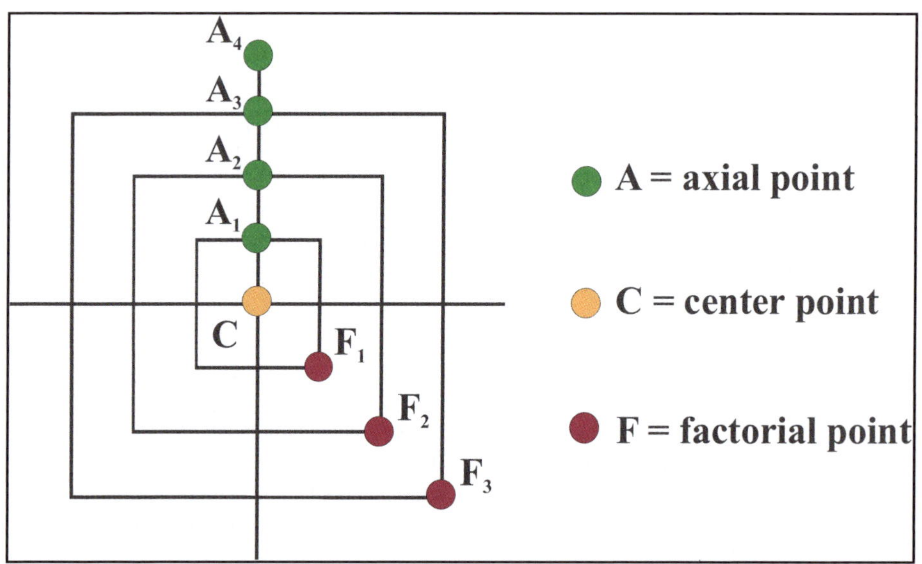

Figure 13.9: *The three types of experimental points that need to be considered when setting up an onion design.*

13.4.1 Some general features of onion design

Here we provide some general advice on how to think when creating an onion design in regular factors from the building blocks of classical designs. The key aspect is that the proposed onion design should be applicable to multivariate calibration. This means that the resulting design should prescribe a variation in each factor (each analyte) encompassing between seven and nine levels. Likewise, it is important to avoid too many runs. Our experience shows that a relatively large number of runs is generally needed in the outer regions of the experimental domain, whilst fewer runs are required the closer we get to the center-point.

From the discussion above it follows that the design intended for the outer layer should preferably be of the RSM type, since such designs generally encode a large number of runs using many levels of the factors being investigated. A classical RSM design in many levels is the CCC design. This design was described in Chapter 6. It has five levels for each factor. Using a CCC in the outer layer, therefore, is beneficial in onion design. The axial point denoted A_4 in Figure 13.9 will be a part of this outer CCC design. Also recall from Chapter 6 that, if we wish to maintain the -1 and +1 settings from the factor definition, a CCF can replace the CCC. However, it is clear that choosing an outer layer CCF will immediately reduce the number of levels in the onion design.

Moreover, there are axial points which are not as essential as A_4 (Figure 13.9). The A_1 point is not needed in a three-layer design. Hence, a two-level factorial design is sufficient as an inner layer protocol in a three layer onion design. In addition, point A_3 is usually not used, unless the outermost design is a CCF. For three or more factors there is yet another opportunity to reduce the number of runs, because the factorial parts of each building block design may well be a fractional factorial design. Whenever fractional factorials are used, different fractions should be selected for the different layers in the onion design.

With these general points in mind, we now outline two simple combinations leading to onion designs with seven and nine levels.

13.4.2 Combinatorial procedures

To accomplish an onion design in seven or nine levels, we propose the following combinatorial schemes:

Seven level onion design (in two layers):

- Inner layer: CCF
- Outer layer: CCC

Nine level onion design (in three layers):

- Inner layer: Two-level full factorial design
- Middle layer: CCF
- Outer layer: CCC

In order to construct such an onion design in MODDE we have to overlay the building block designs sequentially. It is best to start with the outer layer design, since this will make it easier to monitor the necessary changes in the factor settings when switching from one layer to the next. We illustrate this process for a nine level onion design in three factors. As indicated in Figure 13.10, the three example factors are designated as running from -1 to +1.

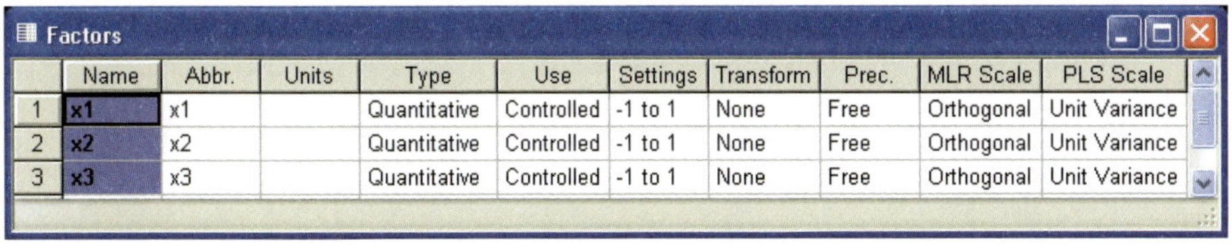

Factors

	Name	Abbr.	Units	Type	Use	Settings	Transform	Prec.	MLR Scale	PLS Scale
1	x1	x1		Quantitative	Controlled	-1 to 1	None	Free	Orthogonal	Unit Variance
2	x2	x2		Quantitative	Controlled	-1 to 1	None	Free	Orthogonal	Unit Variance
3	x3	x3		Quantitative	Controlled	-1 to 1	None	Free	Orthogonal	Unit Variance

Figure 13.10: *The three factors used in the illustration of onion design using classical designs as building blocks. The settings apply to the outer layer CCC design.*

A CCC design augmented with four center-points was generated, resulting in the worksheet in Figure 13.11.

	1	2	3	4	5	6	7
1	Exp No	Exp Name	Run Order	Incl/Excl	x1	x2	x3
2	1	N1	13	Incl	-1	-1	-1
3	2	N2	2	Incl	1	-1	-1
4	3	N3	10	Incl	-1	1	-1
5	4	N4	3	Incl	1	1	-1
6	5	N5	1	Incl	-1	-1	1
7	6	N6	12	Incl	1	-1	1
8	7	N7	8	Incl	-1	1	1
9	8	N8	4	Incl	1	1	1
10	9	N9	5	Incl	-1.682	0	0
11	10	N10	16	Incl	1.682	0	0
12	11	N11	9	Incl	0	-1.682	0
13	12	N12	6	Incl	0	1.682	0
14	13	N13	15	Incl	0	0	-1.682
15	14	N14	14	Incl	0	0	1.682
16	15	N15	11	Incl	0	0	0
17	16	N16	17	Incl	0	0	0
18	17	N17	7	Incl	0	0	0
19	18	N18	18	Incl	0	0	0

Figure 13.11: *The outer layer CCC design in three factors with four center-points.*

In order to specify the middle layer CCF design, we have to initiate a new investigation in MODDE, and in this the factor settings must be modified as shown in Figure 13.12.

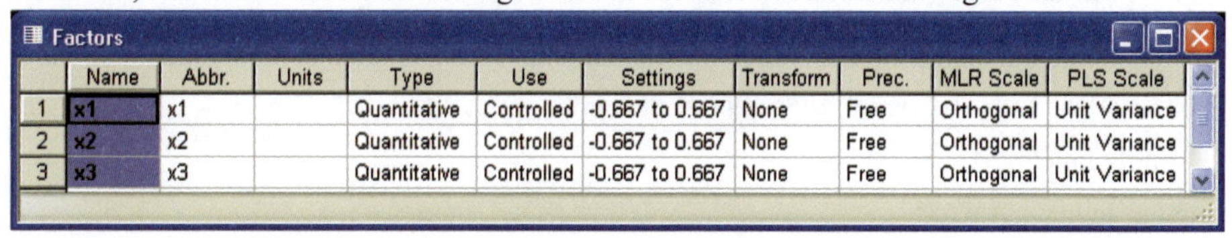

	Name	Abbr.	Units	Type	Use	Settings	Transform	Prec.	MLR Scale	PLS Scale
1	x1	x1		Quantitative	Controlled	-0.667 to 0.667	None	Free	Orthogonal	Unit Variance
2	x2	x2		Quantitative	Controlled	-0.667 to 0.667	None	Free	Orthogonal	Unit Variance
3	x3	x3		Quantitative	Controlled	-0.667 to 0.667	None	Free	Orthogonal	Unit Variance

Figure 13.12: *Modified factor settings for the middle layer CCF design.*

Based on the reduced factor ranges, a CCF design without center-points was constructed. There is no need to include any center-points in this design, since the center-points of the previous CCC are sufficient. Figure 13.13 shows the worksheet for the middle layer CCF design.

Worksheet

	1	2	3	4	5	6	7
1	Exp No	Exp Name	Run Order	Incl/Excl	x1	x2	x3
2	1	N1	17	Incl	-0.667	-0.667	-0.667
3	2	N2	5	Incl	0.667	-0.667	-0.667
4	3	N3	10	Incl	-0.667	0.667	-0.667
5	4	N4	6	Incl	0.667	0.667	-0.667
6	5	N5	2	Incl	-0.667	-0.667	0.667
7	6	N6	15	Incl	0.667	-0.667	0.667
8	7	N7	14	Incl	-0.667	0.667	0.667
9	8	N8	16	Incl	0.667	0.667	0.667
10	9	N9	1	Incl	-0.667	0	0
11	10	N10	13	Incl	0.667	0	0
12	11	N11	3	Incl	0	-0.667	0
13	12	N12	7	Incl	0	0.667	0
14	13	N13	12	Incl	0	0	-0.667
15	14	N14	8	Incl	0	0	0.667

Figure 13.13: *The middle layer CCF design without center-points.*

To facilitate a full factorial design for the inner layer, the factor ranges were again reduced with respect to a new investigation (Figure 13.14).

Factors

	Name	Abbr.	Units	Type	Use	Settings	Transform	Prec.	MLR Scale	PLS Scale
1	x1	x1		Quantitative	Controlled	-0.333 to 0.333	None	Free	Orthogonal	Unit Variance
2	x2	x2		Quantitative	Controlled	-0.333 to 0.333	None	Free	Orthogonal	Unit Variance
3	x3	x3		Quantitative	Controlled	-0.333 to 0.333	None	Free	Orthogonal	Unit Variance

Figure 13.14: *Modified factor settings for the inner layer two-level full factorial design.*

Figure 13.15 shows the worksheet for the factorial design. Again, no center-points are included.

Worksheet

	1	2	3	4	5	6	7
1	Exp No	Exp Name	Run Order	Incl/Excl	x1	x2	x3
2	1	N1	5	Incl	-0.333	-0.333	-0.333
3	2	N2	2	Incl	0.333	-0.333	-0.333
4	3	N3	4	Incl	-0.333	0.333	-0.333
5	4	N4	6	Incl	0.333	0.333	-0.333
6	5	N5	3	Incl	-0.333	-0.333	0.333
7	6	N6	11	Incl	0.333	-0.333	0.333
8	7	N7	1	Incl	-0.333	0.333	0.333
9	8	N8	9	Incl	0.333	0.333	0.333

Figure 13.15: *The inner layer two-level full factorial design without center-points.*

Finally, to produce the nine level onion design, the 14 runs of the middle layer CCF and the eight runs of the inner layer full factorial were copied into the worksheet for the initial outer layer 18 run (including four center-points) CCC. This resulted in the 40 run onion design presented in Figure 13.16. It took us less than five minutes to set it up.

Worksheet

	1	2	3	4	5	6	7
1	Exp No	Exp Name	Run Order	Incl/Excl	x1	x2	x3
2	1	N1	13	Incl	-1	-1	-1
3	2	N2	2	Incl	1	-1	-1
4	3	N3	10	Incl	-1	1	-1
5	4	N4	3	Incl	1	1	-1
6	5	N5	1	Incl	-1	-1	1
7	6	N6	12	Incl	1	-1	1
8	7	N7	8	Incl	-1	1	1
9	8	N8	4	Incl	1	1	1
10	9	N9	5	Incl	-1.682	0	0
11	10	N10	16	Incl	1.682	0	0
12	11	N11	9	Incl	0	-1.682	0
13	12	N12	6	Incl	0	1.682	0
14	13	N13	15	Incl	0	0	-1.682
15	14	N14	14	Incl	0	0	1.682
16	15	N15	11	Incl	0	0	0
17	16	N16	17	Incl	0	0	0
18	17	N17	7	Incl	0	0	0
19	18	N18	18	Incl	0	0	0
20	19	N19	19	Incl	-0.667	-0.667	-0.667
21	20	N20	20	Incl	0.667	-0.667	-0.667
22	21	N21	21	Incl	-0.667	0.667	-0.667
23	22	N22	22	Incl	0.667	0.667	-0.667
24	23	N23	23	Incl	-0.667	-0.667	0.667
25	24	N24	24	Incl	0.667	-0.667	0.667
26	25	N25	25	Incl	-0.667	0.667	0.667
27	26	N26	26	Incl	0.667	0.667	0.667
28	27	N27	27	Incl	-0.667	0	0
29	28	N28	28	Incl	0.667	0	0
30	29	N29	29	Incl	0	-0.667	0
31	30	N30	30	Incl	0	0.667	0
32	31	N31	31	Incl	0	0	-0.667
33	32	N32	32	Incl	0	0	0.667
34	33	N33	33	Incl	-0.333	-0.333	-0.333
35	34	N34	34	Incl	0.333	-0.333	-0.333
36	35	N35	35	Incl	-0.333	0.333	-0.333
37	36	N36	36	Incl	0.333	0.333	-0.333
38	37	N37	37	Incl	-0.333	-0.333	0.333
39	38	N38	38	Incl	0.333	-0.333	0.333
40	39	N39	39	Incl	-0.333	0.333	0.333
41	40	N40	40	Incl	0.333	0.333	0.333

Figure 13.16: *The final nine level onion design in three factors.*

The geometry of the three-layer onion design is presented in Figure 13.17 by means of 2D- and 3D-scatter plots. There is a good spread of the experimental points.

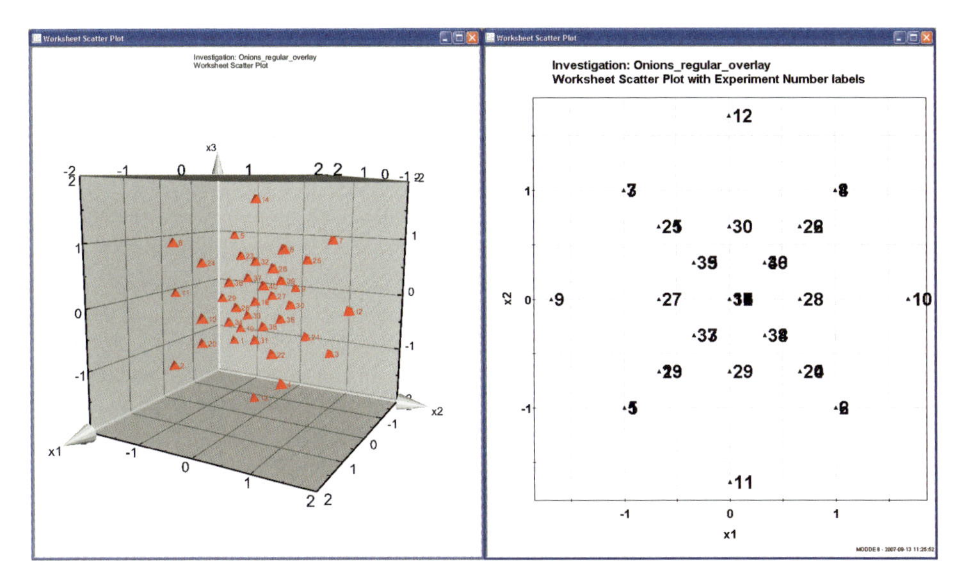

Figure 13.17: *Scatter plots showing how the 40 runs of the onion design spread out across the experimental domain.*

The summary provided in Figure 13.18 indicates how many runs are needed in onion designs for multivariate calibration involving two to five analytes (factors). We have coined tentative names for these designs using the naming principle "number of... factors _ shells _ levels". Hence, the 3_3_9 onion design is the one illustrated graphically in Figure 13.17.

In the table below, the total number of runs includes the number of factorial points, the number of axial points, plus four mandatory center-points. Onion designs involving four and five factors, especially, tend to have a larger number of runs than might be acceptable in an application. In such cases, it is possible to further decrease the number of necessary runs by using fractional factorials instead of full factorials to encode the factorial points. The number of runs achieved with this sort of minimalistic approach is given in the right hand column of Figure 13.18. It should also be noted that when fractional factorials are used, we advocate different fractions being used for the different layers (shells).

Name	Factors	Shells	Levels	Runs	F-points	A-points	C-points	Min.
2_2_7	2	2	7	20	8	8	4	N/A
2_3_9	2	3	9	24	12	8	4	N/A
3_2_7	3	2	7	32	16	12	4	24
3_3_9	3	3	9	40	24	12	4	28
4_2_7	4	2	7	52	32	16	4	36
4_3_9	4	3	9	68	48	16	4	44
5_2_7	5	2	7	88	64	20	4	56
5_3_9	5	3	9	120	96	20	4	72

Figure 13.18: *Overview of some onion designs derived from classical building block designs.*

13.5 Onion design in latent variables (score variables)

Onion design is not necessarily restricted to regular factors. Indeed, onion designs may well be applied to latent variables, or score variables, extracted from a multivariate projection model. When such latent variables relate to properties of molecules they are often called principal properties and therefore the notation statistical molecular design (SMD) is commonly used to refer to designs based on principal properties. For an introduction to SMD see Eriksson, L., et al., Multi- and Megavariate Data Analysis, Part II, Umetrics Academy.

Thus, statistical molecular design (SMD) uses DOE with PCA or PLS scores to select sub-sets of molecules which are representative, informative and diverse. In MODDE, one of the eligible onion designs is based on scores imported from an existing SIMCA-P$^+$ project, using the score vectors as factors. With an onion design based on score variables, the user specifies how many layers are required and, for each layer, its size, number of compounds (runs), number of repetitions and the type of regression model supported.

The purpose of this last tutorial example is to illustrate how an onion design in score variables is set up. It is assumed that an appropriate SIMCA-P$^+$ file, a USP file, is available. The example data set is called ChemGPS, and is used with permission from Johan Gottfries. It deals with a project aimed at defining a stable drug space for orally active drugs (554 compounds). The multivariate data analysis of the ChemGPS data set is detailed in Eriksson, L., et al., Multi- and Megavariate Data Analysis, Part II, Umetrics Academy.

13.5.1 Configuring the Investigation

In the Design Wizard we first have to select 'Advanced Designs' (Figure 13.19).

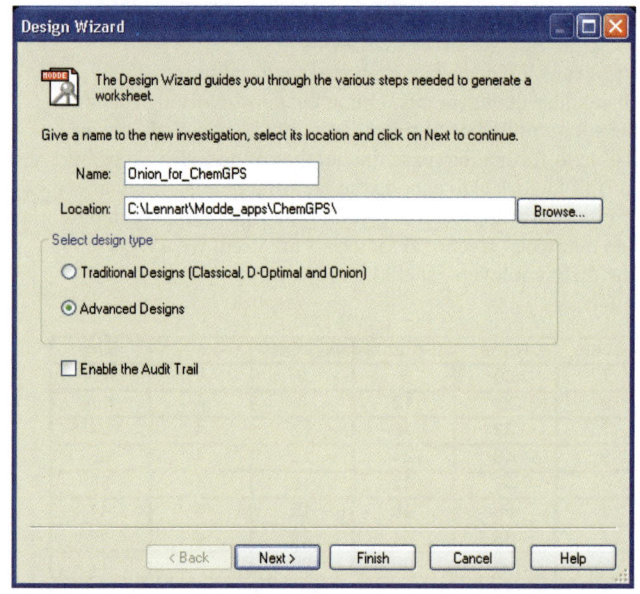

Figure 13.19: To enable onion design in latent variables one must choose the Advanced Designs option.

Then we select the MV Designs option (Figure 13.20), browse to the appropriate SIMCA-P$^+$ project, then select the model number in the corresponding USP-file. We must specify the number of components to consider in the onion design. Here, we work with three significant principal components accounting for 66% of the chemical variance.

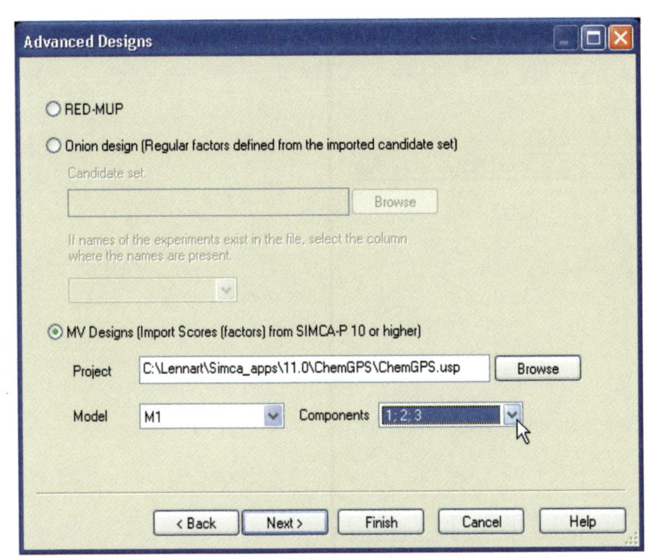

Figure 13.20: *The software must be directed to a USP file from which to import the score vectors.*

In the case of a score variable based onion design, the software will automatically configure the factors according to the information attached to the score variables just imported (Figure 13.21). This means that the Name, Abbreviation and Settings of the factors do *not* need to be entered manually. The low and high values correspond to, respectively, the minimum and maximum score values of the different score variables.

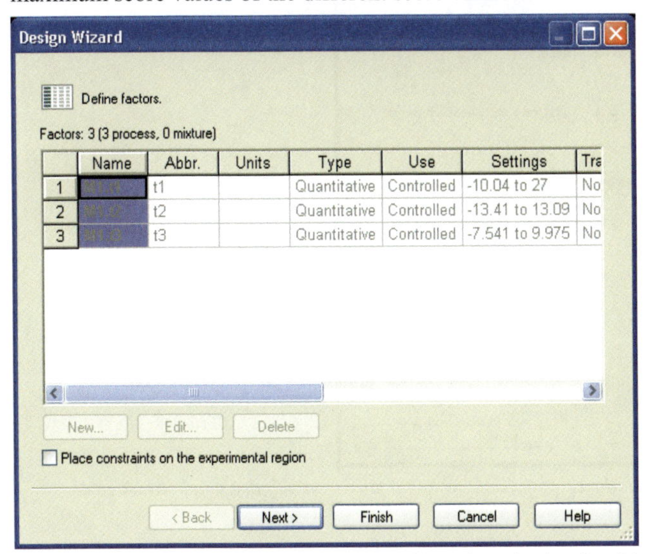

Figure 13.21: *The factors are pre-configured based on the details of the imported score vectors.*

The next step corresponds to defining the responses. However, in most cases involving onion design for sub-set selection of molecular structures, this step can be skipped (Figure 13.22) – there are simply no responses to account for.

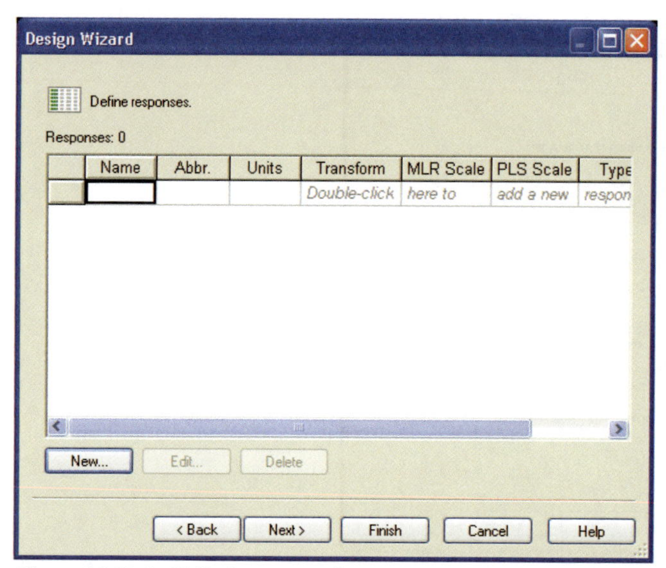

Figure 13.22: *In SMD there is usually no response variable available.*

We then select RSM, since our goal is to select a sub-set with good coverage of the chemical domain of the orally active drugs (Figure 13.23).

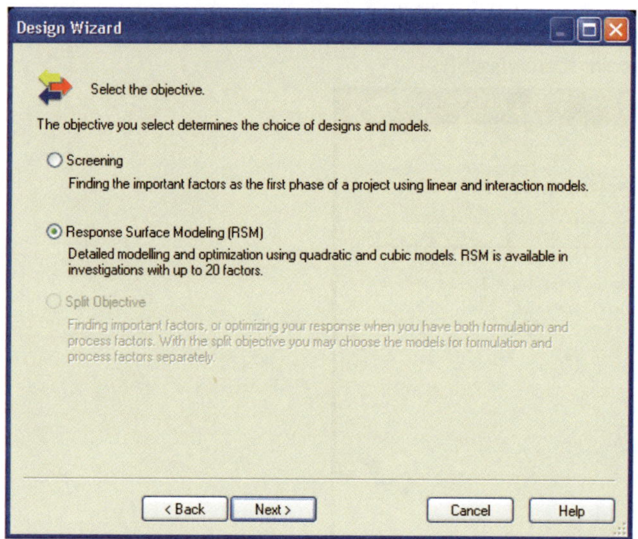

Figure 13.23: *RSM is the appropriate objective in SMD when the aim is to acquire good coverage of the available experimental domain.*

Subsequently, we select Onion D-optimal design (Figure 13.24) and change the defaults Center Points = 3 to 1 and Layers = 3 to 5. By using more than three layers it is possible to increase the number of selected molecules a little, which is advantageous in this case.

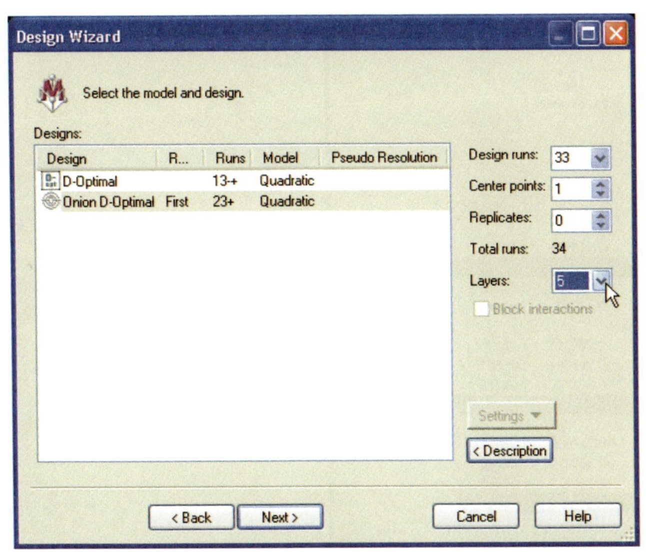

Figure 13.24: *In order to accomplish the desired design an Onion D-optimal design is chosen.*

In the subsequent Layers window (Figure 13.25) one can modify the size of each layer. Note that in MODDE these layers are editable <u>only</u> when working with latent variables as factors. By default, the layer formation is accomplished using a linear sub-division principle. The user may make it non-linear. To avoid compounds with extreme chemical properties, so called outliers, it is advisable to alter the upper limit of the outermost layer (here set to 97% rather than 100%).

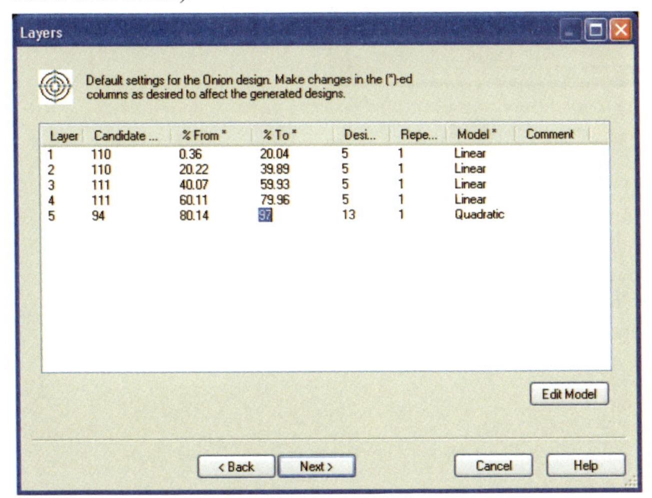

Figure 13.25: *By manipulating the size of the different layers one may exclude outliers and depart from a linearly based layering system.*

In the next stage, MODDE generates a series of D-optimal designs, each restricted to the size of the layer in question (Figure 13.26). Finally, one design for each shell of the onion design is auto-selected according to the selected evaluation criterion (either G-efficiency, Determinant or Condition Number). This step is similar to onion designs generated D-optimally for regular factors.

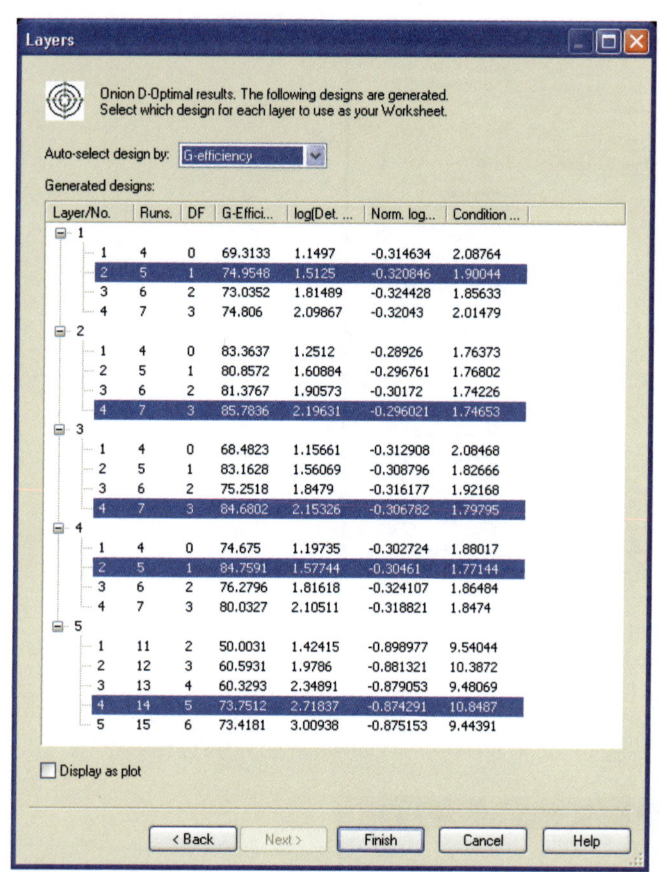

Figure 13.26: *A series of tentative D-optimal designs are generated for each layer. The user has to select which design for each layer will be used as the worksheet.*

The selected sub-set of diverse and representative compounds is shown in the resultant worksheet (Figure 13.27). The column denoted Molecule Name, displays the content of the first Secondary Observation ID used in SIMCA-P^+. (In the current example data set the first Secondary Observation ID corresponds to the Simplified Molecular Input Line Entry Specification – SMILES – code). The numerical values for the three factors correspond to the scores of the compounds for the three principal components.

Worksheet

	1	2	3	4	5	6	7
1	Exp No	Molecule Name	Run Order	Incl/Excl	M1.t1	M1.t2	M1.t3
2	1	CC(Nc1ccc(Br)cn1)n2cncn2	20	Incl	-3.42433	-0.076701	-0.761929
3	2	O=C(Nc1snnc1C(=O)O[n+]2cccc3ccccc23)Nc4ccccc4	10	Incl	2.00901	-1.75997	-1.15524
4	3	O=C(COCC(=O)NN=Cc1ccco1)NN=Cc2ccco2	18	Incl	1.7418	3.10136	0.093803
5	4	CCCCNC(=O)NS(=O)(=O)c1ccc(C)cc1	12	Incl	-1.60212	1.36694	1.39882
6	5	Cc1cccc(OCC(O)CN2CCN(CC2)C(=O)c3ccc(Br)cc3)c1	4	Incl	1.76476	-2.84493	0.662996
7	6	COc1ccccc1OCCNCC(O)c2ccc(C)c(c2)S(=O)(=O)N	9	Incl	3.07046	2.06155	1.38338
8	7	CC(C)(C)NCC(O)COc1ccccc1C2CCCC2	21	Incl	-1.41689	-3.12392	1.40717
9	8	Cc1ncc2C(OCc2c1O)c3ccc(Cl)cc3	39	Incl	-2.72037	-2.51438	-1.39289
10	9	CC(CS)C(=O)N1CCCC1C(=O)O	2	Incl	-3.42389	2.21462	1.26295
11	10	CCCCC1=NC2(CCCC2)C(=O)N1Cc3ccc(cc3)c4ccccc4c5nnn[nH]5	32	Incl	3.43848	-3.36505	-0.190284
12	11	Nc1ccn(C2CSC(CO)O2)c(=O)n1	6	Incl	-2.46199	3.51554	-0.923958
13	12	COc1cc2nc(nc(N)c2cc1OC)N3CCN(CC3)C(=O)c4ccco4	30	Incl	3.56558	1.12066	-1.61098
14	13	NC1C2CN(CC12)c3nc4n(cc(C(=O)O)c(=O)c4cc3F)c5ccc(F)cc5F	37	Incl	3.4549	0.548121	-2.18781
15	14	COc1ccc(cc1)c2cc(=S)ss2	36	Incl	-4.99557	-3.85417	-0.268402
16	15	OC(=O)C=CC(=O)Nc1cccc(C=CC(=O)c2cccc(NC(=O)C=CC(=O)O)c2)c1	35	Incl	5.47618	2.77607	1.11763
17	16	CN(C)S(=O)(=O)NC(O)C(Cl)(Cl)Cl	27	Incl	-3.59391	3.36471	1.86204
18	17	CCOC(=O)N1CCC(=C2c3ccc(Cl)cc3CCc4cccnc24)CC1	16	Incl	1.06648	-5.25788	-1.48906
19	18	S=c1[nH]c2[nH]c(=S)[nH]c(=S)c2[nH]1	19	Incl	-3.51846	3.29835	-2.06673
20	19	CCOC(=O)C1=C(C)NC(=C(C1c2ccccc2C=CC(=O)OC(C)(C)C)C(=O)OCC	8	Incl	3.7152	-3.06722	2.10082
21	20	COc1cc(ccc1Cc2cn(C)c3ccc(NC(=O)OC4CCCC4)cc23)C(=O)NS(=O)(=	17	Incl	8.32001	-3.12651	0.330028
22	21	CN1CCC2=C(C1)C(c3ccccc23)c4ccccc4	14	Incl	-3.06456	-6.40099	-2.17968
23	22	OC1(CC2C3COC1O3)NC(=O)C2N(=O)=O	24	Incl	-1.97193	5.63381	-2.83185
24	23	CN(CCN(C)CCC(=O)NCCN)CCC(=O)NCCN	26	Incl	1.91365	5.00869	3.28435
25	24	CSSCSSC	23	Incl	-8.07294	-2.69295	2.15916
26	25	CC(C)(C)NC(=O)C1CN(Cc2cccnc2)CCN1CC(O)CC(Cc3ccccc3)C(=O)NC	33	Incl	9.51855	-3.04579	1.33887
27	26	ClC=CCl	25	Incl	-10.0365	-3.14501	0.519813
28	27	O=c1oc(=O)c2c1oc3c(=O)oc(=O)c32	7	Incl	-3.07068	5.00359	-3.72482
29	28	OC(=O)CC(SC(CC(=O)O)C(=O)O)C(=O)O	13	Incl	-0.62351	8.71465	1.84722
30	29	c1ccc2c(c1)n3c4ccccc4n2c5ccccc53	22	Incl	-2.88777	-6.68532	-3.79712
31	30	O	28	Incl	-9.82124	7.14439	-0.798358
32	31	C1C2(CC3(CC1(CC(C2)(C3)c4ccccc4)c5ccccc5)c6ccccc6)c7ccccc7	5	Incl	0.428825	-12.0001	-1.25818
33	32	CCCCCCCCCCCCCCCCN(=O)(C)C	1	Incl	-1.51727	-6.07323	5.7327
34	33	OC(=O)CCCCCCC(=O)CCCCCCCCC(=O)CCCCCC(=O)O	34	Incl	3.99185	-0.280972	6.76353
35	34	CC(=O)OC1C2=C(C)C(CC(O)(C(OC(=O)c3ccccc3)C4C5(COC5CC(O)C4(	3	Incl	16.4903	-3.35067	-0.747986
36	35	Cc1cn(C2CC(O)C(COP(=O)(O)OC3CC(OC3COP(=O)(O)O)n4cc(C)c(=O)[	38	Incl	12.0679	9.96732	-1.46403
37	36	COc1cc2CC[N+](C)(C)C3Cc4ccc(O)c(Oc5cc6C(Cc7ccc(Oc(c1O)c32)cc7	29	Incl	10.5198	-4.4294	-5.68526
38	37	NCCNC(=O)CCN(CCN(CCC(=O)NCCN)CCC(=O)NCCN)CCC(=O)NCCN	31	Incl	9.77975	8.81141	4.65301
39	38	NC(=N)NCCCC(NC(=O)N)C(=O)NC(CCCNC(=N)N)C(=O)NC(=O)O	15	Incl	7.9707	13.095	2.3228
40	39	Cn1cc(OCC(O)CNC(C)(C)C)c2ccccc2c1=O	11	Incl	-0.142823	-0.546701	-0.072156

Figure 13.27: *In SMD the worksheet shows the SMILES code of each selected entry, which indicates the structure of each molecule.*

A 3D scatter plot is useful to evaluate the distribution of the substances identified (Figure 13.28). All compounds in the data set, whether selected or not, are plotted. The extent of each layer is indicated by symbol- and color-coding. The selected compounds are indicated by enlarged symbols.

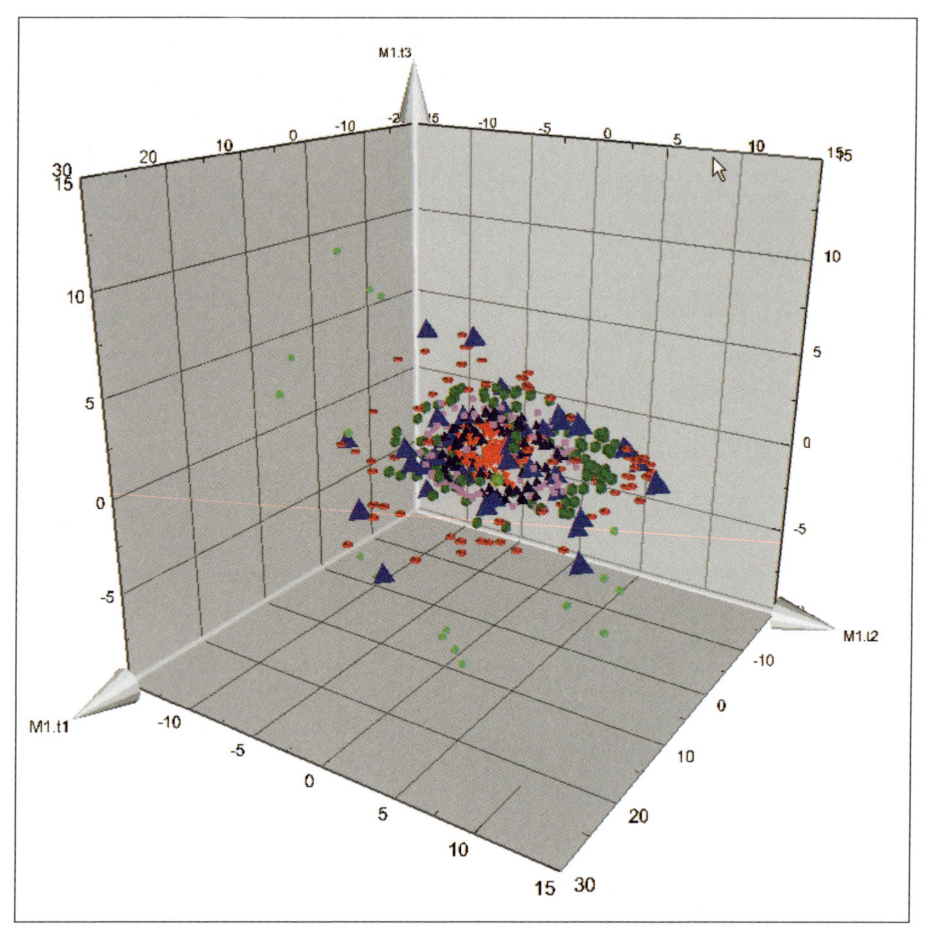

Figure 13.28: An onion 3D scatter plot shows which substances are selected.

13.5.2 Transfer of sub-set selection to SIMCA-P+ (Advanced Section)

This section describes how the contents of the onion design can be brought back into SIMCA-P+ to allow the structures of the selected compounds to be displayed. Note: This is an advanced part of the tutorial, which requires access to EXCEL or a similar spreadsheet program.

- Step 1: In MODDE, Copy the Candidate Set to an EXCEL sheet (here called Sheet 1).
- Step 2: In SIMCA-P+, click on Plot/List | Lists, select Item T2 Range, from components 1–3. Click Add Series and press OK. In the resulting list, mark the T2 Range column, right click and select Sort Ascending.
- Step 3: Copy the sorted T2 Range list to an EXCEL sheet (here called Sheet 2). Note: Now the order of the compounds in Sheet 1 and Sheet 2 is identical.
- Step 4: From EXCEL Sheet 1, copy the column labeled Design Run to EXCEL Sheet 2.
- Step 5: In Sheet 2, re-sort the T2 Range list – now appended with the Design Run column of sheet 1 – according to the Primary Observation ID.

- Step 6: In SIMCA-P+, click on DataSet | Edit to display the raw data. Right click and choose Add Secondary Observation ID, enter a name and press OK.
- Step 7: Copy the column entitled Design Run from EXCEL Sheet 2 to the new Secondary Observation ID column in SIMCA-P+. The new secondary ID is now ready for use.

In order to use the new secondary ID, the Item Information window must be active. Create a score scatter plot and tile it alongside the Item Information window. The plot below (Figure 13.29) illustrates how different symbols, colors, and the SMILES notation can be used to evaluate the onion sub-set selection. Large symbols indicate the onion design content. The molecular structures of two highlighted compounds are displayed.

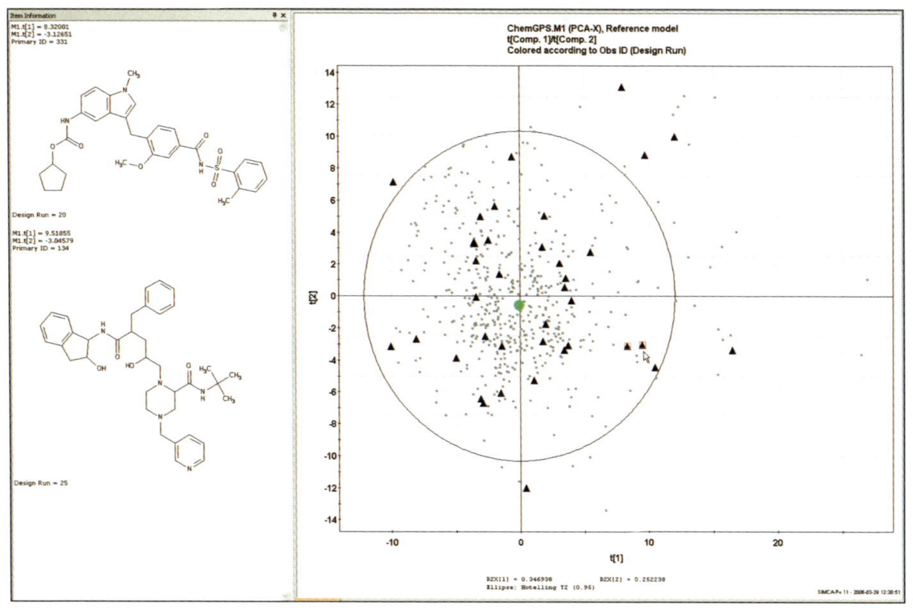

Figure 13.29: Distribution of selected compounds. Through the use of the SMILES code it is possible to view the chemical structures of the selected molecules.

13.6 Questions for Chapter 13

- What is an onion design?
- What is the rationale behind onion design?
- Which classical designs are particularly relevant as building blocks to produce onion designs for regular factors?
- How many levels per factor are recommended for a multivariate calibration design?
- What is statistical molecular design?

13.7 Summary and discussion

Onion design represents a new approach to diversity selection of experiments, where the need to restrict the actual number of trials as much as possible is pivotal. Importantly, onion designs can be laid out both in regular factors and in score variables. A diversity selection encoded by an onion design can be used for two main objectives. First, one may depart from a single key experiment and construct new informative experiments around it. Secondly, one may start from an inventory of existing experiments or trials, and then select a sub-set of diverse and representative experiments.

There are a number of adjustable parameters involved in an onion design, for example the number of factors, the number of layers, the size of each layer, the type of model, the number of runs, the number of repetitions, all of which influence the size and shape of the final protocol. This makes onion designs adaptable to almost any challenge encountered in DOE. Hence, onion design constitutes an interesting and flexible approach to DOE, resulting in parsimonious many-level designs of high quality. The last point is underlined by looking at the correlation matrix of the 3_3_9 onion design presented in Section 13.4; the quadratic terms are essentially uncorrelated (Figure 13.30).

Correlation Matrix

	1	2	3	4	5	6	7	8	9	10
1		x1	x2	x3	x1*x1	x2*x2	x3*x3	x1*x2	x1*x3	x2*x3
2	x1	1	0	0	0	0	0	0	0	0
3	x2	0	1	0	0	0	0	0	0	0
4	x3	0	0	1	0	0	0	0	0	0
5	x1*x1	0	0	0	1	0.0388048	0.0388048	0	0	0
6	x2*x2	0	0	0	0.0388048	1	0.0388048	0	0	0
7	x3*x3	0	0	0	0.0388048	0.0388048	1	0	0	0
8	x1*x2	0	0	0	0	0	0	1	0	0
9	x1*x3	0	0	0	0	0	0	0	1	0
10	x2*x3	0	0	0	0	0	0	0	0	1

Threshold 0.3

Figure 13.30: *Correlation matrix of the 3_3_9 onion design delineated in Section 13.4.*

14 Blocking the experimental design

14.1 Objective

This chapter discusses how to block an experimental design. Blocking is used to supplement randomization. Blocking is needed when one or more external factors change during the period when the experiment is run, so that not all runs of the design are affected by the same level(s) of the external factor(s). For example, a chemical synthesis is investigated by means of DOE. The catalyst comes in bottles of a size that allows only four experimental runs per bottle. Hence, bottle to bottle variations may make runs impossible to compare. If "bottle" is introduced as a blocking factor, the design can be organized in such a way that the results for each bottle are comparable. Hence, blocking of a design makes the block factors appear as ordinary factors in the design, thereby accounting for the influence of the block factor(s), and hence ensuring that all factors are considered in a balanced way. The cost of blocking is an increase in the number of runs, larger designs; as always, it is the user's judgement that determines the route to follow.

14.2 Introduction

Randomization and Blocking are two useful principles to combat and control sources of unwanted variability in measured data. Put simply, it is always recommended that the run order of the experiments in a worksheet is randomized, and that blocking is only performed when full randomization is inappropriate. In this chapter we attempt to shed some light on how to perform blocking. But before we do so, we start by discussing why randomization is necessary and why blocking may sometimes be used to address issues beyond the scope of randomization.

The rationale behind randomization is very simple: randomizing all the runs belonging to an experimental plan will transform any systematic effect(s) of an uncontrolled factor into random, experimental noise. Thus, randomization can be seen as a safeguard against undesired sources of extraneous systematic variability. For example, an environment that often results in uncontrolled factors is associated with the operating conditions of machinery used in a process. Such operating conditions may not remain perfectly consistent over time. Instead, many machines run at a nominal level disturbed by noise, which is correlated from one moment to the next. For example, if a refrigerator has become slightly too cold at one point during the day, it will still be slightly too cold a few minutes later. Analogous phenomena occur with heaters, furnaces, gas flow controllers, and so forth. Statisticians and engineers refer to such behaviour as autoregressive. Randomization in time will eliminate whatever systematic effects this undesired correlation might have on an experiment.

Sometimes, however, due to experimental restrictions, it may not be appropriate to perform all experiments in a completely random order. Let us consider an example that one of the authors experienced a while ago. In a screening experiment in a process industry the temperature needed to be varied between a low level, 700 °C, and a high level, 1000 °C. Due to the unwieldy nature of the process it would take almost 24 hours for any temperature change to equilibrate and thereafter remain stable. Consequently, it was considered impractical to apply full randomization to the temperature factor.

The workaround was to re-arrange the experimental protocol proposed initially, a screening design in 8+4 runs, and instead to sort the runs in ascending order according to temperature. This resulted in three sub-sets of four experiments using temperatures of 700, 850 and 1000 °C. Within each quartet full randomization was used. Moreover, it should be realized that while this way of forming sub-groups of experiments provided a practical solution to the problem of the slow temperature stabilization, there could be other ways of addressing this issue which are more favorable from a theoretical perspective. This is because the sub-sets of experiments formed using the temperature-block approach may not necessarily be orthogonal to one another. The latter aspect is a key point better dealt with using formal blocking procedures.

In general, when you cannot conduct all the experiments in a homogeneous way, randomizing your experiments may not be sufficient to deal with any undesired systematic variability that does occur. Blocking the experiments into synchronized groups may, in such a case, help to decrease the impact of such variability on the estimated factor effects. Suppose we are running a full factorial design in five factors and 32 runs. Further, suppose the batch size of the main raw material only permits eight trials per batch. In such a case, we may want to run our experiments in four blocks, each comprising eight runs, using homogeneous starting material. The method of *Orthogonal Blocking* then makes it possible to divide the 32 experimental trials into four blocks balanced in such a way that the difference between the blocks – the raw material – does not affect estimates of the factor effects. Below, using an example, we illustrate how orthogonal blocking can be accomplished using MODDE.

14.3 Blocking used in a real example

14.3.1 A small chemical investigation

In a chemical experiment, the influence of two factors (time and temperature) on the yield of the main product was investigated (Figure 14.1). Initially a 2^2 factorial design, augmented with two center-points, was performed. Preliminary examination of the results indicated that the design was correctly positioned in the experimental space and hence there was no need to adjust the low and high settings of the factors. However, there was some indication of a non-linear relationship between the factors and the response, so the design was upgraded to a CCC design by adding the star points and two additional center-points. This design comprised 12 experiments: four corner points, four star points and two plus two center-points. This data set is taken from Box, G.E.P., Hunter, W.G., and Hunter, J.S., Statistics for experimenters, John Wiley & Sons, 1978, p. 519.

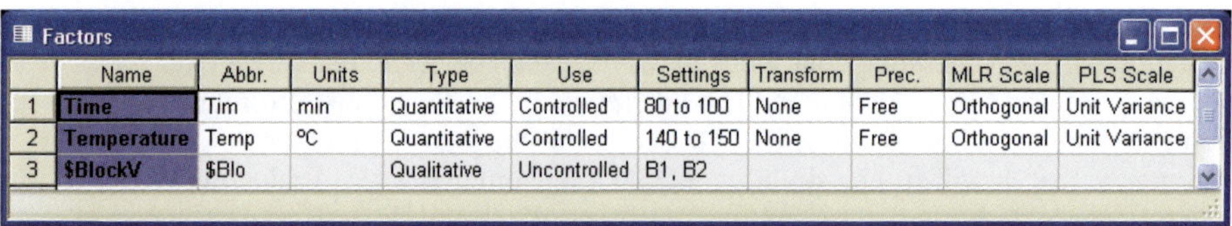
Factors

	Name	Abbr.	Units	Type	Use	Settings	Transform	Prec.	MLR Scale	PLS Scale
1	Time	Tim	min	Quantitative	Controlled	80 to 100	None	Free	Orthogonal	Unit Variance
2	Temperature	Temp	°C	Quantitative	Controlled	140 to 150	None	Free	Orthogonal	Unit Variance
3	$BlockV	$Blo		Qualitative	Uncontrolled	B1, B2				

Figure 14.1: *The two factors varied in the chemical example. The third factor is the block factor, automatically appended by the software when Blocking is invoked.*

The objectives in this example were two-fold:

- to identify the optimal settings for time and temperature,

- to investigate whether there was evidence of a shift in the response data between the two series of experiments (i.e., whether or not there were significant block effects).

14.3.2 Defining the DOE problem using Blocking

The Design Wizard in MODDE supports blocking (Figure 14.2). In the chemical example, the number of blocks was set to two, using two center-points per block (Figure 14.2). The meaning of block interactions will be expanded upon below (Section 14.4.7). The CCC design is blockable, but it should be noted that this is not the case for all the design types implemented in MODDE. For example, the CCF design is not blockable. Whenever a non-blockable design is selected, the blocking functionality is deactivated. An overview of blockable and non-blockable designs is presented in Section 14.4.

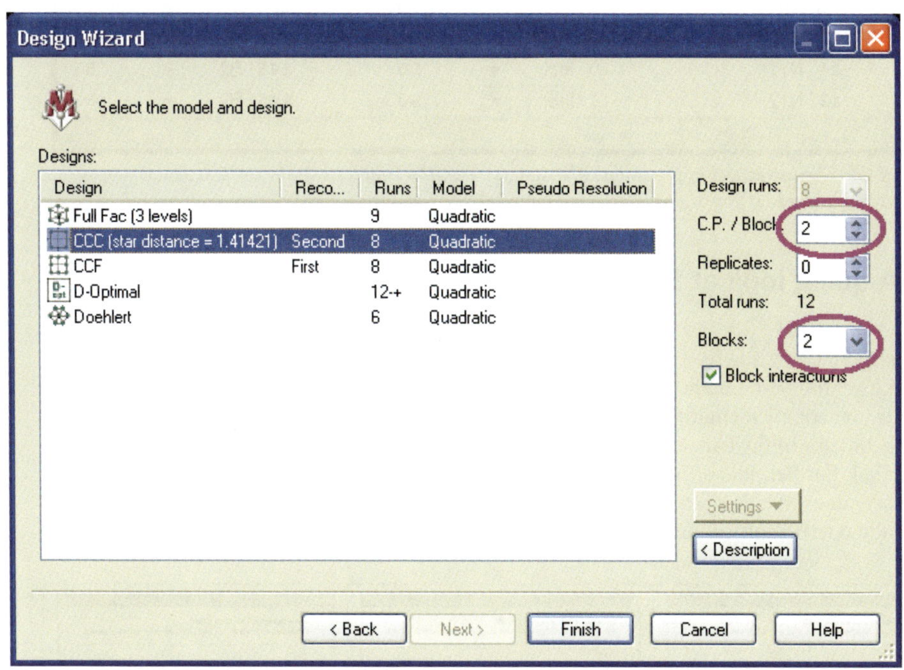

__Figure 14.2:__ To accomplish the experimental design used in the chemical example, two blocks and two center-points per block were chosen.

The resulting worksheet is presented in Figure 14.3. The third factor, denoted $BlockV, has been added by the software. It signifies which experiments were part of the first sampling campaign and which were part of the second. The block-factor gives rise to a main effect term in the resulting regression model. If non-significant, this term can be removed from the model. On the other hand, if it is significant, then there is evidence of a systematic shift in the y-data between the two points in time at which the two blocks of experiments were carried out.

Worksheet

Exp No	Exp Name	Run Order	Incl/Excl	Time	Temperature	$BlockV	Yield
1	N1	3	Incl	80	140	B1	78.8
2	N2	4	Incl	100	140	B1	84.5
3	N3	5	Incl	80	150	B1	91.2
4	N4	2	Incl	100	150	B1	77.4
5	N5	6	Incl	90	145	B1	89.7
6	N6	1	Incl	90	145	B1	86.8
7	N7	11	Incl	76	145	B2	83.3
8	N8	7	Incl	104	145	B2	81.2
9	N9	9	Incl	90	138	B2	81.2
10	N10	8	Incl	90	152	B2	79.5
11	N11	10	Incl	90	145	B2	87
12	N12	12	Incl	90	145	B2	86

Figure 14.3: The worksheet of the chemical example. The Block factor signifies which experiments belong to each block. If non-significant, this factor may be removed from the regression model.

14.3.3 A quick look at the geometric properties of the resultant design

It is instructive to visualize the design region in terms of a scatter plot, to see the CCC design both in its entirety and in the form of the individual blocks (Figure 14.4). With the two sub-set plots, the complementarities of the two blocks are immediately clear. This harmonization of the geometry of the two sub-sets arises as a result of the blocks of the CCC design being orthogonal. Further, this orthogonality means that any arbitrary constant added to the response values in either block, or in both blocks, does not change the estimates of the regression coefficients of the selected regression model.

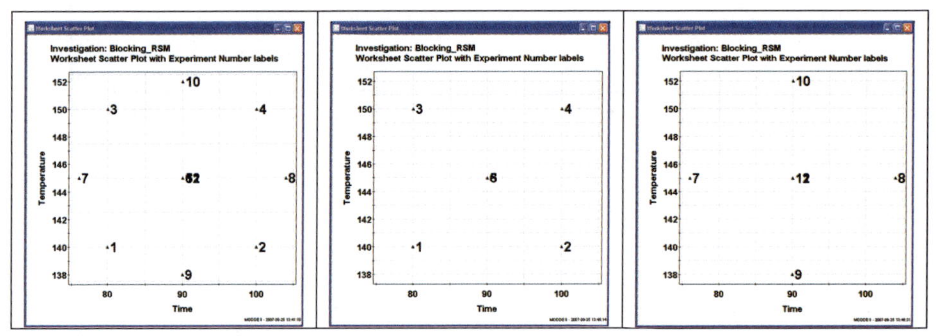

Figure 14.4: (left) Design region of the entire CCC design. (middle) Design region of Block 1. (right) Design region of Block 2.

14.3.4 Evaluation of raw data

We start by evaluating the raw data (Figure 14.5). First we examine the replicate plot, which shows that the replicate error is low; this is good. The histogram plot shows that the response is approximately normally distributed, indicating that we have good data to work with.

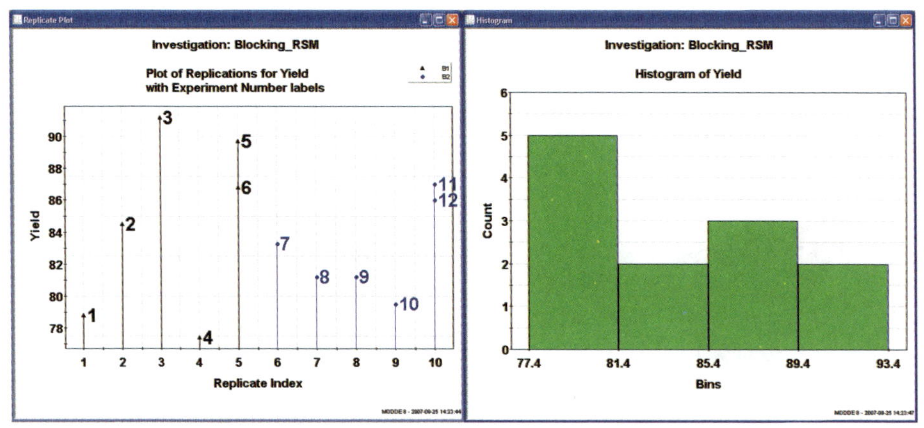

Figure 14.5: (left) Replicate plot for Yield. (right) Histogram plot for Yield.

14.3.5 Data analysis and model interpretation

A strong model was obtained with $R^2 = 0.98$, $Q^2 = 0.95$, Model Validity = 0.99 and Reproducibility = 0.88 (Figure 14.6). The regression coefficients indicate that a low value for time is best, but the linear effect of temperature is not significant (Figure 14.6). However, the quadratic terms for both time and temperature are significant. The block factor and its interactions with time and temperature are not significant. However, the deletion of any of these terms causes the model quality to deteriorate so they are all retained in the model. There is some evidence that slightly lower yields were obtained in the second set of runs.

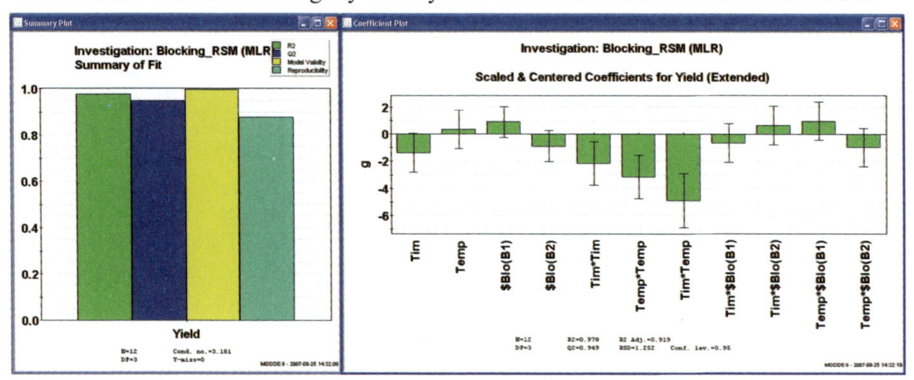

Figure 14.6: (left) Summary of fit plot for Yield. (right) Plot of regression coefficients for Yield.

The two response surface plots below (Figure 14.7) show that higher yields were obtained in the first set of runs (the factorial part of the design). The average difference between the two blocks is 1.76 g. This value is twice the linear coefficient for $Block(B1).

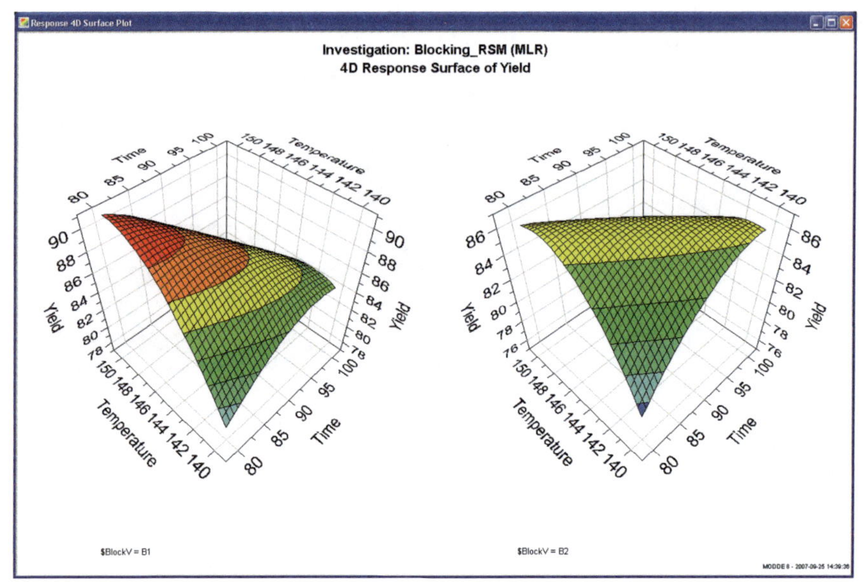

Figure 14.7: Response surface plots demonstrating that slightly higher yields were obtained for the first block.

14.3.6 Conclusion for the chemical example

To maximize yield we should use Time = 76 min and Temperature = 151°C. There is a slight shift in yields between the two blocks of experiments.

14.4 Blocking in MODDE

14.4.1 Blockable and non-blockable designs

MODDE supports orthogonal blocking for two-level full and fractional factorials, CCC-, PB- and BB-designs. Other designs, such as CCF, Rechtschaffner and Doehlert designs, are not blockable. In addition, MODDE allows blocking of D-optimal designs. The latter designs are more flexible with respect to the number of blocks and the block size, but the blocks are usually not orthogonal to the main factors. The only restriction with respect to D-optimal designs is that the number of runs must be a multiple of the block size.

14.4.2 Blocking of two-level full and fractional factorial designs

When blocking two-level factorials, the block size and the number of blocks are always powers of two. The maximum number of blocks supported by MODDE is eight, with a minimum block size of four runs. These designs and the others are blocked by introducing blocking factors, automatically denoted $BlockV. This allows so called "block effects" to be estimable. In this context, block effects are the main effects of the blocking factors and all their interactions. One blocking factor is needed to encode two blocks, two factors for four blocks and three for eight blocks. Hence, with eight blocks and three block factors, it is possible to estimate seven block effects: three main effects, three two-factor interactions and one three-factor interaction. This is equivalent, therefore, to having added seven extra factors to your design. Unfortunately, this also means using up seven degrees of freedom. This high consumption of degrees of freedom is the unfortunate drawback of orthogonal blocking. The latter point is further discussed below (see Section 14.6).

Another negative consequence of the presence of block effects is that the resolution of factorial and fractional factorial designs is usually adversely affected. To indicate the resolution of a factorial design subject to blocking, MODDE specifies its pseudo-resolution. This is the resolution of the design when all the block effects, i.e., all main effects of the block factors and all their interactions, are treated as main effects, under the assumption that there are no further interactions between the blocks and the effects arising from the remaining, "real" factors. MODDE selects the generators of these blocking factors to achieve the highest possible pseudo-resolution of the design.

14.4.3 Blocking of Plackett-Burman designs

Plackett-Burman designs can only be split into two blocks by introducing one block variable, and using its signs to split the design.

14.4.4 Blocking of CCC designs

In general, RSM designs – like the CCCs – are orthogonally blocked when they fulfill the following two conditions:

- Each block must be a first order orthogonal block.
- The fraction of the total sum of squares for each variable contributed by every block must equal the fraction of the total observations (experiments) allotted to the block.

The CCC can be split into two blocks, the cube portion and the star portion, and satisfies the two above conditions when α (the distance from the star points to the center) is equal to

$$\alpha = [\, k\,(1 + p_s\,) \,/\, (1 + p_c)\,]^{1/2}$$

where k = number of factors, p_s = proportion of center-points in the star portion, and p_c = proportion of center-points in the cube portion. This is the value of α implemented in MODDE when you choose to block a CCC design.

Occasionally the cube portion of the CCC can be split into further blocks if (a) the factorial or the fractional factorial part of the design can be split into orthogonal blocks of pseudo-resolution 5, and (b) each one of these blocks has the same number of center-points.

Also note that the CCF design is not blockable.

14.4.5 Blocking of Box-Behnken designs

The Box-Behnken design family is another group of RSM designs that is blockable. However, not all members of this design family can be blocked. The BB design in three factors (BB-3) is not blockable. The BB-4 design is blockable, but only in three blocks, whereas the BB-5–BB-7 are all blockable in two blocks. We recall from Chapter 9 that the BB-8 design does not exist.

14.4.6 Blocking of D-optimal designs

MODDE can block D-Optimal designs, but usually the blocks are *not* orthogonal to the main factors.

The following restrictions apply to blocking D-optimal designs:

- The blocks must be of equal size.
- You cannot have interactions between the block factor and the other factors in the model.
- The selected number of runs for the design must be a multiple of the number of blocks.

MODDE blocks the D-optimal design by generating a qualitative factor, $BlockV, with as many levels as the selected number of blocks. By default, it then selects only balanced designs with respect to the blocking factor.

It is not always possible to generate a balanced D-optimal design with respect to the blocking factor. In this case you may want to change the model, the number of blocks, or generate an unbalanced design.

14.4.7 Block interactions

As was hinted at in Section 14.3.2, it is possible to estimate Block interactions. Block interactions are cross-terms between a main effect and a block effect. When the chosen design supports such additional interactions, the check box labeled "Block interactions" in the Objective dialog becomes active. You can mark the check box if you want to add the block interactions to your model (see Figure 14.8). The example shown in Figure 14.8 relates to a five-factor investigation in 32 runs and using four blocks.

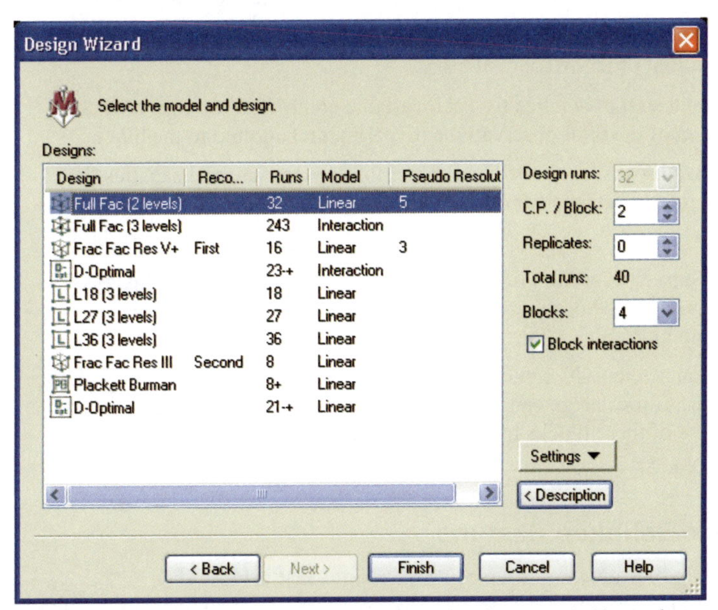

Figure 14.8: *A MODDE investigation involving five factors in a 32-run design divided into four blocks.*

When requesting block interactions, the five-factor regression model will be modified accordingly (Figure 14.9). Figure 14.9 shows the default regression model and its confounding pattern, for the same investigation, with and without block interactions.

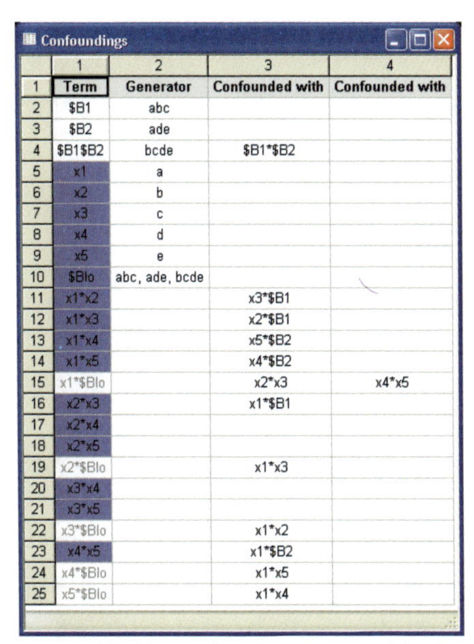

Confoundings

	1	2	3	4
1	Term	Generator	Confounded with	Confounded with
2	$B1	abc		
3	$B2	ade		
4	$B1$B2	bcde	$B1*$B2	
5	x1	a		
6	x2	b		
7	x3	c		
8	x4	d		
9	x5	e		
10	$Blo	abc, ade, bcde		
11	x1*x2		x3*$B1	
12	x1*x3		x2*$B1	
13	x1*x4		x5*$B2	
14	x1*x5		x4*$B2	
15	x1*$Blo		x2*x3	x4*x5
16	x2*x3		x1*$B1	
17	x2*x4			
18	x2*x5			
19	x2*$Blo		x1*x3	
20	x3*x4			
21	x3*x5			
22	x3*$Blo		x1*x2	
23	x4*x5		x1*$B2	
24	x4*$Blo		x1*x5	
25	x5*$Blo		x1*x4	

Confoundings

	1	2	3	4
1	Term	Generator	Confounded with	Confounded with
2	$B1	abc		
3	$B2	ade		
4	$B1$B2	bcde	$B1*$B2	
5	x1	a		
6	x2	b		
7	x3	c		
8	x4	d		
9	x5	e		
10	$Blo	abc, ade, bcde		
11	x1*x2		x3*$B1	
12	x1*x3		x2*$B1	
13	x1*x4		x5*$B2	
14	x1*x5		x4*$B2	
15	x1*$Blo		x2*x3	x4*x5
16	x2*x3		x1*$B1	
17	x2*x4			
18	x2*x5			
19	x2*$Blo		x1*x3	
20	x3*x4			
21	x3*x5			
22	x3*$Blo		x1*x2	
23	x4*x5		x1*$B2	
24	x4*$Blo		x1*x5	
25	x5*$Blo		x1*x4	

Figure 14.9: *The confounding pattern of the five-factor problem in 32 runs defined in Figure 14.8. (left) Without block interactions. (right) With block interactions.*

14.4.8 Recoding the blocking factors

When blocking is introduced into a design, the block to which the individual runs belong is indicated by a separate factor, the block factor. This factor, denoted $BlockV in the worksheet, is a qualitative factor and it assumes as many levels as there are blocks. Figure 14.10 shows the worksheet arising from the design defined in Figure 14.8.

It is important to note that, while the $BlockV factor indicates block membership in a transparent way from a practical perspective, some re-coding is necessary for the regression analysis to work properly. We recall from Section 12.4 that such a recoding operation is also needed when working with regular qualitative factors. For example, to generate the four blocks used above, the following scheme of signs for the blocking factors is used:

$B1	$B2	
–	–	Block 1
+	–	Block 2
–	+	Block 3
+	+	Block 4

	1	2	3	4	5	6	7	8	9	10
1	Exp No	Exp Name	Run Order	Incl/Excl	x1	x2	x3	x4	x5	$BlockV
2	1	N1	6	Incl	-1	-1	-1	-1	-	B1
3	2	N2	4	Incl	-1	1	1	-1	-	B1
4	3	N3	3	Incl	1	1	-1	1	-	B1
5	4	N4	7	Incl	1	-1	1	1	-	B1
6	5	N5	5	Incl	1	1	-1	-1		B1
7	6	N6	10	Incl	1	-1	1	-1		B1
8	7	N7	8	Incl	-1	-1	-1	1		B1
9	8	N8	2	Incl	-1	1	1	1		B1
10	9	N9	1	Incl	0	0	0	0	(	B1
11	10	N10	9	Incl	0	0	0	0	(	B1
12	11	N11	16	Incl	-1	1	-1	-1	-	B2
13	12	N12	11	Incl	-1	-1	1	-1	-	B2
14	13	N13	15	Incl	1	-1	-1	1	-	B2
15	14	N14	17	Incl	1	1	1	1	-	B2
16	15	N15	20	Incl	1	-1	-1	-1		B2
17	16	N16	14	Incl	1	1	1	-1		B2
18	17	N17	19	Incl	-1	1	-1	1		B2
19	18	N18	12	Incl	-1	-1	1	1		B2
20	19	N19	13	Incl	0	0	0	0	(	B2
21	20	N20	18	Incl	0	0	0	0	(	B2
22	21	N21	25	Incl	1	1	-1	-1	-	B3
23	22	N22	28	Incl	1	-1	1	-1	-	B3
24	23	N23	23	Incl	-1	-1	-1	1	-	B3
25	24	N24	24	Incl	-1	1	1	1	-	B3
26	25	N25	30	Incl	-1	-1	-1	-1		B3
27	26	N26	26	Incl	-1	1	1	-1		B3
28	27	N27	22	Incl	1	1	-1	1		B3
29	28	N28	27	Incl	1	-1	1	1		B3
30	29	N29	21	Incl	0	0	0	0	(	B3
31	30	N30	29	Incl	0	0	0	0	(	B3
32	31	N31	34	Incl	1	-1	-1	-1	-	B4
33	32	N32	35	Incl	1	1	1	-1	-	B4
34	33	N33	32	Incl	-1	1	-1	1	-	B4
35	34	N34	37	Incl	-1	-1	1	1	-	B4
36	35	N35	40	Incl	-1	1	-1	-1		B4
37	36	N36	39	Incl	-1	-1	1	-1		B4
38	37	N37	33	Incl	1	-1	-1	1		B4
39	38	N38	36	Incl	1	1	1	1		B4
40	39	N39	38	Incl	0	0	0	0	(	B4
41	40	N40	31	Incl	0	0	0	0	(	B4

Figure 14.10: *To account for the block structure, MODDE will expand the standard worksheet to include an additional factor denoted $BlockV. This factor is a qualitative factor with as many levels as there are blocks.*

In order to display the design in the coded unit, we must select Show | Design Matrix. Figure 14.11 shows the real structure of the design behind the worksheet in Figure 14.10.

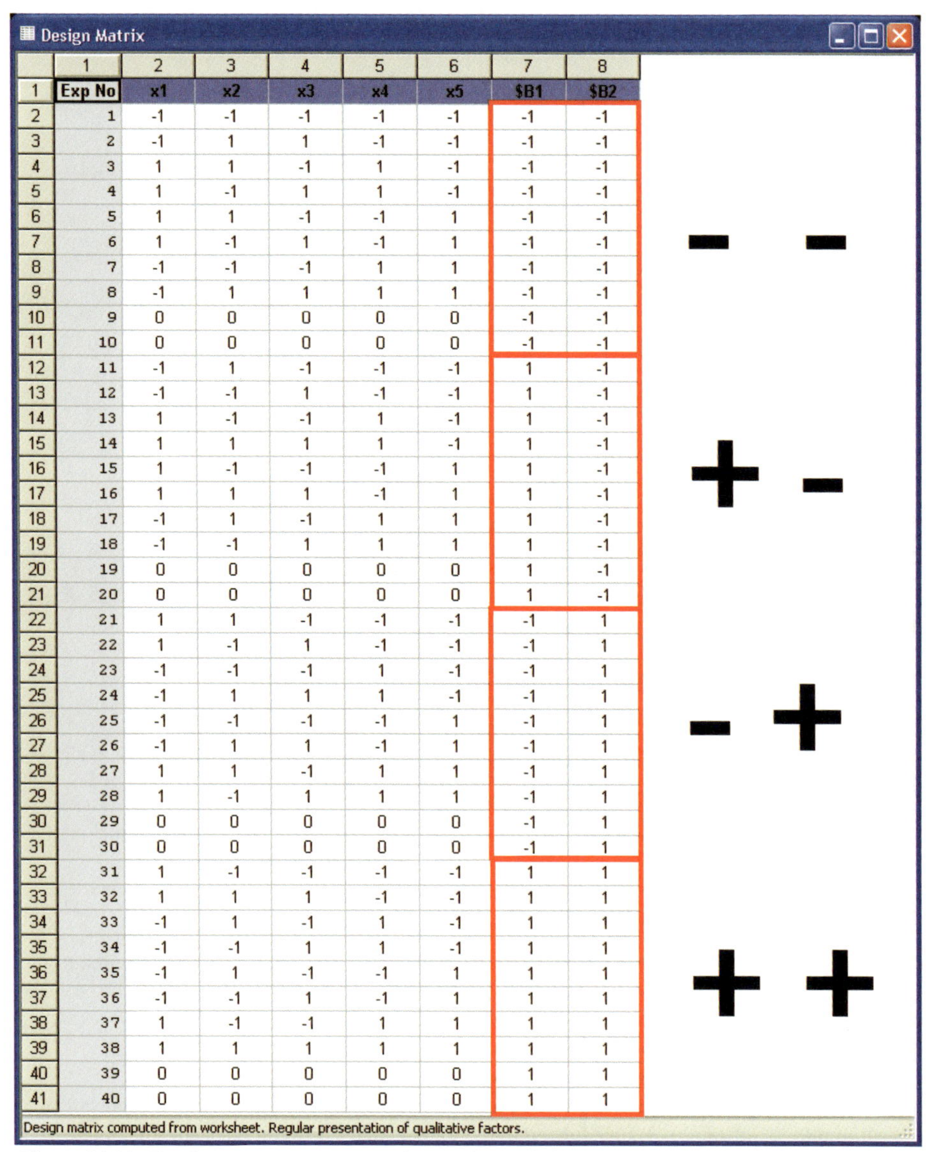

Design Matrix

	1	2	3	4	5	6	7	8
1	Exp No	x1	x2	x3	x4	x5	$B1	$B2
2	1	-1	-1	-1	-1	-1	-1	-1
3	2	-1	1	1	-1	-1	-1	-1
4	3	1	1	-1	1	-1	-1	-1
5	4	1	-1	1	1	-1	-1	-1
6	5	1	1	-1	-1	1	-1	-1
7	6	1	-1	1	-1	1	-1	-1
8	7	-1	-1	-1	1	1	-1	-1
9	8	-1	1	1	1	1	-1	-1
10	9	0	0	0	0	0	-1	-1
11	10	0	0	0	0	0	-1	-1
12	11	-1	1	-1	-1	-1	1	-1
13	12	-1	-1	1	-1	-1	1	-1
14	13	1	-1	-1	1	-1	1	-1
15	14	1	1	1	1	-1	1	-1
16	15	1	-1	-1	-1	1	1	-1
17	16	1	1	1	-1	1	1	-1
18	17	-1	1	-1	1	1	1	-1
19	18	-1	-1	1	1	1	1	-1
20	19	0	0	0	0	0	1	-1
21	20	0	0	0	0	0	1	-1
22	21	1	1	-1	-1	-1	-1	1
23	22	1	-1	1	-1	-1	-1	1
24	23	-1	-1	-1	1	-1	-1	1
25	24	-1	1	1	1	-1	-1	1
26	25	-1	-1	-1	-1	1	-1	1
27	26	-1	1	1	-1	1	-1	1
28	27	1	1	-1	1	1	-1	1
29	28	1	-1	1	1	1	-1	1
30	29	0	0	0	0	0	-1	1
31	30	0	0	0	0	0	-1	1
32	31	1	-1	-1	-1	-1	1	1
33	32	1	1	1	-1	-1	1	1
34	33	-1	1	-1	1	-1	1	1
35	34	-1	-1	1	1	-1	1	1
36	35	-1	1	-1	-1	1	1	1
37	36	-1	-1	1	-1	1	1	1
38	37	1	-1	-1	1	1	1	1
39	38	1	1	1	1	1	1	1
40	39	0	0	0	0	0	1	1
41	40	0	0	0	0	0	1	1

Design matrix computed from worksheet. Regular presentation of qualitative factors.

Figure 14.11: *The design matrix corresponding to the worksheet in Figure 14.10. It is possible to visualize the design matrix in the coded unit. Two -1/+1 factors are needed to encode the four orthogonal blocks.*

Analogously, to encode eight blocks, the recoding of the qualitative block factor involves expansion into three factors:

$B1	$B2	$B3	
–	–	–	Block 1
+	–	–	Block 2
–	+	–	Block 3
+	+	–	Block 4
–	–	+	Block 5
+	–	+	Block 6
–	+	+	Block 7
+	+	+	Block 8

14.4.9 Inclusions and blocks

Accounting for inclusions when undertaking blocking is far from trivial. Consequently, adding inclusions to a blocked design is difficult to accomplish properly, unless the inclusions belong to one of the blocks present.

14.5 Questions for Chapter 14

- What is randomization? Why is it necessary?
- What is blocking? Under what circumstances is it useful?
- Which common designs are blockable? Non-blockable?
- What can the regression coefficient of a block factor be used for?
- What is the main drawback of blocking?

14.6 Summary and discussion

In some DOE investigations, there may be a need, when running an experiment, to eliminate the influence of extraneous factors. As discussed in this chapter, such disturbance can be counteracted either by randomization or by blocking. For example, let us assume we anticipate that predictable shifts will occur whilst an experiment is being run. This might happen, for example, when one has to change to a new batch of raw materials halfway through the experiment. The effect of the change in raw materials is well known, and we want to eliminate its influence on any subsequent data analysis. This can be accomplished by dividing the parent design into two mathematically equivalent sub-sets, two so-called orthogonal blocks.

This sub-division has to balance out the effect of the materials change in such a way as to eliminate its influence on the data analysis. The simple quandary is just how to spread the experiments making up the different blocks so that this balancing is achieved. Fortunately, this is taken care of by MODDE, and in the case of factorial designs, for example, it selects the generators of the blocking factors that will achieve the highest possible pseudo-resolution for the design.

In a general sense, the basic working principle is that each level of any regular factor should be tested in each block, and this should be achieved by using the same number of experiments in each block. As an illustration of this phenomenon, we can study the distribution of the experiments in Figure 14.12. Here, all the non-center-point trials contained in the worksheet of Figure 14.10 are plotted using text- and color-coding to indicate block-membership. Each block, B1–B4, occurs twice in each "corner-cube", i.e., sub-space spanned by the factors $x_1 - x_3$, and all levels of the five factors are tested using the same number of experiments.

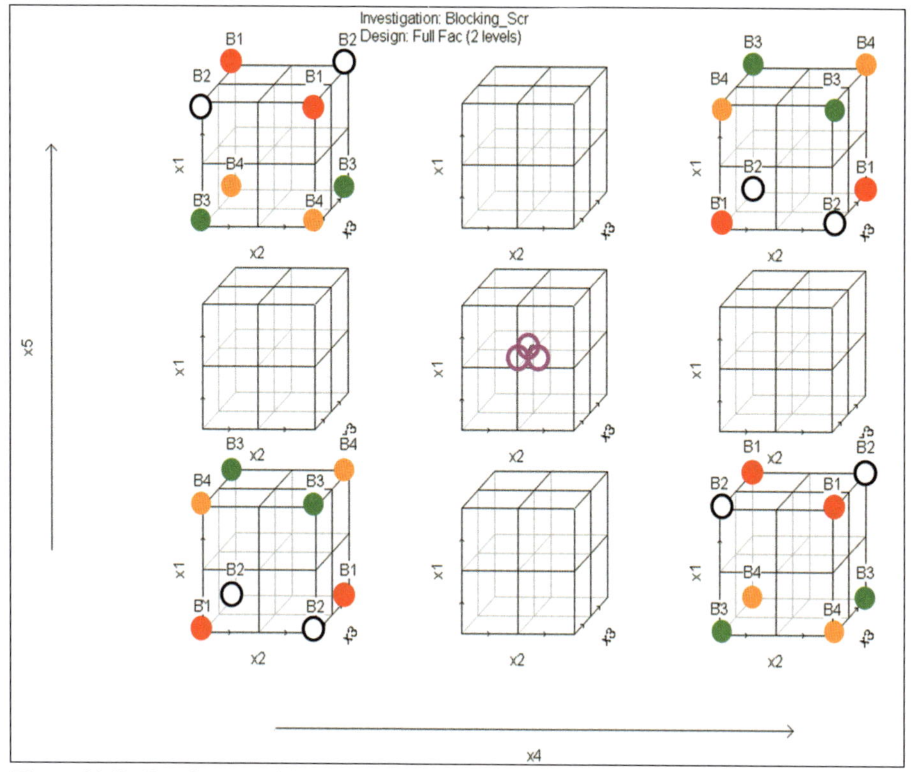

Figure 14.12: *Visualization of the design region of the five-factor example in 32 runs. Block membership is indicated using text- and color-coding.*

The main drawback of blocking is that blocked designs can create a large number of additional main and interaction effects. This increase in the number of terms reduces the resolution of the design. More importantly, however, the degrees of freedom are decreased, which, in turn, weakens the reliability of residual analysis and similar modelling diagnostics. *Therefore, it is our firm belief that one should only block when the extraneous source of variability is high and cannot be dealt with by randomizing the run order!*

14.6.1 Random versus fixed effects

We end this chapter by discussing analysis and prediction aspects of blocked designs. For prediction, in particular, it is necessary to determine whether the block factors are to be regarded as random or fixed factors. The block factors should be considered to be fixed when the external variability can be set by the experimenter, i.e., is controllable, and the primary purpose of blocking is to eliminate that source of variability. A fixed block can be modelled as a controlled qualitative factor with a limited number of levels. All predictions of the responses and contour plots will be made for a selected block level. The block is a fixed effect, for example, if you are running 32 experiments, and each sub-set of eight runs is undertaken on one of four different machines. No machine other than those four is or will be used in the experiment. We may want to specify four blocks to eliminate the variability introduced by the machines, but all predictions of the response(s) are made for a specific, individual machine.

Conversely, the block must be regarded as a random effect when the external variability cannot be controlled and set by the experimenter, and the primary objective is to make predictions without specifying the block level, and without taking into account the external variability. Then, since the block level of future experiments is unknown, all predictions for the responses to random block effects are made without specifying the block level. The confidence intervals for the responses are increased to account for the uncontrolled external block variability. For example, the block is a random effect if you are conducting 32 experiments, and each block of eight runs is made using a different batch of raw material. Our primary objective is to make predictions for the next unknown batch of raw material.

Whether the block effects are considered to be fixed or random also has a bearing on the data analysis phase. When the block factor is treated as a random effect, it is often desirable to investigate the consistency of the factor effects by including in the model, if possible, all the interactions of the block factor with the main factors. In MODDE the model is always fitted as a fixed effect model, that is with the block factor treated as a controlled qualitative variable, even when the blocks are specified as being random.

If the random block interactions with the main factors are large and significant, the effect of the main factors varies from block to block, and the confidence intervals for the prediction will be large due to this uncontrolled variability. However, if the random block interaction effects are small and insignificant, the effects of the main factors will be consistent from block to block and the uncertainty of the predictions is greatly reduced. To produce confidence intervals of a realistic size, trim the model and remove all insignificant block interaction effects. If the block factor is specified as a fixed effect, the interactions between the main factors with the block factor are of less interest.

15 The Taguchi approach to robust design

15.1 Objective

The objective of this chapter is to describe the Taguchi approach to robust design. "Robust design" is a term with a confusing similarity to "robustness testing", discussed in Chapter 7. In the Taguchi approach, robustness has a different connotation and objective. The objective is to find conditions where simultaneously the responses have values close to target and low variability. Hence, factors are often varied in large intervals and with designs very different from those discussed in Chapter 7. This contrasts with robustness testing where small factor intervals are used. The Taguchi philosophy aims to reduce variation in the process outcomes by finding operating conditions under which uncontrollable variation in some process factors has minimal impact on the product quality. It is the aim of this chapter to convey the essence of this approach. Taguchi distinguishes between *design factors* which are easy-to-control factors, and *noise factors* which are hard-to-control factors. These factors are usually varied in so called inner and outer arrays. We will use the full CakeMix application, and an additional example from the pharmaceutical industry to illustrate the steps of the Taguchi approach. Two ways to analyze Taguchi arrayed data, the classical analysis approach and the interaction analysis approach, are considered.

15.2 The Taguchi approach – Introduction

The original ideas of robust design were formulated by the Japanese engineer Genichi Taguchi. The Taguchi approach made rapid progress in Japanese industry, and partly explains the success of the Japanese electronics and car industries in the 1970s and 1980s. Taguchi divides robust design into three major blocks, the *product* design, the *parameter* design, and the *tolerance* design phases.

Product design involves the use of off-line quality improvement schemes. Taguchi advocates a philosophy where quality is measured in terms of loss suffered by society as a result of product variability around a specified target. This includes losses to the manufacturer during production, and losses to the consumer after the release of the product. Every contribution is quantified in monetary terms and the total loss is computed. A desirable product is one for which the total loss is acceptably small. This means that high quality is coupled to low loss. In this context, Taguchi recommends the use of a loss function for estimating the loss. The problem is that the computation of such a loss expression is often obscured by hurdles in the definition of relevant target values and an appropriate form of the polynomial model. Often wide-ranging assessments of the product impact on society must be made. If successful, this may lead to the specification of an acceptable region within which the final design can lie. It is in this region that the second part of the Taguchi approach, the parameter design, is applied.

The parameter design is equivalent to using design of experiments (DOE) for finding optimal settings of the process variables. The additional feature of the Taguchi approach is to make a distinction between *design* factors and *noise* factors. The former are easy-to-control and are supposed to affect the mean output of the process. The latter are hard-to-control and may or may not affect the process mean, and the spread around the mean. The subsequent data analysis tries to identify design factors which affect only the mean, design factors which affect only the spread around the mean, and factors which affect both properties. The idea is then to choose levels of the design factors so that the process output is high and relatively insensitive to the noise factors. However, one limitation of the Taguchi approach concerning this second stage, is that the prescribed strategy is weak on the modelling-of-data aspect.

Finally, the third part of the Taguchi approach, the tolerance design, takes place when optimal factor settings have been specified. Should the variability in the product quality still be unacceptably high, the tolerances on the factors have to be further adjusted. Usually, this is accomplished by using a derived mathematical model of the process, and the loss function belonging to the product property of interest.

15.3 Arranging factors in inner and outer arrays

In the Taguchi approach, *design factors* and *noise factors* are varied systematically according to an architecture consisting of inner and outer factor arrays. We will now study this arrangement. For this purpose, we will use an extended version of the CakeMix application. We recall that this is an industrial pilot plant investigation aimed at designing a cake mix giving tasty products. Previously, we studied how changes of the factors Flour, Shortening, and Eggpowder affected the taste. In reality, this example is nothing less than a robust design application, where the final goal is to design a cake mix which will produce a good cake even if the customer does not follow the baking instructions. To explore whether this was feasible, the factors Flour, Shortening, and Eggpowder were used as design factors and varied in a cubic inner array (see Figure 15.1). In addition, two noise factors were incorporated in the experimental design as a square outer array. These factors were baking temperature, varied between 175 and 225°C, and time spent in oven, varied between 30 and 50 minutes.

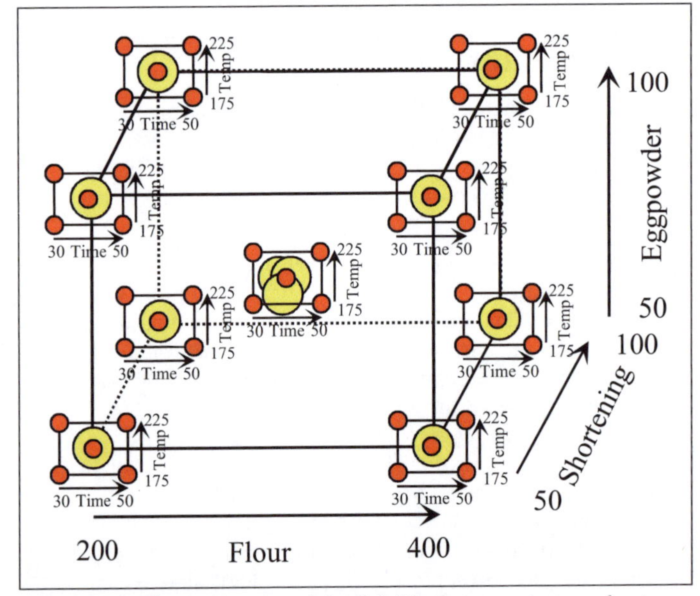

Figure 15.1: *The arrangement of the CakeMix factors as inner and outer arrays.*

We can see in Figure 15.1 that for each design point in the inner cubic array, a square outer array is laid out. The number of experiments according to the inner array is eleven, and according to the outer array five, which makes 11*5 = 55 experiments necessary. With this arrangement of the experiments, the experimental goal was to find levels of the three ingredients producing a good cake (a) when the noise factors temperature and time were correctly set according to the instructions on the box, and (b) when deviations from these specifications occur. Hence, in this kind of testing, the producer has to consider worst-case scenarios corresponding to what the consumer might do with the product, and let these considerations regulate the low and high levels of the noise factors. Finally, we have another important aspect to observe in this context – the fact that the noise factors are often controllable in the experiment, but not in full-scale production.

15.4 The classical analysis approach

The classical way to analyze Taguchi-like DOE data is to form, for each experimental point in the inner array, two responses. The first response is the average response value and the second the standard deviation around this average. These experimental values are shown in Figure 15.2. The Taste response is in the same shape as before. Thus, this is the average taste for the five outer array experiments at each inner array point. Similarly, the second response, denoted StDev, is the standard deviation among the five outer array experiments at each inner array point. Because such standard deviation responses tend to be non-normally distributed, it is common practice to model the log-transformed variable. Thus, in the following, the StDev response will be log-transformed. With the classical analysis approach the goal is to find which design factors affect the variation (StDev) only, which affect the mean level (Taste) only, and which affect both. Note that with this approach, there will be no model terms related to the noise factors (Time and Temperature). We will now study the results of the regression modelling.

Worksheet

	1	2	3	4	5	6	7	8	9
1	Exp No	Exp Name	Run Order	Incl/Excl	Flour	Shortening	Eggpowder	Taste	LogStD
2	1	N1	4	Incl	200	50	50	3.52	0.389
3	2	N2	5	Incl	400	50	50	3.66	0.031
4	3	N3	11	Incl	200	100	50	4.74	0.17
5	4	N4	6	Incl	400	100	50	5.2	-0.029
6	5	N5	7	Incl	200	50	100	5.38	0.163
7	6	N6	9	Incl	400	50	100	5.9	-0.166
8	7	N7	2	Incl	200	100	100	4.36	0.259
9	8	N8	8	Incl	400	100	100	4.86	-0.177
10	9	N9	10	Incl	300	75	75	4.73	0.079
11	10	N10	1	Incl	300	75	75	4.61	0.105
12	11	N11	3	Incl	300	75	75	4.68	0.132

Figure 15.2: *Data arrangement for the classical analysis approach.*

15.4.1 Regression analysis

It is instructive to first consider the raw experimental data. Figures 15.3 and 15.4 show the replicate plots of the responses. We can see that for both responses the replicate error is satisfactorily small. It is also of interest that the responses are inversely correlated, as is evidenced by Figure 15.5. We recall that the experimental goal is a factor combination producing a tasty cake with low variation. Hence, it seems as if experiment number 6 is the most promising.

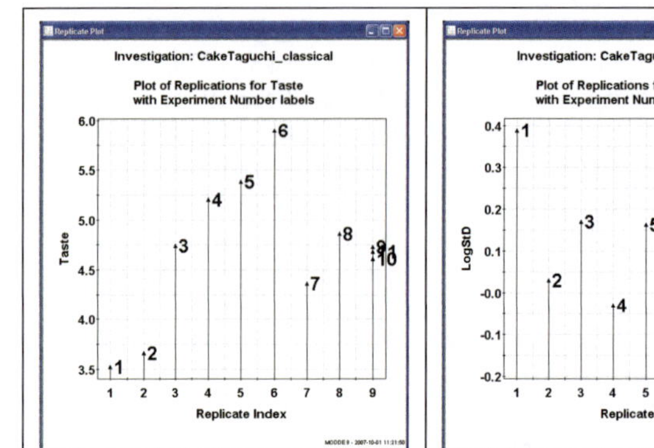

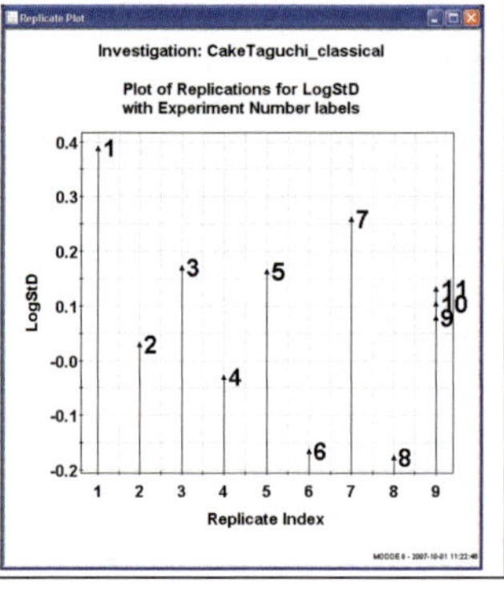

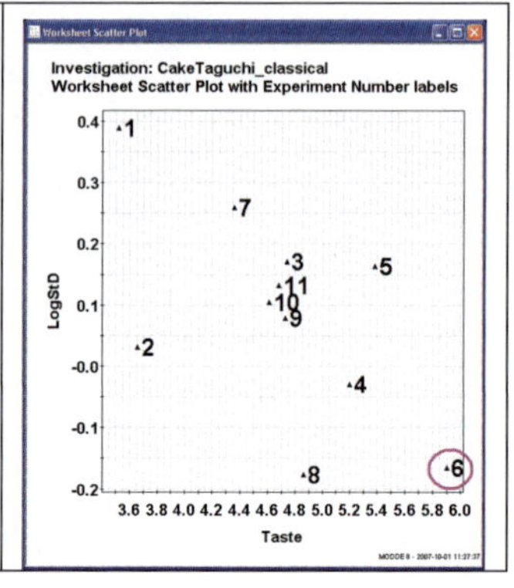

Figure 15.3: *(left) Replicate plot of Taste.*
Figure 15.4: *(middle) Replicate plot of LogStDev.*
Figure 15.5: *(right) Raw data scatter plot of LogStDev versus Taste.*

Figures 15.6 – 15.8 show the modelling results obtained from fitting an interaction model to each response. Noteworthy is the negative Q^2 of LogStDev, indicating model problems. The model for Taste is better, but we remember from previous modelling attempts that even better results are possible if the two non-significant two-factor interactions are omitted.

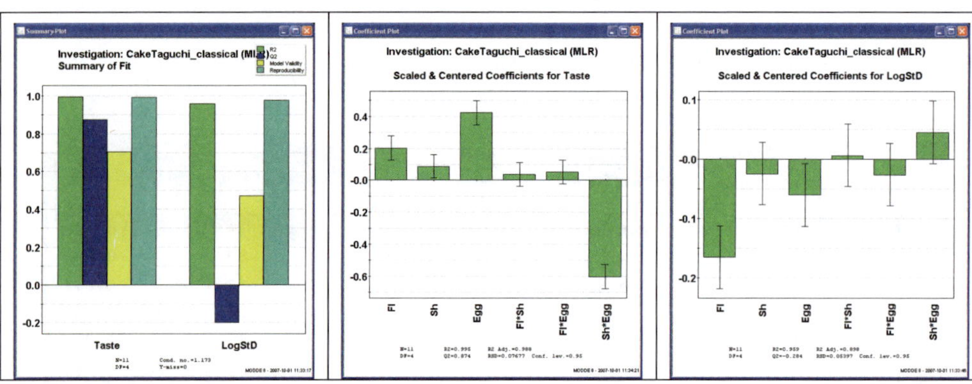

Figure 15.6: *(left) Summary of fit plot of the interaction model.*
Figure 15.7: *(middle) Regression coefficient plot for Taste.*
Figure 15.8: *(right) Regression coefficient plot for LogStDev.*

The results from the fitting of a refined model to each response are seen in Figures 15.9 – 15.11. Obviously, the model for LogStDev has improved a lot thanks to the model pruning. Two interesting observations can now be made. The first is related to the Sh*Egg interaction, which is much smaller for LogStDev than for Taste. The second observation concerns the Fl main effect, which shows that Flour is the factor causing most spread around the average Taste. Hence, this is a factor which must be better controlled if robustness is to be achieved. The models that we have derived will now be used to try to achieve the experimental goal.

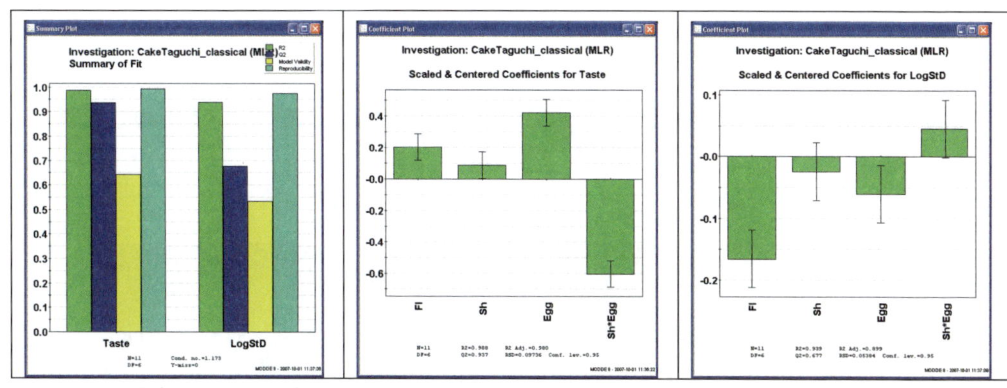

Figure 15.9: (left) Summary of fit plot of the refined interaction model.
Figure 15.10: (middle) Regression coefficient plot for Taste – refined model.
Figure 15.11: (right) Regression coefficient plot forLog StDev – refined model.

15.4.2 Interpretation of model

One way to understand the impact of the surviving two-factor interaction is to make interaction plots of the type shown in Figures 15.12 and 15.13. Evidently, the impact of this model term is greater for Taste than for LogStDev. This is inferred from the fact that the two lines cross each other in the plot related to Taste, but do not cross in the other interaction plot. Both plots indicate that a low level of Shortening and a high level of Eggpowder are favorable for high Taste and low LogStDev.

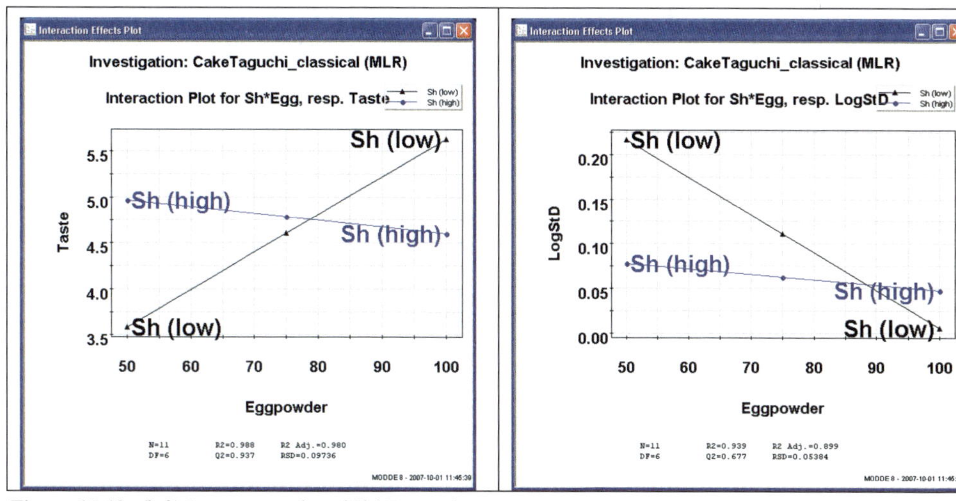

*Figure 15.12: (left) Interaction plot of Sh*Egg with respect to Taste.*
*Figure 15.13: (right) Interaction plot of Sh*Egg with respect to LogStDev.*

An alternative procedure for understanding the modelled system consists of making response contour plots. Such response contour plots are shown in Figure 15.14. These contours were created by setting Flour to its high level, as this was found favorable in the modelling. The two contour plots convey an unambiguous message. The best cake mix conditions are found in the upper left-hand corner, where the highest taste is predicted to be found, and at the same time the lowest standard deviation. This location corresponds to the factor settings Flour = 400, Shortening = 50, and Eggpowder = 100. At this factor combination, Taste is predicted at

5.84 ± 0.18, and LogStDev at -0.16 ± 0.10. Bearing in mind that the highest registered experimental value of Taste is 6.9, and the lowest value of LogStDev -0.18, these predictions appear reasonable.

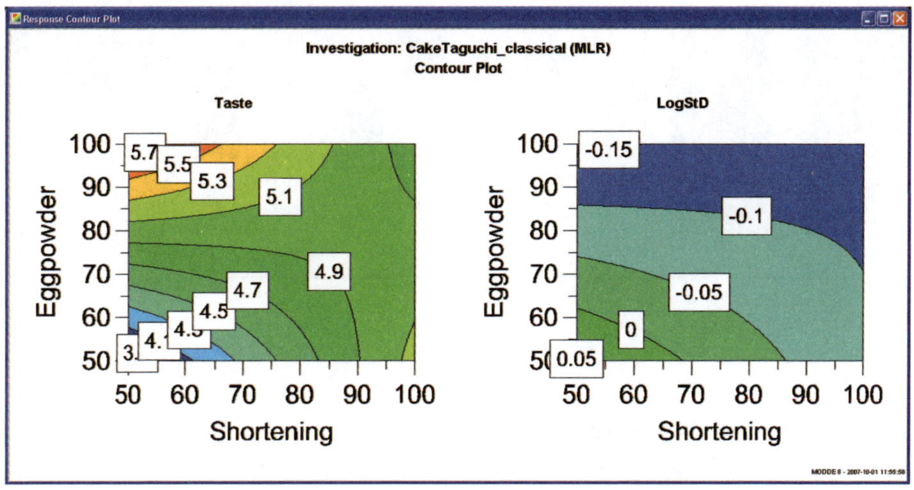

Figure 15.14: Response contour plots of Taste and LogStDev (Flour = 400g).

15.5 The interaction analysis approach

One drawback of the classical approach to data analysis is that it does not allow the user to identify which noise factors may be affecting the variability of the responses. For the Taguchi method to be really successful, we would need to be able to estimate the impact of the noise factors *and* possible interactions between the design and the noise factors. Clearly, by definition, the success of the Taguchi approach critically depends on the existence of such noise-design factor interactions. Otherwise, the noise (variability) cannot be reduced by changing any design factors.

This kind of noise/design factor interaction information can be extracted if the design factors and the noise factors are combined into one single design, and a regression model is fitted which contains both types of factors, as well as their interactions. In such a case, what in the classical analysis approach were design factor effects on the variation around the mean, would in this alternative *interaction analysis* approach correspond to noise-design factor cross-terms. We will now use the full CakeMix application to illustrate the interaction analysis approach. For this purpose, we need to rearrange the experimental worksheet, so that it comprises 55 rows. An excerpt of this worksheet is shown in Figure 15.15. As seen, there is only one response, the raw value of Taste. One additional benefit obtained by ordering the experiments in this latter way, is that it enables single deviating experiments to be identified.

No	Flour	Shortening	Eggpowder	Temp	Time	Taste	No	Flour	Shortening	Eggpowder	Temp	Time	Taste
1	200	50	50	175	30	1.1	34	200	50	50	225	50	1.3
2	400	50	50	175	30	3.8	35	400	50	50	225	50	2.1
3	200	100	50	175	30	3.7	36	200	100	50	225	50	2.9
4	400	100	50	175	30	4.5	37	400	100	50	225	50	5.2
5	200	50	100	175	30	4.2	38	200	50	100	225	50	3.5
6	400	50	100	175	30	5	39	400	50	100	225	50	5.7
7	200	100	100	175	30	3.1	40	200	100	100	225	50	3
8	400	100	100	175	30	3.9	41	400	100	100	225	50	5.4
9	300	75	75	175	30	3.5	42	300	75	75	225	50	4.1
10	300	75	75	175	30	3.4	43	300	75	75	225	50	3.8
11	300	75	75	175	30	3.4	44	300	75	75	225	50	3.8
12	200	50	50	225	30	5.7	45	200	50	50	200	40	3.1
13	400	50	50	225	30	4.9	46	400	50	50	200	40	3.2
14	200	100	50	225	30	5.1	47	200	100	50	200	40	5.3
15	400	100	50	225	30	6.4	48	400	100	50	200	40	4.1
16	200	50	100	225	30	6.8	49	200	50	100	200	40	5.9
17	400	50	100	225	30	6	50	400	50	100	200	40	6.9
18	200	100	100	225	30	6.3	51	200	100	100	200	40	3
19	400	100	100	225	30	5.5	52	400	100	100	200	40	4.5
20	300	75	75	225	30	5.15	53	300	75	75	200	40	6.6
21	300	75	75	225	30	5.3	54	300	75	75	200	40	6.5
22	300	75	75	225	30	5.4	55	300	75	75	200	40	6.7
23	200	50	50	175	50	6.4							
24	400	50	50	175	50	4.3							
25	200	100	50	175	50	6.7							
26	400	100	50	175	50	5.8							
27	200	50	100	175	50	6.5							
28	400	50	100	175	50	5.9							
29	200	100	100	175	50	6.4							
30	400	100	100	175	50	5							
31	300	75	75	175	50	4.3							
32	300	75	75	175	50	4.05							
33	300	75	75	175	50	4.1							

Figure 15.15: *Data arrangement for the interaction analysis approach.*

15.5.1 Regression analysis

As usual, we commence the data analysis by evaluating the raw data. Figure 15.16 suggests that the replicate error is small, and Figure 15.17 that the response is approximately normally distributed. Hence, we may proceed to the regression analysis phase, without further preprocessing of the data.

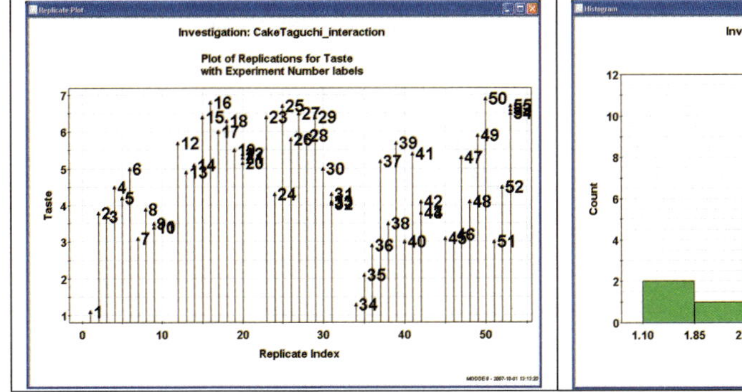

Figure 15.16: *(left) Replicate plot of Taste.*
Figure 15.17: *(right) Histogram plot of Taste.*

As seen in Figure 15.18, the regression analysis gives a poor model with $R^2 = 0.60$ and $Q^2 = 0.18$. Such a large gap between R^2 and Q^2 is undesirable and indicates model inadequacy.

The N-plot of residuals in Figure 15.19 reveals no clues which could explain the poor modelling performance. However, the regression coefficient plot in Figure 15.20 does reveal two plausible causes. Firstly, the model contains many irrelevant two-factor interactions. Secondly, it is surprising that the Fl*Te and Fl*Ti two-factor interactions are so weak. Since in the previous analysis we observed the strong impact of Flour on StDev, we would now expect much stronger noise-design factor interactions. In principle, this means that there must be a crucial higher-order term missing in the model, the Fl*Te*Ti three-factor interaction. Consequently, in the model revision, we decided to add this three-factor interaction and remove six unnecessary two-factor interactions.

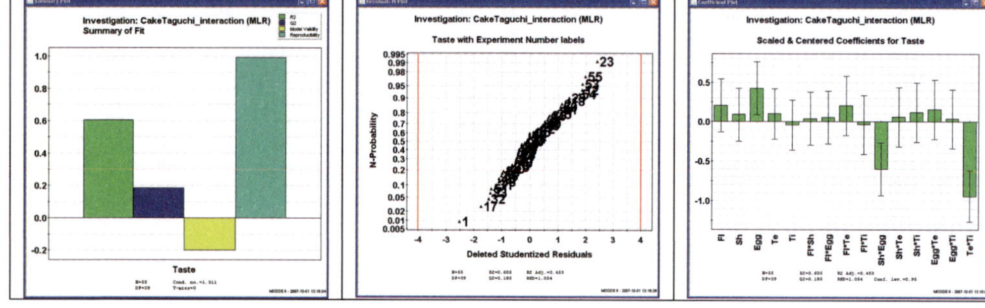

Figure 15.18: *(left) Summary of fit of Taste.*
Figure 15.19: *(middle) N-plot of residuals for Taste.*
Figure 15.20: *(right) Regression coefficients for Taste.*

When re-analyzing the data, a more stable model with the reasonable statistics $R^2 = 0.69$ and $Q^2 = 0.57$ was the result (Figure 15.21). An interesting aspect is that the R^2 obtained is lower than in the classical analysis approach. This is due to the stabilizing effect of averaging Taste over five trials in the classical analysis approach. Concerning the current model, we are unable to detect significant outliers. The relevant N-plot of residuals is displayed in Figure 15.22. Therefore, it is appropriate to consider the regression coefficients, which are displayed in Figure 15.23. We can see the significance of the included three-factor interaction. This is in line with the previous finding regarding the impact of Flour on StDev. Some smaller two-factor interactions are also kept in the model to make the three-factor interaction more interpretable.

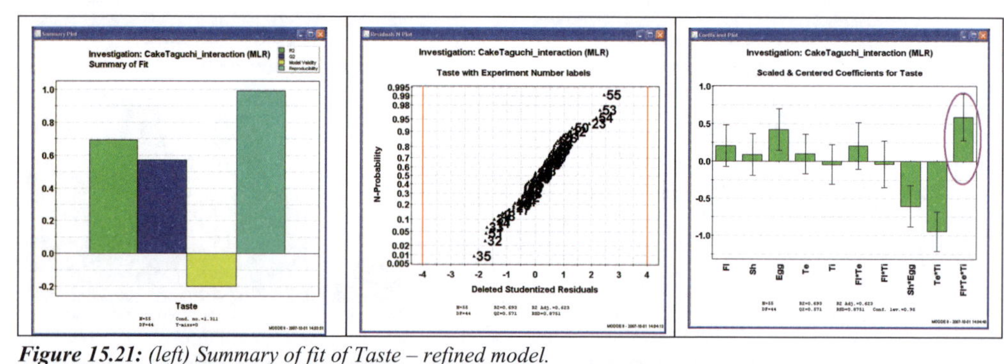

Figure 15.21: *(left) Summary of fit of Taste – refined model.*
Figure 15.22: *(middle) N-plot of residuals for Taste – refined model.*
Figure 15.23: *(right) Regression coefficients for Taste – refined model.*

15.5.2 An important three-factor interaction

The meaning of the three-factor interaction is most easily understood by constructing an interaction plot. Figure 15.24 displays the impact of the three-factor interaction. Now, one may wonder, what is it that we should look for in this kind of plot? The answer is that we want to understand how to adjust the controllable factor Flour, so that the impact of variations in the uncontrollable factors Temperature and Time is minimized. It is seen in Figure 15.24 that by adjusting Flour to 400g the spread in Taste due to variations in Temperature and Time is limited.

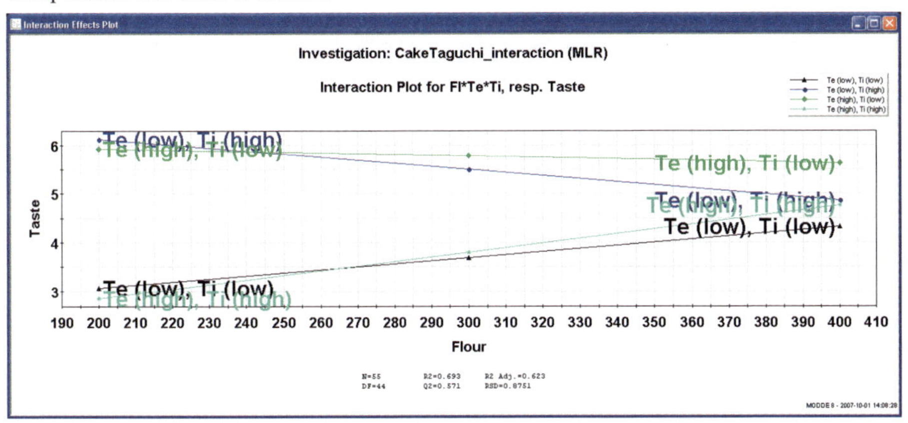

*Figure 15.24: Interaction plot showing the significance of the Fl*Te*Ti three-factor interaction.*

Furthermore, in solving the problem, we must not forget the significance of the Sh*Egg two-factor interaction. We know from the initial analysis that the combination of low Shortening and high Eggpowder produces the best cakes. These considerations lead to the construction of the response contour triplet shown in Figure 15.25. Because these contours are flat, especially when Flour = 400, we can infer robustness. Hence, when producing the cake mix industrially with the composition Flour 400g, Shortening 50g, and Eggpowder 100g, together with a recommendation on the box of using the temperature 200°C and time 40 min, sufficient robustness towards consumer misuse ought to be the result.

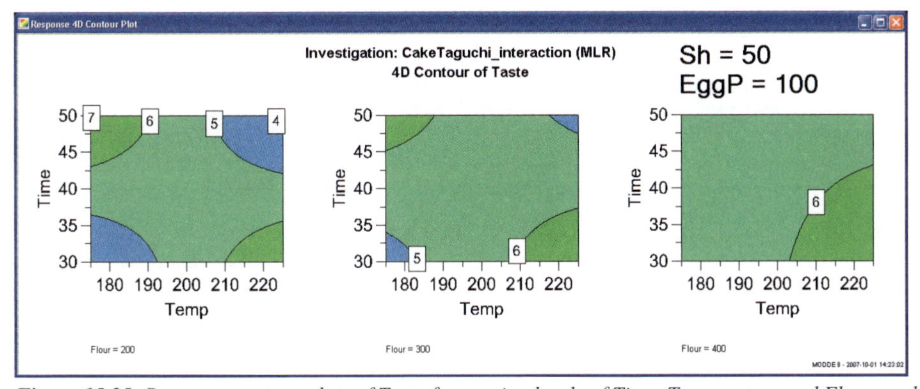

Figure 15.25: Response contour plots of Taste for varying levels of Time, Temperature and Flour, and with Shortening and Eggpowder fixed at 50 and 100g, respectively.

15.6 A second example – DrugD

We will now repeat the two analytical approaches to Taguchi planned experiments with an application taken from pharmaceutical industry, which we call DrugD. The background to the example is as follows. The manufacturer of a drug to be administered in tablet form wanted to test the robustness of the tablet for patients with varying conditions. Four factors were varied in an inner array setup to simulate a human stomach. These are seen in Figure 15.26 together with their settings.

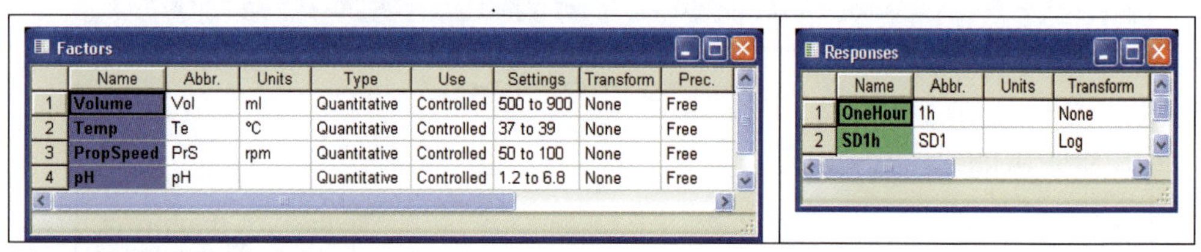
Factors

	Name	Abbr.	Units	Type	Use	Settings	Transform	Prec.
1	Volume	Vol	ml	Quantitative	Controlled	500 to 900	None	Free
2	Temp	Te	°C	Quantitative	Controlled	37 to 39	None	Free
3	PropSpeed	PrS	rpm	Quantitative	Controlled	50 to 100	None	Free
4	pH	pH		Quantitative	Controlled	1.2 to 6.8	None	Free

Responses

	Name	Abbr.	Units	Transform
1	OneHour	1h		None
2	SD1h	SD1		Log

Figure 15.26: *(left) The factors varied in the DrugD example – classical analysis setup.*
Figure 15.27: *(right) The responses registered in the DrugD example – classical analysis setup.*

Volume simulates the size of the stomach, Temperature whether the patient has fever or not, PropSpeed a calm or a stressed human being, and pH different levels of acidity in the gastrointestinal tract. The inner array used was a CCF design with 27 runs. This design was tested in 6 parallel baths, here labeled B1 – B6, and the measured response was the release after 1h of the active ingredient. In the written documentation the manufacturer declared that after 1h the release should be between 20 and 40%. Hence, the experimental goal was to assess whether the variation in the release rates across the entire design was consistent with this claim.

In the classical analysis approach one would regard the qualitative Bath factor as the outer array, and use the average release and its standard deviation across the six baths as responses. These responses are summarized in Figure 15.27. Note the log-transform of the SD response. In the interaction analysis approach, one would combine the inner and outer array experiments in one single design, and fit a regression model with the qualitative factor also included. As seen in Figures 15.28 and 15.29 this means five factors and one response. The combined worksheet then has 27*6 =162 rows. In the following we will contrast the results obtained with the classical and the interaction analysis approaches.

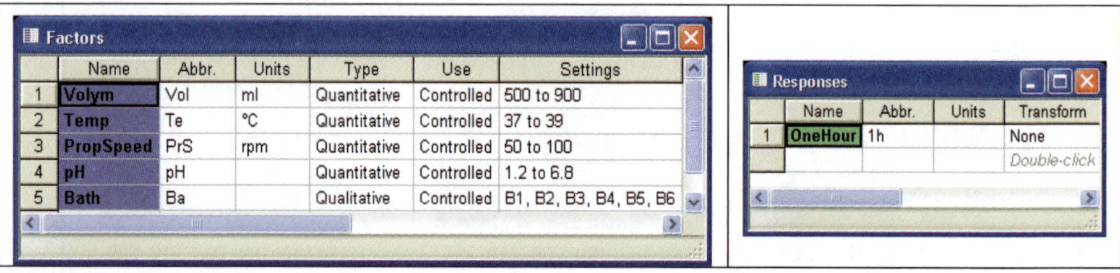
Factors

	Name	Abbr.	Units	Type	Use	Settings
1	Volym	Vol	ml	Quantitative	Controlled	500 to 900
2	Temp	Te	°C	Quantitative	Controlled	37 to 39
3	PropSpeed	PrS	rpm	Quantitative	Controlled	50 to 100
4	pH	pH		Quantitative	Controlled	1.2 to 6.8
5	Bath	Ba		Qualitative	Controlled	B1, B2, B3, B4, B5, B6

Responses

	Name	Abbr.	Units	Transform
1	OneHour	1h		None
				Double-click

Figure 15.28: *(left) The factors varied in the DrugD example – interaction analysis setup.*
Figure 15.29: *(right) The response registered in the DrugD example – interaction analysis setup.*

15.6.1 DrugD – The classical analysis approach

The simultaneous analysis of the average release, denoted 1h, and its standard deviation, denoted SD1, using a full quadratic model, gave the modelling results rendered in Figure 15.30. After some model refinement, that is, exclusion of six non-significant cross- and square terms, we obtained two refined models with performance statistics as displayed in Figure 15.31.

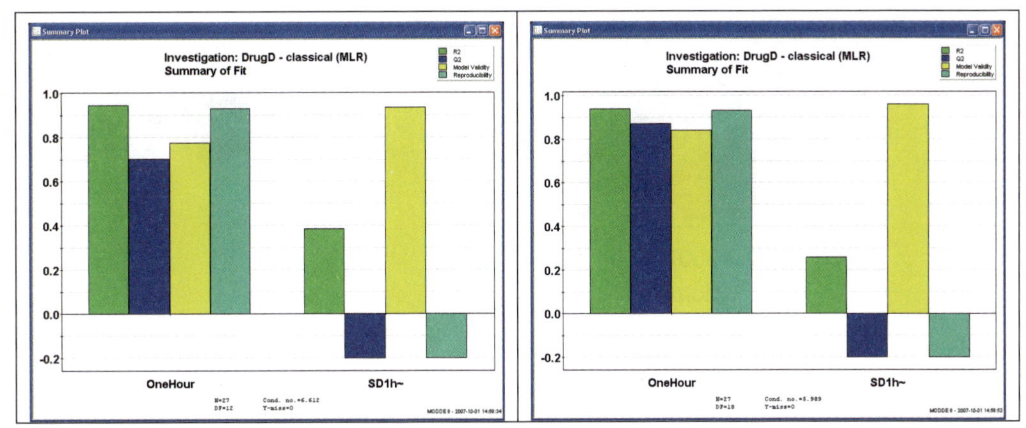

Figure 15.30: *(left) Summary of fit of initial models.*
Figure 15.31: *(right) Summary of fit of revised models.*

As seen from Figure 15.31, the model refinement resulted in an excellent model for 1h, but no model at all for SD1. This suggests that the variation in the four factors significantly affect the average release, but not the standard deviation. We can see from Figure 15.32 that all factors but Volume influence the average release. In contrast, Figure 15.33 indicates that there are no significant effects with regard to the standard deviation. Hence, the latter response is robust to changes in the factors.

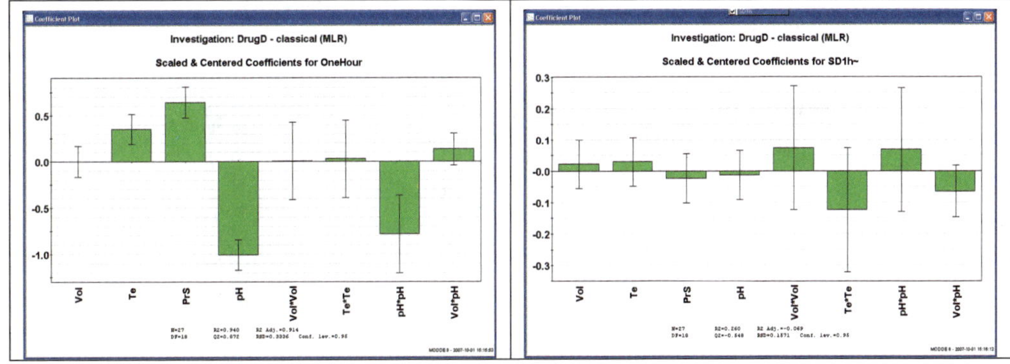

Figure 15.32: *(left) Regression coefficients of the revised model for 1h.*
Figure 15.33: *(right) Regression coefficients of the revised model for SD1.*

To better understand the features of the first response we created the response contour plot shown in Figure 15.34. From this figure it is easy to get the impression that the response 1h changes dramatically as a result of altered factor settings. However, this is not the case. We are fooled by the way this plot was constructed. A more appropriate plot is given in Figure 15.35. This is a response surface plot in which the z-axis, the release axis, has been re-expressed to go between 20 and 40%, which corresponds to the promised variation range. Now we can see the flatness of the response surface. Remarkably, the difference between the highest and the lowest measured values is as low as 4.1%. Hence, we can conclude that the average release response is robust, because it is inside the given specification.

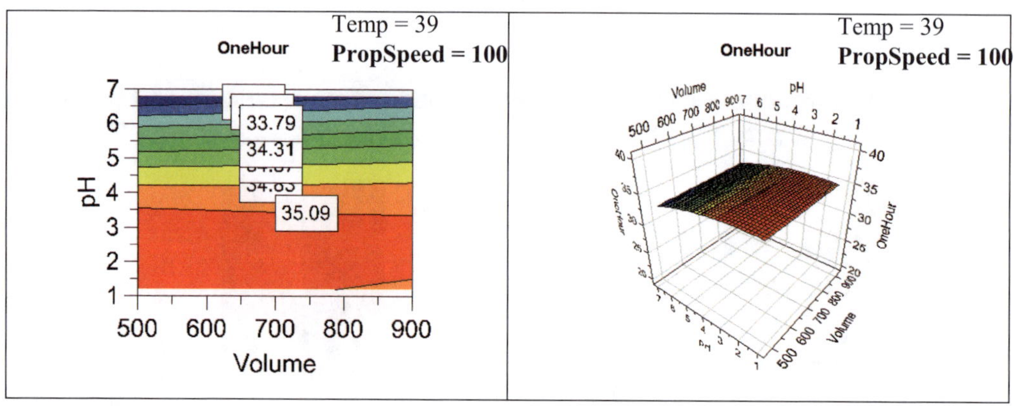

Figure 15.34: *(left) Response contour plot for 1h.*
Figure 15.35: *(right) Response surface plot for 1h.*

15.6.2 DrugD – The interaction analysis approach

For comparative purposes we now turn to the interaction analysis approach of the DrugD data. Figure 15.36 is a replicate plot over the entire set of 162 experiments, which shows the robustness of the tablet. In this case, we fitted a partially quadratic model, and, after some model refinement, a model with $R^2 = 0.79$ and $Q^2 = 0.74$ was obtained. The summary of fit plot is provided in Figure 15.37.

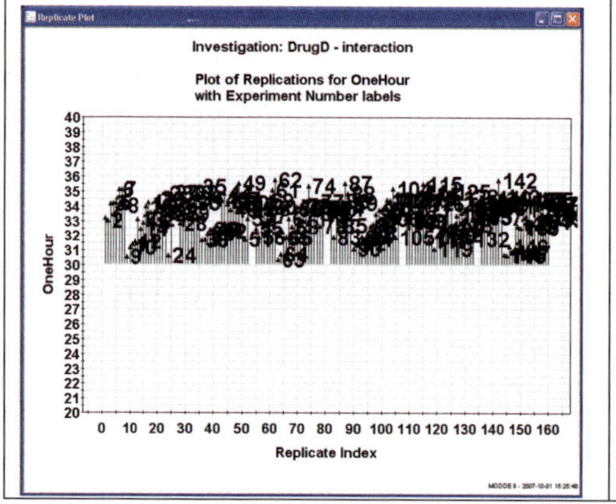

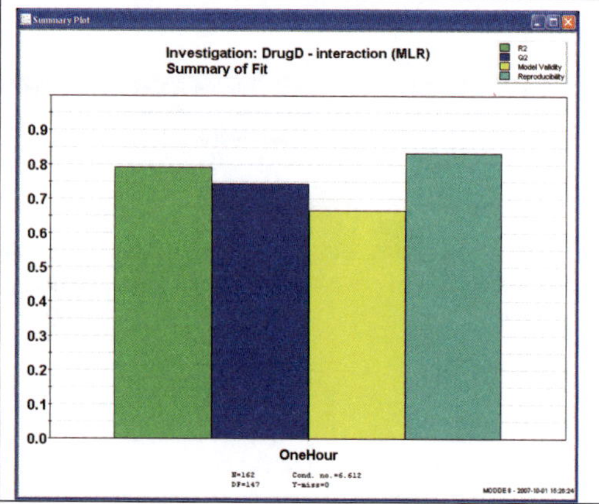

Figure 15.36: *(left) Replicate plot of 1h.*
Figure 15.37: *(right) Summary of fit plot of 1h.*

The N-plot of residuals in Figure 15.38 reveals no strange experiments, and the ANOVA table listed in Figure 15.39 also suggests that the model is adequate. The model itself is given in Figure 15.40. Because there were no strong model terms influencing the standard deviation in the foregoing analysis, we would not expect any significant interactions among the inner array factors and the Bath factor. This is indeed the case.

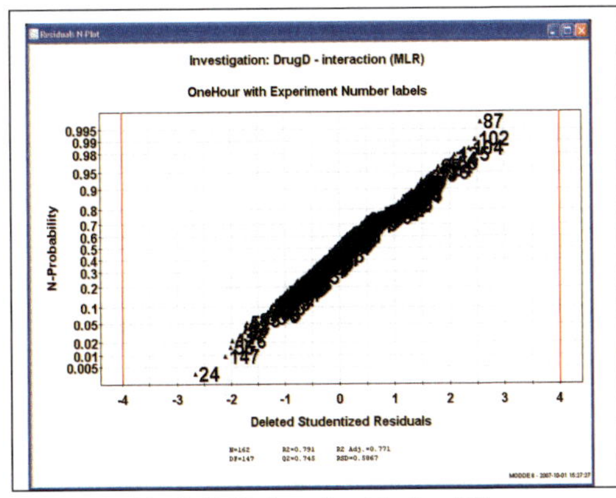

ANOVA Table - Response OneHour

	1	2	3	4	5	6	7
1	**OneHour**	**DF**	**SS**	**MS (variance)**	**F**	**p**	**SD**
2	Total	162	180958	1117.02			
3	Constant	1	180715	180715			
4							
5	Total Corrected	161	242.203	1.50437			1.22653
6	Regression	14	191.6	13.6857	39.7565	**0.000**	3.69942
7	Residual	147	50.6031	0.344239			0.586719
8							
9	Lack of Fit	135	47.581	0.352452	1.39952	**0.265**	0.593677
10	(Model Error)						
11	Pure Error	12	3.02206	0.251838			0.501835
12	(Replicate Error)						
13							
14		N = 162	Q2 =	0.745	Cond. no. =	6.612	
15		DF = 147	R2 =	0.791	Y-miss =	0	
16			R2 Adj. =	0.771	RSD =	0.5867	

The selected response 'OneHour'

Figure 15.38: *(left) N-plot of residuals of 1h.*
Figure 15.39: *(right) ANOVA of 1h.*

Moreover, with the interaction analysis approach we are able to uncover one little piece of new information with regard to the six baths used. The regression coefficients suggest that with bath B2 the experimenters might obtain slightly higher numerical results than with the other five baths. Observe, though, that this effect of B2 is statistically weak and must not be over-interpreted. We show in Figure 15.41 an alternative version of Figure 15.40, in which the vertical scale is re-expressed to go between –5 and +5. Any model term of large magnitude in this plot would constitute a serious problem for achieving the experimental goal. This is not the case, and hence we understand that the effect of B2 is not large. However, the recognition of this small term is important for further fine-tuning of the experimental equipment.

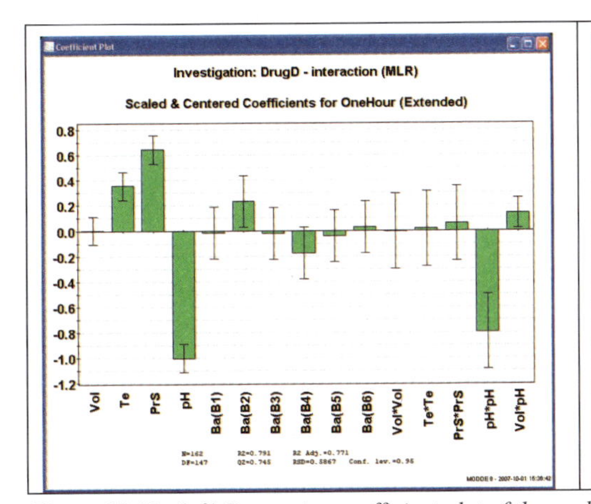

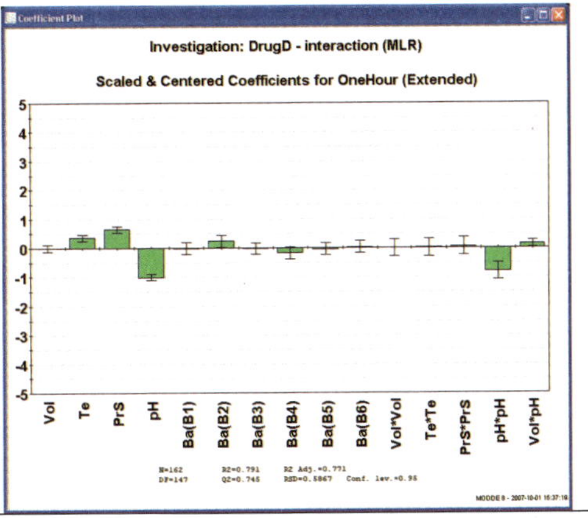

Figure 15.40: *(left) Regression coefficient plot of the model for 1h.*
Figure 15.41: *(right) Same as Figure 15.40 but with a re-expressed y-axis.*

15.7 An additional element of robust design

Sometimes it is of interest to perform a robust design test where it is less relevant to differentiate between controllable and uncontrollable factors, and more appropriate to distinguish among factors which are *expensive* and *inexpensive* to vary. We will now do a thought experiment to illustrate this situation. Let us assume that we want to increase the durability and reliability of drills to be used for drilling stainless steel. Further, assume that our goal is to measure the lifetime of a drill under different conditions. This is a response which is cheap to measure and we wish to maximize it. In this case, we have two types of factors, those that are expensive to vary, and those that are inexpensive to change. Examples belonging to the costly category are factors related to the features of the drill, such as, diameter, length, and geometry. Examples of the cheaper type are factors related to the machine conditions, such as, cutting speed, feed rate, cooling, and so on.

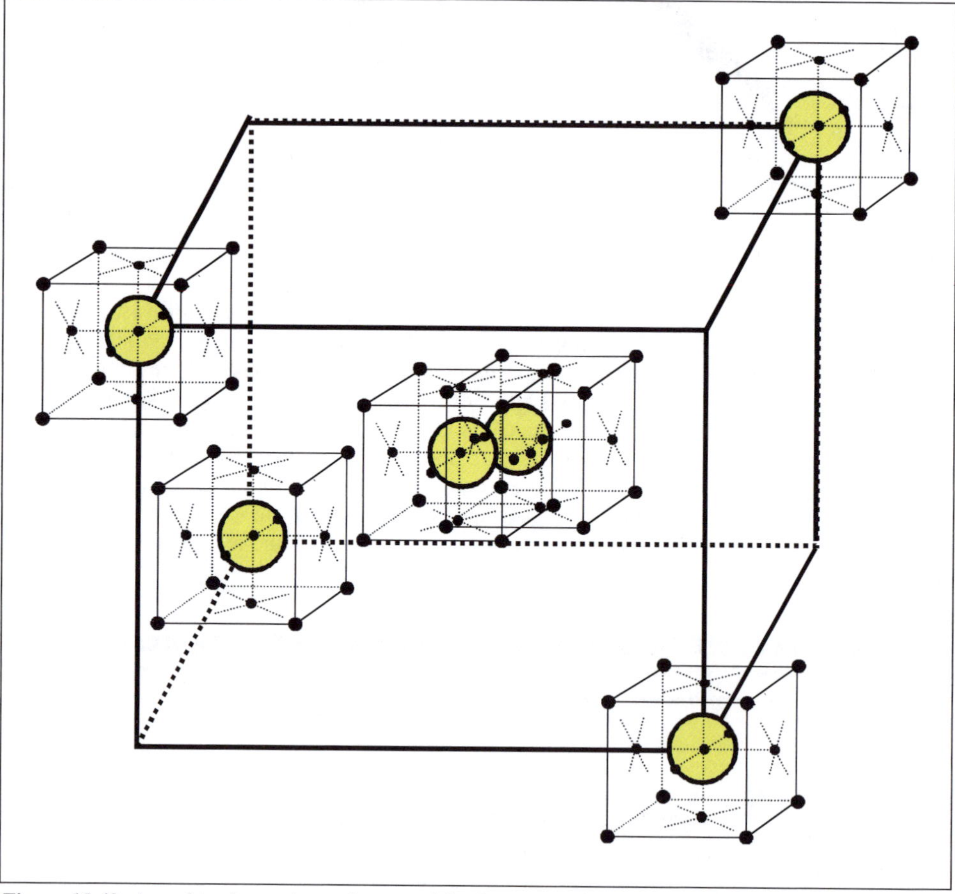

Figure 15.42: *A combined experimental protocol for the testing of expensive and inexpensive factors in drilling.*

In this kind of application it is relevant to try to minimize the number of alterations made with the most expensive factors, without sacrificing the quality of the work. Thus, to adequately address this problem we propose the following: A combined experimental design of the type shown Figure 15.42 is created, where the inner array is a medium resolution fractional factorial design laid out in the expensive factors, and the outer array is a composite design defined for the inexpensive factors. For instance, if we have three expensive and three inexpensive factors, we could lay out a combined design in 102 experiments, or 17 experiments per drill. This would be one way of assessing the robustness of a drill towards different customer practices, without draining the available test resources.

15.8 Questions for Chapter 15

- What is a design factor in the Taguchi approach? A noise factor?
- How are design and noise factors varied in a Taguchi design?
- How does the classical analysis approach work?
- What are its advantages and disadvantages?
- Why is it usually necessary to log-transform a standard deviation response?
- How does the interaction analysis approach work?
- What are its advantages and disadvantages?
- Which procedure may be used for robust design of expensive and inexpensive factors?

15.9 Summary and discussion

In this chapter, we have described the Taguchi approach to robust design. Briefly, in this approach a distinction is made between design factors and noise factors. The former are easy-to-control and correspond to the factors that we normally vary in a statistical experimental design. The latter are hard-to-control and correspond to factors that in principle should have no effect on the responses. Taguchi proposed that design factors and noise factors be co-varied together by means of the inner and outer array system.

Experimental data generated with the inner and outer array planning can be analyzed in two ways. In the classical analysis approach, the average response and the standard deviation are used as response variables. One is then interested in finding out which design factors mainly affect the signal, the average, and which affect the noise, the standard deviation. Note that with this approach no effect estimates are obtainable for the noise factors. In the alternative approach, the interaction analysis approach, a combined design is created, which includes all experiments of the inner and outer arrays. This makes it possible to fit a regression model in which model terms of the noise factors are included. It also enables the detection of single deviating experiments. It is important to realize that what in the classical analysis approach were design factor effects on the standard deviation response, now correspond to noise-design factor interactions. Recall that we encountered a strong three-factor interaction in the CakeMix study.

16 Models of variability: Distributions

16.1 Objective

The objective of this chapter is to provide an introduction to some basic univariate statistical concepts, which are useful for the evaluation of DOE data. We will first focus on the concept of variability, which was partly introduced in Chapter 1. In order to deal with variability, we will need models of variability. Such models are called distributions, and the first part of this chapter is devoted to the discussion of some common distributions. The distributions that will be treated are the normal distribution, the t-distribution, the log-normal distribution and the F-distribution. These distributions give rise to two diagnostic tools which are useful in DOE, the confidence interval and the F-test. Here, we will demonstrate how to compute confidence intervals for regression coefficients.

16.2 Models of variability - Distributions

Suppose we have measured the yield of a product ten times, and that these measurements were registered under identical experimental conditions. Then we might end up with a situation resembling the data displayed in Figure 16.1. Apparently, these data vary, despite the fact that they were obtained using identical conditions. The reason for this variation is that every measurement and every experiment is influenced by noise. This happens in the laboratory, in the pilot-plant, and in the full-scale production.

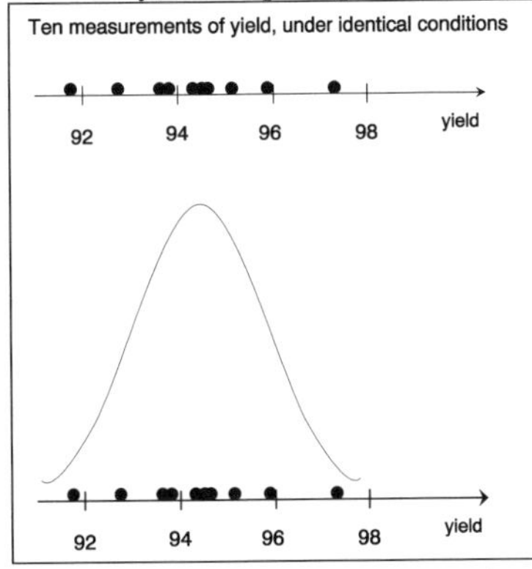

- Data vary,
- but vary in a *limited interval,*
- and with a central tendency

average:

$$\bar{x} = \sum (x_i) / N = 94.4$$

standard deviation:

$$S = \sqrt{\sum (x_i - \bar{x})^2 /(N-1)} = 1.56$$

variance: S^2

Figure 16.1: *(left) Data vary even when acquired under identical conditions.*
Figure 16.2: *(right) The average and the standard deviation of the data displayed in Figure 16.1.*

In order to draw correct conclusions, an experimenter must be able to determine the size of the experimental variation. To accomplish this, however, the experimenter needs tools to handle variability. Such tools, or models, of variability exist and they are called *distributions*.

Figure 16.1 illustrates what a distribution might look like in the context of the 10 replicated experiments. We can see that the data vary in a limited interval and with a tendency to group around a central value. The *size* of this interval, usually measured with the *standard deviation*, and the *location* of this central value, usually estimated with the *average*, may be used to characterize the properties of the occurring variability. Figure 16.2 lists the standard deviation and the average value of the ten experiments.

16.3 The normal distribution

The most commonly used theoretical distribution is the Normal or Gaussian distribution, which corresponds to a bell-shaped probability curve. A schematic example for a variable x is provided in Figure 16.3. In this probability graph, the y-axis gives the probability count and the x-axis the measured values of the studied variable. For convenience, this variable may be re-expressed to become a *standardized normal deviate* z, which is a variable standardized such that the mean is 0 and the standard deviation 1. Figure 16.4 shows how this standardization is conducted.

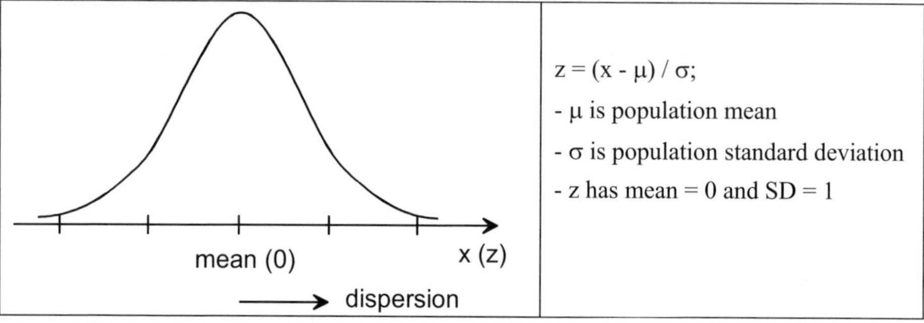

Figure 16.3: *(left) Normal probability curve of a variable x or a standardized normal variate z.*
Figure 16.4: *(right) How to standardize a variable to get average 0 and standard deviation 1.*

One interesting property of the normal distribution is that the area under the curve corresponds to a probability level, and that the area under the whole curve is 1 or 100% probability. This property means that we may now conduct an assessment of the probability that |z| will not exceed a given reference value z_0. Such probability levels are tabulated in Figure 16.5.

z_0	P
0.5	0.383
1	0.683
1.5	0.866
2	0.955
2.5	0.988
3	0.9973
4	0.9999

Figure 16.5: *Probability, P, that |z| will not exceed a given value,* z_0

It is seen that a normally distributed variable will seldom give a value of |z| greater than 3. An alternative and perhaps easier way to understand this phenomenon consists of the following: Consider the mean value of 0. From this mean and out to $-1\ \sigma$ and $+1\ \sigma$ an area corresponding to 68.3% of the total area under the curve is covered. Similarly, in the range mean $\pm$ 2 σ's the covered area is 95.5% of the total area, and when using the range mean $\pm$ 3 σ's 99.7% of the total area is covered. Hence, when working with the normal distribution, almost all data are found in the interval mean $\pm 3\ \sigma$.

In summary, the normal distribution makes it possible for the experimentalist to quantify variability. We will now describe how such information may be used for statistical inference.

16.4 The t-distribution

A common statistical operation is to calculate the *precision* of some estimated statistic, for instance, a mean value. Usually, such a precision argument is known as a *confidence interval*. One term contributing to such a precision estimate is the standard deviation of the data series distribution. If the form of this distribution is well-established, and if its true mean, μ, and true standard deviation, σ, are both known, it is possible to make straightforward probability statements about the mean value of a number of observations. In most practical circumstances, however, the true standard deviation σ is not known, and only an estimate, s, of σ, based on a limited number of degrees of freedom, is available.

Whenever the standard deviation s is estimated from the data, a distribution known as the *t-distribution* has to be employed. A simplified illustration of the t-distribution is given in Figure 16.6. As seen, the t-distribution is less peaked than the normal distribution and has heavier tails. These special features of the t-distribution arise because the estimated standard deviation, s, is itself subject to some uncertainty. Observe that with sufficiently many degrees of freedom, say, more than 30, the uncertainty in s is comparatively small, and the t-distribution is practically identical to the normal distribution. We will now examine how the t-distribution may be utilized to accomplish a precision estimate of a mean value.

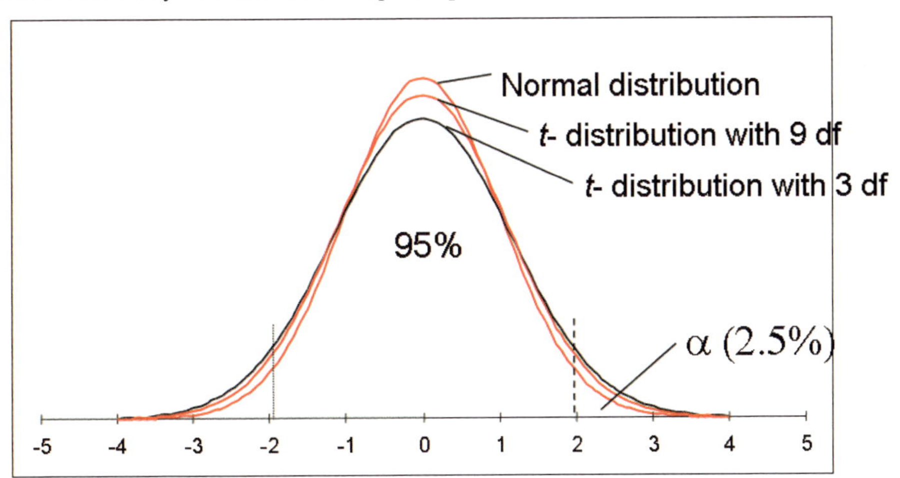

Figure 16.6: *A comparison of the normal distribution and the t-distribution.*

16.5 Confidence intervals

When estimating a mean value we would like to know the significance of this parameter, that is, we would like to know not only the estimated value of the statistic, but also how precise it is. In other words, we want to be able to state some reference limits within which it may reasonably be declared that the true value of the statistic lies. Such statements may assert that the true value is unlikely to exceed some upper limit, or that it is unlikely to be less than some lower limit, or that it is unlikely to lie outside a pair of limits. Such a pair of limits is often known as *confidence limits*, or a *confidence interval*, and is just as important as the estimated statistic itself. The degree of confidence used is usually set at 95%, but higher or lower levels may be chosen by the user.

We will now illustrate how to compute a confidence interval of a mean value. Consider the series of ten replicated experiments. The mean value of this series is 94.4 and the standard deviation 1.56. In order to compute the confidence interval around this mean, one proceeds as depicted in Figure 16.7. The first step is to calculate the standard error, SE, which is the standard deviation of the mean. Hence, SE is expressed in the original measurement unit and constitutes a dispersion estimate around the mean. Secondly, in order to convert the standard error into a confidence expression, we need a reference t-value acquired from an appropriate t-distribution. The numerical value of t reflects the size of the data set, that is, the available degrees of freedom, and the chosen level of confidence. By inspecting a table of t-values, we can see that the relevant t-value for nine degrees of freedom is 2.262 if a 95% confidence interval is sought. Observe that the t-distribution is listed as a single-sided argument, whereas the confidence interval is a double-sided argument, and hence a t-value of confidence level 0.025 is used. Finally, in the third step, the SE is multiplied by t and attached to the estimated mean value. Figure 16.8 shows how the confidence interval of the mean value may be interpreted graphically. The numerical range from 93.3 to 95.5 corresponds to the interval within which we can be 95% confident that the true mean will be found.

SE (standard error = SD of the mean) = $1.56/\sqrt{10}$ = 0.493

t (9, 0.025) = 2.262; from a t-table with 9 (10-1) degrees of freedom

The 95% confidence interval for our average is:
$94.4 \pm (2.262 * 0.493) = 94.4 \pm 1.1$

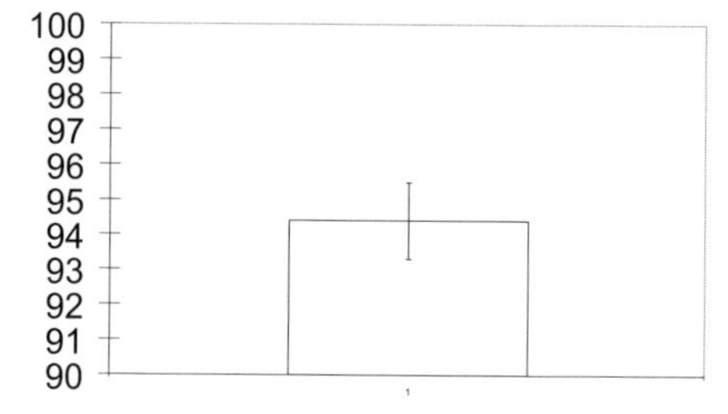

100
99
98
97
96
95
94
93
92
91
90

Figure 16.7: *(left) How to compute a confidence interval of a mean value.*
Figure 16.8: *(right) How to display a confidence interval of a mean value.*

16.5.1 Confidence intervals of regression coefficients

The use of a mean value and its confidence interval closely resembles how the statistical significance of regression coefficients is assessed. We recall from Chapter 1, that a regression coefficient is estimated in terms of averaging several differences in response values between high and low levels of the varied factors. Hence, regression coefficients may be interpreted as mean values, and their precision understood through the calculation of 95% confidence intervals. As an example, consider the regression coefficients of the CakeMix example plotted in Figure 16.9.

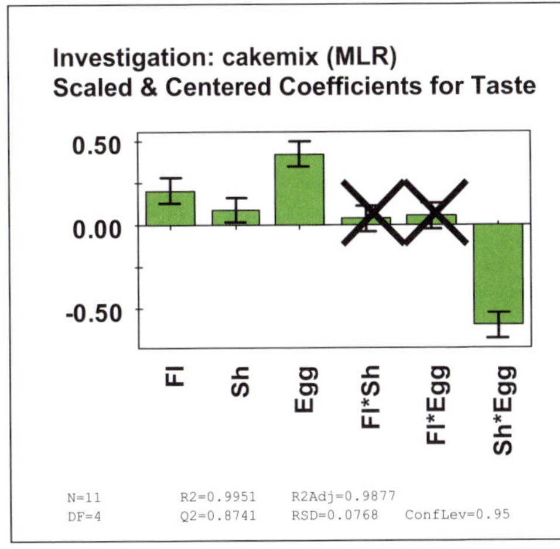

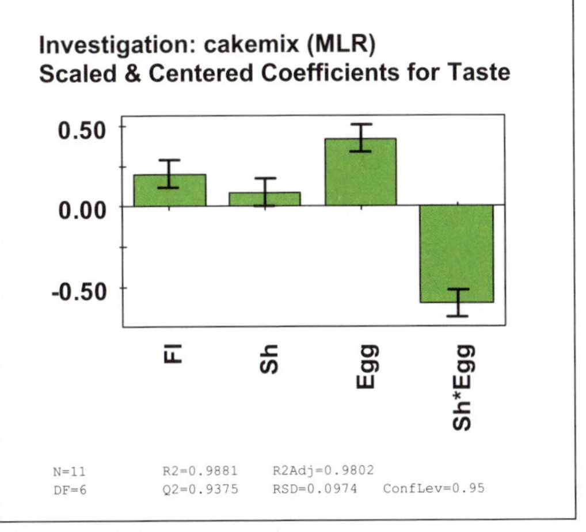

Figure 16.9. *(left) Confidence intervals of regression coefficients – CakeMix example.*
Figure 16.10: *(right) Confidence intervals and regression coefficients after model refinement.*

The appearance of the confidence intervals reveal that the value of zero is plausible for two of the interaction terms. Because of this, we can conclude that these terms are not statistically significant, and accordingly they can be removed from the model. The result after the removal of these terms and refitting of the model is displayed in Figure 16.10. It is interesting to note that Q^2 has increased from 0.87 to 0.94, which vindicates the model pruning.

The example outlined indicates that the confidence interval of a regression coefficient is an informative tool for obtaining better models. The computation of such confidence intervals is slightly more complex than for an ordinary mean value. Figure 16.11 shows how this is done for well-designed data, and the essence of the given expression is explained in the statistical appendix. At this stage, it suffices to say that three terms are involved in the computation of confidence intervals, and that the sharpest precisions are obtained with (i) a good design with low condition number, (ii) a good model with low residual standard deviation, and (iii) sufficiently many degrees of freedom.

$$\pm\sqrt{(X'X)^{-1}} * RSD * t(\alpha/2, DF_{resid})$$

Figure 16.11: *How to compute confidence intervals of regression coefficients.*

A good regression model is obtained with:

- a good design of low condition number.
- a good model with low RSD.
- sufficiently many degrees of freedom.

16.6 The log-normal distribution

We will now consider another distribution which is commonly encountered in DOE, the log-normal distribution. Sometimes it will be found that the distribution of a variable departs from normality, and it would be desirable if, by a simple transformation of the variable, an approximately normal distribution could be obtained. The principles for this are illustrated in Figures 16.12 and 16.13.

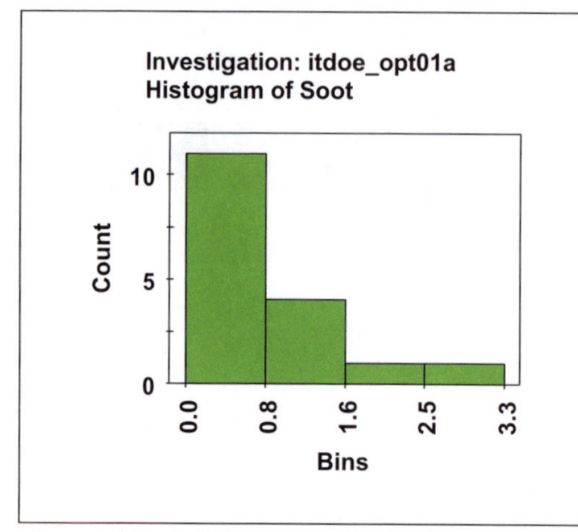

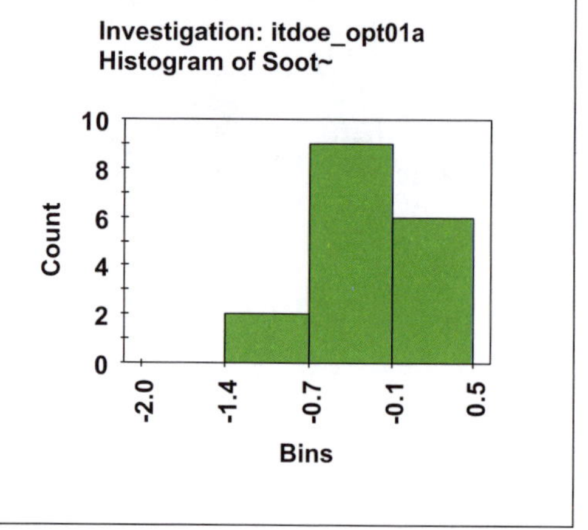

Figure 16.12: *(left) Histogram plot for the response Soot of the truck engine data.*
Figure 16.13: *(right) Histogram of the log-transformed response.*

Figure 16.12 shows the histogram plot for the Soot response of the truck engine application. The appearance of this histogram suggests that Soot adheres to the log-normal distribution. This distribution is characterized by a heavy tail to the right, that is, a few large numerical values and a majority of low measured values. By applying a logarithmic transformation to the raw data of Soot, a distribution which is much closer to normality is obtained (Figure 16.13). This transformed variable is better suited for regression analysis, and will make the modelling easier. Other examples of transformations, which are often used for transferring a skew distribution into an approximate normal distribution, are the square root, the fourth root, and the reciprocal of a variable.

16.7 The F-distribution

The last distribution which will be considered here is the *F-distribution*. In contrast to the t-distribution, which is used in conjunction with statistical testing of, for instance, mean values, the F-distribution is used for comparing variances with each other. It is possible to test whether two estimated variances differ significantly. This testing is carried out by forming the *ratio* between two compared variances. The larger variance is used in the numerator and the smaller variance in the denominator. When the underlying measured values are normally distributed, a variance ratio formed in this manner is distributed according to a distribution known as the F-distribution. Further, this ratio only depends on the degrees of freedom associated with the two variance estimates. In DOE, the F-test is ubiquitous in analysis of variance, ANOVA, which is explained in Chapter 17.

16.8 Questions for Chapter 16

- What is a distribution?
- What is the normal distribution?
- When is the t-distribution appropriate?
- What is the essence of a confidence interval?
- How can a confidence interval be used for model refinement?
- What is the log-normal distribution?

- Which transformations can be used to obtain approximately normally distributed data?
- Which distribution is used for comparing variances?

16.9 Summary and discussion

In this chapter, we have considered variability and models for variability. Such models are called distributions, and they enable uncertainty to be handled in a rational manner. Initially, the most commonly used distribution, the normal distribution, was scrutinized. Often, however, in DOE, the number of runs in an experimental plan is rather small, rendering statistical reasoning based on the normal distribution inappropriate. As a consequence, much use is made of the t-distribution, for example in the estimation of confidence intervals of regression coefficients. Another common distribution is the log-normal distribution. Response variables whose numerical values extend over more than one order of magnitude often display a log-normal distribution. This skewness can be altered by using an appropriate transformation, like the logarithmic transformation. Other common transformations are the square root, the fourth root, and the reciprocal of the variable. Finally, we discussed the F-distribution, which is used to make assessments between two compared variances. Variance testing is done in analysis of variance, ANOVA.

17 Analysis of variance, ANOVA

17.1 Objective

Analysis of variance, ANOVA, is used as a basis for regression model evaluation. This technique was briefly introduced in Chapter 4. It is the objective of the current chapter to provide more details of ANOVA and its use in connection with regression analysis. Since much of the statistical testing in ANOVA is based on the F-test, we will also explain this test.

17.2 Introduction to ANOVA

Multiple linear regression, MLR, is based on finding the regression model which minimizes the residual sum of squares of the response variable. We have seen in the earlier part of the course, that with designed data it is possible to add terms or to remove terms from an MLR model. The question which arises then is *how can we be sure that any revised model is better than the model originally postulated*? Fortunately, we can use analysis of variance, ANOVA, R^2/Q^2 and residual plots to shed some light on this question. ANOVA makes it possible to formally evaluate the performance of alternative models. We will now study how this is carried out.

Consider the CakeMix application, and recall that the performed experiments support the interaction model $y = \beta_0 + \beta_1 x_1 + \beta_2 x_2 + \beta_3 x_3 + \beta_{12} x_1 x_2 + \beta_{13} x_1 x_3 + \beta_{23} x_2 x_3 + \varepsilon$. We found in Chapter 4 that when applying MLR to the CakeMix data, the model shown in Figure 17.1 was the result. However, we also found that two model terms could be removed from the model and the model thus refined. The revised model is shown in Figure 17.2.

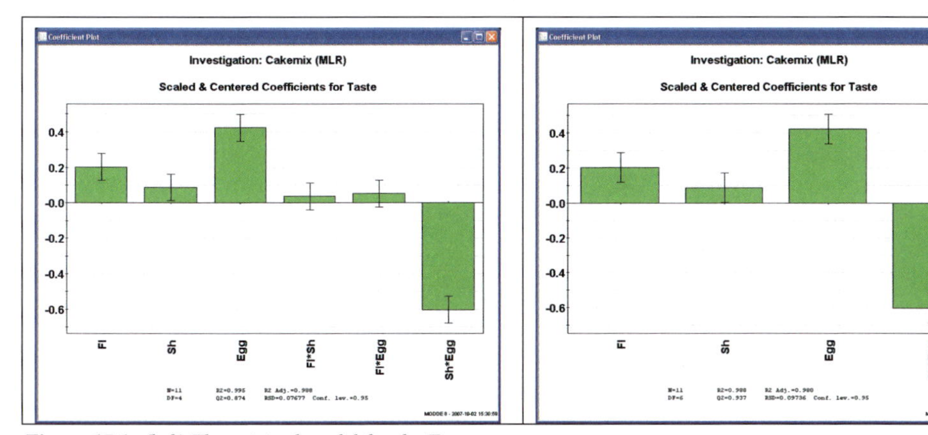

Figure 17.1: *(left) The original model for the Taste response.*
Figure 17.2: *(right) The revised model for the Taste response.*

Now, it is of interest to compare these models and this may be done with ANOVA (figures 17.3 and 17.4). ANOVA is based on partitioning the total variation of a selected response into one part due to the regression model and another part due to the residuals. At times, when replicated experiments are available, ANOVA also decomposes the residual variation into one part related to the model error and another part linked to the replicate error. Subsequently, the numerical sizes of these variance estimates are formally compared by means of F-tests. How to interpret such tables is explained below.

ANOVA Table - Response Taste

	1	2	3	4	5	6	7
1	Taste	DF	SS	MS (variance)	F	p	SD
2	Total	11	247.205	22.4731			
3	Constant	1	242.426	242.426			
4							
5	Total Corrected	10	4.77829	0.477829			0.691252
6	Regression	6	4.75472	0.792453	134.468	0.000	0.890198
7	Residual	4	0.023573	0.00589324			0.0767674
8							
9	Lack of Fit	2	0.0163063	0.00815315	2.24399	0.308	0.0902948
10	(Model Error)						
11	Pure Error	2	0.00726666	0.00363333			0.0602771
12	(Replicate Error)						
13							
14		N = 11	Q2 =	0.874	Cond. no. =	1.173	
15		DF = 4	R2 =	0.995	Y-miss =	0	
16			R2 Adj. =	0.988	RSD =	0.07677	

The selected response 'Taste'

Taste	DF	SS	MS (variance)	F	p	SD
Total	11	247.205	22.473			
Constant	1	242.426	242.426			
Total Corrected	10	4.778	0.478			0.691
Regression	6	4.755	0.792	134.469	1.45E-04	0.89
Residual	4	0.024	0.006			0.077
Lack of Fit (Model Error)	2	0.016	0.008	2.244	0.308	0.09
Pure Error (Replicate Error)	2	0.007	3.63E-03			0.06
N = 11	Q2 =	0.8741	CondNo =	1.1726		
DF = 4	R2 =	0.9951	Y-miss =	0		
	R2Adj =	0.9877	RSD =	0.0768		

ANOVA Table - Response Taste

	1	2	3	4	5	6	7
1	Taste	DF	SS	MS (variance)	F	p	SD
2	Total	11	247.205	22.4731			
3	Constant	1	242.426	242.426			
4							
5	Total Corrected	10	4.77829	0.477829			0.691252
6	Regression	4	4.72142	1.18035	124.525	0.000	1.08644
7	Residual	6	0.056873	0.00947883			0.0973593
8							
9	Lack of Fit	4	0.0496064	0.0124016	3.41329	0.239	0.111362
10	(Model Error)						
11	Pure Error	2	0.00726666	0.00363333			0.0602771
12	(Replicate Error)						
13							
14		N = 11	Q2 =	0.937	Cond. no. =	1.173	
15		DF = 6	R2 =	0.988	Y-miss =	0	
16			R2 Adj. =	0.980	RSD =	0.09736	

The selected response 'Taste'

Taste	DF	SS	MS (variance)	F	p	SD
Total	11	247.205	22.473			
Constant	1	242.426	242.426			
Total Corrected	10	4.778	0.478			0.691
Regression	4	4.721	1.18	124.525	6.68E-06	1.086
Residual	6	0.057	0.009			0.097
Lack of Fit (Model Error)	4	0.05	0.012	3.413	0.239	0.111
Pure Error (Replicate Error)	2	0.007	3.63E-03			0.06
N = 11	Q2 =	0.9375	CondNo =	1.1726		
DF = 6	R2 =	0.9881	Y-miss =	0		
	R2Adj =	0.9802	RSD =	0.0974		
			ConfLev=	0.95		

Figure 17.3: *(left) ANOVA of the original model for Taste.*
Figure 17.4: *(right) ANOVA of the refined model for Taste.*

17.3 ANOVA – Regression model significance test

In ANOVA, a common term is *sum of squares*, SS. Sums of squares are useful to quantify variability, and can be decomposed into smaller constituents with ANOVA. In ANOVA, the first decomposition is $SS_{total\ corrected} = SS_{regression} + SS_{residual}$. The first term, $SS_{total\ corrected}$, is the total variation in the response, corrected for the average. With least squares analysis we want to create a mathematical model, which can describe as much as possible of this total variation. The amount of variation that we can model, "describe", is given by the second term, $SS_{regression}$. Consequently, the amount of variation that we can *not* model is given by the third term, $SS_{residual}$.

Usually, a model is good when the modellable variation, $SS_{regression}$, is high and the unmodellable variation, $SS_{residual}$, is low. It is possible to formulate a formal test to check this. In doing so, one first converts these two *variation* estimates into their *mean squares (variance)* counterparts, that is, $MS_{regression}$ and $MS_{residual}$. This is accomplished by dividing the SS estimates with the corresponding degrees of freedom, DF. Subsequently, the sizes of these two variances are formally compared by an F-test. This is accomplished by forming the ratio $MS_{regression}/MS_{residual}$ and then retrieving the probability, p, that these two variances originate from the same distribution. It is common practice to set p = 0.05 as the critical limit. Figures 17.3 and 17.4 show that the relevant p-value for the original CakeMix model is

1.45E-04 and for the revised model 6.68E-06. These values are well below 0.05, indicating that each model is good. However, the p-value is lower for the revised model and this suggests that the model refinement was appropriate. In conclusion, the variance explained by either version of the model is significantly larger than the unexplained variance.

17.4 ANOVA – Lack of fit test

Furthermore, in ANOVA, a second decomposition of sums of squares may be made. This decomposition is done according to $SS_{residual} = SS_{model\ error} + SS_{replicate\ error}$. Thus, the unmodellable variation, $SS_{residual}$, has two components, one arising from the fact that the model is imperfect, the model error, and one arising from the fact that there is always variation when doing replicated experiments, the replicate error. In the computation of the replicate error one considers the replicated experiments and their deviation around the local replicate mean value. Once the replicate error has been determined it is subtracted from the residual sum of squares, and the remainder then corresponds to the model error.

In the ideal case, the model error and the replicate error are small and of similar size. Whether this is the case may be formally tested with an F-test. In this test, the sizes of the two variances $MS_{model\ error}$ and $MS_{replicate\ error}$ are compared by forming their ratio and F-testing this ratio. We can infer from the ANOVA table in Figure 17.3 that the model error of the original model is of the same magnitude as the replicate error, because the p-value 0.308 is *larger* than the critical reference value of 0.05. Hence, our model has small model error and good fitting power, that is, shows no *lack of fit*. As seen Figure 17.4, the p-value for the alternative model is slightly lower, 0.239, indicating that the original model is preferable to the revised model. However, the important fact is that both models pass the lack of fit test, and the fact that the p-value is slightly lower for the revised model is of only marginal relevance.

17.5 Summary of ANOVA

We have now seen that with ANOVA pairs of variances are formally compared by conducting F-tests. The first F-test, comparing modellable and unmodellable variance, is satisfied when p is < 0.05. The second F-test, comparing model and replicate errors with each other, is satisfied when p is > 0.05. The second F-test is called the *lack of fit* test and is an important diagnostic test, but it cannot be carried out if replicates are lacking. This test is the basis for the Model Validity bar seen in the Summary of Fit plot.

In addition, the ANOVA table contains other statistical measures, which are useful in model evaluation. One parameter is the *explained variation*, R^2, which is calculated as $1-SS_{residual}/SS_{total\ corrected}$, and lies between 0 and 1. This parameter is the classical quantity used for model evaluation. Unfortunately, R^2 is very sensitive to the degrees of freedom, and by including more terms in the model it is possible to fool R^2 and make it arbitrarily close to 1. Recognizing this deficiency of R^2, one may choose to work with the *explained variance,* R^2_{adj}, which is a goodness of fit measure adjusted for degrees of freedom. The adjusted R^2 is computed as $1-MS_{residual}/MS_{total\ corrected}$, and this parameter is always lower than R^2.

A third parameter often employed in this context is the *predicted variation*, Q^2, which is computed in the same way as R^2, the only difference being that $SS_{predictive\ residual}$ is used in place of $SS_{residual}$. Q^2 estimates the predictive power of a model, and is therefore of primary interest in regression modelling. We can see in Figures 17.3 and 17.4, that R^2 and R^2_{adj} have decreased a little as a result of the model revision. Normally, R^2 decreases when less useful model terms are removed, whereas R^2_{adj} should essentially remain unchanged. Also, if irrelevant model terms are omitted from the model, it is expected that Q^2 should increase. We can see that Q^2 has increased from 0.87 to 0.94, that is, we have obtained a verification that the model revision was appropriate.

17.6 The F-test

We will end this chapter by considering the F-test. The F-test compares the ratio of two variances and returns the probability that these originate from the same distribution, that is, the probability that these are not significantly different. As an example, consider the upper F-test of the original CakeMix model (Figure 17.3). The larger variance, $MS_{regression}$, is 0.792, and the smaller variance, $MS_{residual}$, is 0.006, and they have six and four degrees of freedom, respectively. The F-value is found by forming the ratio of these two variances, which is 134.468. Figure 17.5 shows an F-table pertaining to $p = 0.05$, and it shows that with six and four degrees of freedom the critical F-value is 6.2. Because the obtained F-value of 134.468 is larger than the critical F-value, the conclusion is that the two variances are unequal and are not drawn from the same distribution, that is, they are significantly different.

	1	**2**	**3**	**4**	**5**	**6**	**7**	**8**	**9**	**10**	**11**	**12**
1	161.4	199.5	215.7	224.6	230.2	234.0	236.8	238.9	240.5	241.9	243.0	243.9
2	18.5	19.0	19.2	19.2	19.3	19.3	19.4	19.4	19.4	19.4	19.4	19.4
3	10.1	9.6	9.3	9.1	9.0	8.9	8.9	8.8	8.8	8.8	8.8	8.7
4	7.7	6.9	6.6	6.4	6.3	**6.2**	6.1	6.0	6.0	6.0	5.9	5.9
5	6.6	5.8	5.4	5.2	5.1	5.0	4.9	4.8	4.8	4.7	4.7	4.7
6	6.0	5.1	4.8	4.5	4.4	4.3	4.2	4.1	4.1	4.1	4.0	4.0
7	5.6	4.7	4.3	4.1	4.0	3.9	3.8	3.7	3.7	3.6	3.6	3.6
8	5.3	4.5	4.1	3.8	3.7	3.6	3.5	3.4	3.4	3.3	3.3	3.3
9	5.1	4.3	3.9	3.6	3.5	3.4	3.3	3.2	3.2	3.1	3.1	3.1
10	5.0	4.1	3.7	3.5	3.3	3.2	3.1	3.1	3.0	3.0	2.9	2.9
11	4.8	4.0	3.6	3.4	3.2	3.1	3.0	2.9	2.9	2.9	2.8	2.8
12	4.7	3.9	3.5	3.3	3.1	3.0	2.9	2.8	2.8	2.8	2.7	2.7

Figure 17.5: *An excerpt of the p = 0.05 F-table used for assessing the significance of the CakeMix regression model.*

The p-value given in the ANOVA table (Figure 17.3) of 1.45E-04 states the real probability at which a critical reference value would be equal to the tested ratio. This is shown by the excerpt of the F-table corresponding to p = 1.45E-04 in Figure 17.6.

	3	**4**	**5**	**6**	**7**
3	514.1	499.2	489.9	483.4	478.7
4	149.9	142.5	137.8	**134.5**	132.0
5	74.0	69.2	66.1	63.9	62.4
6	47.1	43.4	41.1	39.5	38.3
7	34.5	31.5	29.5	28.2	27.2

Figure 17.6: *Excerpt of the p = 1.45E-04 F-table.*

17.7 Questions for Chapter 17

- What is the basic intention of ANOVA?
- Which two tests are carried out in ANOVA?
- What is an F-test?
- What is R^2? R^2_{adj}? Q^2?

17.8 Summary and discussion

Analysis of variance, ANOVA, is an important diagnostic tool in regression analysis. ANOVA partitions the total variation of a response variable into one component due to the regression model and another component due to the residuals. Furthermore, when replicated experiments are available, ANOVA also decomposes the residual variation into one part related to the model error and another part linked to the replicate error. Then, the numerical sizes of these variance estimates are formally compared by means of F-tests. The first F-test, comparing modellable and unmodellable variances, is satisfied when p is smaller than 0.05. The second F-test, comparing model and replicate errors, is satisfied when p exceeds 0.05. The second F-test is called the *lack of fit* test and is an important diagnostic test, but it cannot be carried out if replicates are lacking. It constitutes the basis for the Model Validity diagnostic presented in the Summary of Fit plot. In this context, we also examined the properties of the F-test.

Moreover, ANOVA gives rise to the goodness of fit parameters R^2 and R^2_{adj}, which are called *explained variation* and *explained variance*, respectively. The former parameter is sensitive to degrees of freedom, whereas for the latter this sensitivity has been reduced, thereby creating a more useful parameter. In the regression analysis it is also possible to compute the goodness of prediction, Q^2, which is the most realistic parameter of the three. Observe that Q^2 is not directly given by ANOVA, but is derivable through similar mathematical operations.

18 PLS

18.1 Objective

In this chapter, the objective is to describe PLS. PLS is an alternative to MLR, which can be used for the evaluation of complex experimental design data, or even non-designed data. PLS is an acronym of partial least squares projections to latent structures, and is a multivariate regression method. We will begin this chapter with a polymer example containing 14 responses and show how PLS may be applied to this data set. This example will help us to position PLS appropriately in the DOE framework, and highlight its essential modelling features. We will then try to introduce PLS from a geometrical perspective, but also provide relevant equations. A statistical account of PLS is given in the statistical appendix.

18.2 When to use PLS

When several responses have been measured, it is useful to fit a model simultaneously representing the variation of all responses to the variation of the factors. This is possible with PLS, because PLS deals with many responses by taking their covariances into account. MLR is not as efficient in this kind of situation, because separate regression models are fitted for each response.

Another situation in which PLS is appealing is when the experimental design is distorted. PLS handles distorted designs more reliably than MLR, since MLR in principle assumes perfect orthogonality. A design distortion may, for instance, occur because one corner of the design was inaccessible experimentally, or because some critical experiment failed. Usually, a distorted design has a high condition number, that is, it has lost some of its sphericity. As a rough rule of thumb, as soon as the condition number exceeds 10-20, it is recommended to use PLS. MLR should not be used with such high condition numbers, because the interpretability of the regression coefficients breaks down. Some regression coefficients become larger than expected, some smaller, and some may even have the wrong sign.

A third argument in favor of PLS is when there are missing data in the response matrix. MLR cannot handle missing data efficiently, and therefore each experiment for which data are missing, must be omitted from the analysis. PLS can handle missing data, as long as they are missing in a random fashion. In summary, PLS is a pertinent choice, if (i) there are several correlated responses in the data set, (ii) the experimental design has a high condition number, or (iii) there are small amounts of missing data in the response matrix.

18.3 The LOWARP application

One nice feature with PLS is that all the model diagnostic tools which we have considered so far, that is, R^2/Q^2, ANOVA, residual plots, and so on, are retained. But, in addition to these diagnostic tools, PLS provides other model parameters, which are useful for model evaluation and interpretation. We will now describe some of these PLS parameters and illustrate how they are used in the interpretation of a PLS regression model.

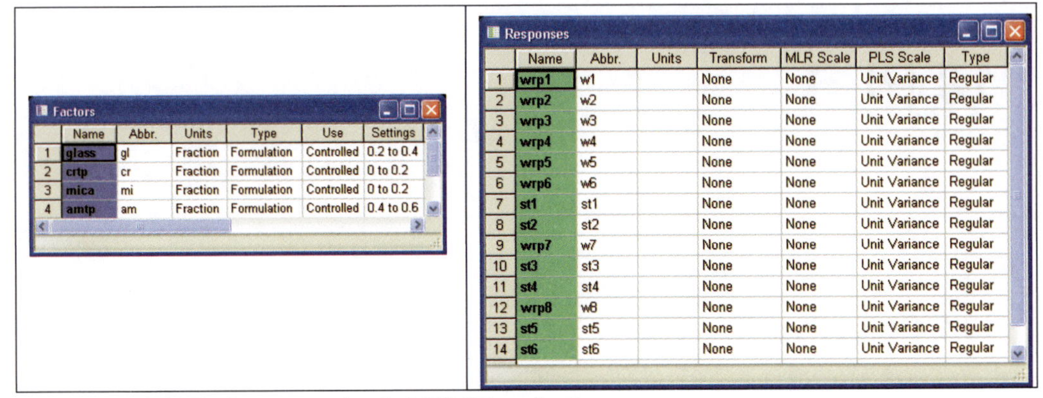

	Name	Abbr.	Units	Type	Use	Settings
1	glass	gl	Fraction	Formulation	Controlled	0.2 to 0.4
2	crtp	cr	Fraction	Formulation	Controlled	0 to 0.2
3	mica	mi	Fraction	Formulation	Controlled	0 to 0.2
4	amtp	am	Fraction	Formulation	Controlled	0.4 to 0.6

	Name	Abbr.	Units	Transform	MLR Scale	PLS Scale	Type
1	wrp1	w1		None	None	Unit Variance	Regular
2	wrp2	w2		None	None	Unit Variance	Regular
3	wrp3	w3		None	None	Unit Variance	Regular
4	wrp4	w4		None	None	Unit Variance	Regular
5	wrp5	w5		None	None	Unit Variance	Regular
6	wrp6	w6		None	None	Unit Variance	Regular
7	st1	st1		None	None	Unit Variance	Regular
8	st2	st2		None	None	Unit Variance	Regular
9	wrp7	w7		None	None	Unit Variance	Regular
10	st3	st3		None	None	Unit Variance	Regular
11	st4	st4		None	None	Unit Variance	Regular
12	wrp8	w8		None	None	Unit Variance	Regular
13	st5	st5		None	None	Unit Variance	Regular
14	st6	st6		None	None	Unit Variance	Regular

Figure 18.1: *(left) The factors varied in the LOWARP application.*
Figure 18.2: *(right) The measured responses in the LOWARP application.*

The application that we will present concerns the production of a polymer with certain desired properties. Four factors, ingredients, were varied according to a 17 run mixture design, that is, each run of the design corresponded to one polymer. These factors are presented in Figure 18.1. To map the properties of the polymers produced, 14 responses were measured. Eight responses reflecting the warp and shrinkage of the polymers, and six responses expressing the strength of the polymers, were measured. These responses are summarized in Figure 18.2. The desired combination was low warp/shrinkage and high strength. The worksheet used is listed in Figure 18.3. As seen, small amounts of missing data are found among the responses, but this is not a problem when employing PLS.

	1	2	3	4	5	6	7	8	9	10	11	12	13	14	15	16	17	18	19	20	21	22
1	Exp I	Exp N	Run	Incl/Excl	glass	crtp	mica	amtp	wrp1	wrp2	wrp3	wrp4	wrp5	wrp6	st1	st2	wrp7	st3	st4	wrp8	st5	st6
2	1	N1	16	Incl	0.4	0.1	0.1	0.4	0.9	5	0.2	1	0.3	4.2	232	15120	1.2	2190	26390	1.3	2400	0.7
3	2	N2	1	Incl	0.2	0.2	0	0.6	3.7	7.3	0.7	1.8	2.5	5.4	150	12230	1.8	905	20270	2.1	1020	0.6
4	3	N3	11	Incl	0.4	0.2	0	0.4	3.6	6.9	0.9	2.1	4.8	9.4	243	15550	1.2	1740	21180	1.4	1640	
5	4	N4	6	Incl	0.2	0.2	0.2	0.4	0.6	3.1	0.3	0.4	0.4	1.1	188	11080	1	1700	17630	1	1860	0.5
6	5	N5	13	Incl	0.2	0.1	0.2	0.5	0.3	2.1	0.3	0.3	0.8	1.1	172	11960	1.2	1810	21070	1.3	1970	0.5
7	6	N6	3	Incl	0.4	0	0.2	0.4	1.2	5					245	15600	1.1	2590	25310	1.3	2490	0.6
8	7	N7	4	Incl	0.2	0	0.2	0.6	2.3	3.9	0.3	0.4	0.7	1.4	242	13900	1.5	1890	21370	1.6	1780	
9	8	N8	7	Incl	0.4	0	0.1	0.5	2.6	5.9	0.4	0.2	0.7	1.2	243	17290	1.6	2130	30530	1.6	2320	0.7
10	9	N9	14	Incl	0.3	0.2	0.1	0.4	2.2	5.3	0.2	0.7	0.6	2	204	11170	1	1670	19070	1.1	1890	0.6
11	10	N10	5	Incl	0.4	0	0	0.6	5.8	7	0.9	1	5.6	11.8	262	20160	1.6	1930	29830	1.8	1890	
12	11	N11	12	Incl	0.3	0	0.2	0.5	0.8	2.9	0.5	0.6	1.1	2	225	14140	1.3	2140	22850	1.3	2110	0.7
13	12	N12	15	Incl	0.3	0.1	0	0.6	2.8	5.1	1	1.2	2.7	6.1	184	15170	1.9	1230	23400	2.1	1250	0.6
14	13	N13	9	Incl	0.3	0.1	0.1	0.5	1.1	4.7	0.6	0.9	1.3	3.5	198	13420	1.4	1750	23790	1.4	1930	0.7
15	14	N14	8	Incl	0.3	0.1	0.1	0.5	1.9	4.7	1	1	2.8	5.4	234	16970	1.5	1920	25010	1.6	1790	0.7
16	15	N15	10	Incl	0.3	0.1	0.1	0.5	2.9	5.9	0.5	0.6	1	6.6	239	15480	1.5	1800	23140	1.6	1730	
17	16	N16	2	Incl	0.4	0.1	0	0.5	5.5	7.9	0.8	2.4	5.5	9.3	256	18870	1.5	1880	28440	1.8	1790	
18	17	N17	17	Incl	0.3	0	0.1	0.6	3.2	6	0.3	0.5	1.5	5.2	249	16310	1.5	1860	24710	1.7	1780	

Figure 18.3: *The LOWARP experimental data.*

18.3.1 PLS model interpretation – Scores

In the analysis of the LOWARP data it is recommended to start by evaluating the raw experimental data. We made a histogram and a replicate plot for each one of the 14 responses, but could detect no anomalies in the data. No plots of this phase are provided. We will now describe the results of the PLS analysis. Conceptually, PLS may be understood as computing pairs of new variables, known as *latent variables* or *scores*, which summarize the variation in the responses and the factors. When modelling the LOWARP data, three such pairs of latent variables, three *PLS components*, were obtained. In Figure 18.4, we can see how R^2 and Q^2 evolve as a result of calculating each consecutive pair of latent variables.

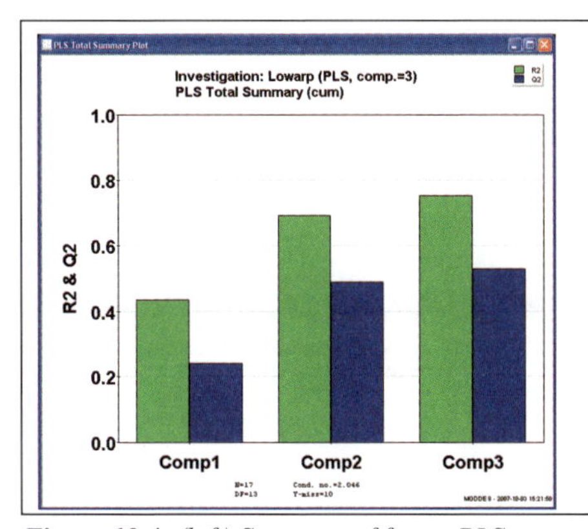

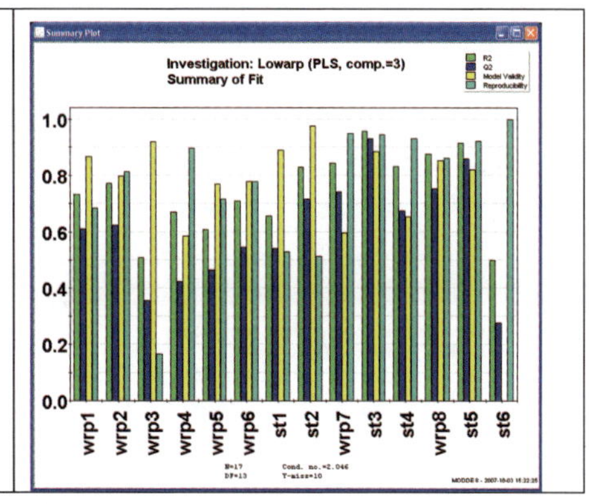

Figure 18.4: *(left) Summary of fit per PLS component.*
Figure 18.5: *(right) Summary of fit per response variable, after 3 components.*

After three components $R^2 = 0.75$ and $Q^2 = 0.53$. These are excellent values considering the fact that 14 responses are modelled at the same time. It is possible to fractionate these overall performance statistics into R^2's and Q^2's related to the individual responses. This is shown in Figure 18.5. The individual R^2-values range between 0.50 – 0.96 and the Q^2-values between 0.28 – 0.93, but the interesting feature is that for no response are the individual R^2/Q^2-values separated by more than 0.25. This constitutes a strong indication that the multivariate model is well-founded and warranted.

Furthermore, the three pairs of latent variables offer the possibility to visually inspect the correlation structure between the factors and the responses. Figure 18.6 provides a scatter plot of the first pair of latent variables. Here, the first linear summary of Y, called u_1, is plotted against the first linear summary of X, called t_1, and each plot mark depicts one polymer, one experiment in the design. With a perfect match between factors (X) and responses (Y), all samples would line up on the diagonal going from the lower left-hand corner to the upper right-hand corner. Conversely, the weaker the correlation structure between factors and responses, the more scattered the samples are expected to be around this ideal diagonal. Any point remote from this diagonal may be an outlier. We can see in Figures 18.6 and 18.7 that for the two first PLS components, there is a strong correlation structure. The correlation band for the third PLS component, as shown in Figure 18.8, is not as distinct, but still reasonably strong, considering it is a third model component. In conclusion, these plots indicate that the factor changes are influential for the monitored responses, and that there are no outliers in the data. This is vital information.

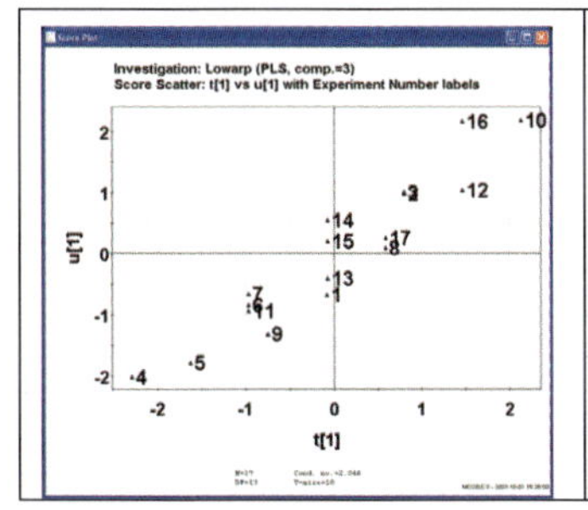

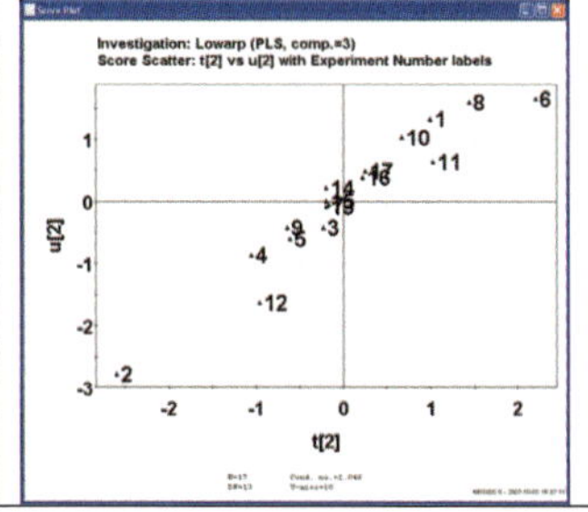

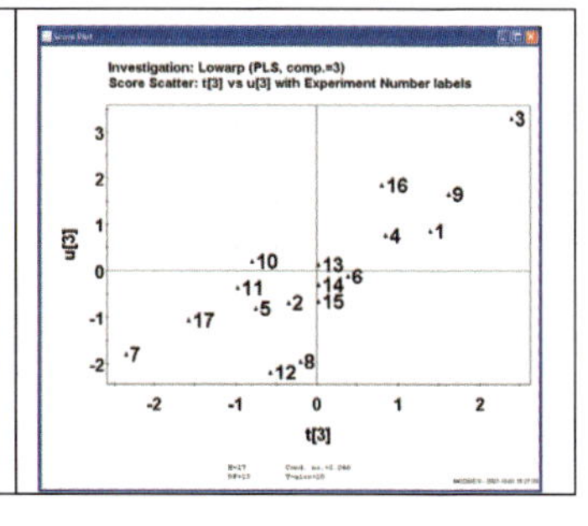

Figure 18.6: *(left) PLS score plot of the first model dimension.*
Figure 18.7: *(middle) PLS score plot of the second model dimension.*
Figure 18.8: *(right) PLS score plot of the third model dimension.*

18.3.2 PLS model interpretation – Loadings

Because PLS results in only one model, it is possible to overview the relationships between *all* factors and *all* responses at the same time. An efficient means of interpreting the PLS model is the loading plot displayed in Figure 18.9. We have plotted the loadings of the second model dimension, denoted wc_2, against the loadings of the first model dimension, denoted wc_1.

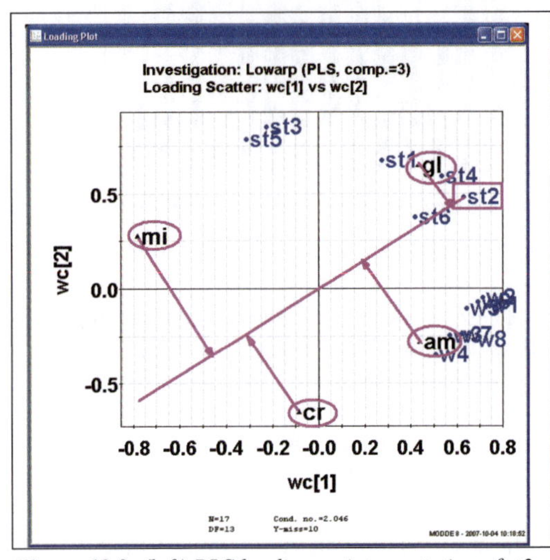

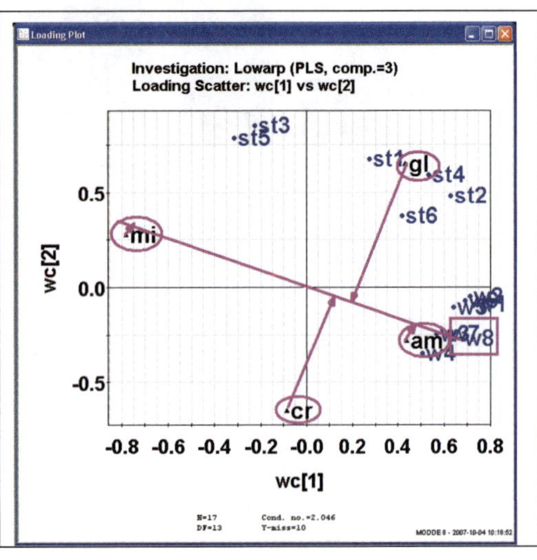

Figure 18.9: *(left) PLS loadings – interpretation of st2.*
Figure 18.10: *(right) PLS loadings – interpretation of w8.*

In the model interpretation one considers the distance to the plot origin. The further away from the plot origin an X- or Y-variable lies, the stronger the model impact that particular variable has. In addition, we must also consider the sign of the PLS loading, which informs about the correlation among the variables. For instance, the X-variable (factor) glas is influential for the Y-variable (response) st4. This is inferred from their closeness in the loading plot. Hence, when glas increases st4 increases.

Figure 18.9 contains some extra lines which are meant to support the interpretation process. First, the interpreter has to focus on an interesting response, for example the boxed st2. From the location of this response one then draws a reference base-line through the origin. Subsequently, orthogonal projections are carried out from the position of each factor and onto this first reference line. The resulting intercepts will help us in the interpretation; the further away from the plot origin the more potent the factor. Hence, we understand that glas and mica are most important for st2, albeit with different relationships to the response. The latter factor is negatively correlated with st2, since the intercept with the reference line is located on the opposite side of the plot origin.

Now, if we want to interpret the model with regard to another response, say w8, we have to draw another reference base-line and make new orthogonal factor projections. This is displayed in Figure 18.10. For the w8 response, mica is the most important factor, and exerts a negative influence, that is, to decrease w8 we should increase mica. The factors glas and amtp are positively correlated with w8, but their impact is weaker than mica. The last factor, crtp, is the least influential in the first two model dimensions. This is inferred from the fact that the projection intercept is almost at the plot origin.

18.3.3 PLS model interpretation – Coefficients

As mentioned before, it is possible to transfer the PLS solution into an expression based on regression coefficients. We will now demonstrate the practical part of this, not the matrix algebra involved. Consider Figure 18.11, in which four responses of the LOWARP application have been marked.

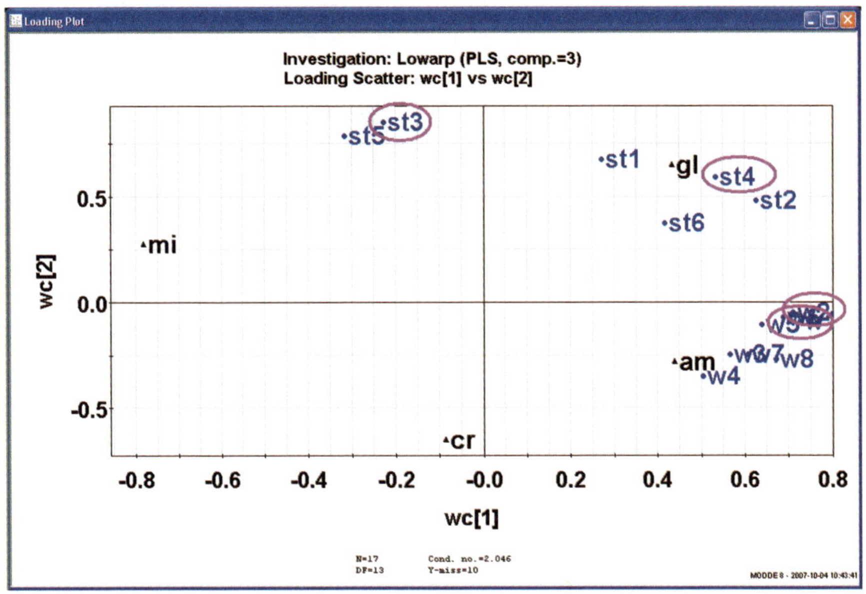

Figure 18.11: The PLS loadings of the two first model dimensions. The four responses w2, w6, st3, and st4 are marked for comparative purposes (see text).

The responses denoted w2 and w6 lie almost on top of each other and their labels are hard to discern. These represent a pair of strongly correlated responses. Based on their orientation in the loading plot, we expect them to have similar regression coefficient profiles. That this is indeed the case is evidenced by Figures 18.12 and 18.13.

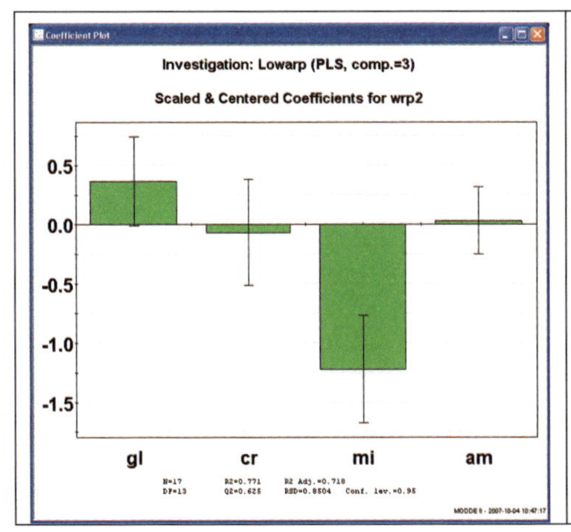

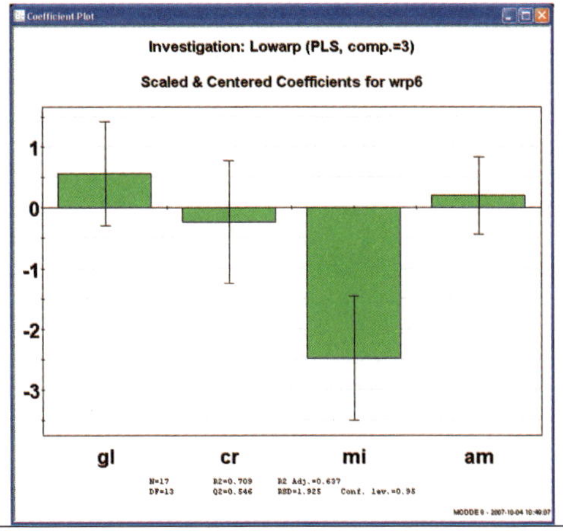

Figure 18.12: (left) Regression coefficients for w2.
Figure 18.13: (right) Regression coefficients for w6.

The other variable pair is made up of st3 and st4. Since these are located in different regions of the loading plot, they should not have similar regression coefficient profiles. Figures 18.14 and 18.15 indicate that these profiles mainly differ with regard to the third and fourth factors.

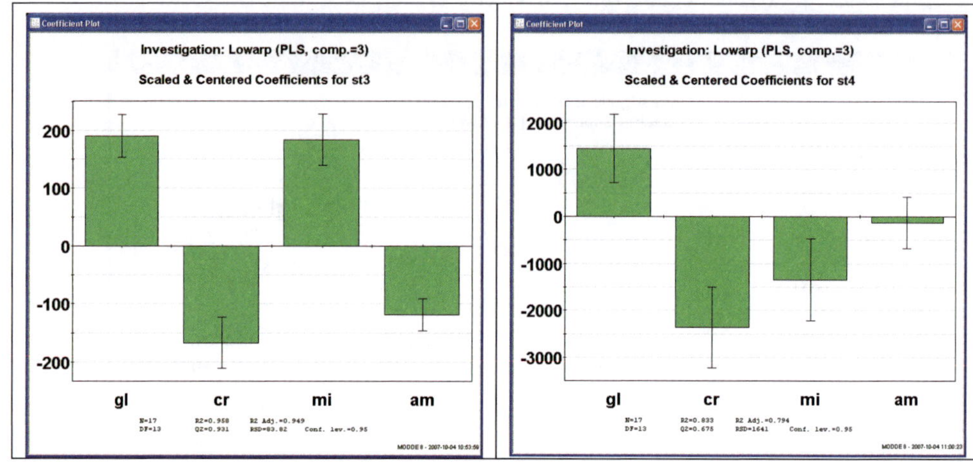

Figure 18.14: *(left) Regression coefficients for st3.*
Figure 18.15: *(right) Regression coefficients for st4.*

This ability of the loading plot to overview the relationships among the responses is very useful in the model interpretation. It will facilitate the understanding of which responses provide similar information about the experimental runs, and which provide unique information. The implication of a loading plot like the one in Figure 18.11 is that it is not necessary to measure 14 responses in the future. It is sufficient to select a sub-set of responses representing a good spread in the loading plot. Both categories of responses are grouped in two clusters. Hence, a proper selection might be st1 and st2 to represent the st-grouping in the upper right-hand quadrant, st3, w1-w2 for one w-group, and w3-w4 for the other w-group. This corresponds to a reduction by 50%.

18.3.4 PLS model interpretation – VIP

So far, we have only considered the PLS loadings associated with the two first model components. However, in order to conduct the model interpretation stringently, we ought to consider the loadings of the third component as well. With three PLS loading vectors, three bivariate scatter plots of loadings are conceivable, and these are plotted in Figures 18.16 – 18.18.

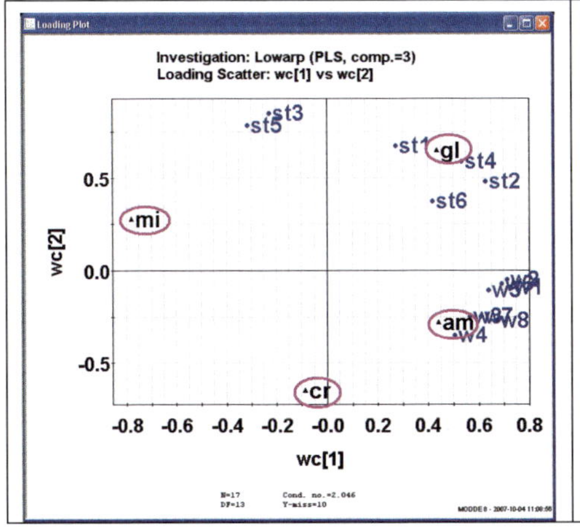

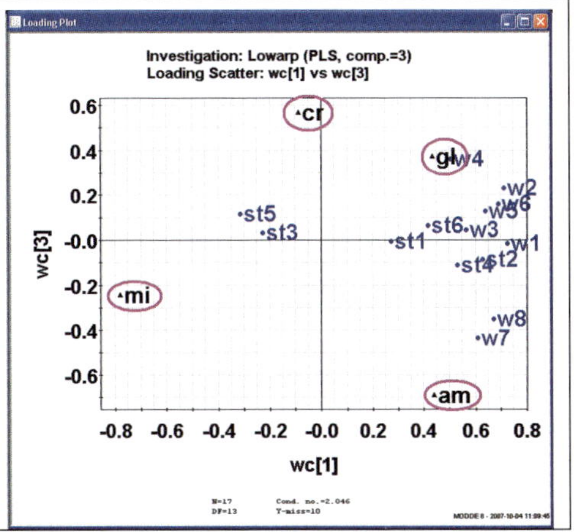

Figure 18.16: *(left) Scatter plot with loading vector 2 versus loading vector 1.*
Figure 18.17: *(right) Scatter plot with loading vector 3 versus loading vector 1.*

From a model overview perspective, it is quite natural to primarily consider the loading plot displayed in Figure 18.16, as this plot represents most of the modelled response variation. However, if details are sought, one must also consider the loadings of the higher-order model dimensions. With many PLS components, say four or five, the interpretation of loadings in terms of scatter plots becomes impractical. Also, with as many as 14 responses, the model interpretation based on regression coefficients is cumbersome, since 14 plots of coefficients must be created. In this context of many PLS components and many modelled responses, PLS provides another parameter which is useful for model interpretation, the VIP parameter (see Figure 18.19).

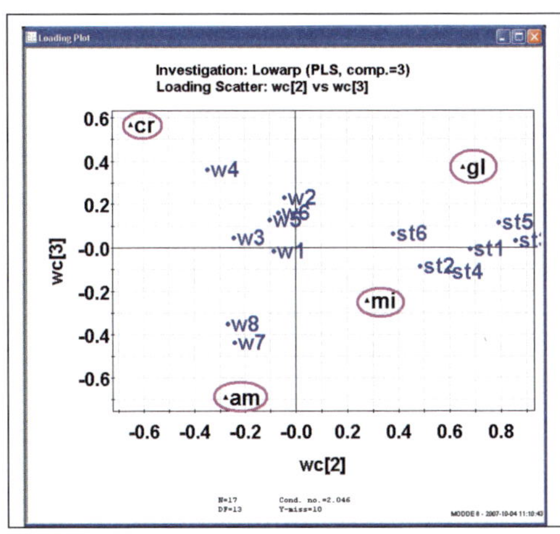

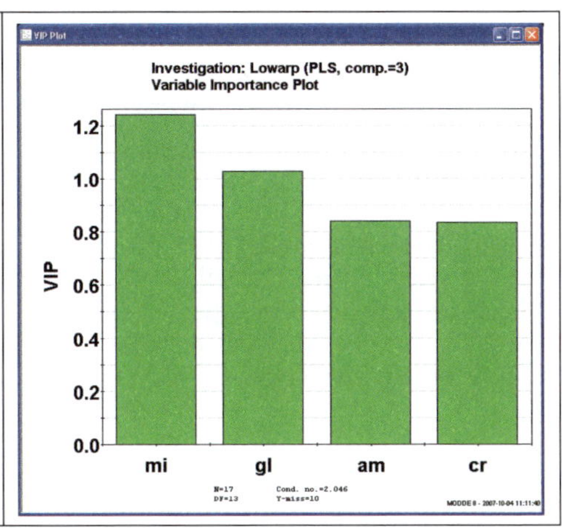

Figure 18.18: *(left) Scatter plot with loading vector 3 versus loading vector 2.*
Figure 18.19: *(right) VIP plot.*

VIP is the acronym for *v*ariable *i*mportance in the *p*rojection. This parameter represents the most condensed way of expressing the PLS model information. VIP is a weighted summary of all loadings and across all responses, and hence there can be only one VIP-expression per PLS model. Model terms with high VIP-values, often higher than 0.7 – 0.8, are the most influential in the model. We can see in Figure 18.19 that mica and glas are the two most important factors. In the case of many model terms, many more than four, the VIP plot often displays discrete jumps, like stairs, and it is possible to define a suitable threshold below which factors might be discarded from the model. Therefore, in complicated PLS applications, a plot of the VIP parameter is often a useful start in the model interpretation for revealing which are the dominating terms. Such information is subsequently further explored by making appropriate plots of loadings or coefficients.

18.4 The linear PLS model – Matrix and geometric representation

Now, that we have seen how PLS operates from a practical point of view, it is appropriate to try to understand how the method works technically. The development of a PLS model can be described as follows: For a certain set of observations – experimental runs in DOE – appropriate response variables are monitored. These form the N x M response data matrix Y, where N and M are the number of runs and responses, respectively. This is shown in Figure 18.20. Moreover, for the same set of runs, relevant predictor variables are gathered to constitute the N x K factor matrix X, where N is the same as above and K the number of factors. The Y-data are then modelled by the X-data using PLS.

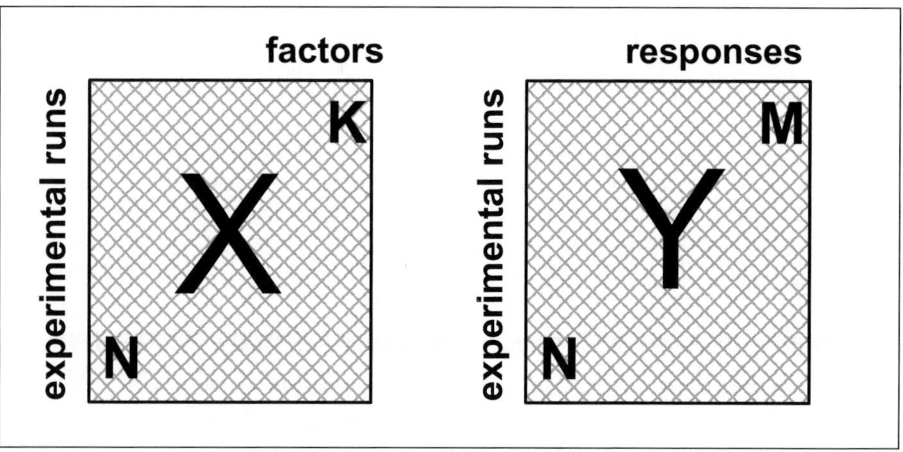

Figure 18.20: *Some notation used in PLS.*

A geometric representation of PLS is given in Figure 18.21. The experimental runs can be seen as points in two spaces, that of X with K dimensions and that of Y with M dimensions. PLS finds lines, planes or hyperplanes in X and Y that map the shapes of the point-swarms as closely as possible. The drawing in Figure 18.21 is made such that PLS has found only one-dimensional models in the X- and Y-spaces.

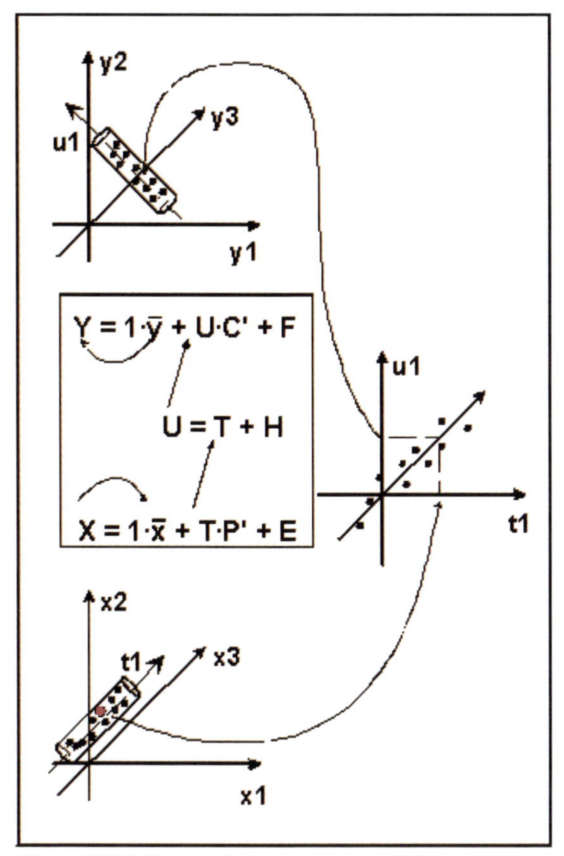

Figure 18.21: *Geometric representation of PLS.*

PLS has two primary objectives, namely to well approximate X and Y and to model the relationship between X and Y. This is accomplished by making the bilinear projections

$$\mathbf{X} = \mathbf{TP'} + \mathbf{E} \qquad \text{(eqn 18.1)}$$

$$\mathbf{Y} = \mathbf{UC'} + \mathbf{G} \qquad \text{(eqn 18.2)}$$

and connecting X and Y through the inner relation

$$\mathbf{U} = \mathbf{T} + \mathbf{H} \qquad \text{(eqn 18.3)}$$

where E, G and H are residual matrices. A more detailed account of the PLS algorithm is given in the statistical appendix. PLS simultaneously projects the X- and Y-variables onto the same subspace, T, in such a manner that there is a good relation between the position of one observation on the X-plane and its corresponding position on the Y-plane. Moreover, this relation is asymmetric (X ➔ Y), which follows from equation (3). In this respect, PLS differs from for instance canonical correlation where this relation is symmetric.

18.5 The linear PLS model – Overview of algorithm

In essence, each PLS model dimension consists of the X score vector t, the Y score vector u, the X loading vector p', the X weight vector w' and the Y weight vector c'. This is illustrated in Figure 18.22, which provides a matrix representation of PLS. Here, the arrows show the order of calculations within one round of the iteration procedure. Solid parts of the arrows indicate data "participating" in the computations and dashed portions "inactive" data. For clarity, the X score matrix T, the X loading matrix P', the X weight matrix W', the Y score matrix U, and the Y weight matrix C' are represented as matrices, though in each iteration round only the last vector is updated.

It is the PLS vectors w' and c', formally known as *weights* but in everyday language known as *loadings*, that are used in the interpretation of each PLS component. Based on this argumentation one may see that another way to understand PLS is that it forms "new x-variables", t, as linear combinations of the old ones, and thereafter uses these new t's as predictors of Y. Only as many new t's are formed as are needed, and this is assessed from their predictive power using a technique called cross-validation. Cross-validation is described in the statistical appendix.

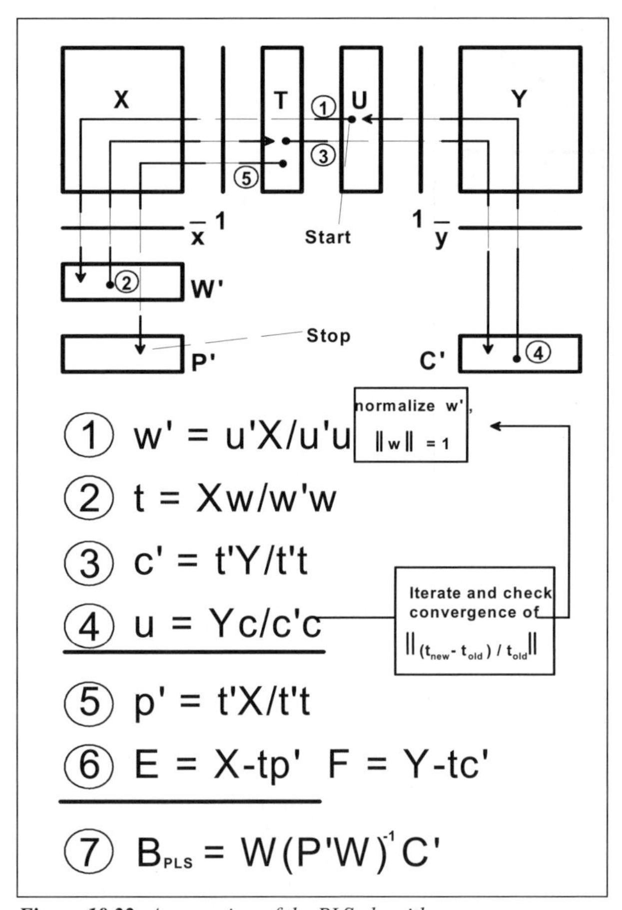

Figure 18.22: *An overview of the PLS algorithm.*

18.6 Summary of PLS

Once a PLS model has been derived, it is important to interpret its meaning. For this, the scores t and u can be considered. They contain information about the experimental runs and their similarities / dissimilarities in the X- and the Y-space with respect to the given problem and model. To understand which factors and responses are well described by the model, PLS provides three complementary interpretational parameters. These are the loadings, the coefficients, and VIP, which are exemplified in Figures 18.23 – 18.25.

The loadings wc provide information about how the variables combine to form t and u, which, in turn, expresses the quantitative relation between X and Y. Hence, these loadings are essential for the understanding of which X-variables are important for modelling Y (numerically large w-values), for understanding of which X-variables provide common information (similar profiles of w-values), and for the interpretation of the scores t.

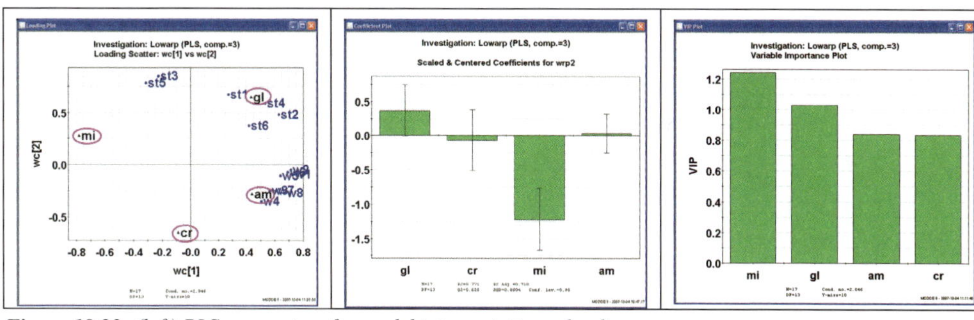

Figure 18.23: *(left) PLS parameters for model interpretation – loadings.*
Figure 18.24: *(middle) PLS parameters for model interpretation – coefficients.*
Figure 18.25: *(right) PLS parameters for model interpretation – VIP.*

Sometimes it may be quite taxing to overview the PLS loadings and in such circumstances the VIP parameter is useful. VIP in square is a weighted sum of squares of the PLS loadings, w, taking into account the amount of Y-variance explained. Alternatively, the PLS solution may be transferred into a regression-like model:

$$Y = X B_{PLS} + F \qquad \text{(eqn 18.4)}$$

Here B_{PLS} corresponds to the regression coefficients. Thus, these coefficients are determined from the underlying PLS model and can be used for interpretation, in the same way as coefficients originating from MLR. However, with collinear variables, which occur for instance in mixture design applications, we must remember that these coefficients are *not* independent. Further, the parts of the data that are not explained by the model, the residuals, are of diagnostic interest. Large Y-residuals indicate that the model is inadequate, and a normal probability plot of the residuals of a single Y-variable is useful for identifying outliers.

18.7 Questions for Chapter 18

- When is it appropriate to use PLS?
- Which diagnostic tools are available through PLS?
- What is a latent variable?

18.8 Summary and discussion

PLS is a multivariate regression method which is useful for handling complex DOE problems. This method is particularly useful when (i) there are several correlated responses in the data set, (ii) the experimental design has a high condition number, or (iii) there are small amounts of missing data in the response matrix. We used the LOWARP application with 14 responses and some missing Y-data to exemplify the tractable properties of PLS. Interestingly, PLS contains the multiple linear regression solution as a special case, when one response is modelled and the condition number is unity. Model interpretation and evaluation is carried out analogously to MLR, the difference being that with PLS more informative model parameters are available. Particularly, we think here of the PLS loadings, which inform about the correlation structure among all factors and responses, and the PLS scores, which are useful for finding trends, groups, and outliers.

Part IV

Chapters 19-22

Applications of DOE in Pharma

19 DOE in organic chemistry

19.1 Objective

Many reactions in organic chemistry are complex processes and the outcome (e.g. yield, purity, cost, …) typically depends on a number of factors acting together. Hence, the optimization of an organic reaction can benefit substantially from the use of DOE. The objective of this chapter is to discuss the use of DOE from the viewpoint of the organic chemist.

19.2 Introduction

Organic chemistry is a discipline within chemistry which involves the study of chemical compounds consisting primarily of carbon and hydrogen. Such compounds may also contain any number of other elements, including nitrogen, oxygen and halogens, as well as phosphorus, silicon and sulfur. Because of their unique properties, multi-carbon compounds exhibit wide structural variety and the range of applications of organic compounds is enormous. They form the basis, or are important constituents, of many products, such as paints, plastics, food, explosives, drugs, petrochemicals, and so on.

Typically, an organic reaction is affected by a number of potentially adjustable factors, both quantitative and qualitative in nature. The choice of which quantitative factors to vary and their ranges naturally depends on the particular reaction in question, but it is generally agreed that temperature, addition rates, concentrations, stoichiometries, reaction times, and pH are the factors which have the greatest impact on the result. With respect to the qualitative factors, the obvious prime aspect to consider is the choice of solvent. Additional discretely varying factors might involve the selection of e.g. Lewis acids, catalysts, amines, additives, etc., some of which may be varied using multiple settings.

Because of the large number of factors that can potentially influence the outcome of an organic synthesis, any optimization task may rapidly become quite a complex endeavor. As within other manufacturing industries, it is paramount to minimize cost, time and waste, while maximizing product yield, quality and duration. Given the multifaceted nature of organic chemistry, the DOE approach represents a useful and economical way of studying these problems, minimizing the number of experimental trials necessary.

The proper use of DOE in organic chemistry may not only provide a better mechanistic insight into the investigated reaction system, but also usually results in significantly improved cost-efficiency. In fact, the Lanxess company, in their 2005 annual report, noted this and stated: "Efficiency – thanks to MODDE. The Fine Chemicals business uses the new MODDE program for its process development in the field of customer synthesis. This enables us to systematically plan and evaluate tests series with a minimum number of experiments. All in all, it significantly boosts the efficiency of process development." (URL: http://reports.westernreal.de/reports/lxs_ar_2005_en/pdf_cache/lxs_ar_2005_en_extract_30.pdf).

In order to illustrate the applicability of DOE to organic chemistry problems, we start by examining a small chemical example, where the objective is to increase the yield of the desired product. Subsequently, there is a short discussion about which factors should be varied and which responses measured. This chapter also addresses the possibilities for applying multivariate design to organic chemistry.

19.3 An organic chemistry example: WILLGE

This example data-set concerns a chemical synthesis using the Willgerodt-Kindler reaction. This process is a re-arrangement reaction which occurs when aryl-alkyl-ketones are heated in the presence of sulfur and an amine (Figure 19.1). The general objective of the investigation was to elucidate the reaction mechanism by means of DOE and, in particular, to determine which factors have the greatest influence on the reaction. The original literature source is: Torbjörn Lundstedt, PhD Thesis, Umeå University, Sweden, 1986.

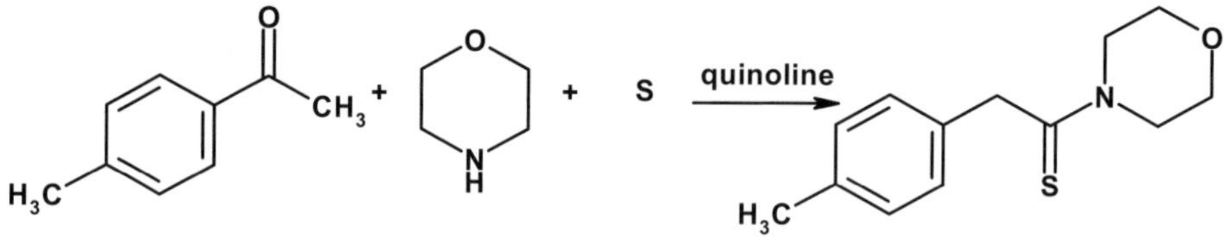

O
CH_3 +
O
N
H
+ S
quinoline
H_3C
O
N
S
H_3C

Figure 19.1: *Schematic of the Willgerodt-Kindler reaction.*

19.3.1 Screening phase

The objective of the first stage of the investigation was screening, in order to identify the most important factors. The screening study included five factors (Figure 19.2). The single response variable was the percentage yield of the oxidized aryl-alkyl-ketone.

	Factors	Levels		
		-	0	+
SK	The proportion Sulphur/Ketone (mole/mole)	5	8	11
MK	The proportion Morpholine/Ketone (mole/mole)	6	8	10
Te	Temperature (°C)	100	120	140
Pa	Particle size of Sulphur (mm)	0.1	0.15	0.2
Sti	Stirring speed	300	500	700

Figure 19.2: *Factors varied in the study of the Willgerodt-Kindler reaction.*

A two-level fractional factorial design in five factors, a so called 2^{5-1} design, was constructed. As shown by Figure 19.3, this design consists of 16 factorial experiments excluding center-points (in this study no center-points were included). The design supports an interaction model.

Worksheet

	1	2	3	4	5	6	7	8	9	10
1	Exp No	Exp Name	Run Order	Incl/Excl	SulfKet	MorfKet	Temp	PartSize	Stirring	Yield
2	1	N1	10	Incl	5	6	100	0.1	700	11.5
3	2	N2	1	Incl	11	6	100	0.1	300	55.8
4	3	N3	13	Incl	5	10	100	0.1	300	55.7
5	4	N4	16	Incl	11	10	100	0.1	700	75.1
6	5	N5	9	Incl	5	6	140	0.1	300	78.1
7	6	N6	4	Incl	11	6	140	0.1	700	88.9
8	7	N7	14	Incl	5	10	140	0.1	700	77.6
9	8	N8	5	Incl	11	10	140	0.1	300	84.5
10	9	N9	2	Incl	5	6	100	0.2	300	16.5
11	10	N10	8	Incl	11	6	100	0.2	700	43.7
12	11	N11	11	Incl	5	10	100	0.2	700	38
13	12	N12	3	Incl	11	10	100	0.2	300	72.6
14	13	N13	15	Incl	5	6	140	0.2	700	79.5
15	14	N14	6	Incl	11	6	140	0.2	300	91.4
16	15	N15	12	Incl	5	10	140	0.2	300	86.2
17	16	N16	7	Incl	11	10	140	0.2	700	78.6

Figure 19.3: *Worksheet for the 2^{5-1} fractional factorial design used in the screening phase of the Willgerodt-Kindler reaction.*

Analysis of the screening data indicated a very strong relationship between the factors and the response variable (Figure 19.4). The Q^2 of the model is as high as 0.95. Note that the Summary of Fit plot has been reduced to contain just two bars, since there are no replicates in the worksheet. The default interaction model has been revised by omitting six non-significant cross-terms (Figure 19.4).

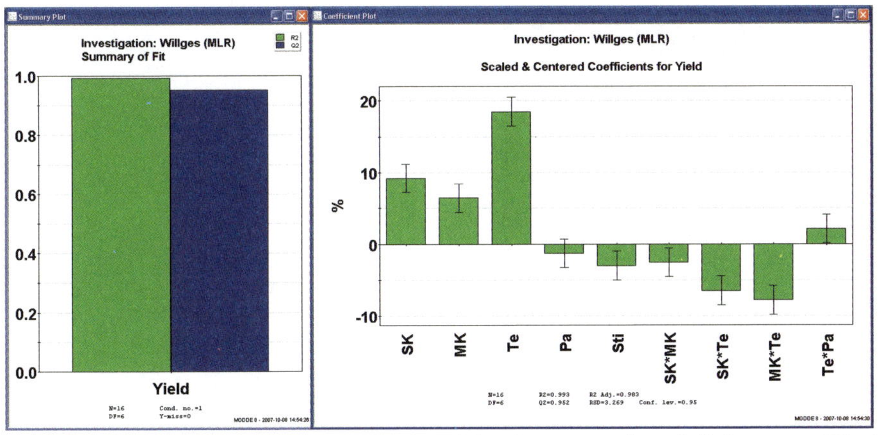

Figure 19.4: *Outcome of fitting the interaction model to the response data. (left) Summary of Fit plot. (right) Regression coefficient plot, which shows that three factors have more influence on the Yield than the others.*

The principal conclusion of the screening design is that three out of the five factors (SK, MK and Te) have the greatest influence on the Yield; there are also significant interactions between these three. Below, these three factors are further explored using an RSM design.

19.3.2 Optimization design

In the second phase, the initial screening design was upgraded to an optimization design. Design augmentation is easily accomplished in MODDE using the Complement Design

Wizard, which offers different options depending on what the user wants to do (Figure 19.5). In our example, the original investigator decided to upgrade the screening design to an RSM design.

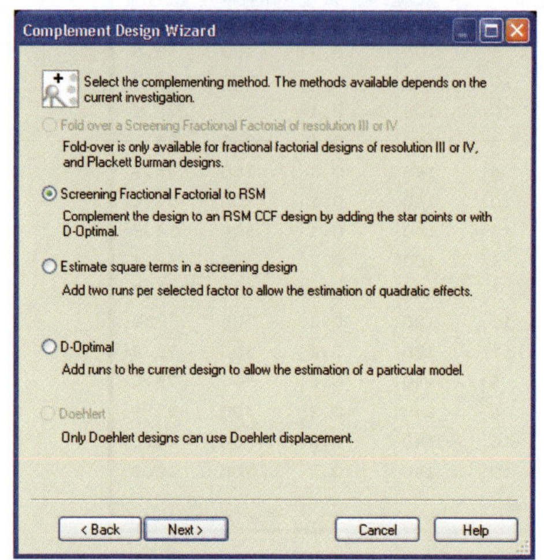

Figure 19.5: *With the Complement Design Wizard in MODDE, design augmentation is quickly accomplished. In the WILLGE example it was decided to upgrade the initial screening design to an RSM design.*

As shown by Figure 19.6, the screening design was extended by including additional trials in the three most important factors. The star distance was set at 1.68, which implies that the original design, together with its complement, corresponds to a CCC design in the three selected factors.

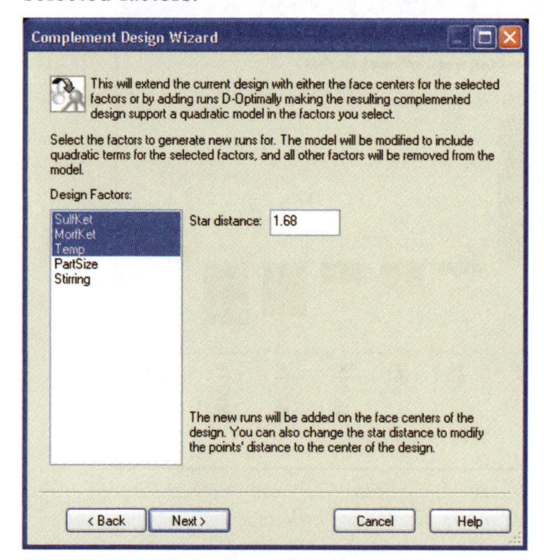

Figure 19.6: *Only three of the factors were considered. In the screening design, these factors were found to be the most important.*

In order to allow an estimate to be made of the replicate error, six replicated center-points were included in the design complement. The final worksheet, consisting of the 16 original runs and the 12 new runs, is presented in Figure 19.7. Note that MODDE calculates the star distance to five decimal places, so it is necessary to manually adjust some of the factor values (Figure 19.7). In addition, sulfur with a particle size of 0.15 mm was not available, so all experiments in the second phase were conducted using 0.1 mm sulfur.

The updated design supports a quadratic model in the three first factors. Furthermore, we can see that a new factor, denoted $Block, has been appended by the software. As discussed in Chapter 14, the addition of a block factor makes it possible to quantify whether there are systematic differences between the two blocks of experiments. (However, in contrast to the Blocking example shown in Chapter 14, the two blocks formed in the WILLGE application are NOT orthogonal.)

Worksheet

	1	2	3	4	5	6	7	8	9	10	11
1	Exp No	Exp Name	Run Order	Incl/Excl	SulfKet	MorfKet	Temp	PartSize	Stirring	$Block	Yield
2	1	N1	10	Incl	5	6	100	0.1	700	-1	11.5
3	2	N2	1	Incl	11	6	100	0.1	300	-1	55.8
4	3	N3	13	Incl	5	10	100	0.1	300	-1	55.7
5	4	N4	16	Incl	11	10	100	0.1	700	-1	75.1
6	5	N5	9	Incl	5	6	140	0.1	300	-1	78.1
7	6	N6	4	Incl	11	6	140	0.1	700	-1	88.9
8	7	N7	14	Incl	5	10	140	0.1	700	-1	77.6
9	8	N8	5	Incl	11	10	140	0.1	300	-1	84.5
10	9	N9	2	Incl	5	6	100	0.2	300	-1	16.5
11	10	N10	8	Incl	11	6	100	0.2	700	-1	43.7
12	11	N11	11	Incl	5	10	100	0.2	700	-1	38
13	12	N12	3	Incl	11	10	100	0.2	300	-1	72.6
14	13	N13	15	Incl	5	6	140	0.2	700	-1	79.5
15	14	N14	6	Incl	11	6	140	0.2	300	-1	91.4
16	15	N15	12	Incl	5	10	140	0.2	300	-1	86.2
17	16	N16	7	Incl	11	10	140	0.2	700	-1	78.6
18	17	C17	28	Incl	3	8	120	0.1	500	1	48.5
19	18	C18	21	Incl	13	8	120	0.1	500	1	91.5
20	19	C19	18	Incl	8	4.6	120	0.1	500	1	58.5
21	20	C20	25	Incl	8	11.2	120	0.1	500	1	94.7
22	21	C21	19	Incl	8	8	86	0.1	500	1	14.4
23	22	C22	22	Incl	8	8	154	0.1	500	1	94.1
24	23	C23	26	Incl	8	8	120	0.1	500	1	83.9
25	24	C24	27	Incl	8	8	120	0.1	500	1	84.2
26	25	C25	20	Incl	8	8	120	0.1	500	1	85.6
27	26	C26	24	Incl	8	8	120	0.1	500	1	82.6
28	27	C27	23	Incl	8	8	120	0.1	500	1	83.2
29	28	C28	17	Incl	8	8	120	0.1	500	1	84.9

Figure 19.7: *The WILLGE RSM worksheet. The 12 runs appended to the original design are highlighted.*

Figure 19.8 shows a scatter plot of SulfKet and MorfKet and a histogram plot of the response variable. The CCC design structure, which is valid across the first three factors, is fully visible. From the histogram plot, two interesting observations can be made. First, the variation among the six replicates is exceptionally small. The likely consequence of this is that no matter what we do in the model revision phase, there will almost certainly be a low Model Validity bar in the Summary of Fit plot. Secondly, the response values of the replicates are at the high end of the numerical range of the response variable, suggesting quadratic dependence of the response on the factors.

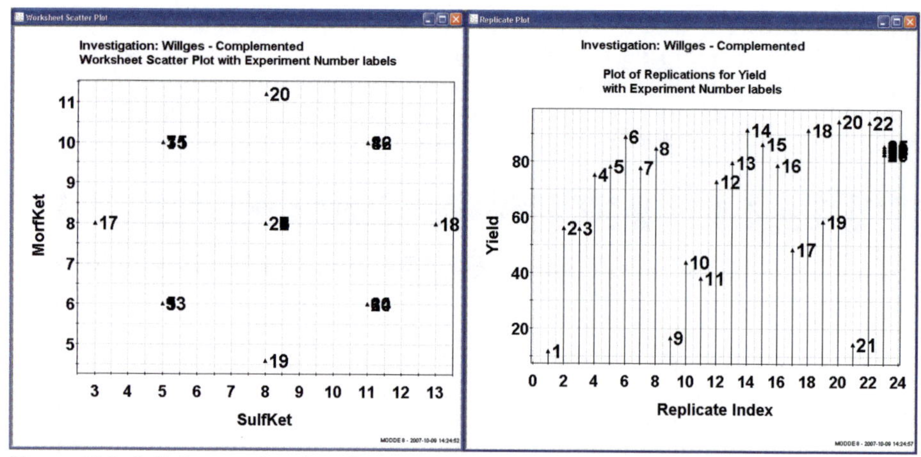

Figure 19.8: (left) Scatter plot of MorfKet and SulfKet. The CCC structure is apparent. (right) Histogram plot of Yield.

The analysis of the RSM data also indicated a very strong relationship between the factors and the response variable (Figure 19.9). In the first preliminary model, the linear terms for PartSize, Stirring and $Block were incorporated into the model, however, they were not meaningful and were therefore excluded immediately. Hence, the final model corresponds to the full quadratic model in the three factors SK, MK and Te (Figure 19.9). The order of factors parallels the findings of the screening design and the reaction temperature is by far the most important factor. A higher temperature gives a higher Yield. Given the distribution of the replicated center-points in the replicate plot, the occurrence of quadratic effects is as expected.

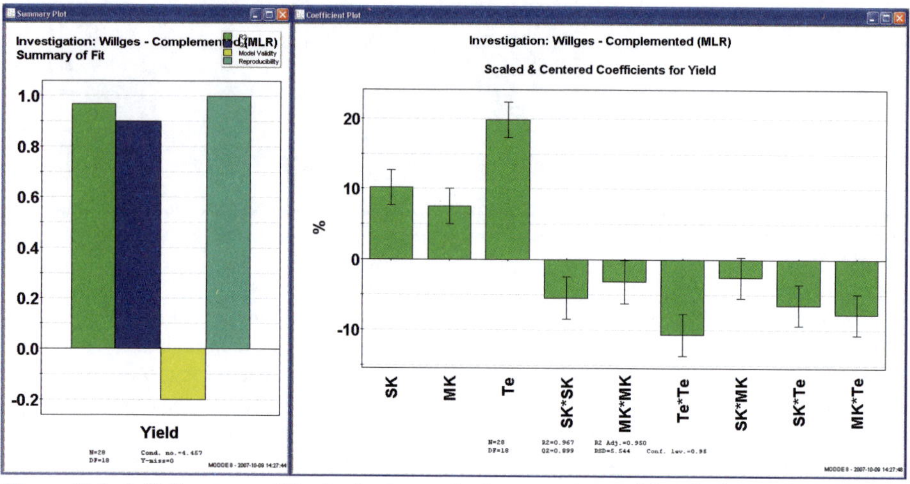

Figure 19.9: (left) Summary of Fit plot for the quadratic regression model in the factors SK, MK and Te. (right) Regression coefficients of the quadratic model.

The response surface plot presented in Figure 19.10 helps us to appreciate that a Yield of approximately 95% is not unrealistic within the experimental domain under consideration. A more precise co-ordinate, determined using the software optimizer, is SK = 9.5, MK = 8.2 and Temp = 134, where the Yield is predicted to be ≈ 94.3%.

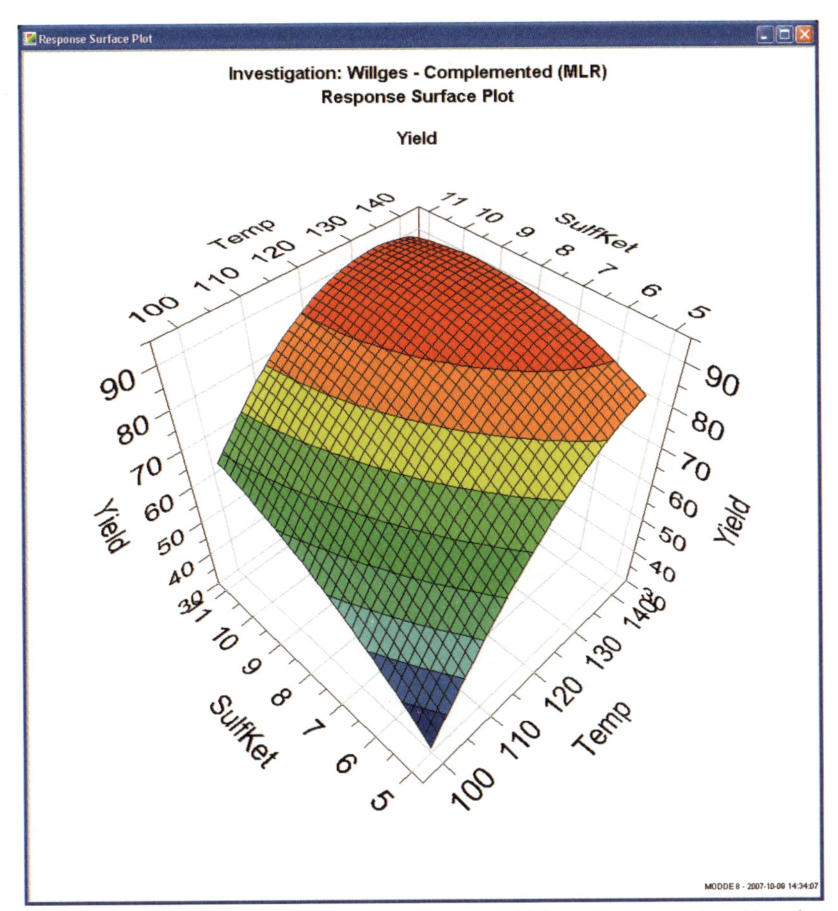

Figure 19.10: *Response surface plot suggesting that a Yield of approximately 95% is obtainable within the experimental region investigated.*

19.3.3 Conclusion of the study

This example illustrates the value of the working principle of first conducting a careful, but shallow screening investigation and thereafter a detailed optimization study. The screening phase identified three factors as being more influential than the others. When examining these three factors in the optimization stage, an area of very high yield was identified. The appropriateness of this region of operability was later checked experimentally by Torbjörn Lundstedt, who obtained yields of 94.3% and 96.7% in the verification experiments. The yields obtained were much higher than those previously reported in the literature (< 70%). Interestingly, these experiments led to an increased understanding of the reaction mechanism of the Willgerodt-Kindler reaction pathway.

19.4 Commonly chosen factors

With the WILLGE example in mind, the idea now is to discuss, briefly, some factors (Section 19.4), responses (Section 19.5) and designs (Section 19.6), that are commonly used in the literature on DOE and synthetic organic chemistry. For a detailed treatment of this topic, the leading book in the field is: Carlson, R. and Carlson, J., *Design and Optimization in Organic Synthesis*, Volume 24: Second Revised and Enlarged Edition, Data Handling in

Science and Technology, Elsevier, 2005. An interesting and general observation is that there is a tendency in many papers to restrict the number of factors to, perhaps, between two and five, even though a higher number could easily be accommodated in a screening study without the number of experiments becoming prohibitively large.

An obvious group of factors to consider is those that regulate the experimental conditions. This group includes factors such as temperature, addition rates, concentrations, reaction times, and pH. For some reactions, additional factors such as pressure, gas flow rate, stirring speed, order of addition of reagents, etc., may turn out to be influential. Moreover, the experimenter should not overlook the possible occurrence of uncontrollable factors. We recall from Chapter 2, that an uncontrolled factor is one which cannot be set at a fixed value during the experiment, but which still might have a profound influence on the responses measured. For example, the pH in an aqueous solution can, in some cases, be affected by changes in the proportions of the reagents, and may therefore be specified as an uncontrolled factor.

Another main group of factors relates to the reagents and other constituents added to the reaction vessel. To regulate such features we may, for instance, specify factors such as "type of ketone", a qualitative factor, and "amount of ketone", a quantitative factor, or we may simply specify factors in terms of stoichiometric relationships (molar ratios) between the various reagents. Usually, the solvent has a major impact on the reaction system, and one way to investigate this is to define a qualitative "type of solvent" factor. Note, however, that aspects such as solvents, reagents, catalysts, etc., may also be varied according to the philosophy of multivariate design, as discussed below (see Section 19.7).

A third group of factors comprises equipment and performance related characteristics. For instance, there are situations where systematic differences between reaction compartments might be expected, and where it may be appropriate to define factors such as reactor 1/reactor 2 or vessel 1/2/3, and so on. Sometimes there are restrictions with respect to the raw material, and it might be necessary to specify factors such as "type of raw material" or "supplier of raw material". Also remember that it might be possible to deal with some of the latter issues by using orthogonal blocking, as outlined in Chapter 14.

Last, but certainly not least, a fourth group of factors concerns the work-up after the completion of an organic synthesis, where the objective is usually to isolate and purify the desired product, but may also include aspects such as recovery of solvent and re-generation of unconsumed chemicals. Work-up involves processes such as filtering, (re-) crystallization, precipitation, extraction, distillation, etc., and typically there are a number of additional factors to consider. It is entirely possible to include work-up factors as part of the screening design, however, the drawback is that the design protocol often becomes unreasonably large. Usually, a better strategy, therefore, is to focus first on finding appropriate experimental conditions leading to a good result, and only thereafter to address the work-up in terms of a separate and dedicated experimental design.

19.5 Commonly chosen responses

It is important to remember that many responses can be used in DOE; we are not statically locked into considering just the same old yield response over and over again. Of course, by far the most frequently used response *is* the yield of the desired product, usually expressed as a percentage, but there are certainly other important responses to consider too. Some syntheses produce a profile of products and the yield of, or the selectivity for, each compound can be stated. One way to define selectivity is as the number of moles of the side-product produced per 100 moles of starting material, corrected for unreacted starting material. A similar response measure for enantiomers is denoted enantiomeric excess, similarly the term diastereomeric excess is applied to diastereomers.

Apart from yield-related responses, the most commonly used responses in organic synthesis relate to aspects of purity, quality and cost. Impurity levels in the final product depend on the reaction conditions, but may also be a function of work-up procedures, so several measures of impurities might have to be considered. Quality indices, which of course differ from application to application, can also be numerous and are typically linked to properties like density, rheology, particle size distribution, filterability, and so on. Cost measures may include consideration of the cost of reagents, energy consumption, environmental load, and procedural duration (e.g., time for reaction completion or timing of a quiescent step).

19.6 Commonly chosen design protocols

For screening purposes, two-level fractional factorial designs of resolution IV or V (see Chapter 5), Plackett-Burman designs or Rechtschaffner designs are often selected (see Chapter 8). D-optimal designs (see Chapter 10) are becoming increasingly widespread, particularly when the synthetic problem consists of one or more qualitative factors examined at multiple settings. The RSM phase is dominated by the central composite design family (see Chapter 6), however, Doehlert designs and Box-Behnken designs (see Chapter 9) are not uncommon. To accomplish robustness testing (see Chapter 7), low-resolution fractional factorials or Plackett-Burman designs are commonly used. The interested reader is again referred to the seminal book by Rolf Carlson.

19.7 Multivariate design

Multivariate design can supplement the construction of DOE in regular quantitative and qualitative factors. It is based on creating a statistical experimental design in latent variables. As explained in Section 13.5, latent variables are score vectors from a PCA or PLS model of descriptor variables, reflecting molecular properties of the molecular “items” under investigation. Multivariate design is a versatile and flexible approach that may conveniently be used to select sub-sets of representative molecular “items”, for example solvents, reagents and catalysts. In the context of drug design, multivariate design is also known as statistical molecular design. An example of the latter was outlined in Section 13.5 using onion design in PCA score variables. For a more detailed discussion of multivariate design we refer to the companion book on multivariate data analysis (Eriksson, L., et al., *Multi- and Megavariate Data Analysis*, Part I, Chap. 5, Umetrics Academy). Below, we use two examples to illustrate (i) multivariate design to select a representative sub-set of solvents, and (ii) multivariate design used in conjunction with regular DOE in experimental conditions.

19.8 A multivariate design example: SOLVENT

The illustration that we have selected is called SOLVENT and deals with how to choose representative solvents for synthetic organic chemistry. Professor Rolf Carlson at the University of Tromsö, Norway, is the originator of this application. The assumption was that intrinsic properties of the studied solvents, could, at least in part, be seen in well chosen chemical model systems. Rolf Carlson and co-workers compiled a series of 103 solvents described by nine chemical property values. These descriptors were: (1) melting point; (2) boiling point; (3) dielectric constant; (4) dipole moment; (5) refractive index; (6) ET30, a transition energy from UV-spectroscopy of an organic dye; (7) density; (8) log P, octanol–water partition coefficient; and (9) water solubility.

When PCA was applied to the 103 (solvents) × 9 (chemical descriptors) SOLVENT data set, a two-component model resulted. It should be noted that PCA cannot be conducted in MODDE but must be performed in SIMCA-P$^+$. The PCA model produced an R^2X of 0.68 and a Q^2X of 0.46. The first component explains 40% of the variation and the second 28%.

This suggests that the nine chemical descriptors encode two intrinsic molecular properties. The loadings and scores of the two components are plotted in Figure 19.11.

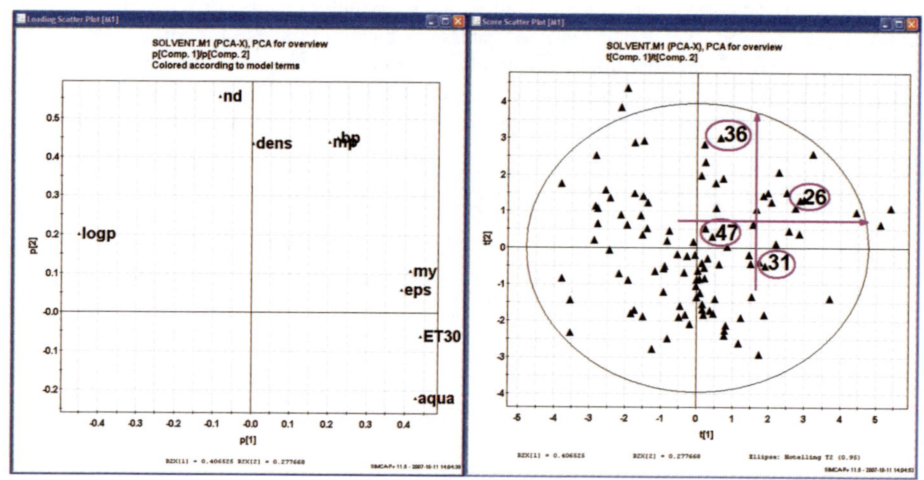

Figure 19.11: Plot of PCA model parameters. (left) PCA loadings expressing chemical descriptor related information. Each point is one chemical descriptor. (right) PCA scores showing the relationships among the solvents, i.e., this is a chemical property map of the solvents. Each point represents one solvent.

Figure 19.11 shows a score plot with a rather uniform distribution of the solvents, indicating a homogenous class of compounds. In order to ascribe a physico-chemical meaning to the two principal components (or latent variables), we look at the loading plot. The first component is dominated by lipophilicity (log P), water solubility, and the UV-related variable ET30. The investigators interpreted the first component as mainly mirroring polarity. The second component is primarily influenced by refractive index, boiling point, and melting point, and was, therefore, interpreted as reflecting polarizability.

The interesting feature of the latent variables is that they are mathematically independent of each other (orthogonal) and limited in number (often between two and four). This implies that they can readily be used in statistical experimental design schemes, such as fractional factorial, D-optimal or onion designs, to encode representative and diverse molecules. If there are no restrictions related to e.g. synthetic feasibility, cost, toxicity, ..., such a design can be applied across the whole score space (chemical property map). This is similar in scope to the onion design example outlined in Section 13.5. Quite frequently, however, there are restrictions of some sort, rendering the accessible search space smaller than the full chemical domain. The important point is still the same, namely that DOE can efficiently be used to accomplish good mapping of the restricted chemical domain.

In the next section, we discuss another DOE example, in which the choice of the solvents is an integral part of the design protocol. In that particular application, it was necessary to focus the choice of solvent within and near the first quadrant of the t_1/t_2 score plot presented in Figure 19.11, i.e., the quadrant corresponding to moderate-to-high polarity and moderate-to-high polarizability. Inside this localized area, a two-level factorial design was approximated onto the distribution of the solvents (this is indicated in the score plot, see Figure 19.11). Subsequently, four solvents with good diversity across that limited area in the score plot were selected as follows:

Coded t_1	Coded t_2	Real t_1	Real t_2	No	Solvent
-	-	0.43	0.29	47	Pyridine
+	-	1.89	-0.50	31	DMFA
-	+	0.65	3.00	36	Nitrobenzene
+	+	2.86	1.29	26	DMSO

19.9 A combined design example: Nenitzescu reaction

The application reviewed here originates from Torbjörn Lundstedt and co-workers at BMC, Uppsala, Sweden (Katkevica, D., Lundstedt, T., et al., *J. Chemometrics*, 18, 183-187, 2004), who studied the so called Nenitzescu reaction in a little more detail. In this reaction indoles are produced by heating benzoquinones with enamines in a solvent. The indole structure is a biologically active scaffold that occurs in several applications. In order to better understand the properties of this reaction, Lundstedt and co-workers used the reaction between benzoquinone and methyl-3-aminocrotonate as a model reaction (Figure 19.12).

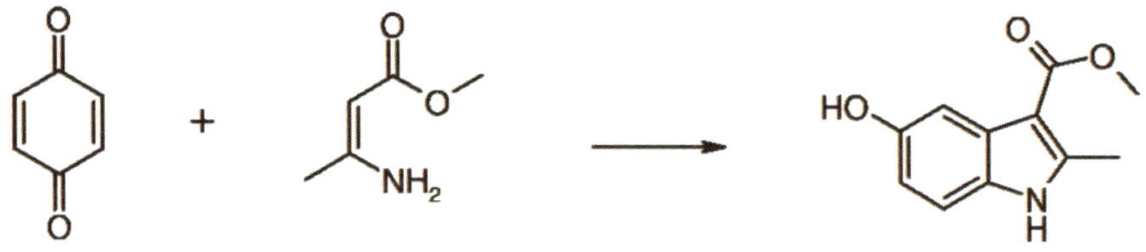

Figure 19.12: *As a model reaction, that between benzoquinone and methyl-3-amonicrotonate was used.*

19.9.1 Initial screening design

Initially, four solvents were identified, as described in the previous section. In addition, it was decided to vary three experimental factors, namely (i) the molar ratio benzoquinone / enamine, (ii) the volume of solvent, and (iii) the reaction temperature. An overview of the factors is presented in Figure 19.13. The measured response was the yield of the main product after synthesis for 120 minutes.

Factors

	Name	Abbr.	Units	Type	Use	Settings	Transform	Prec.	MLR Scale	PLS Scale
1	t1	t1		Quantitative	Controlled	0.2 to 2.86	None	Free	Orthogonal	Unit Variance
2	t2	t2		Quantitative	Controlled	-0.5 to 2.83	None	Free	Orthogonal	Unit Variance
3	Q/EA	Q/EA	-	Quantitative	Controlled	0.67 to 1.5	None	Free	Orthogonal	Unit Variance
4	Volume	Vol	-	Quantitative	Controlled	0.5 to 1	None	Free	Orthogonal	Unit Variance
5	Temperature	Temp	°C	Quantitative	Controlled	50 to 90	None	Free	Orthogonal	Unit Variance

Figure 19.13: *An overview of the five factors of the Nenitzescu reaction example. Note that the first two factors are used to encode the four solvents.*

The experimental settings were varied using a 2^{5-1} fractional factorial design in 16 + 4 runs, with a center-point for each solvent. The worksheet for this design is shown in Figure 19.14. Note that we used the Exp Name column to indicate the solvent used. The numerical range of the response variable is between 1 and 50%.

Worksheet

	1	2	3	4	5	6	7	8	9	10
1	Exp No	Exp Name	Run Order	Incl/Excl	t1	t2	Q/EA	Volume	Temperature	Yield
2	1	N1:Pyridine	16	Incl	0.43	0.29	0.67	0.5	90	11
3	2	N2:Nitrobenzene	7	Incl	0.65	3	0.67	0.5	50	8
4	3	N3:DMFA	15	Incl	1.89	-0.5	0.67	0.5	50	10
5	4	N4:DMSO	4	Incl	2.86	1.29	0.67	0.5	90	47
6	5	N5:Pyridine	17	Incl	0.43	0.29	1.5	0.5	50	25
7	6	N6:Nitrobenzene	11	Incl	0.65	3	1.5	0.5	90	19
8	7	N7:DMFA	6	Incl	1.89	-0.5	1.5	0.5	90	31
9	8	N8:DMSO	3	Incl	2.86	1.29	1.5	0.5	50	31
10	9	N9:Pyridine	2	Incl	0.43	0.29	0.67	1	50	19
11	10	N10:Nitrobenzene	1	Incl	0.65	3	0.67	1	90	6
12	11	N11:DMFA	20	Incl	1.89	-0.5	0.67	1	90	19
13	12	N12:DMSO	13	Incl	2.86	1.29	0.67	1	50	18
14	13	N13:Pyridine	10	Incl	0.43	0.29	1.5	1	90	12
15	14	N14:Nitrobenzene	5	Incl	0.65	3	1.5	1	50	1
16	15	N15:DMFA	18	Incl	1.89	-0.5	1.5	1	50	6
17	16	N16:DMSO	12	Incl	2.86	1.29	1.5	1	90	50
18	17	N17:Pyridine	8	Incl	0.43	0.29	1	0.75	70	17
19	18	N18:Nitrobenzene	19	Incl	0.65	3	1	0.75	70	5
20	19	N19:DMFA	9	Incl	1.89	-0.5	1	0.75	70	7
21	20	N20:DMSO	14	Incl	2.86	1.29	1	0.75	70	29

Figure 19.14: *Worksheet of the 2^{5-1} screening design. The name of the solvent is given in the Exp Name column.*

The PLS method (see Chapter 18) was used to establish a multivariate model between the solvent properties, the three experimental condition variables and the yield. A two-component PLS model was produced, with $R^2 = 0.91$ and $Q^2 = 0.61$; these can be considered to be excellent model performance indicators (Figure 19.15). The influence of the different descriptors can be seen in the plot of the PLS regression coefficients (Figure 19.15). As seen, the model consists of the five linear terms plus three two-factor interactions. These coefficients indicate that a high t_1 (high polarity) for the solvent, and a high value for the interaction between the solvent descriptors t_1 (polarity) and t_2 (polarizability), together with a high temperature, are associated with a high yield. In addition, there is a strong interaction between the solvent polarity and the temperature. Hence, the provisional conclusion is that the solvent has a strong impact on this reaction system.

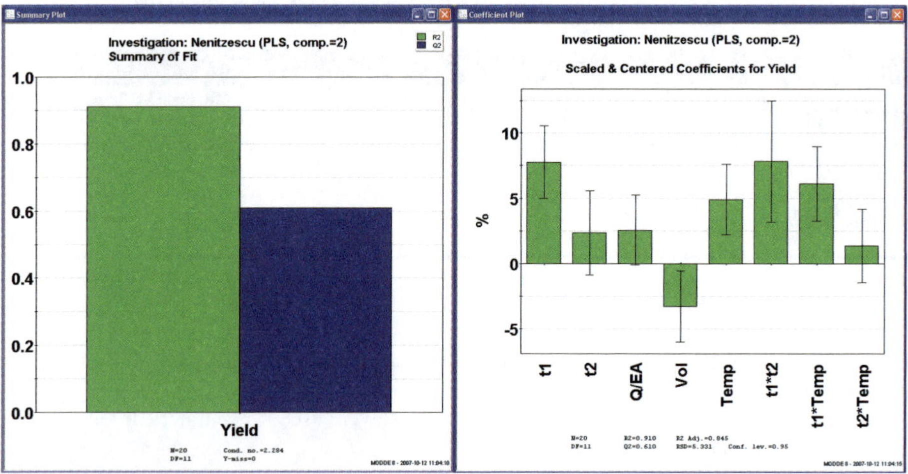

Figure 19.15: *(left) Summary of Fit plot of the PLS model for the Nenitzescu reaction data. $R^2 = 0.91$ and $Q^2 = 0.61$. (right) Plot of regression coefficients of the PLS model.*

One way of further exploring the cross-terms consists of creating interaction plots. The meaning of the two largest cross-terms is represented graphically in Figure 19.16. With respect to the polarity*temp (t_1*temp) interaction, we can see that at a low polarity it does not matter what temperature is used, the yield remains pretty much the same. However, when the polarity of the solvent is high, there is a considerable increase in yield when moving from a low to a high temperature. An analogous phenomenon is valid for the slightly weaker t_1*t_2 cross-term. At high polarity (t_1) there is a strong effect on the yield when switching from a low- to a highly-polarizable (low to high t_2) solvent, whereas at the low end of the polarity scale, a much smaller impact on the yield occurs when manipulating the solvent polarizability.

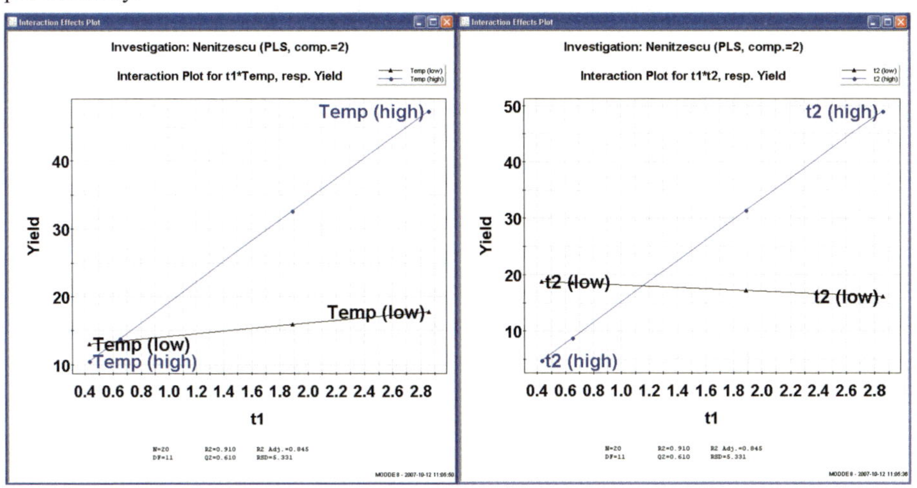

*Figure 19.16: Interaction plots for two cross-terms. (left) The polarity*temp cross-term. (right) The polarity*polarizability cross term.*

19.9.2 An efficient re-parameterization of the solvent factors

When the solvent scales t_1 and t_2 are used as design variables, all we can see is the impact of the solvent on the yield from an overall perspective. With this representation it is hard to deduce the effect of individual solvents. In order to capture the influence of each of the four solvents, we must carry out a small re-parameterization of the solvent factors. As shown in Figure 19.17, the way to accomplish this is to define a qualitative factor in four levels (because there are four solvents). This also results in a drop in the number of design factors, from five to four.

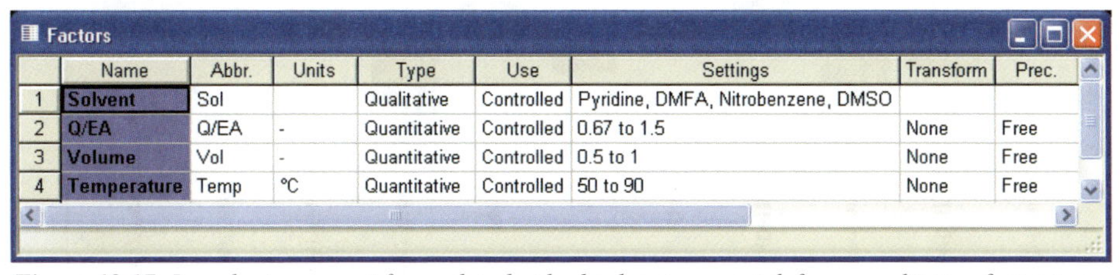

Factors

	Name	Abbr.	Units	Type	Use	Settings	Transform	Prec.
1	Solvent	Sol		Qualitative	Controlled	Pyridine, DMFA, Nitrobenzene, DMSO		
2	Q/EA	Q/EA	-	Quantitative	Controlled	0.67 to 1.5	None	Free
3	Volume	Vol	-	Quantitative	Controlled	0.5 to 1	None	Free
4	Temperature	Temp	°C	Quantitative	Controlled	50 to 90	None	Free

Figure 19.17: In order to account for each individual solvent we must define a qualitative factor in as many levels as there are solvents. This results in a four-factor problem.

The resulting experimental scheme is presented in Figure 19.18. There are no other changes besides the introduction of the four-level qualitative solvent factor.

Worksheet

	1	2	3	4	5	6	7	8	9
1	Exp No	Exp Name	Run Order	Incl/Excl	Solvent	Q/EA	Volume	Temperature	Yield
2	1	N1	19	Incl	Pyridine	0.67	0.5	90	11
3	2	N2	6	Incl	Nitrobenzene	0.67	0.5	50	8
4	3	N3	20	Incl	DMFA	0.67	0.5	50	10
5	4	N4	8	Incl	DMSO	0.67	0.5	90	47
6	5	N5	16	Incl	Pyridine	1.5	0.5	50	25
7	6	N6	5	Incl	Nitrobenzene	1.5	0.5	90	19
8	7	N7	13	Incl	DMFA	1.5	0.5	90	31
9	8	N8	14	Incl	DMSO	1.5	0.5	50	31
10	9	N9	18	Incl	Pyridine	0.67	1	50	19
11	10	N10	2	Incl	Nitrobenzene	0.67	1	90	6
12	11	N11	11	Incl	DMFA	0.67	1	90	19
13	12	N12	17	Incl	DMSO	0.67	1	50	18
14	13	N13	10	Incl	Pyridine	1.5	1	90	12
15	14	N14	7	Incl	Nitrobenzene	1.5	1	50	1
16	15	N15	1	Incl	DMFA	1.5	1	50	6
17	16	N16	4	Incl	DMSO	1.5	1	90	50
18	17	N17	3	Incl	Pyridine	1	0.75	70	17
19	18	N18	12	Incl	Nitrobenzene	1	0.75	70	5
20	19	N19	15	Incl	DMFA	1	0.75	70	7
21	20	N20	9	Incl	DMSO	1	0.75	70	29

Figure 19.18: *The modified worksheet containing one four-level qualitative factor.*

A new PLS model was fitted to the response data using the same model configuration as above. The regression coefficients of the new model are plotted in Figure 19.19. The remarkable difference is that, due to the factor re-parameterization, we are now able to resolve the influence of each solvent. Clearly, with DMSO we obtain the highest yields whereas with nitrobenzene we obtain the lowest yields. A look at the raw data corroborates this finding. The average yield across the five experiments using DMSO as the solvent is 35%. The corresponding number for nitrobenzene is 7.8%.

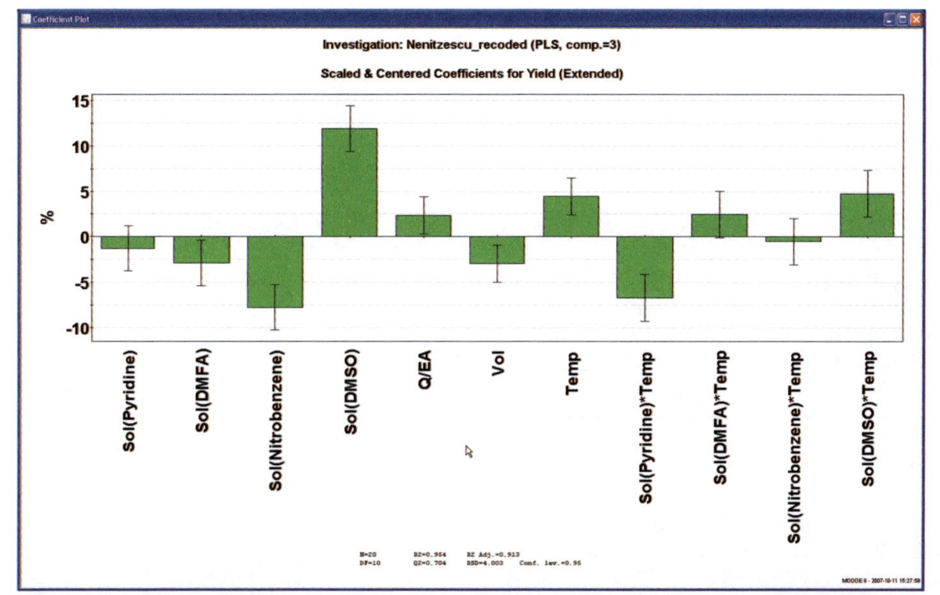

Figure 19.19: *Regression coefficients of the new PLS model based on four factors rather than five.*

Since the presence of the qualitative factor results in there being one coefficient for each solvent, the important t_1*Temp cross-term discussed above is also split into four Sol*Temp interaction coefficients. We can see that a high temperature corresponds to an increased yield in the DMFA and DMSO reaction systems, while a low temperature is favourable for the yield when pyridine is the solvent. Moreover, a slight excess of benzoquinone (a high molar ratio) and a small solvent volume is important to achieve higher yields.

19.9.3 Conclusion of the study

The dataset from Lundstedt and co-workers beautifully illustrates the option of combining – in the same design – regular experimental factors and latent variables. There is no doubt that the solvent exerts a profound impact on the Nenitzescu reaction, at least in the cited study. This strong influence must be attributed to the fact that systematic changes in important solvent properties were imposed on the experimental system using DOE. With an alternative, arbitrary selection of solvents it is not at all certain that the reaction system would have been affected and responded in the same characteristic manner.

From the regression coefficient profile seen in Figure 19.15 it is concluded that the desired profile for "future" solvents would, most importantly, be one that results in a high t_1 and a high t_1*t_2 interaction, and possibly also a high t_2, although the latter condition is not as important as the first two. By limiting the selection of additional solvents to those identified by Rolf Carlson (see Section 19.8), it appears that formamide, sulfolane, and N-methylacetamide would be interesting candidate solvents for future work to determine whether even higher yields can be achieved (Figure 19.20). In the original work, Lundstedt and co-workers indicate that such studies with sulfolane are underway, but no results have been released so far.

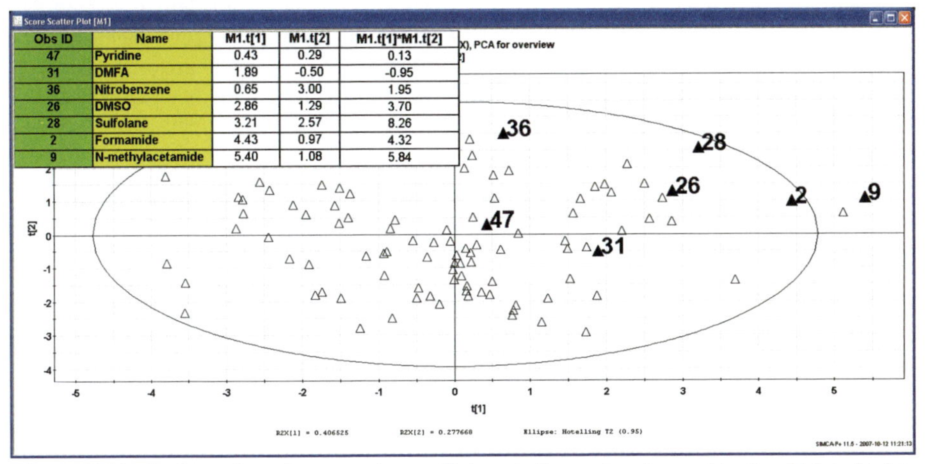

Obs ID	Name	M1.t[1]	M1.t[2]	M1.t[1]*M1.t[2]
47	Pyridine	0.43	0.29	0.13
31	DMFA	1.89	-0.50	-0.95
36	Nitrobenzene	0.65	3.00	1.95
26	DMSO	2.86	1.29	3.70
28	Sulfolane	3.21	2.57	8.26
2	Formamide	4.43	0.97	4.32
9	N-methylacetamide	5.40	1.08	5.84

Figure 19.20: *Guidance from the regression coefficients indicates that the combination high t_1, high t_2 and high t_1*t_2 interaction is an appropriate property profile for new solvents. The plot provides the t_1 and t_2 scores, and their interaction, for the four solvents actually tested plus three promising candidate solvents. This means that by interpreting the model parameters we have identified a direction in the solvent score space in which to move to enhance the possibility of obtaining even higher yields of the final indole structure.*

19.10 Questions for Chapter 19

- List some factors commonly used for DOE investigations in synthetic organic chemistry.
- List some frequently used responses.
- Which designs are often used in screening? In RSM? In robustness testing?
- What is multivariate design?
- What is a latent variable?
- How can multivariate design be used to enrich DOE in regular factors?

19.11 Summary and discussion

As shown by the examples above, and certainly by numerous studies in the literature, DOE is an ideal tool to optimize the performance of reactions in organic chemistry. In addition to higher yields and a better mechanistic understanding of the reaction system, DOE often reveals fine details and unexpected effects which would not have been anticipated on the basis of purely mechanistic considerations. Likewise, decisive evidence regarding a failing reaction, or an otherwise unsuccessful application, where the only realistic option is termination of the ongoing project, can also be identified.

There is no doubt that we are currently witnessing an increase in the use of DOE in organic chemistry. There are several reasons for this. First and foremost, there is a pressing need to produce the right product at the right time, and here DOE comes in very handy. DOE is also a very timely concept in the era of PAT (see Chapter 22) and other legislative initiatives, and such good, general manufacturing principles will inevitably spill over into organic chemistry, as well as other fields related to process chemistry and engineering. A second important reason concerns the recent advances in laboratory automation, i.e. robots and other automatic device. Using these, more experiments can be conducted more rapidly and reliably than ever before. Finally, the spread of DOE in organic chemistry may, in part, be due to improved DOE software.

However, we want to emphasize very strongly that the DOE methodology should not be viewed or interpreted as a black-box approach. Of course it matters greatly how the experiments within any experimental series are configured, and the DOE plan chosen will take care of that. However, the selection of factors is also very important along with the levels defined for them, the responses are to be measured, etc. Hence, the chemist *must* think carefully about the problem at hand before launching any design. DOE will never be a replacement for sound, critical chemical thinking, nor will it ever become a magic wand with which a lousy series of experiments can be salvaged.

Moreover, an often overlooked aspect is the issue of scale-up. Results obtained at the laboratory scale are usually not directly or proportionally transferable to the pilot plant or full scale production. One reason for this is that optimum conditions in a test tube do not remain the same once the reaction is carried out at full scale. Another reason is that experimental and mechanical restrictions apply at full scale, which may not have even been considered at the laboratory bench. Nonetheless, if a factor is found to be important at the small scale, it will almost certainly turn out to be influential at the large scale, although it may be that its optimal value is scale-sensitive. DOE is a vital resource in pinpointing, addressing and solving such scaling-up challenges.

20 DOE in assay development

20.1 Objective

Multiwell microtiter plates are key pieces of modern technology that facilitate high-throughput screening and are used in myriads of biological assays in the pharmaceutical industry, other biotechnological sectors and diverse academic applications. This chapter introduces a new set of designs for conducting DOE using multiwell plates, known as Rectangular Experimental Designs for Multi-Unit Platforms (RED-MUPs). RED-MUPs are built from two separate precursor designs which are combined to form the final RED-MUP protocol. RED-MUPs have been developed for 96-, 384-, and 1536-well plates.

20.2 Introduction

20.2.1 DOE and microtiter plates

Multiwell microtiter plates have become essential pieces of equipment for the pharmaceutical industry, in which huge numbers of biological and pharmacological assays have to be performed, both for routine testing and various other purposes, notably high throughput screening (HTS), where the objective is to identify potential drug candidates from a large library of chemical compounds. Important examples of the assays in which they are often used include reporter gene assays, and similar biological systems, which provide information on gene regulation.

Users of multiwell plate technology have been slow to adopt the ideas of DOE. In part, this may be due to a general lack of awareness of the DOE concept. More likely, however, the main reason for not using DOE has been that classical experimental designs are in a format that is incompatible with the requirements of the multipipettes and robotized equipment used in laboratories today. It has – simply speaking – not been practical to use DOE, and the approach has been viewed by many people as more of a hindrance than a help.

Until recently, there were no experimental designs that were directly applicable to multiwell microtiter plates. Thus, in order to facilitate the use of DOE in assay development and optimization, a new family of experimental designs was recently introduced (Olsson, I.-M., et al., Chemometrics and Intelligent Laboratory Systems 83, 2006, 66-74). This group of designs is known as the Rectangular Experimental Designs for Multi-Unit Platforms, or RED-MUPs for short. The objective of the RED-MUPs is to enable straightforward DOE by minimizing the workload associated with the preparation of a multi-unit plate according to a given experimental design.

20.2.2 Main components of a RED-MUP protocol

A RED-MUP design consists of two building blocks – two partial designs that are multiplied together. Within MODDE, these two designs are referred to as the *vertical design* and the *horizontal design*. Figure 20.1 shows the general principle of a RED-MUP protocol. This particular design relates to a reporter gene assay, the analysis of which is reviewed in Section 20.3. As shown by Figure 20.1, the vertical design dictates the lay-out of the columns of the 96-well plate used in the example, whilst the horizontal design encodes the arrangement of the rows of the same plate. Thus, the vertical design contains as many runs as there are rows and the horizontal design as many runs as there are columns on the plate. With this arrangement of two combined building block designs, the practical consequence is that the individual factors are *not* varied within a given row or column, that is, they are kept constant. This is the key to ensuring the efficient utilization of multi-pipettes and other automatic laboratory equipment.

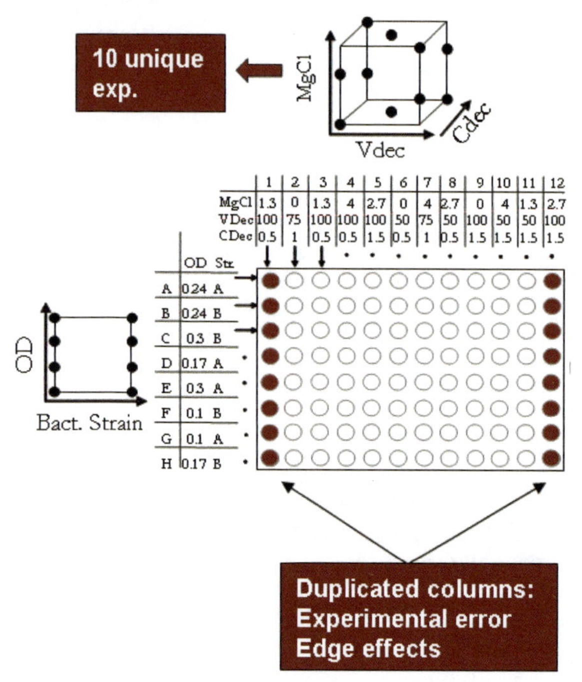

Figure 20.1: *Schematic of a RED-MUP design for a 96-well plate. The vertical design shows the arrangement of the columns. It has the same number of experiments as there are rows, i.e., eight. The horizontal design shows the arrangement of the rows. It has the same number of runs as there are columns, i.e., twelve. More details are found in the text.*

In MODDE, RED-MUPs have been implemented for 96-, 384- and 1536-well plates, which comprise 8 (rows) *12 (columns), 16 * 24 and 32 * 48 rows and columns, respectively. Unfortunately, there are *few* classical experimental designs containing exactly these numbers of runs, particularly when they incorporate center-points. As discussed in earlier chapters, center-points are highly recommended since their presence (i) enables the replicate error to be determined, (ii) helps to reveal curvature, and (iii) indicates where response transformations may be needed. In order to construct building block designs with the appropriate number of runs (8, 12, 16, 24 etc.), the work-around has been to exclude experiments that produce the minimum geometric distortion. However, one cannot mechanically and arbitrarily delete some points in a design just to get the desired number of design runs. In the construction of RED-MUP building block designs there will always be a trade-off; it is preferable to have a high number of unique experiments, but it is also vital to have a few center-points to account for the experimental error. For a given application (type of plate), number of factors, and experimental objective (screening or RSM) MODDE will recommend two building block designs that, once combined, will form a good RED-MUP design. Of course, an experienced user may ignore these proposals.

20.3 Example I: A reporter gene assay

The first example presented here concerns a reporter gene assay. The experimental objective of this investigation was to increase the signal from the assay. Here, our data analysis objective is to demonstrate how RED-MUPs are implemented in MODDE and how data generated according to such a design may be visualized and analyzed.

20.3.1 Background and objective

Reporter gene assays are used, *inter alia*, in mechanistic studies of gene regulation. They also have great potential utility in toxicology and drug development. A reporter gene is associated with an easily measurable phenotype, the transcription of which is controlled by an appropriate promoter. Reporter gene assays provide important information on gene regulation relating to expression (i.e. numbers of copies) and when and where a particular protein is formed.

The objective of this investigation was to maximize the signal in a reporter gene assay conducted under different conditions on 96-well plates (Olsson, I-M., *et al.*, 2006, Chemometrics and Intelligent Laboratory Systems, 83, 2006, 66-74). This specific assay is used to identify inhibitors of Type III Secretion (TTS) in *Yersinia* by measuring luminescence. If the TTS apparatus is inhibited, no light is emitted from the well. It was of particular interest to examine how the experimental factors influenced the response, in relative light units (RLU), when no inhibitors were added. Five factors, four quantitative and one qualitative, were investigated:

- Optical density (OD, relates to the number of cells)
- Amount of $MgCl_2$
- Volume of decanal (VDec)
- Concentration of decanal solution (CDec)
- Bacterial strain; qualitative factor on two levels

The first and last factors were investigated in the eight run vertical design. The remaining three factors were studied using the twelve run horizontal design. In total 192 experiments were performed, i.e., the design was performed twice, using two 96-well plates. The design lay-out is presented above in Figure 20.1, which shows that in the vertical eight-experiment design the two factors bacterial strain and OD were varied according to a complete 2*4 factorial design. The remaining three factors were varied in a reduced 3*3*4 factorial design (the horizontal component). This set-up enables the estimation of the linear terms of all factors and the interaction effects between factors, as well as quadratic effects of two of the factors. The highlighted columns represent duplicated experiments that were placed in the outer columns to investigate edge-effects. The final configuration of experiments is shown below (Figure 20.2).

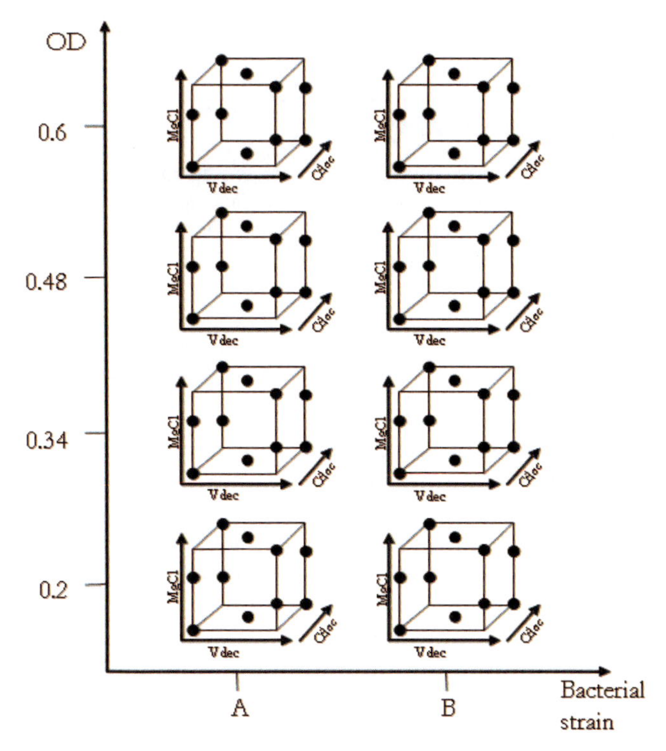

Figure 20.2: *The configuration of experiments used to investigate which factors affect the signal in the reporter gene assay.*

20.3.2 Configuring the investigation

The RED-MUP design type is available in MODDE within the Advanced Designs option (Figure 20.3).

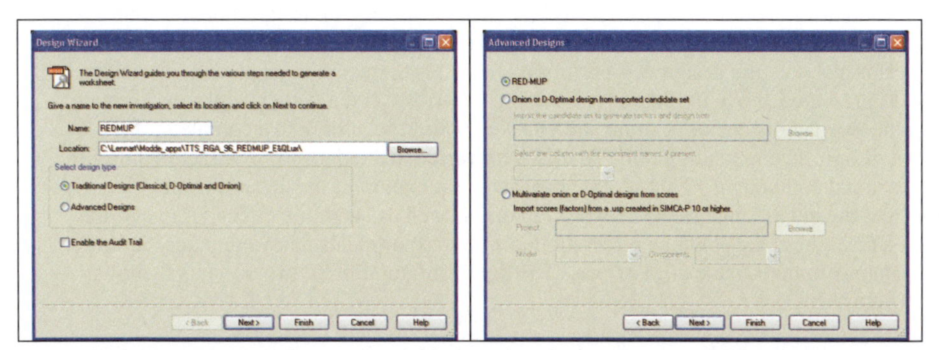

Figure 20.3: *The RED-MUP design is available from the Advanced Designs dialogue box.*

After choosing the RED-MUP option, the factors are entered in the usual way in MODDE. Figure 20.4 shows the settings of the five factors used. Note that the last factor is qualitative.

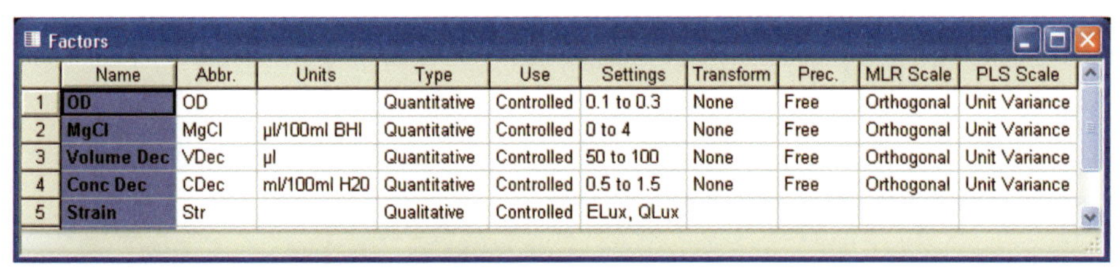

Factors

	Name	Abbr.	Units	Type	Use	Settings	Transform	Prec.	MLR Scale	PLS Scale
1	OD	OD		Quantitative	Controlled	0.1 to 0.3	None	Free	Orthogonal	Unit Variance
2	MgCl	MgCl	µl/100ml BHI	Quantitative	Controlled	0 to 4	None	Free	Orthogonal	Unit Variance
3	Volume Dec	VDec	µl	Quantitative	Controlled	50 to 100	None	Free	Orthogonal	Unit Variance
4	Conc Dec	CDec	ml/100ml H20	Quantitative	Controlled	0.5 to 1.5	None	Free	Orthogonal	Unit Variance
5	Strain	Str		Qualitative	Controlled	ELux, QLux				

Figure 20.4: *The factors and their settings.*

Only one response was specified; this needed to be maximized (Figure 20.5).

Responses

	Name	Abbr.	Units	Transform	MLR Scale	PLS Scale	Type
1	Relative Light Unit	RLU		None	None	Unit Variance	Regular

Figure 20.5: *The RLU response.*

The next step is to assign which factors should be varied according to the vertical and horizontal designs (Figure 20.6). Here, Optical density and Bacterial Strain are vertical design factors. Magnesium-chloride, Volume Decanal and Concentration Decanal are horizontal design factors. Since, in total, there are 192 experiments, we need to select two plates, the Block Use option, and ensure that the Plate/Block factor interactions box is checked.

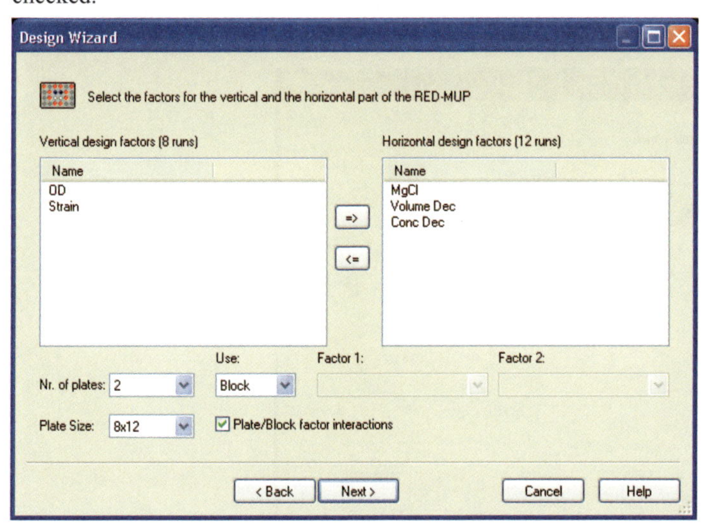

Figure 20.6: *The five factors must be assigned to either the vertical or the horizontal design.*

In the study considered here, the experimenters then selected screening as the objective. This was because the majority of estimable model terms were either linear or interaction types. Square terms that are estimable can be added manually to the model at a later stage of the investigation. MODDE then proposes a design protocol (Figure 20.7). We select the first design choice for the vertical design. It is important to note that the total number of runs in this design is eight. Note that the design presented below was not the one performed in reality. The settings were changed to the settings of the real protocol at a later stage.

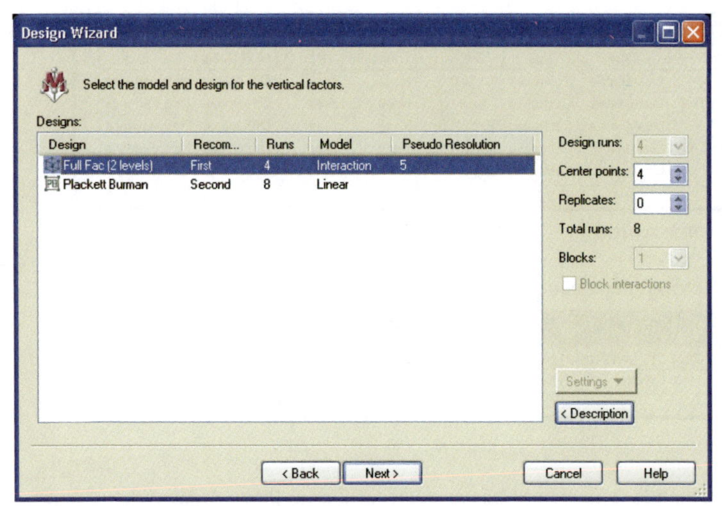

Figure 20.7: A tentative design protocol must be selected for the vertical factors. The settings of the chosen design may, if necessary, be changed at a later stage.

Similarly, the first design option was chosen for the three horizontal factors (Figure 20.8). Note that the total number of runs is equal to twelve. It should also be noted that the design presented below is not the one performed in reality. The settings were changed to the settings of the actual protocol at a later stage.

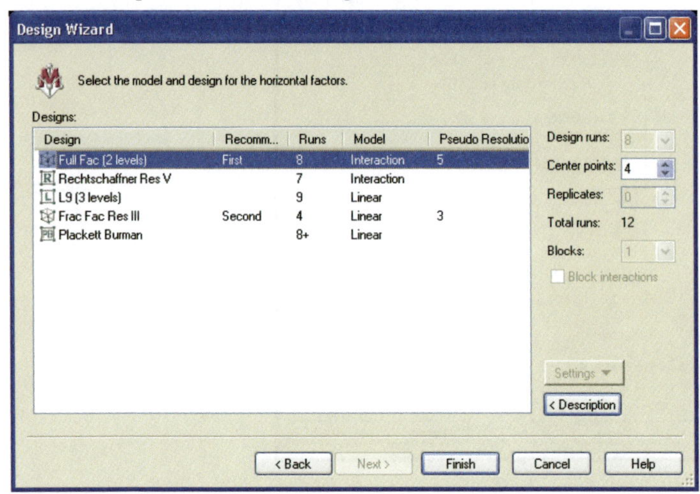

Figure 20.8: A tentative design protocol must be selected for the horizontal factors. The settings of the chosen design may, if necessary, be changed at a later stage.

In order to create the final RED-MUP protocol, the settings for the original vertical design must be changed to correspond to the settings for the real design. This is accomplished by manually editing the partial worksheet describing the vertical factors (Figure 20.9).

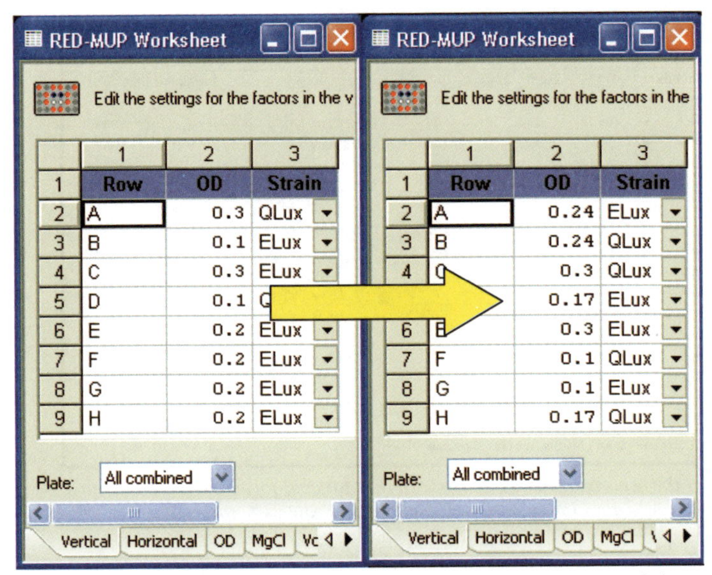

Row	OD	Strain
A	0.3	QLux
B	0.1	ELux
C	0.3	ELux
D	0.1	[illegible]
E	0.2	ELux
F	0.2	ELux
G	0.2	ELux
H	0.2	ELux

Row	OD	Strain
A	0.24	ELux
B	0.24	QLux
[illegible]	0.3	QLux
[illegible]	0.17	ELux
[illegible]	0.3	ELux
F	0.1	QLux
G	0.1	ELux
H	0.17	QLux

Figure 20.9: To get the correct vertical design we must change its worksheet manually.

Likewise, the required settings for the horizontal design are obtained after manually editing the corresponding partial worksheet (Figure 20.10).

Column	MgCl	Volume	Conc De
1	2	75	1
2	2	75	1
3	4	100	1.5
4	4	100	0.5
5	2	75	1
6	0	100	[illegible]
7	4	50	0.5
8	0	50	1.5
9	0	100	0.5
10	2	75	1
11	4	50	1.5
12	0	50	0.5

Column	MgCl	Volume	Conc De
1	1.3	100	0.5
2	0	75	1
3	1.3	100	0.5
4	4	100	0.5
5	2.7	100	1.5
[illegible]	0	50	0.5
[illegible]	4	75	1
8	2.7	50	0.5
9	0	100	1.5
10	4	50	1.5
11	1.3	50	1.5
12	2.7	100	1.5

Figure 20.10: To get the correct horizontal design we must change its worksheet manually.

Following the editing of the two partial worksheets, it is possible to construct the complete worksheet. In so doing, there are two options: the final worksheet can be presented using the classical view (Figure 20.11) or the plate view (Figure 20.12). The former format lists all the runs from first to last, whereas the latter format sorts the runs according to the row-and-column structure of the plate to be used (i.e., 8 by 12 in our case).

RED-MUP Worksheet

	1	2	3	4	5	6	7	8	9	10	11
1	Exp No	Exp Name	Run Order	Incl/Excl	OD	MgCl	Volume Dec	Conc Dec	Strain	$Plate	Relative Light Unit
2	1	N1	1	Incl	0.24	1.3	100	0.5	ELux	1	6678
3	2	N2	2	Incl	0.24	1.3	100	0.5	QLux	1	4856
4	3	N3	3	Incl	0.3	1.3	100	0.5	QLux	1	5017
5	4	N4	4	Incl	0.17	1.3	100	0.5	ELux	1	10282
6	5	N5	5	Incl	0.3	1.3	100	0.5	ELux	1	7551
7	6	N6	6	Incl	0.1	1.3	100	0.5	QLux	1	2592
8	7	N7	7	Incl	0.1	1.3	100	0.5	ELux	1	4148
9	8	N8	8	Incl	0.17	1.3	100	0.5	QLux	1	4237
10	9	N9	9	Incl	0.24	0	75	1	ELux	1	10052
11	10	N10	10	Incl	0.24	0	75	1	QLux	1	6228
12	11	N11	11	Incl	0.3	0	75	1	QLux	1	7058
13	12	N12	12	Incl	0.17	0	75	1	ELux	1	16947
14	13	N13	13	Incl	0.3	0	75	1	ELux	1	8051
15	14	N14	14	Incl	0.1	0	75	1	QLux	1	8372
16	15	N15	15	Incl	0.1	0	75	1	ELux	1	10043
17	16	N16	16	Incl	0.17	0	75	1	QLux	1	7023

Vertical | Horizontal | OD | MgCl | Volume Dec | Conc Dec | Strain | Relative Light Unit | Worksheet

Figure 20.11: *The worksheet of the reporter gene assay listed in the classical format, from first to last run. Only a small part of the total investigation is shown. The entire worksheet contains 192 experiments.*

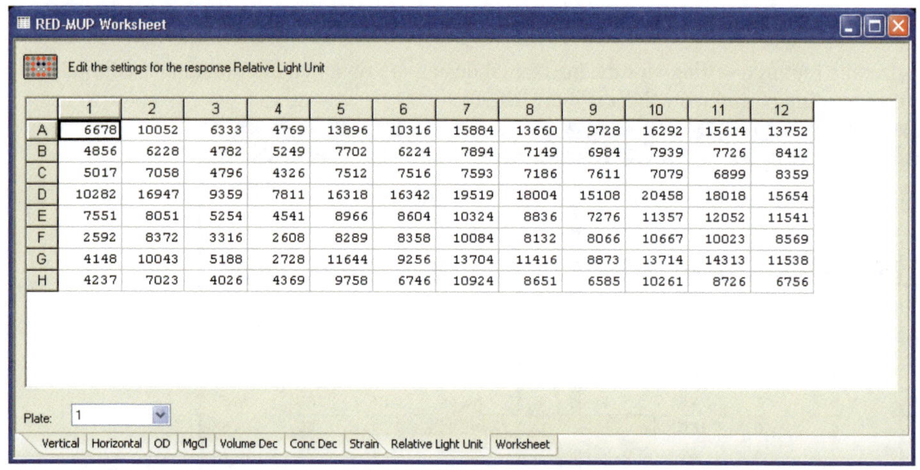

RED-MUP Worksheet

Edit the settings for the response Relative Light Unit

	1	2	3	4	5	6	7	8	9	10	11	12
A	6678	10052	6333	4769	13896	10316	15884	13660	9728	16292	15614	13752
B	4856	6228	4782	5249	7702	6224	7894	7149	6984	7939	7726	8412
C	5017	7058	4796	4326	7512	7516	7593	7186	7611	7079	6899	8359
D	10282	16947	9359	7811	16318	16342	19519	18004	15108	20458	18018	15654
E	7551	8051	5254	4541	8966	8604	10324	8836	7276	11357	12052	11541
F	2592	8372	3316	2608	8289	8358	10084	8132	8066	10667	10023	8569
G	4148	10043	5188	2728	11644	9256	13704	11416	8873	13714	14313	11538
H	4237	7023	4026	4369	9758	6746	10924	8651	6585	10261	8726	6756

Plate: 1

Vertical | Horizontal | OD | MgCl | Volume Dec | Conc Dec | Strain | Relative Light Unit | Worksheet

Figure 20.12: *Alternatively, the data in the worksheet may be presented using the plate view. Here, the response data from the first of the two 96-well plates are presented.*

This concludes the design generation phase and we now turn our attention to the analysis of these data in order to pinpoint a factor combination where the RLU response is likely to be maximized.

20.3.3 Evaluation of raw data

In order to evaluate the raw data we start by inspecting the histogram plot and the replicate plot presented in Figure 20.13. The appearance of the histogram suggests no transformation is needed (no great asymmetry). The replicate plot suggests the spread among the replicates (a total of 32) is small in relation to the variation in RLU values across the entire design. Hence, we have satisfactory response data to work with.

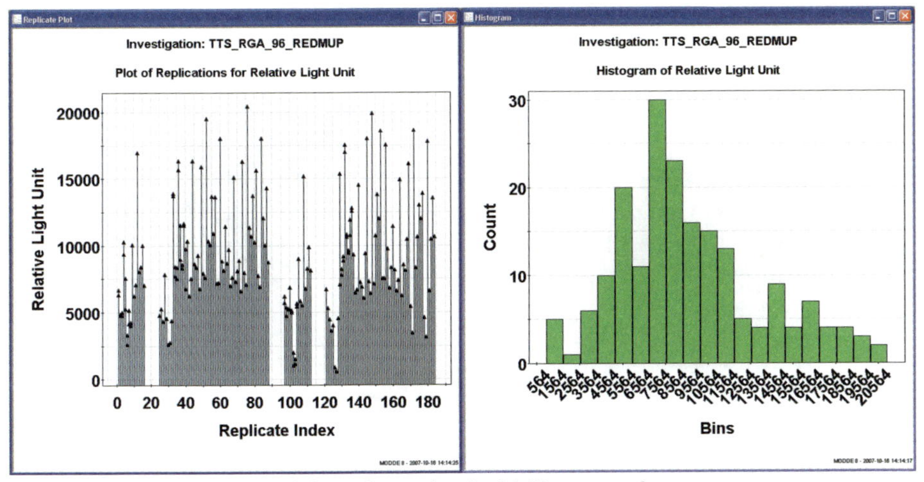

Figure 20.13: *Raw data plots. (left) Replicate plot. (right) Histogram plot.*

An important part of this stage is to consider the design geometry. This is especially relevant in the current application, since the real RED-MUP design is different from the one first proposed by MODDE. The geometry can be examined by constructing scatter plots of the factors. The scatter plot below (Figure 20.14) shows that MgCl and OD are varied in four levels. Hence, it is possible to estimate their quadratic terms ($MgCl^2$ & OD^2).

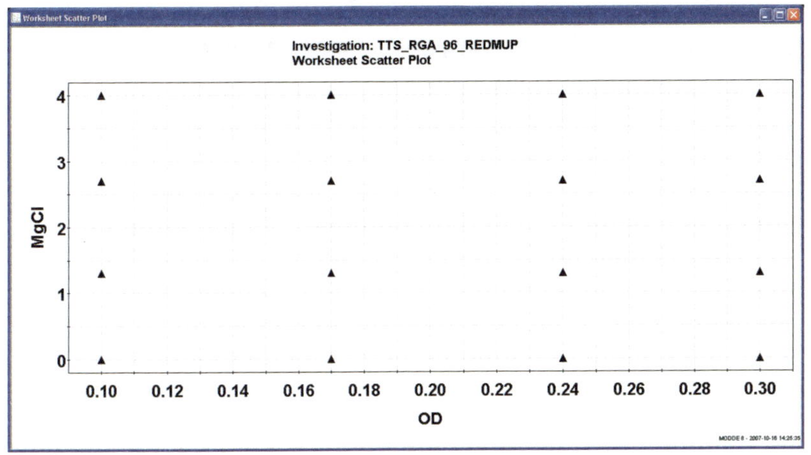

Figure 20.14: *Scatter plot showing the design geometry in the subspace spanned by OD and MgCl. Since both factors are explored at four levels, their quadratic terms can be estimated in the regression analysis.*

20.3.4 Data analysis and model interpretation

When fitting the regression model to the RLU response, the summary of fit plot presented below was produced (Figure 20.15). The negative Model Validity is an artifact arising from the extremely small replicate error (Figure 20.13). No matter how the model is revised, the model error will still remain significantly larger than the replicate error. The R^2 for the model is 0.77 and Q^2 is 0.71. Hence, we have a very good model.

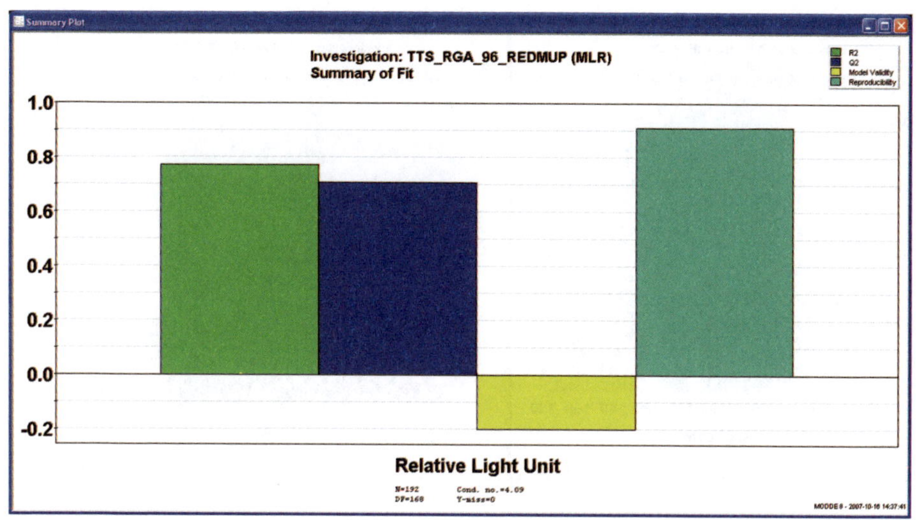

Figure 20.15: *Summary of fit plot with $R^2 = 0.77$, $Q^2 = 0.71$, Model Validity = -0.20 and Reproducibility = 0.91.*

Examining the regression coefficients (Figure 20.16) we can see that the Block factor (for the two plates) is not significant. This is encouraging. Hence, there are no systematic or undesirable differences between the two plates. Consequently, the measured data for the two plates can be investigated in a single MODDE project.

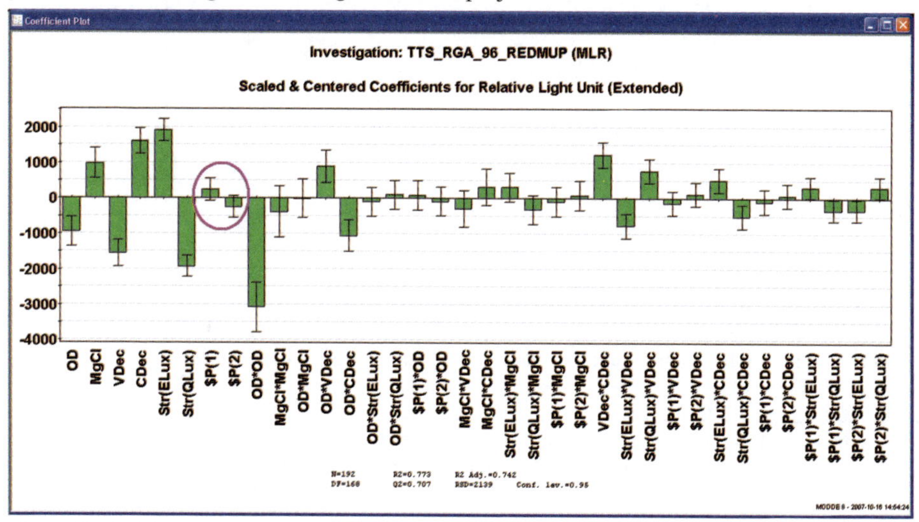

Figure 20.16: *Plot of regression coefficients. The Block factor representing the two plates is* ***not*** *significant.*

As there are several non-significant terms in the model, we have to revise it by excluding the unnecessary terms. We removed the following terms: $P(1); $P(2); MgCl*MgCl; OD*MgCl; OD*Str(ELux); OD*Str(QLux); $P(1)*OD; $P(2)*OD; MgCl*VDec; MgCl*CDec; Str(ELux)*MgCl; Str(QLux)*MgCl; $P(1)*MgCl; $P(2)*MgCl; $P(1)*VDec; $P(2)*VDec; $P(1)*CDec; $P(2)*CDec; $P(1)*Str(ELux); $P(1)*Str(QLux); $P(2)*Str(ELux) & $P(2)*Str(QLux). The revised model has similar performance statistics to the initial model ($R^2 = 0.75$ / $Q^2 = 0.72$), but it is a lot easier to interpret (Figure 20.17).

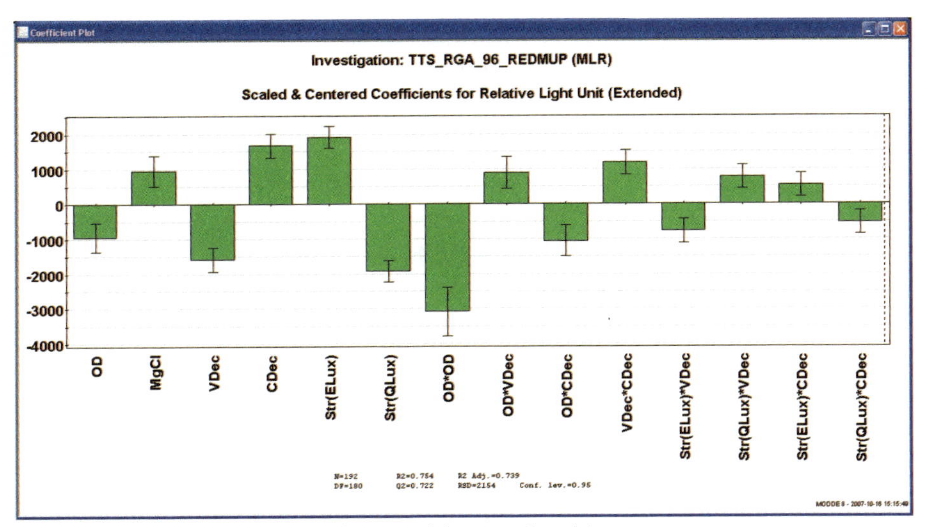

Figure 20.17: Plot of regression coefficients of the revised model.

From Figure 20.17 it can be deduced that an increased RLU will occur at low OD and VDec and high MgCl and CDec. There is a significant non-linear relationship between RLU and OD. In addition, the signal strength is generally higher for the bacterial strain ELux.

A final diagnostic check of the model, by inspecting the N-plot of residuals, shows that there are no outliers (Figure 20.18). Hence, we have a reliable model.

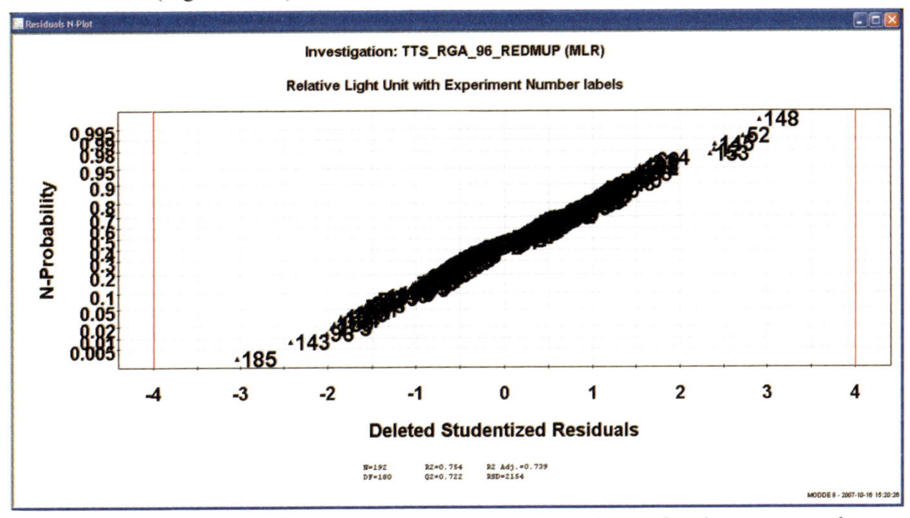

Figure 20.18: Normal probability of model residuals. This plot indicates that there are no outliers.

20.3.5 Using the model for predictions

Response contour and surface plots are informative when we wish to overview the impact of multiple factor combinations. Figure 20.19 shows a three-by-three grid of response contour plots pertaining to ELux. The arrow indicates the region where maximum RLU is achieved. The filled ellipse indicates the region where the biological assay was normally conducted, prior to the current study.

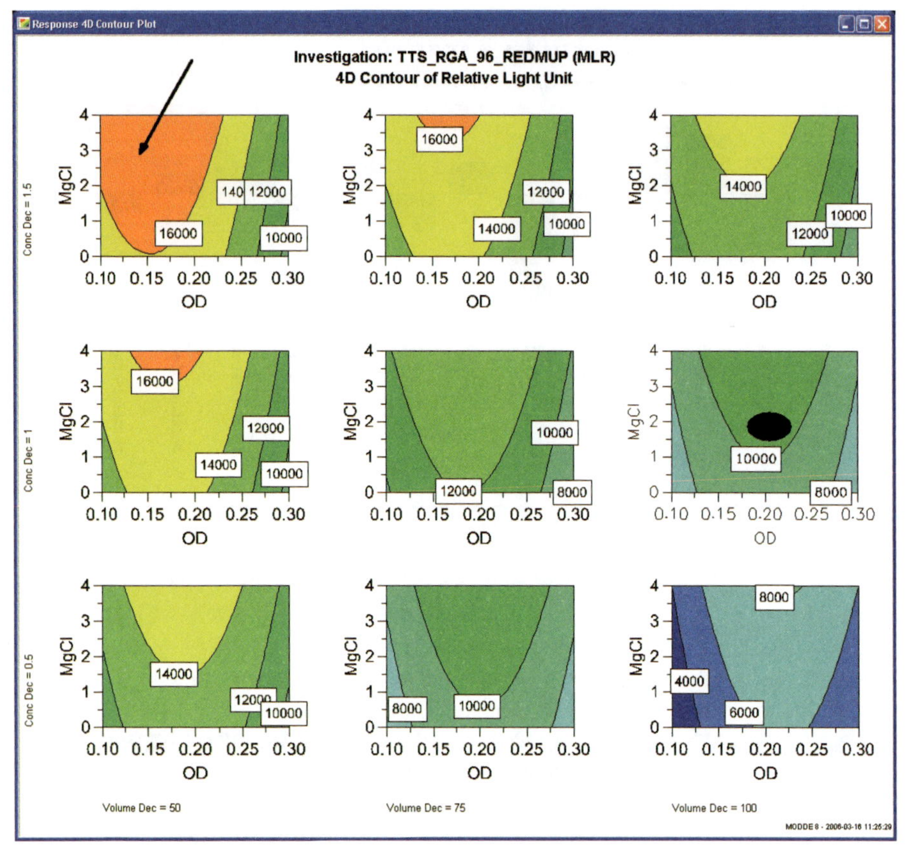

Figure 20.19: A 4D response contour plot showing the change in RLU value following changes to four of the five factors. The results are valid for the ELux bacterial strain. The solid ellipse shows the factor combination where the assay was normally operated. The arrow shows the optimum as suggested by the current modelling efforts. The final result was a nearly 70% increase in the RLU signal.

20.3.6 Conclusions of the study

The main conclusion from this application was that the concept of RED-MUP design represents a powerful, flexible and efficient approach for optimizing experimental systems and biological assays performed on multiwell plates. The current investigation led to (i) an improved understanding of the biological system under study and (ii) a greatly increased signal strength (70% higher RLU value) compared with earlier experimental campaigns.

20.4 Example II: An HTS assay measuring tryptase activity

Our second example deals with an assay measuring the activity of the enzyme tryptase, for which we are grateful to Per Rosén for giving us access to the data (P. Rosén, Experimental Design in Assay Development, Masters Thesis, Dept. of Chemical and Biological Engineering, Chalmers University of Technology, Gothenburg, Sweden, 2006). In comparison with the previous example, the scope of the tryptase data-set is much broader. First and foremost, the complete study reported by Per Rosén contained four chained RED-MUPs, one screening design, two optimization designs and one final robustness testing design. Second, the experiments were performed using 384-well plates. Third, three responses were measured (rather than one) related to signal, stability and linearity. Here, we review the results obtained utilizing the initial screening RED-MUP, focusing on the signal response.

20.4.1 Background and objective

High throughput screening (HTS) is used by the pharmaceutical industry to identify potential candidate drugs from large inventories of chemical substances. These compounds are tested in multiple *in vitro* screenings (biological assays) that monitor the effects of the compounds on the protein targets of interest. Ideally, such assays are set-up so that they can be performed on 384- or 1536-multiwell plates on a routine basis.

In order to evaluate the applicability of RED-MUP designs to HTS assays, Per Rosén choose to work with a model assay measuring the activity of the enzyme tryptase by means of fluorescence spectroscopy. The effects of seven factors were explored using four RED-MUPs. For more experimental details see the original literature, the reference for which is given in the previous section. Below, we review the main results of the screening phase.

20.4.2 Setting up the screening design

The screening phase involved seven factors: five quantitative and two qualitative. One qualitative factor, Stabilizer, was defined at four levels and the other, Buffer, at three levels (Figure 20.20). The factors Heparin, Conc $MgCl_2$ and Stabilizer were assigned to the vertical design with 16 experiments (because there are 16 rows on the 384 plate). The remaining four factors were assigned to the 24-run horizontal design. One qualitative factor was allotted to each of the two building block designs.

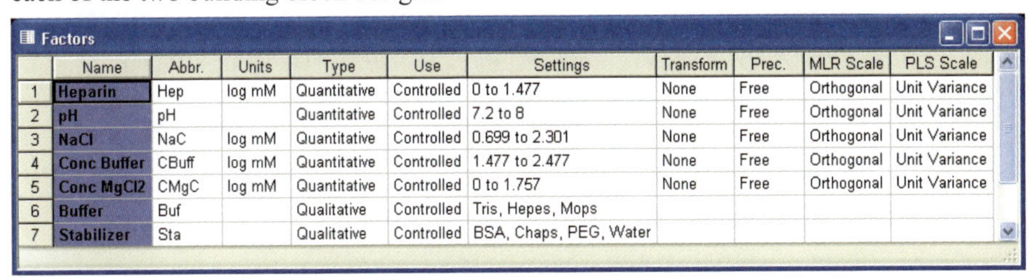

Factors

	Name	Abbr.	Units	Type	Use	Settings	Transform	Prec.	MLR Scale	PLS Scale
1	Heparin	Hep	log mM	Quantitative	Controlled	0 to 1.477	None	Free	Orthogonal	Unit Variance
2	pH	pH		Quantitative	Controlled	7.2 to 8	None	Free	Orthogonal	Unit Variance
3	NaCl	NaC	log mM	Quantitative	Controlled	0.699 to 2.301	None	Free	Orthogonal	Unit Variance
4	Conc Buffer	CBuff	log mM	Quantitative	Controlled	1.477 to 2.477	None	Free	Orthogonal	Unit Variance
5	Conc MgCl2	CMgC	log mM	Quantitative	Controlled	0 to 1.757	None	Free	Orthogonal	Unit Variance
6	Buffer	Buf		Qualitative	Controlled	Tris, Hepes, Mops				
7	Stabilizer	Sta		Qualitative	Controlled	BSA, Chaps, PEG, Water				

Figure 20.20: *The seven factors examined in the tryptase screening design.*

The worksheet for the vertical design is presented in Figure 20.21. The core part of this building block design is a 2^2 design in Heparin and Conc $MgCl_2$ with one center-point, but where the +1/+1 setting has been omitted (Figure 20.21). In this way, the number of runs is reduced from five to four. This four-run pattern is then preserved across all four levels of the Stabilizer factor, i.e., four runs at four levels giving 16 experiments.

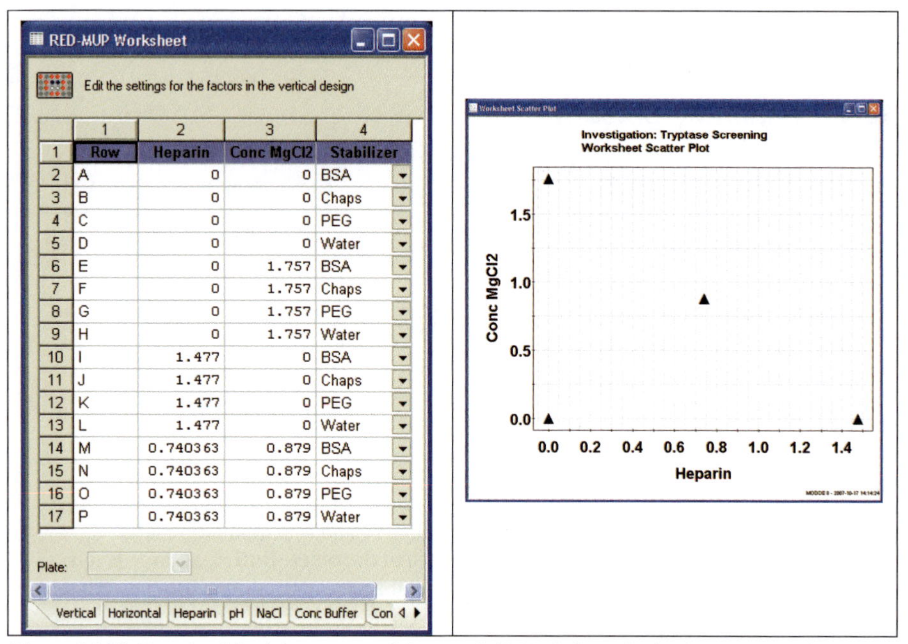

1	2	3	4
Row	Heparin	Conc MgCl2	Stabilizer
A	0	0	BSA
B	0	0	Chaps
C	0	0	PEG
D	0	0	Water
E	0	1.757	BSA
F	0	1.757	Chaps
G	0	1.757	PEG
H	0	1.757	Water
I	1.477	0	BSA
J	1.477	0	Chaps
K	1.477	0	PEG
L	1.477	0	Water
M	0.740363	0.879	BSA
N	0.740363	0.879	Chaps
O	0.740363	0.879	PEG
P	0.740363	0.879	Water

Figure 20.21: *Details of the 16-run vertical design. (left) Worksheet. (right) Scatter plot of the two quantitative factors.*

The worksheet for the horizontal design is presented in Figure 20.22. The basic structure of this building block design is a 2^3 design in pH, NaCl and Conc Buffer with one center-point, but where the -1/-1/-1 factorial point has been omitted (Figure 20.22). In this way, the number of runs is reduced from nine to eight. This eight-run pattern is then preserved across all the three levels of the Buffer factor, except in one case where another corner experiment is exchanged for an additional center-point to allow space for a column with replicates.

Combining the two building block designs results in a 384-run RED-MUP, in which the number of unique experiments is 368. It has a condition number of 5.10 and a maximum correlation between any pair of model terms of 0.52. These performance statistics, although somewhat high for a conventional DOE protocol, are acceptable for a RED-MUP and thus indicate that the design is fit-for-purpose (i.e. for maximizing the signal from the assay). The model supported by the RED-MUP protocol is a full interaction model, containing 29 model terms.

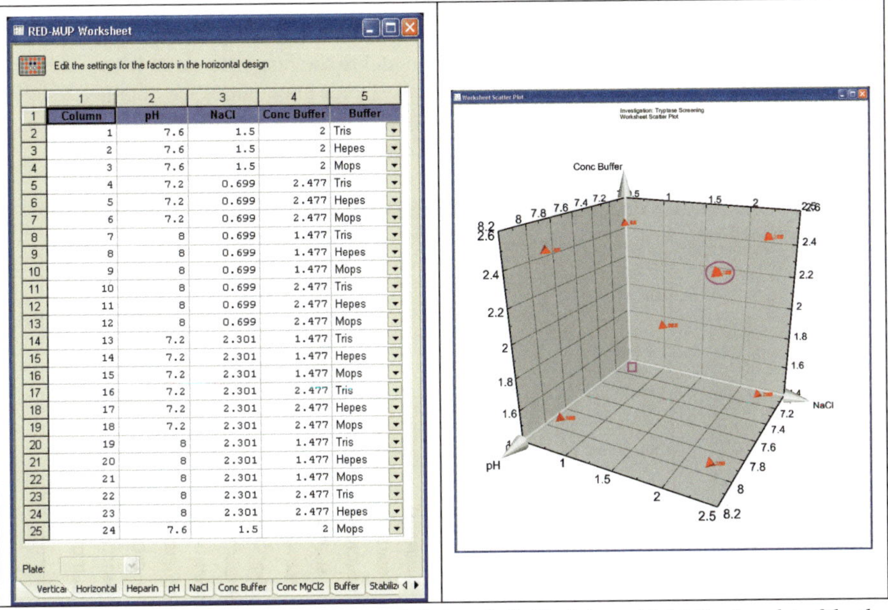

RED-MUP Worksheet

Edit the settings for the factors in the horizontal design

	1	2	3	4	5
1	Column	pH	NaCl	Conc Buffer	Buffer
2	1	7.6	1.5	2	Tris
3	2	7.6	1.5	2	Hepes
4	3	7.6	1.5	2	Mops
5	4	7.2	0.699	2.477	Tris
6	5	7.2	0.699	2.477	Hepes
7	6	7.2	0.699	2.477	Mops
8	7	8	0.699	1.477	Tris
9	8	8	0.699	1.477	Hepes
10	9	8	0.699	1.477	Mops
11	10	8	0.699	2.477	Tris
12	11	8	0.699	2.477	Hepes
13	12	8	0.699	2.477	Mops
14	13	7.2	2.301	1.477	Tris
15	14	7.2	2.301	1.477	Hepes
16	15	7.2	2.301	1.477	Mops
17	16	7.2	2.301	2.477	Tris
18	17	7.2	2.301	2.477	Hepes
19	18	7.2	2.301	2.477	Mops
20	19	8	2.301	1.477	Tris
21	20	8	2.301	1.477	Hepes
22	21	8	2.301	1.477	Mops
23	22	8	2.301	2.477	Tris
24	23	8	2.301	2.477	Hepes
25	24	7.6	1.5	2	Mops

Plate:

Vertical | Horizontal | Heparin | pH | NaCl | Conc Buffer | Conc MgCl2 | Buffer | Stabiliz

Figure 20.22: *Details of the 24-run horizontal design. (left) Worksheet. (right) Scatter plot of the three quantitative factors. The corner marked with the square was not tested. The corner marked with the ellipse was omitted for the Mops buffer to allow a column for replication to be included.*

20.4.3 Evaluation of raw data

Figure 20.23 shows two plots of the response data: the replicate plot and the histogram plot. Although it is hard to see, the replicate error is relatively small. The numerical range of the signal is between 7 and 100. The histogram suggests that there is a small tail to the right, but this structure is not sufficiently strong to justify any response transformation.

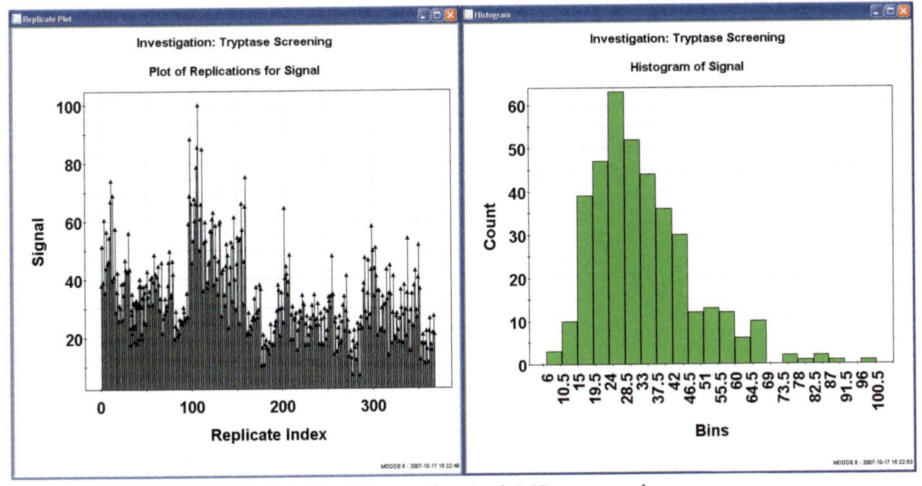

Figure 20.23: *Raw data plots. (left) Replicate plot. (right) Histogram plot.*

20.4.4 Data analysis and model interpretation

The data analysis based on fitting the full interaction model to the response variable revealed a number of irrelevant cross-terms. After deleting 16 two-factor interactions we obtained a sound model with $R^2 = 0.80$, $Q^2 = 0.78$, Model Validity = 0.51 and Reproducibility = 0.86 (Figure 20.24). In comparison with the original model, the revised model is a lot easier to examine and interpret (Figure 20.24).

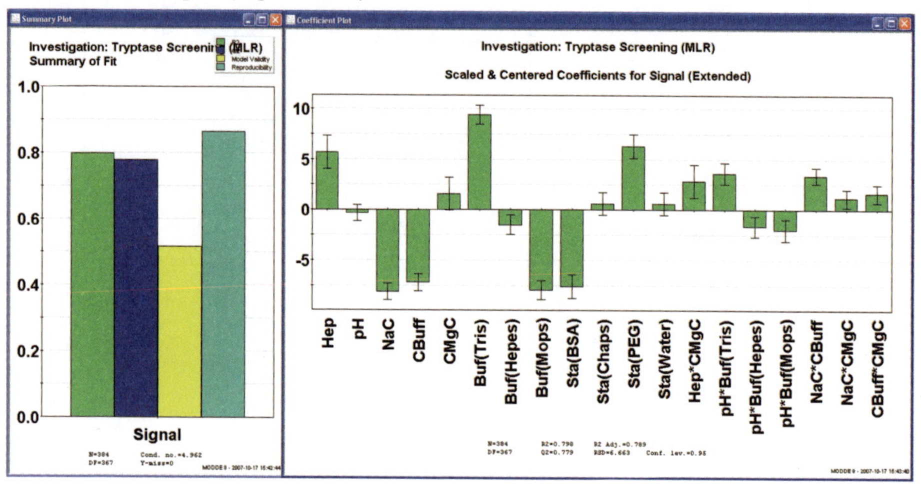

Figure 20.24: Results from fitting the refined interaction model to the signal response. (left) Summary of fit plot. (right) Regression coefficient plot.

The regression coefficients suggest that an increased signal is connected to the profile with high Heparin, low NaCl, low Conc Buffer, Tris as the buffer and PEG as the stabilizer. pH and Conc $MgCl_2$ have relatively little impact on the signal. Further, it can also be concluded that the combination Mops (buffer) and BSA (stabilizer) should be completely avoided. A final diagnostic check using the N-plot of residuals corroborates the soundness of the model (plot not provided).

20.4.5 Using the model for predictions

According to Figure 20.25, which shows a three-by-three grid of response contour plots, the model estimates that the profile high Heparin, intermediate pH, low NaCl, low Conc Buffer, high Conc $MgCl_2$, Buffer TRIS and Stabilizer PEG will produce a signal of approximately 80 (an arrow on the plot indicates this factor combination).

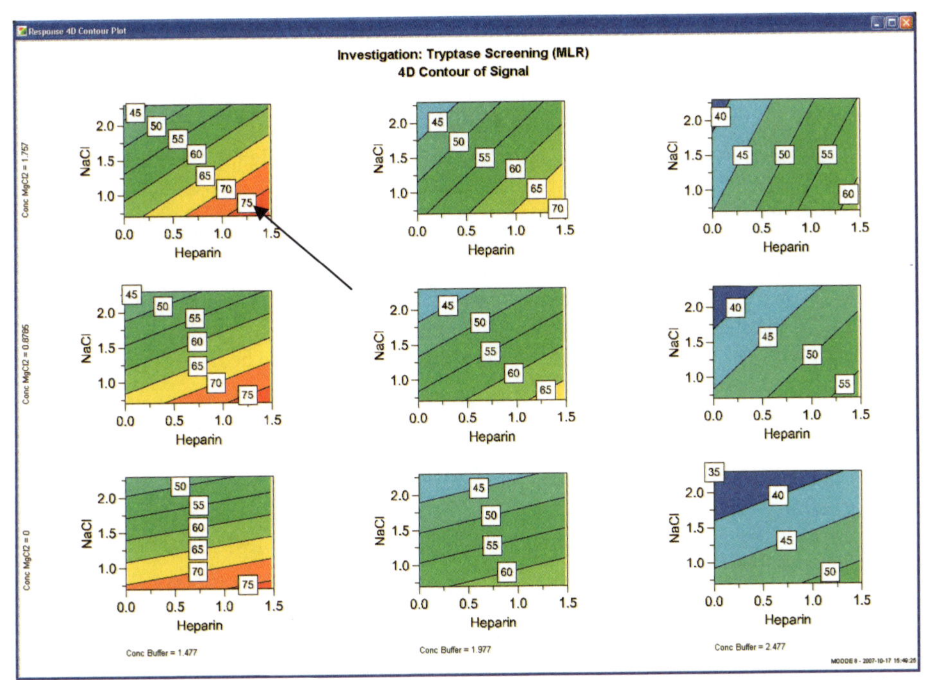

Figure 20.25: 4D-response contour plot of signal when varying Heparin, NaCl, Conc Buffer and Conc $MgCl_2$. The other three factors have been fixed as follows: pH = intermediate, Buffer = TRIS and Stabilizer = PEG. The arrow indicates where the highest signal should be encountered, according to the regression model.

20.4.6 Concluding remarks

The screening design clearly demonstrated that the best settings for the qualitative factors were Buffer TRIS and Stabilizer PEG. It was decided to preserve this combination in the subsequent optimization RED-MUPs, although the concentration of PEG was included as a new factor. The ranges of the quantitative factors were adjusted in accordance with the results of the refined interaction regression model.

We will not review all the details of the two optimization RED-MUPs and the RED-MUP for the robustness testing, but reference is made to the Masters thesis of Per Rosén (P. Rosén, Experimental Design in Assay Development, Masters Thesis, Dept. of Chemical and Biological Engineering, Chalmers University of Technology, Gothenburg, Sweden, 2006). The final result after the four chained RED-MUPs was that the tryptase assay became more reliable and stable, and produced an enhanced signal. More precisely, the signal was increased four-fold and the loss of activity was reduced from a 40% decrease to a 5% decrease within a 24 hour period. A more general conclusion was that the RED-MUP approach provides a versatile and valuable tool for assay development and optimization. The RED-MUP designs were found to make manual sample preparatory work easier and to prepare the ground for more efficient use of the microtiter plates.

20.5 Some technical aspects

The two cases presented above illustrate the REDMUP concept, but without going into technical details. In this section, some technical aspects are considered. A general dilemma when constructing a RED-MUP is that any two precursor designs (the vertical design plus the horizontal design) that are to be combined must comply with the constraints of 8, 12, 16 or

24 runs (numbers valid for 96-well and 384-well plates). However, most of the classical designs intended for screening and optimization do *not* meet these requirements, particularly if center-points are to be considered. Hence, prior to creating the RED-MUP, the building block designs must first be pruned so that they contain the correct number of runs.

Whenever distortions are introduced into a traditional experimental design, irrespective of whether the design is a screening or optimization scheme, care must be exercised so that the statistical properties of the resulting modified designs are monitored properly. As reported in the pioneering work of IngMarie Olsson (Olsson, I., Experimental Designs at the Crossroads of Drug Discovery, Ph.D. Thesis, Umeå University, 2006), it was found that "combining precursor designs with good statistical properties resulted in RED-MUP combinations with good statistical properties". In other words, if we combine inappropriately pruned precursor designs of low statistical quality, their product, the RED-MUP, will also be of low quality.

In her doctorate work, Olsson lists a whole series of building block designs that are useful for setting up pure screening and optimization RED-MUPs, as well as RED-MUPs with a mixed screening–optimization objective. A total of 162 precursor designs are listed, which are sub-divided into four groups depending on the number of runs (8, 12, 16 or 24). Depending on the experimental objective and the number of runs, these building block designs cover anything between 1 and 23 factors, including qualitative factors at multiple levels. A sub-set of these designs has been implemented in MODDE.

When browsing through the work of Olsson, it becomes evident that sometimes only small adjustments are required to tailor many classical design families to fit the RED-MUP context. The inventories of RED-MUP primers have entries derived from factorial, fractional factorial, Rechtschaffner, Plackett-Burman, Doehlert and central composite designs, with and without center-points. All these building block designs have been quality assured by considering a number of statistical performance criteria, such as (i) the number of unique experiments, (ii) the condition number, (iii) the logarithm of the normalized determinant of the variance–covariance matrix and (iv) the maximum pairwise correlation between any two terms in the regression model supported for that particular precursor design.

Another interesting observation in Olsson's work was the discovery that the orthogonal combination of two resolution III fractional factorial designs results in RED-MUPs of a higher resolution (Res III/IV). This means that in a 96-run RED-MUP it is possible to estimate all interaction effects *between* factors varied according to the vertical and horizontal designs. Hence, if a specific two-factor interaction is deemed to be important it can be estimated by positioning one factor in the vertical design and the other in the horizontal design. As a consequence, when planning a RED-MUP protocol it is essential to have a clear understanding of the investigated assay or system, so that factors are allocated to the correct RED-MUP primer. In particular, some factors may be linked together for practical reasons, and may therefore have to be manipulated coherently in terms of being part of the same building block design.

Finally, it should be realized that while randomization of the run order is generally very important to counteract the possible influence of an uncontrollable factor, it is hardly a principle that can be easily superimposed onto the technical demands associated with multiwell technology. Thus, in order to make the RED-MUPs as user-friendly as possible, randomization of the run order has been downgraded in relation to other key technical features. However, should for instance edge effects occur and be significant, the possibilities of run order randomization and column/row-replication should not be forgotten.

20.6 Questions for Chapter 20

- What is the objective of the RED-MUP approach?
- What is the role of the vertical design?
- What is the role of the horizontal design?
- Which plate formats are usually used?
- Which statistical performance criteria should be considered when setting up a RED-MUP protocol?

20.7 Summary and discussion

In this chapter, we have discussed and exemplified RED-MUP designs. The RED-MUP protocols have proved to be highly efficient for assay development and optimization. The driving force underpinning the RED-MUPs is customization of DOE to the formats and demands of multiwell microtiter plate technology. Some of the notable advantages of the RED-MUPs are:

- An efficient framework for assay development and optimization.
- Decreased workload for the experimenter.
- Decreased assay optimization times.
- Increased data quality.
- Enhanced understanding of the system under investigation.
- A standardized communication platform for information exchange between experimenters.
- Improved documentation and reporting format for the results obtained.
- Optimized information flow between various steps in the drug discovery process.

21 DOE in formulation development

21.1 Objective

Formulation development represents a complex chain of technical challenges. The great benefit of using DOE to develop and enhance formulations is that it allows all potential factors – factors related to formulation and factors related to manufacturing – to be investigated systematically and simultaneously. This makes it possible to identify the most critical factors and any likely interactions between them. The objective of Chapter 21 is to discuss DOE from the viewpoint of a formulator. Some of the concepts are illustrated using an example dealing with tablet formulation.

21.2 Introduction

A *formulation* is the constitution of a pharmaceutical product. It comprises the active pharmaceutical ingredient (API) *and* a multitude of other excipients. All pharmaceutical products need to be formulated to a specific dosage form in order to be conveniently and efficiently administered to users. A good formulation must be easy to manufacture and, over time, must consistently yield the desired functional product. There are many types of dosage forms, the most common being tablets, ointments, capsules, suspensions, and gels. The highly variable properties of the different dosage forms present a manufacturing challenge, in the sense that different production strategies and technologies are usually required. Moreover, even for just a single dosage form, there are normally a number of steps in the production chain, making the whole manufacturing process a complex task. In tablet manufacturing, for instance, typical steps include weighing, milling, granulation, drying, blending, tabletting and coating. Each one of these individual operations involves a number of factors that have the potential to influence the performance of the final product.

A key issue in formulation development concerns identifying the constituents (excipients) to deploy in the actual formulation. Apart from the API, common excipients in tablet formulation are fillers, binders, disintegrants, lubricants, colorants and sometimes surfactants; these are added to serve specific purposes in order to guarantee good product performance, consistency and appearance. For instance, a filler is used to influence the size of the tablet and a binder is included to affect the binding capacity and therefore, indirectly, the hardness of the tablet. A mandatory step in the early phase of formulation development is to verify that the types of excipient selected are physically and chemically compatible with each other and the API. A final formulation must remain stable throughout the manufacturing process and the product's shelf-life.

At the start of a formulation process it is advantageous to consider more than one alternative compound for each main type of constituent, for example more than one type of binder. It is rarely the best approach to limit the exploratory work and focus on just one single binder. Our recommendation, in fact, is that the choice should be made from a diverse sub-set of binders that are representative of a larger group of untested candidate binders. Deciding which binders (or fillers, or lubricants, …) to evaluate may be based on chemical knowledge and previous experience, and may even rely on intuition, but in the long-term using DOE as a supplementary tool is a more efficient strategy. It should be emphasized that a sub-set containing a few, diverse and representative binders (fillers, lubricants, …) may well be identified using the principles of multivariate design. Multivariate design was reviewed in Chapter 19 in connection with the choice of solvent in organic synthetic chemistry. Of course, other aspects like the cost of excipients may need to be considered, especially in situations where many, very similar excipients are available. Section 21.4 below is devoted to a discussion of how to change the type of excipient and their levels at the same time.

Furthermore, in addition to pinpointing which combination of excipients corresponds to the optimal composition, it is equally important to determine the optimal proportion of each excipient. The optimal proportion of each excipient in the formulation blend will depend not only on the properties of the API but also on the specific production environment. Depending on how the formulation problem is addressed, variations in the excipient recipe can be induced by means of a process design such as a fractional factorial, or through a mixture design. This is discussed in more detail in Section 21.4.

In conclusion to this introduction, it can be stated that because of the wide variety of technical challenges that formulators encounter during formulation development, it is crucial to use an effective experimental methodology. DOE is one such methodology and it has been widely used in formulation development. The major advantage of using DOE in formulation development is that it permits all potential factors and their possible interactions to be estimated in an informationally optimal manner. In order to shed some light on the value of DOE in formulation development, we will review a tablet formulation example in which many factors and responses are considered simultaneously. As will become apparent, there will be some conflicting results as far as the responses are concerned. Hence, we need to consider the linked responses functionality in MODDE to achieve the final optimization.

21.3 A tablet formulation example

We want to provide a relevant example early in this chapter. Hence, we shall use this section to review an application originating from AstraZeneca R&D in Södertälje. The original publication is: Andersson, M., Ringberg, A., and Gustafsson, C., Multivariate methods in tablet formulation suitable for early drug development: Predictive models from a screening design of several linked responses, Chemometrics and Intelligent Laboratory Systems, 87, 2007, 151-156. The aim of the study was to define a formulation resulting in a tablet of a desired strength with a fast drug release profile. More details are given below.

The tablet formulation data-set is interesting for several reasons. First and foremost, it includes multiple factors and responses. Second, the whole study was planned and managed using MODDE; the optimizer functionality was found to be particularly important for identifying follow-up experiments. Third, the linked responses option was invoked to enable optimization of partially uncorrelated responses. In addition, PLS was used as the regression engine because of the presence of one uncontrollable factor.

21.3.1 Configuring the investigation

The study involves six excipients, namely two fillers, a binder, a disintegrant, a surfactant and a lubricant (Figure 21.1). As shown in Figure 21.1, six quantitative factors were specified to take account of the changes in the proportions of the excipients. Two of these were used to express the amount of the two fillers, i.e. two ratio factors defined as API/filler and filler1/filler2. A seventh factor was compression force, for which the target value was 10 kN.

This factor was set as being uncontrolled, since there was a small variation that could not be satisfactorily controlled within the experiment.

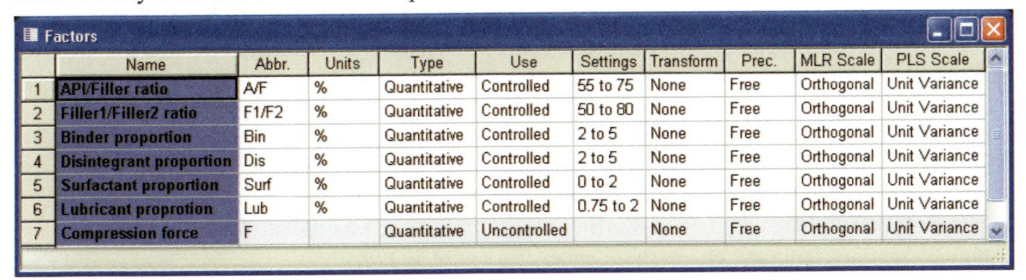

Factors

	Name	Abbr.	Units	Type	Use	Settings	Transform	Prec.	MLR Scale	PLS Scale
1	API/Filler ratio	A/F	%	Quantitative	Controlled	55 to 75	None	Free	Orthogonal	Unit Variance
2	Filler1/Filler2 ratio	F1/F2	%	Quantitative	Controlled	50 to 80	None	Free	Orthogonal	Unit Variance
3	Binder proportion	Bin	%	Quantitative	Controlled	2 to 5	None	Free	Orthogonal	Unit Variance
4	Disintegrant proportion	Dis	%	Quantitative	Controlled	2 to 5	None	Free	Orthogonal	Unit Variance
5	Surfactant proportion	Surf	%	Quantitative	Controlled	0 to 2	None	Free	Orthogonal	Unit Variance
6	Lubricant proprotion	Lub	%	Quantitative	Controlled	0.75 to 2	None	Free	Orthogonal	Unit Variance
7	Compression force	F		Quantitative	Uncontrolled		None	Free	Orthogonal	Unit Variance

Figure 21.1: The seven factors defined in the Tablet Formulation data-set.

It was deemed important to measure responses reflecting tablet strength and dissolution profile. A total of nine response variables were defined (Figure 21.2), of which a quintet reflect *in vitro* dissolution values sampled at regular time intervals (5, 15, 30, 45, and 60 minutes). For experimental details of the response measurement methods, reference is made to the original publication (see above). We shall refer to the first four responses as the "strength responses" and the last five as the "dissolution responses".

Responses

	Name	Abbr.	Units	Transform	MLR Scale	PLS Scale	Type
1	True Density	Trde	g/cm3	None	None	Unit Variance	Regular
2	Tablet Hardness	TH	N	None	None	Unit Variance	Regular
3	Tensile Strength 1	TS1	MPa	None	None	Unit Variance	Regular
4	Tensile Strength 2	TS2	MPa	None	None	Unit Variance	Regular
5	T5	T5	%	None	None	Unit Variance	Regular
6	T15	T15	%	None	None	Unit Variance	Regular
7	T30	T30	%	None	None	Unit Variance	Regular
8	T45	T45	%	None	None	Unit Variance	Regular
9	T60	T60	%	None	None	Unit Variance	Regular

Figure 21.2: The nine responses defined in the Tablet Formulation data-set.

In order to screen the six controlled factors, the authors decided to use a two-level fractional factorial design in 19 runs, 16 factorial points and three replicated center-points. This design supports a linear regression model. The worksheet is shown in Figure 21.3. Using standard nomenclature, the chosen design is identified as a 2^{6-2} design of resolution IV. Remember that a discussion of fractional factorial designs and the concept of their resolution was presented in Chapter 5.

Worksheet

	Exp No	Exp Na	Run O	Incl/Excl	API/Fille	Filler1/F	Binder p	Disinteg	Surfacta	Lubricar	Compre
2	1	N1	1	Incl	55	50	2	2	0	0.75	7.75
3	2	N2	6	Incl	75	50	2	2	2	0.75	11.06
4	3	N3	17	Incl	55	80	2	2	2	2	10.01
5	4	N4	15	Incl	75	80	2	2	0	2	10.72
6	5	N5	19	Incl	55	50	5	2	2	2	9.7
7	6	N6	11	Incl	75	50	5	2	0	2	9.66
8	7	N7	16	Incl	55	80	5	2	0	0.75	11.05
9	8	N8	18	Incl	75	80	5	2	2	0.75	10.26
10	9	N9	12	Incl	55	50	2	5	0	2	9.7
11	10	N10	8	Incl	75	50	2	5	2	2	10.29
12	11	N11	5	Incl	55	80	2	5	2	0.75	10.47
13	12	N12	4	Incl	75	80	2	5	0	0.75	12.02
14	13	N13	9	Incl	55	50	5	5	2	0.75	11.19
15	14	N14	14	Incl	75	50	5	5	0	0.75	10.27
16	15	N15	13	Incl	55	80	5	5	0	2	8.6
17	16	N16	3	Incl	75	80	5	5	2	2	9.7
18	17	N17	2	Incl	65	65	3.5	3.5	1	1.375	9.55
19	18	N18	10	Incl	65	65	3.5	3.5	1	1.375	9.68
20	19	N19	7	Incl	65	65	3.5	3.5	1	1.375	9.13

	True Density	Tablet Hardı	Tensile Strength 1	Tensile Strength 2	T5	T15	T30	T45	T60
2	1.483	164	4.112	13.97	12	45	69	79	85
3	1.47	234	5.716	22.1	20	74	92	97	98
4	1.462	186	4.563	19.2	31	87	97	98	97
5	1.464	208	5.128	19.66	2	10	33	59	73
6	1.466	122	2.878	11.07	27	81	98	99	98
7	1.455	213	5.101	19.72	9	42	77	88	93
8	1.462	201	4.822	18.89	10	40	79	92	96
9	1.451	204	4.799	18.88	21	68	98	100	99
10	1.481	216	5.181	19.1	33	84	94	97	96
11	1.461	251	5.939	23.52	35	85	96	97	97
12	1.469	210	5.059	19.93	49	97	99	98	97
13	1.474	256	6.254	24.47	19	67	89	94	95
14	1.466	252	5.901	24.08	36	94	99	99	99
15	1.462	280	6.548	25.73	31	86	97	99	99
16	1.463	193	4.513	17.32	35	90	97	97	97
17	1.45	234	5.425	22.31	23	66	90	92	93
18	1.462	214	5.133	19.01	36	86	96	97	97
19	1.465	232	5.499	20.81	40	90	97	98	98
20	1.465	208	5.075	19.24	37	85	94	95	95

Figure 21.3: The worksheet of the screening design used in the Tablet Formulation example.

21.3.2 Evaluation of raw data

Initial histogram plots of the nine responses revealed skewed distributions for the five dissolution responses, while the four strength responses were approximately normally distributed. As a consequence, the negative logarithmic transformation ("NegLog" in MODDE) was applied to the last five responses. Figure 21.4 shows the histograms of the nine responses after transformation of the five dissolution responses. The latter five are now much closer to being normally distributed than they were before. Thus, in the following, we will analyze the four strength responses in their untransformed metric and the five dissolution responses in their transformed metric. Further inspection of the raw data in terms of replicate plots indicated small replicate errors in all cases (plots not shown).

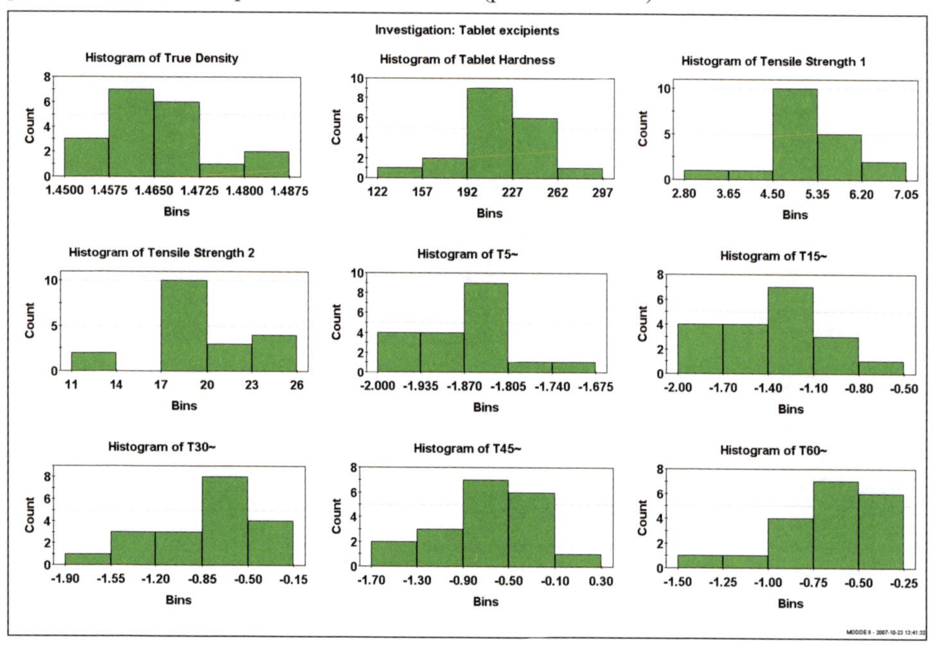

Figure 21.4: Histogram plots of the nine responses, after Neg-Log transformation of the last five.

Another tool that is available in the MODDE worksheet toolbox is the correlation matrix (see Figure 21.5). This tool is particularly relevant when there are uncontrolled factors. We can see from Figure 21.5 that the uncontrollable factor (compression force) is moderately correlated with the six designed factors. Hence, these correlations will not violate the relevance of the data analysis and the conclusions thereof.

Correlation Matrix

	1	2	3	4	5	6	7	8
1		A/F	F1/F2	Bin	Dis	Surf	Lub	F
2	A/F	1	0	0	0	0	0	0.332901
3	F1/F2	0	1	0	0	0	0	0.19394
4	Bin	0	0	1	0	0	0	-0.096064
5	Dis	0	0	0	1	0	0	0.122648
6	Surf	0	0	0	0	1	0	0.175815
7	Lub	0	0	0	0	0	1	-0.343776
8	F	0.332901	0.19394	-0.096064	0.122648	0.175815	-0.343776	1

Threshold 0.3

Figure 21.5: Correlation matrix of the model for the Tablet Formulation example. Note that only the part related to the factors is displayed.

21.3.3 Data analysis and model interpretation

Because of the uncontrollable factor, the PLS method was chosen for the regression analysis. A PLS model with four significant components was obtained with the following performance indicators: $R^2 = 0.72$; $Q^2 = 0.27$; Model Validity = 0.69; Reproducibility = 0.92. Figure 21.6 shows the corresponding values for the individual response variables. The general observation is that the strength responses fit better and are better predicted by the model than the dissolution responses. The worst fits are for the dissolutions recorded after 45 and 60 minutes. This is quite natural since in almost all runs most of the tablets have been dissolved by this stage, meaning that the numerical values of these responses are approaching the upper limit of 100. This, in turn, implies that there is comparatively low degree of variability in these two responses upon which to base the model.

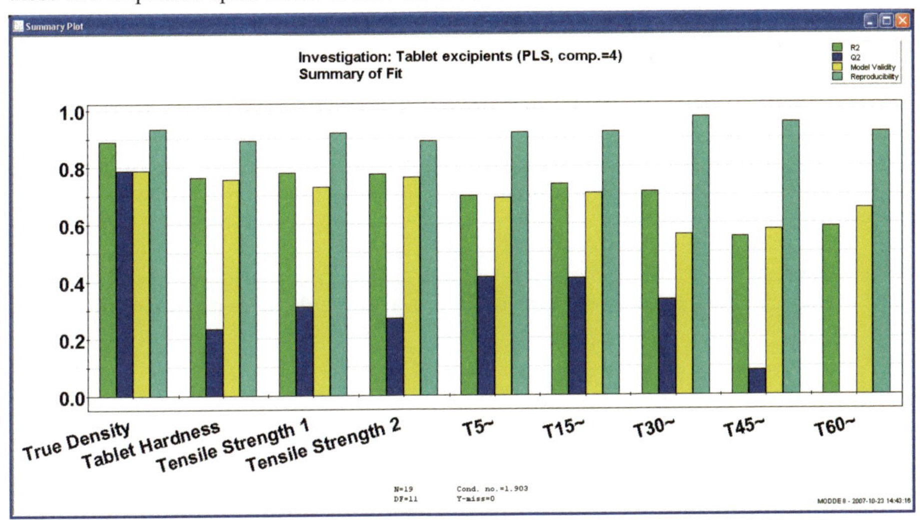

Figure 21.6: *Summary of fit plot for the Tablet Formulation application. Initial model based on nine responses.*

The relatively high number of components in the PLS model suggests that the responses are fairly unrelated and disparate (i.e., providing different information on the runs in the experimental design). Looking at the loading plot shown in Figure 21.7, it is apparent that the nine responses are clustered into two groups. One cluster is located in the first quadrant and comprises all the dissolution responses. The other cluster is situated to the far right along the first model component. This latter group contains strength related responses. This group of responses includes the Trde response, although it is inversely correlated to the other three. Moreover, it can be tentatively concluded that the amounts of disintegrant and surfactant are very influential with respect to the dissolution responses. The greater the amount of these excipients, the faster the dissolution. For the strength responses, the factors compression force, API/filler ratio and amount of disintegrant exert the greatest influence.

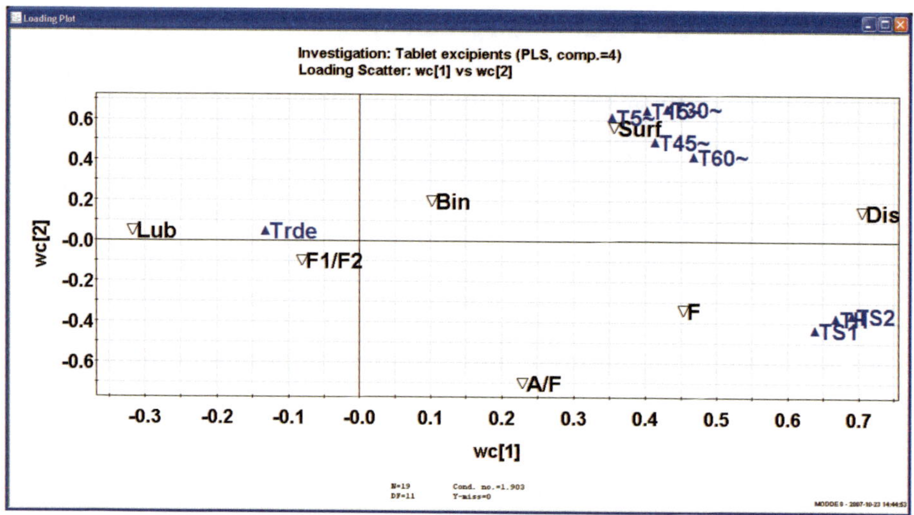

Figure 21.7: PLS loading plot of the first two components of the model. Open inverted triangles denote the factors. Solid triangles represent the responses.

When the responses are clustered into two sub-sets it is recommended that they be split and separate PLS models constructed for each group. Below, we describe what happened when this was done.

21.3.4 Separate PLS model for the strength responses

The PLS model for the strength responses (the first four responses of the original project) consisted of three components, i.e., one less than before. This indicates that the responses encompassed by the model are now more coherent. The performance statistics of the strength model are $R^2 = 0.82$; $Q^2 = 0.51$; Model Validity = 0.77; Reproducibility = 0.91, indicating a highly valid model (plot not shown). Figure 21.8 provides the corresponding information for the individual responses. The highest predictivity is found for true density, which can be explained because this response has no outliers. Experiment number 5 is a moderate outlier for the other three responses, accounting for a slight decay in the predictive ability. According to the original report, experiment number 5 had an unusually low tablet strength. Accordingly, it was re-run twice, only to demonstrate that the initial value was reasonable.

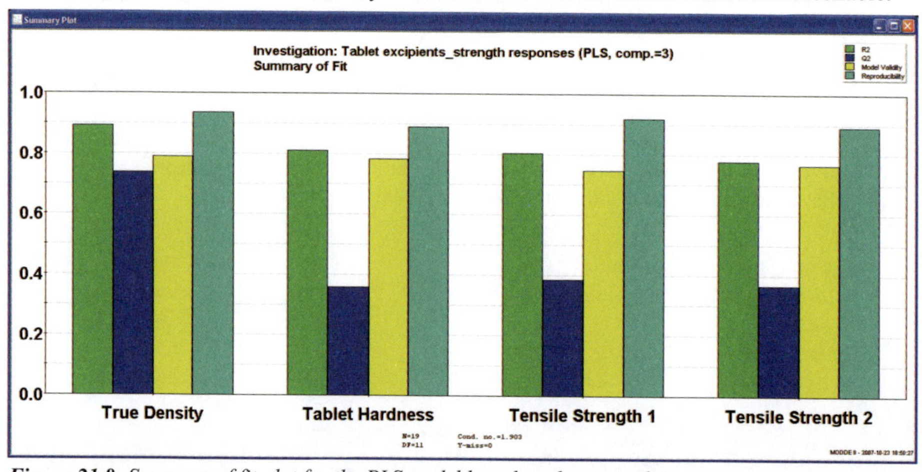

Figure 21.8: Summary of fit plot for the PLS model based on the strength responses.

Figure 21.9 shows the regression coefficients. The responses tablet hardness and tensile strengths 1 and 2 are highly correlated and, therefore, are affected by the factors in a similar way. In all three cases, the influences of the API/filler ratio and the amount of disintegrant are statistically significant, and higher factor values correspond to higher tablet strength. The compression force is the third strongest factor, but its effect on the responses is not statistically significant. For the first response (true density) the pattern is somewhat different. The API/filler ratio is still very important, but now, in addition, the filler 1/filler 2 ratio and the levels of binder, surfactant and lubricant are significant. The amount of disintegrant is no longer of relevance.

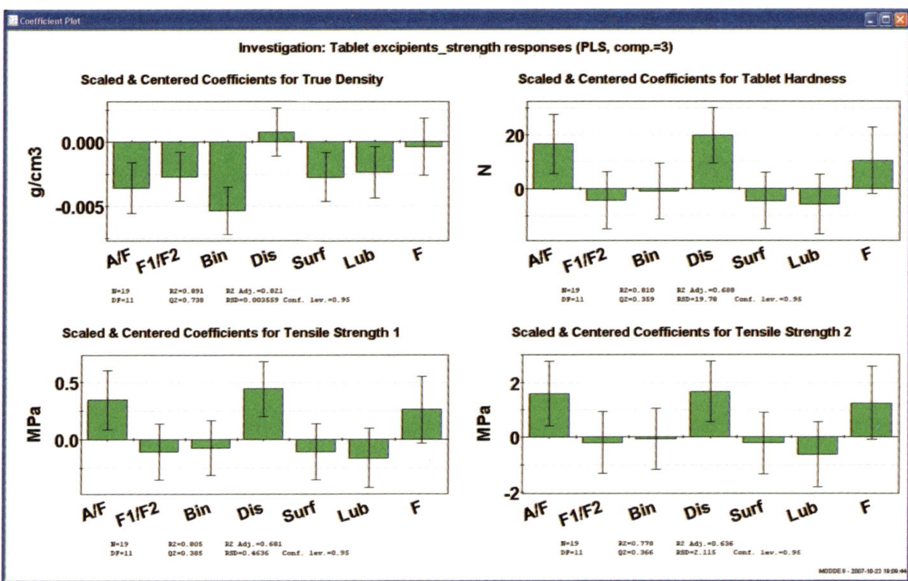

Figure 21.9: Regression coefficients of the PLS model tailored to the strength responses.

21.3.5 Separate PLS model for the dissolution responses

The PLS model for the dissolution responses (the last five responses in the original project) consisted of two components, the performance statistics of which are $R^2 = 0.60$; $Q^2 = 0.09$; Model Validity = 0.40; Reproducibility = 0.98 (plot not shown). Figure 21.10 provides the corresponding values for the individual responses. The highest predictivity is found for the T5 and T15 responses, where Q^2 is 0.43 and 0.26, respectively. A closer examination of the residuals does not indicate any outliers, so there must be some other reason lurking in the background to explain the comparatively poor performance of the PLS model.

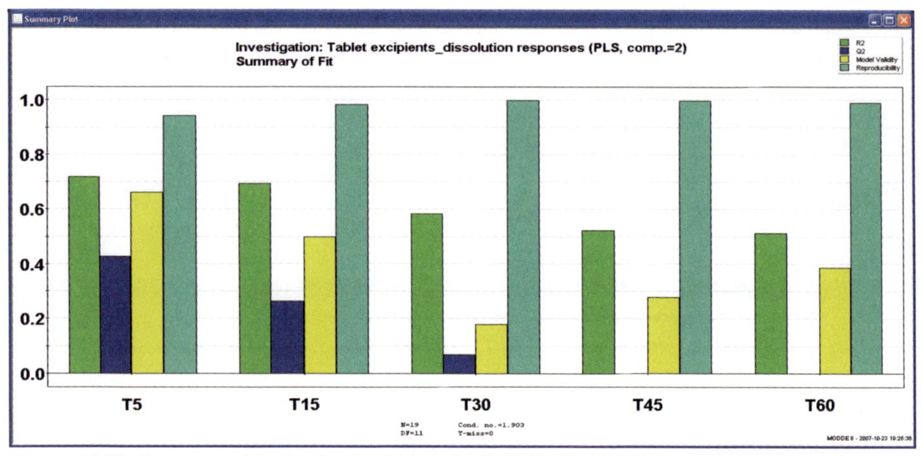

Figure 21.10: *Summary of fit plot for the PLS model based on the dissolution responses.*

An important clue about the weakness of the PLS model is found in the scatter plots of the PLS scores (Figure 21.11). There is a clear indication that the t_1/u_1 inner relationship for the first PLS component is not linear. In the second component, the relationship between the scores, t_2 and u_2, is predominantly linear. This suggests that the second component is merely a compensation component addressing the non-linearity that the first component was unable to accommodate.

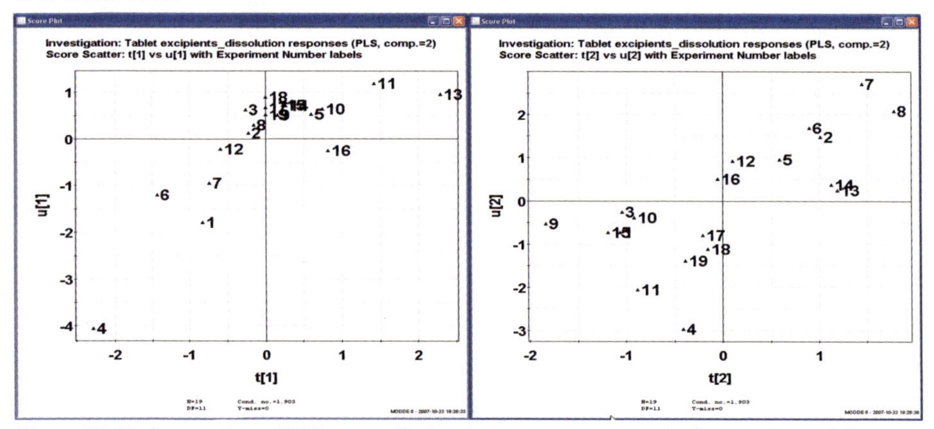

Figure 21.11: *Scatter plots of PLS scores. Each point represents one run in the experimental design.*

We suggested above that one likely reason for the model's shortcomings is non-linearity. Nonetheless, it is instructive to examine the regression coefficients. Figure 21.12 provides these for the two best modelled responses (T5 and T15) with the best predictions. Similar profiles were achieved for the other three responses, because of the strong inter-relatedness between the responses. Evidently, important factors for this model are the amount of disintegrant, the amount of surfactant and to some extent the API/filler ratio. To obtain rapid dissolution, a low ratio is preferable, associated with large amounts of the two excipients. However, remember that a low API/filler ratio will impact negatively on the tablet strength.

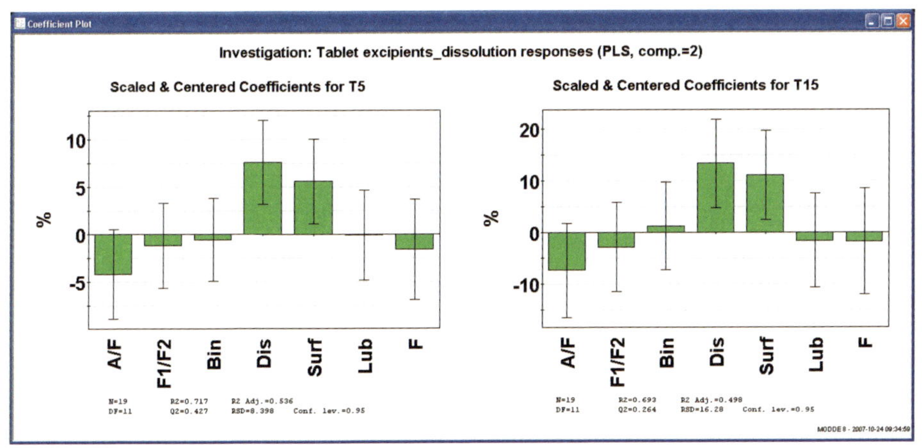

Figure 21.12: *Regression coefficients for the T5 and T15 dissolution responses.*

Figure 21.13 shows two scatter plots of the relationships between T5/T15 and the most influential factor, the amount of disintegrant. These two plots indicate a curved relationship. Such a curvature cannot be adequately handled by the linear model. The model needs to be augmented by more terms, i.e. quadratic terms, and possibly also some selected cross-terms. Thus, the experimenters decided to append more runs to the original screening design.

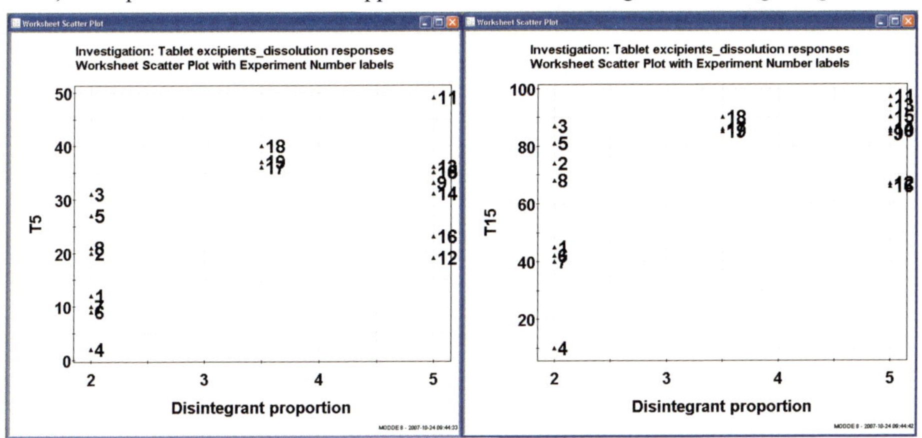

Figure 21.13: *Raw data plots for the T5 and T15 dissolution responses.*

21.3.6 Using linked responses to identify where to position new runs

Before conducting any new experiments it is necessary to identify appropriate settings for the factors. The investigators chose to use the optimizer, as there were conflicting demands with respect to the settings for the first factor, the API/filler ratio. A high ratio is beneficial for tablet strength, but a low ratio is favorable for rapid dissolution. Moreover, we must use the linked responses option in MODDE to enable a simultaneous optimization of all the nine responses fitted by the two separate PLS models. Remember that the functionality of the optimizer was described in Chapter 6.

In preparing the optimizer it is necessary to define acceptance criteria. Such criteria, also called desirabilities, were defined for two responses, one from each sub-cluster. For the strength responses, the target value of tablet hardness was set at 270 N, with lower and upper limits of 260 and 280 N. For the dissolution responses, the target value of T15 was set to 90% release, with 80% and 100% as the lower and upper limits. The other seven responses were not considered in the optimization.

Once completed, the clear message from the simplex optimizations was that the amount of disintegrant should be kept high. This is because the eight simplexes converged towards a point where the amount of disintegrant was at the upper end of its numerical range. For the other factors, the outcome was variable and not at all as clear-cut. Consequently, in the second optimization cycle, the settings for the proportion of disintegrant were modified by increasing the upper limit to 6.5% (instead of 5%). The outcome of the second optimization confirmed earlier findings, i.e., a high value of the proportion of the disintegrant was the most critical factor for maintaining rapid dissolution and high strength.

Based on the optimization results, it was decided to add two new experiments to the design (Figure 21.14). These were added to allow an estimate of Dis^2 and to resolve Dis*Surf from the cross-term with which it is confounded. In the original model, the Dis*Surf cross-term is confounded with the A/F*Lub cross-term. In the two extra runs, all factors were maintained at their center-level, except for the disintegrant proportion, which was allocated the new high setting, and the surfactant proportion, which was tested at both low and high levels (Figure 21.14). One interesting feature is that the factor disintegrant proportion is now examined at four levels, enabling its quadratic term to be rigorously determined.

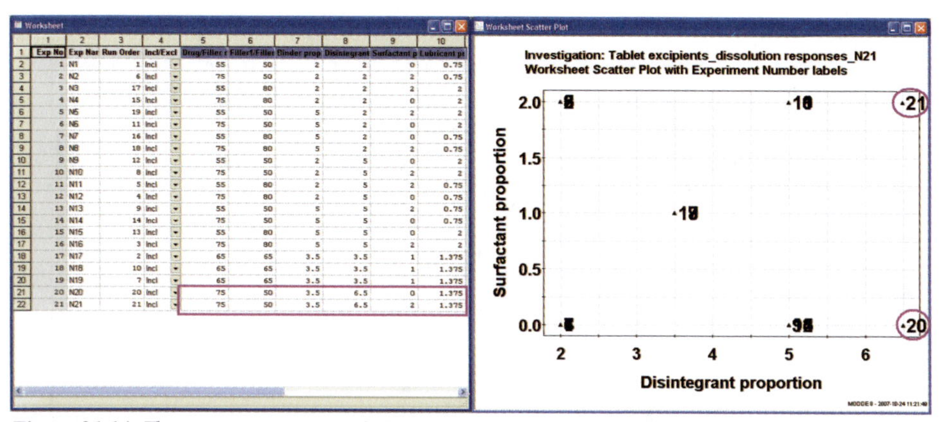
Worksheet

	1	2	3	4	5	6	7	8	9	10
1	Exp No	Exp Nar	Run Order	Incl/Excl	Drug/Filler r	Filler1/Filler	Binder prop	Disintegrant	Surfactant p	Lubricant pr
2	1	N1	1	Incl	55	50	2	2	0	0.75
3	2	N2	6	Incl	75	50	2	2	2	0.75
4	3	N3	17	Incl	55	80	2	2	2	2
5	4	N4	15	Incl	75	80	2	2	0	2
6	5	N5	19	Incl	55	50	5	2	2	2
7	6	N6	11	Incl	75	50	5	2	0	2
8	7	N7	16	Incl	55	80	5	2	0	0.75
9	8	N8	18	Incl	75	80	5	2	2	0.75
10	9	N9	12	Incl	55	50	2	5	0	2
11	10	N10	8	Incl	75	50	2	5	2	2
12	11	N11	5	Incl	55	80	2	5	2	0.75
13	12	N12	4	Incl	75	80	2	5	0	0.75
14	13	N13	9	Incl	55	50	5	5	2	0.75
15	14	N14	14	Incl	75	50	5	5	0	0.75
16	15	N15	13	Incl	55	80	5	5	0	2
17	16	N16	3	Incl	75	80	5	5	2	2
18	17	N17	2	Incl	65	65	3.5	3.5	1	1.375
19	18	N18	10	Incl	65	65	3.5	3.5	1	1.375
20	19	N19	7	Incl	65	65	3.5	3.5	1	1.375
21	20	N20	20	Incl	75	50	3.5	6.5	0	1.375
22	21	N21	21	Incl	75	50	3.5	6.5	2	1.375

*Figure 21.14: The two new runs appended to the screening design. With these added, it becomes possible to calculate Dis^2 and the Dis*Surf two-factor interaction, the latter with less disturbance from A/F*Lub.*

21.3.7 Updating the model for the dissolution responses

Following the addition of the two new experiments, the PLS model for the dissolution responses was refined by including Dis^2 and Dis*Surf, and removing all linear terms except A/F, Dis and Surf (Figure 21.15). The net result is a much enhanced model for all five Y-variables (compare Figures 21.10 and 21.15).

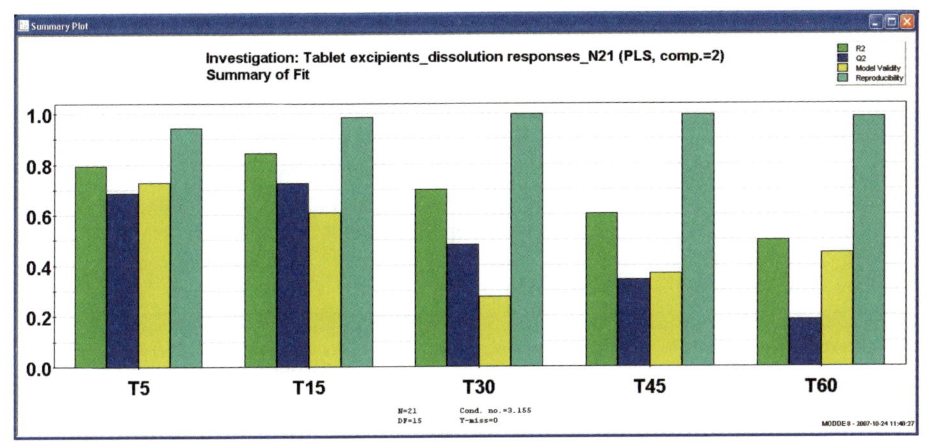

Figure 21.15: Summary of fit plot for the refined PLS model. Compare with Figure 21.10.

The relevance of the cross-term and quadratic term added to the model of the dissolution responses can be seen from the coefficients plotted in Figure 21.16. The confidence intervals for the coefficients are narrower, compare Figures 21.12 and 21.16, indicating that a stabilization of the model has taken place.

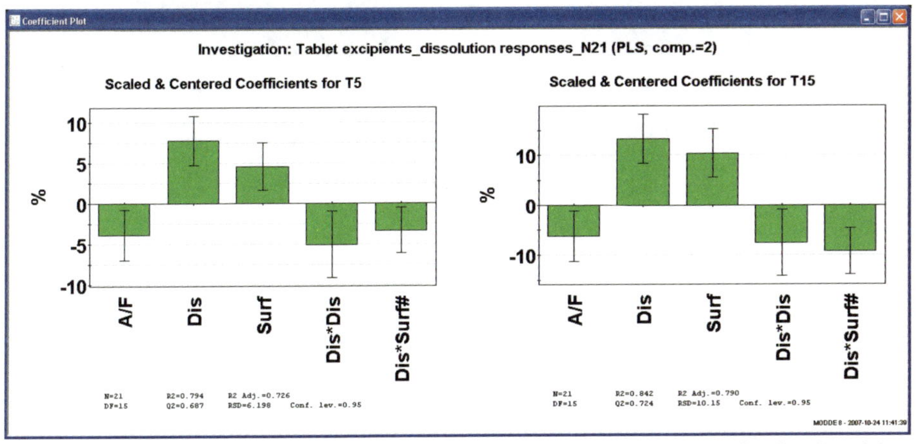

Figure 21.16: Regression coefficients for the T5 and T15 dissolution responses. Refined PLS model. Compare with Figure 21.12.

21.3.8 Conclusions from the study

The goal of the study was to screen a number of formulation factors in order to (a) identify the most important ones and (b) propose a factor combination resulting in a strong tablet with rapid dissolution. Two key responses were highlighted by specifying desirabilities for them: for tablet hardness the target was 270 N and for T15 the desired dissolution was 90%. One main conclusion drawn from the initial PLS model for all responses was that the responses were grouped into two clusters. Accordingly, the responses were divided into two sub-sets, which were modelled separately. For the four responses mirroring tablet strength, the default linear model was found to be adequate. However, this was not the case for the five dissolution responses, the main reason for this being a suspected quadratic dependence between the factors and the responses. The two models helped to uncover conflicting demands with respect to the first factor, the API/filler ratio.

The simultaneous optimization of all nine responses, using the linked responses functionality in MODDE, suggested that a key to achieving a compromise between the conflicting demands would be to increase the amount of disintegrant beyond its upper level of 5%, using 6.5% instead. Consequently, two new experiments were performed with the goal of eventually improving the dissolution model. The refined dissolution model *was* also improved, due mainly to the introduction of the Dis^2 and Dis*Surf terms, which could not be estimated from the original design. Figure 21.17 shows contour plots of the final models, based on the extended range for the disintegrant proportion factor. It can be seen that by avoiding the maximum value for API/filler, but still staying on the high side, there is a region where we are close to the goal (Tablet hardness around 270, T15 around 90%).

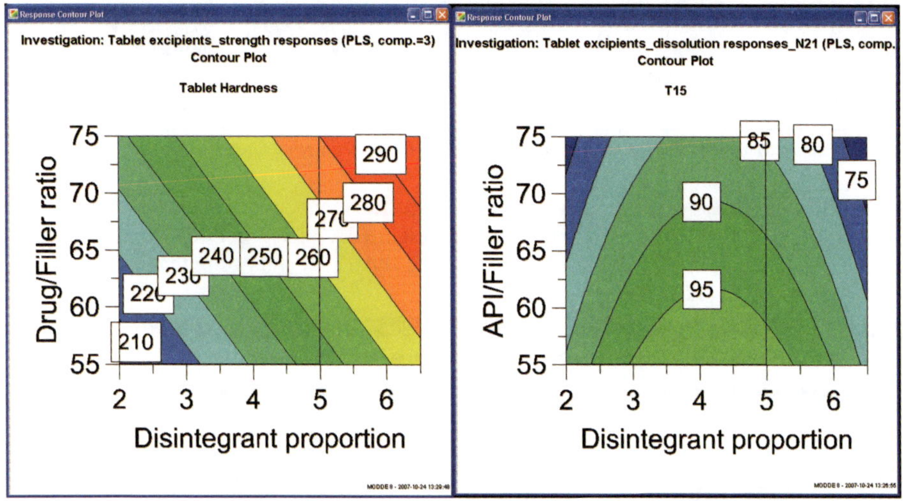

Figure 21.17: *Contour plots showing the change in two responses as a function of changing values for the factors disintegrant proportion and API/filler ratio. (left) Tablet Hardness. (right) T15, dissolution after 15 minutes.*

A final observation was that the uncontrollable factor, compression force, was only slightly correlated with the six properly designed factors. Hence, its presence in the design would perturb an otherwise ideal configuration of the experiments only a little. This means we can be highly confident in the results obtained.

21.4 DOE – A smorgasbord of opportunities for the formulator

When planning a formulation application we recommend that a fishbone or Ishikawa diagram is constructed. This tool, introduced in Section 5.3.2, is a system diagram that helps to provide an overview of the factors that might *potentially* influence the critical quality attributes of the product. Figure 21.18 provides an example fishbone diagram related to a formulation investigation. Correctly used, the fishbone diagram will decrease the risk of overlooking important factors. Ideally, all DOE projects in formulation development should begin with a complete fishbone diagram with all factors and responses listed for all steps. Thus, DOE requires the investigator to select factors and responses that are relevant for the stated purposes. When all factors have been listed, their ranges must be defined. This is accomplished by determining the lower and upper limits for the investigation values for each factor. Usually, comparatively large ranges are investigated in screening, because one does not want to run the risk of overlooking a meaningful factor.

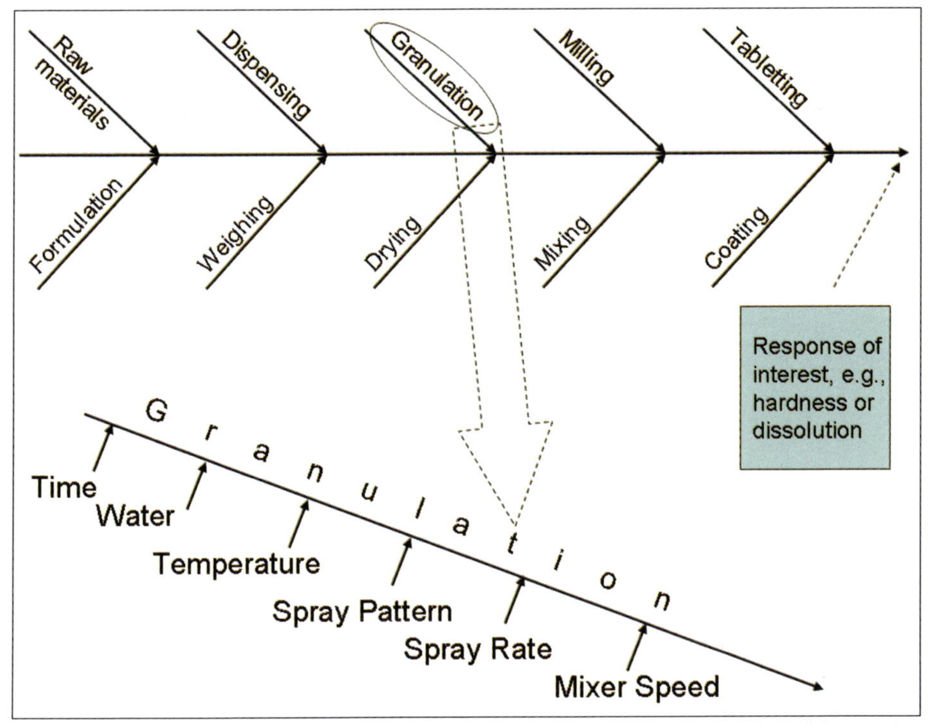

Figure 21.18: *A fishbone diagram is a system diagram within which it is possible to list all factors that might affect a response variable. This tool was introduced in Section 5.3.2.*

Most material attributes and process factors that are identified via the fishbone diagram can be varied using classical design protocols for screening, optimization and robustness testing. However, for the formulation recipe, the design problem can be specified either as a mixture or a process design problem. Mixture design, as applied to a tablet formulation example, was discussed in Chapter 11. In a mixture design, the amount of each excipient is expressed as a proportion between 0 and 1. With this approach we are interested in the effect of each excipient on the responses. Alternatively, if there is a bulk excipient or a solvent in a mixture, a so-called filler, whose impact on the responses is less interesting *per se*, the mixture problem can be re-expressed as a problem to which some kind of process design is applicable. The amount of each important excipient is then expressed in relation to the filler in terms of one or more ratio variables. Thus, the presence of the designated filler allows the mixture factors to be manipulated independently. We emphasize that this is one of a number of possible ways to switch from the mixture design situation to the process design situation. Such ratio variables were used in the above example.

Moreover, setting our sights slightly higher, it is possible to develop experimental designs that at the same time encode both type *and* amount of excipient (see illustration in Figure 21.19). This approach, pioneered by Jon Gabrielsson (Multivariate Methods in Tablet Formulation, Ph.D. Thesis, Umeå University, Sweden, 2004), is based on multivariate design, and yields a diverse excipient selection among a multitude of classes of excipients. Multivariate design was discussed in Chapter 19.

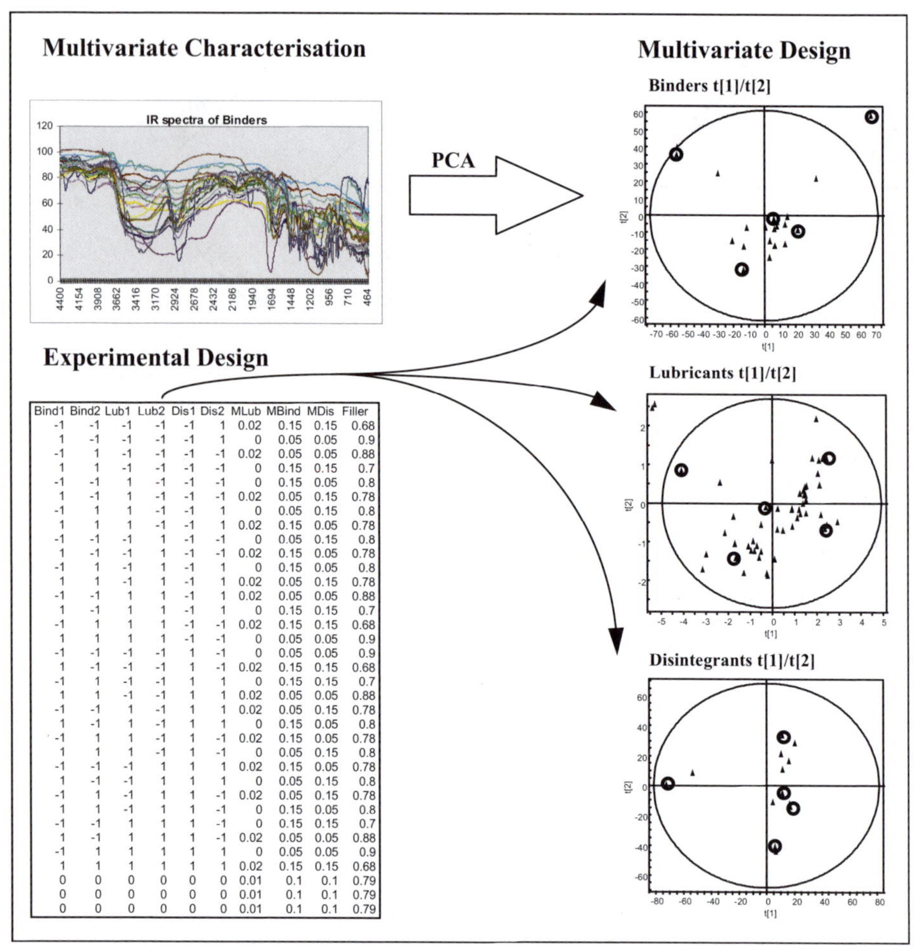

Bind1	Bind2	Lub1	Lub2	Dis1	Dis2	MLub	MBind	MDis	Filler
-1	-1	-1	-1	-1	1	0.02	0.15	0.15	0.68
1	-1	-1	-1	-1	1	0	0.05	0.05	0.9
-1	1	-1	-1	-1	-1	0.02	0.05	0.05	0.88
1	1	-1	-1	-1	-1	0	0.15	0.15	0.7
-1	-1	1	-1	-1	-1	0	0.15	0.05	0.8
1	-1	1	-1	-1	-1	0.02	0.05	0.15	0.78
-1	1	1	-1	-1	1	0	0.05	0.15	0.8
1	1	1	-1	-1	1	0.02	0.15	0.05	0.78
-1	-1	-1	1	-1	-1	0	0.05	0.15	0.8
1	-1	-1	1	-1	-1	0.02	0.15	0.05	0.78
-1	1	-1	1	-1	1	0	0.15	0.05	0.8
1	1	-1	1	-1	1	0.02	0.05	0.15	0.78
-1	-1	1	1	-1	1	0.02	0.05	0.05	0.88
1	-1	1	1	-1	1	0	0.15	0.15	0.7
-1	1	1	1	-1	-1	0.02	0.15	0.15	0.68
1	1	1	1	-1	-1	0	0.05	0.05	0.9
-1	-1	-1	-1	1	-1	0	0.05	0.05	0.9
1	-1	-1	-1	1	-1	0.02	0.15	0.15	0.68
-1	1	-1	-1	1	1	0	0.15	0.15	0.7
1	1	-1	-1	1	1	0.02	0.05	0.05	0.88
-1	-1	1	-1	1	1	0.02	0.05	0.15	0.78
1	-1	1	-1	1	1	0	0.15	0.05	0.8
-1	1	1	-1	1	-1	0.02	0.15	0.05	0.78
1	1	1	-1	1	-1	0	0.05	0.15	0.8
-1	-1	-1	1	1	1	0.02	0.15	0.05	0.78
1	-1	-1	1	1	1	0	0.05	0.15	0.8
-1	1	-1	1	1	-1	0.02	0.05	0.15	0.78
1	1	-1	1	1	-1	0	0.15	0.05	0.8
-1	-1	1	1	1	-1	0	0.15	0.15	0.7
1	-1	1	1	1	-1	0.02	0.05	0.05	0.88
-1	1	1	1	1	1	0	0.05	0.05	0.9
1	1	1	1	1	1	0.02	0.15	0.15	0.68
0	0	0	0	0	0	0.01	0.1	0.1	0.79
0	0	0	0	0	0	0.01	0.1	0.1	0.79
0	0	0	0	0	0	0.01	0.1	0.1	0.79

Figure 21.19: *Schematic of a combined design, simultaneously encoding type and amount of excipient. This illustration was provided by Jon Gabrielsson, for which we are grateful.*

An interesting aspect of the Gabrielsson's approach is that the data underlying the excipient selection can reflect both physical properties and spectroscopic properties of the excipients. The joint information content in these descriptor variables is summarized by means of the principal properties derived from a Principal Component Analysis (PCA, see Chapters 13 and 19). As concluded by Gabrielsson, "using principal properties and multivariate design is a viable alternative if many excipients are to be included in a screening design".

Moreover, not only the type and the amount of excipient will impact on the final product quality, but also the raw material characteristics might influence the product performance. Figure 21.20 shows an example from the pharmaceutical industry. In this particular case, there were three vendors of the same excipient. All incoming batches during a period of approximately two years were characterized using particle size distribution measurements. The resulting curves, based on standard particle size distributions between 2.5 and 150 μm, are plotted in Figure 21.20. As shown in the figure, there is considerable variation between the different batches and suppliers of the raw material.

In total, there are 303 variables, from the curves, that convey information about the physical properties of the raw material batches. PCA can be used to summarize such data. In this case, PCA reveals strong groupings among the three vendors. In particular, it shows the batches from the vendor denoted L2 – marked by the solid line in Figure 21.20 – to have been highly variable, and thus the quality of the raw material delivered by this vendor to the client (the pharmaceutical company) was unreliable.

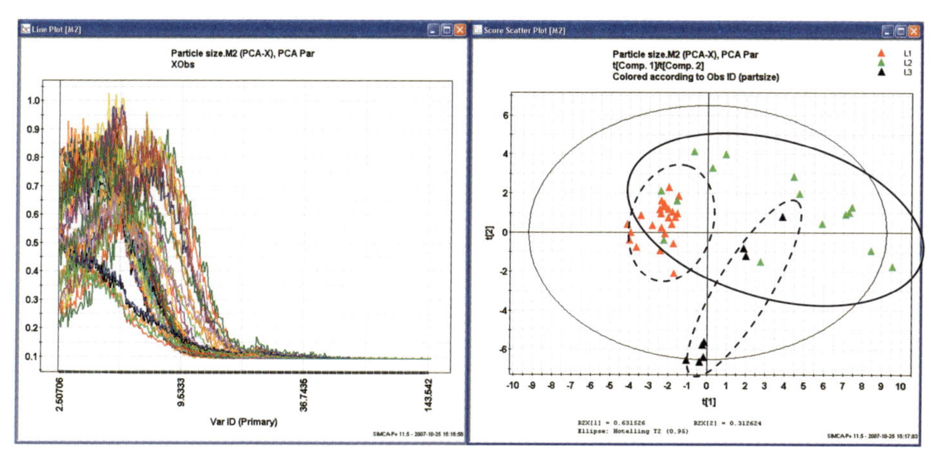

Figure 21.20: *Example of how to characterize and account for raw material variability. (left) Particle size distribution curves. (right) PCA score plot from the analysis of the curves. The three different vendors are indicated by ellipses. Raw material from vendor L2 (solid line) has the highest variability in quality. Representative raw material batches can be selected using e.g. onion design, which was discussed in Chapter 13.*

The approach we recommend here is to use the score values from the PCA as factors in an experimental design to take account of this variability in raw material properties. Such factors can then be combined with other factors in terms of a D-optimal design. Due to the flexibility of the D-optimal approach, the user may select any desired number of samples, with a good spread of the raw material properties and other factors, for further optimization work. It should also be realized that the physical and chemical characterization of the raw materials can be accomplished using other types of data as well, for instance NIR or measurements of viscosity, rheological properties, and so on. The crucial aspect is perhaps not so much which technique is selected for raw material characterization, but rather the fact that this is done in some way and that it allows raw material variability to be quantitatively encoded into the DOE protocol of the formulation development process.

21.5 Questions for Chapter 21

- How can changes in a formulation blend be encoded using mixture design?
- How can changes in a formulation blend be encoded using process design?
- Which designs allow simultaneous changes among both the type and amount of excipients?
- How can variability in raw materials be taken into account?
- How can variability among suppliers of the same raw material be taken into account?
- What is the role of the fishbone diagram?

21.6 Summary and discussion

Design of experiments is a viable technique in formulation development. Using DOE technology, the formulator is guided to address the problem at hand in a sequential manner, thereby helping to build quality into the products, so-called quality by design (QBD). As discussed in Chapter 2, this sequential approach to experimental work and changes to the production environment involves three stages: (i) screening, (ii) optimization and (iii) robustness testing. In the first stage, the main objective is to sort out the dominating factors and their appropriate ranges. Thereafter a detailed optimization design is constructed, so that the sweet spot area of the system or process can be rapidly identified (see Chapter 7 for a discussion on using the sweet spot plot). Conceptually, this sweet spot area can be thought of as representing a window of operation for the system or process, in which all quality demands are met. The final stage then consists of performing a robustness testing design on the sweet spot area. The latter can be seen as a final test to verify that the system or process is running in a stable manner and is not sensitive to small changes in the critical factors.

As mentioned in the introduction, the major advantage of DOE is that it enables all potential factors to be explored in a systematic fashion. With this approach the formulator can investigate the effect of all factors and their interactions. Once the critical factors have been identified through regression analysis, the optimal formulation can be identified, consisting of the best mix of excipients, all available in just the right proportions. This is one of the greatest assets of DOE, i.e., that it is possible for the formulator, in a single experimental design, to determine the optimal choice in each excipient category and the optimal level of each excipient. Furthermore, the manufacturing process can also be enhanced and optimized in the same manner. When the formulation recipe and the manufacturing process have been screened, optimized and fine-tuned using a systematic approach based on DOE and MVDA, issues such as scale-up and process validation may be addressed very efficiently because of the comprehensive understanding that will be acquired of the entire formulation development environment.

22 DOE in PAT

22.1 Objective

Chapter 22 discusses the use of DOE in Process Analytical Technology, PAT. PAT has its origins in an initiative of the US Food and Drug Administration, FDA, first launched in mid-2002 (http://www.fda.gov/cder/ops/pat.htm). The goal of PAT is to improve the understanding and control of the entire pharmaceutical manufacturing process. One way of achieving this is through timely measurements of critical quality and performance attributes of raw and in-process materials and processes, combined with multivariate data analysis (MVDA). This needs to be coupled with Design of Experiments (DOE) to maximize the information content in the measured data. A central concept within the PAT paradigm is that quality should arise as a result of design-based understanding of the processes, rather than merely by aiming to generate products that meet minimum criteria within defined confidence limits, and rejecting those that fail to meet the criteria. This, in turn, implies the need for DOE.

22.2 Introduction

Chapters 19-21 described the use of DOE in synthetic organic chemistry (19), assay improvement (20) and formulation development (21). As shown by the examples reviewed in these chapters, there is no doubt about the relevance and usefulness of DOE in research and development within the pharmaceutical and biotechnological industries. If they are to maintain a competitive edge, pharmaceutical and biotechnological companies are strongly adviced to use properly designed experiments that provide the maximum amount of information as cost-effectively as possible. DOE is a key element of rational scientific and technological progress.

In the last main chapter of this textbook we examine DOE from a broader, overall perspective, considering the potential use of DOE across major stages of pharmaceutical and biotechnological manufacturing; a series of potential applications collectively incorporated under the umbrella name of Process Analytical Technology, PAT.

22.2.1 The PAT initiative

There are many objectives associated with the PAT concept, but the foremost goal is to improve the *understanding* and *control* over the entire pharmaceutical or biotechnological manufacturing process. Briefly, PAT can be understood as a framework of tools and technologies for accomplishing this goal. Interestingly, the US FDA defines *process understanding* as: the identification of critical sources of variability, management of this variability by the manufacturing process, and the ability to accurately and reliably predict quality attributes. DOE and MVDA are clearly needed to achieve this, together with insight, process knowledge, and relevant measurements.

22.2.2 What can PAT accomplish?

A common question is "what can be accomplished with PAT?" The answer to this is multi-faceted. In part, this is because PAT means different things to different people. Its answer may also depend on whether one is involved in academia, industry or a regulatory agency. Moreover, the connotations associated with PAT depend to a great extent on the context, e.g. whether it is to be applied to an existing process (continuous or batch) or a process still under development. The manner in which we (i.e. the authors) interpret PAT and the use of MVDA and DOE in PAT is outlined below in Section 22.3.

Recently, Johnson and co-authors published an article dealing with applications of MVDA in the biotechnology industry (Johnson, R., et al., Applications of Multivariate Data Analysis in Biotech Processing, BioPharm International, October 2007, pages 130-144). Primarily, the article discusses and illustrates how MVDA tools can aid in solving problems encountered in biotechnological processing. However, the content and suggestions presented in this article also have a bearing on the pharmaceutical industry. The authors highlight four application areas that are typical of PAT in general, and where MVDA combined with DOE has added value to production:

- *Optimization of large-scale production.* In a commercial antibody production process there were problems with reductions in productivity and increases in process variability. A combined analysis of historical and current data led to the identification of several influential process variables (initial concentrations, timings, etc.). Appropriate changes to these process variables were implemented incrementally via a new experimental campaign – DOE was of course an ideal approach for planning such changes. The final result was a 10% increase in productivity, no loss of quality and a 15% decrease in process variability.
- *Establishment of process comparability and trouble-shooting.* MVDA was used to model data from small-scale (2 L) and large-scale (2000 L) cell culture batches. The data-set included many inputs (pCO_2, PO_2, glucose, pH, lactate, ...) – varied according to DOE – and outputs (purity, cell density, viability, osmolality, …) which were related using PLS. Separate models were constructed for the 2L and 2000L data in order to evaluate process comparability across scales. PLS loadings and VIP plots were used to identify variables contributing similar information at the two scales, and to highlight critical variables for which the impact on the process differed at the two scales. It was discovered that a few inputs influenced the responses differently at the two scales. Therefore, the models developed resulted in not just a better understanding of the production process, but assisted in trouble-shooting issues encountered during scale-up.
- *Routine monitoring of manufacturing processes.* In a protein purification process, eight process variables were identified (harvest volume, harvest amount, β-sepharose recovery, overall recovery, …). A training set consisting of 100 observations was used together with PCA to define a process model. After outlier detection and removal, the updated process model was used for monitoring the manufacturing process. This monitoring indicated how the current process conditions compared with the previous conditions and provided early warning of changes in process behaviour. Understanding the root causes of such process drifts provides opportunities to further improve the process, i.e., provides the basis for a subsequent DOE.
- *Raw material characterization and screening.* In a study involving cell growth media, lipoprotein fraction batches from three vendors were characterized analytically: 16 analytes were determined reflecting lipid profiles, fatty acids and lipoproteins, then PCA was applied to the raw material characterization data. Each supplier was found to

deliver material with stable quality, which did not vary much from one batch to another. However, the composition of the raw material supplied by the three vendors was found to differ substantially, since the individual batches from the three vendors were strongly grouped, forming a triangular pattern in the resultant PCA score plot. The loading plot indicated that the main determinants for separating the three vendors were fatty acid composition, phosphatidyl content and the lipoprotein profile. Moreover, cell culture tests were performed and significantly less titer was produced from batches from one of the suppliers. PLS analysis of the relationships between the analytical characterization data and the titer resulted in a better understanding of the chemical composition of the lipoprotein batches that produced more titer.

As with the biotechnological applications briefly described above, MVDA and PAT may be applied to important unit operations in the pharmaceutical industry. In our experience, the most common applications of PAT relate to on-line monitoring of the blending, drying and granulation steps. Here, the measurement and analysis of multivariate spectroscopic data are of central importance. These spectroscopic data form the so-called X-matrix, and if there are response data (Y-data), the former can be related to the latter using PLS or OPLS to establish a multivariate calibration model (see also Section 22.3). However, MVDA should not be regarded as the only important approach; DOE should also be considered. DOE is very useful for defining optimal and robust conditions in such applications.

22.2.3 What are the benefits of using DOE?

DOE is complementary to MVDA. They form two strong links in a chain. In DOE the findings of a multivariate PCA, PLS or OPLS model are often used as the point of departure, because such models highlight which process variables have been important in the past. Systematic and simultaneous changes in such variables may then be induced using an informative DOE protocol. In addition, the controlled changes to these important process factors can be supplemented with measurements of more process variables which cannot be controlled. Hence, when DOE and MVDA are used in conjunction, there is the possibility of analyzing both designed and non-designed process factors at the same time, along with their putative interactions. This allows us to see how they jointly influence key production attributes, such as the cost-effectiveness of the production process, and the amounts and quality of the products obtained.

Generally, DOE is used for three main experimental objectives, regardless of scale. The first of these is screening. Screening is used to identify the most influential factors, and to determine the ranges across which they should be investigated. This is a straightforward aim, so screening designs require relatively few experiments in relation to the number of factors. Sometimes more than one screening design is needed.

The second experimental objective is optimization. Here, the interest lies in defining which approved combination of the important factors will result in optimal operating conditions. Since optimization is more complex than screening, optimization designs demand more experiments per factor. With such an enriched body of data it is possible to derive quadratic regression models.

The third experimental objective is robustness testing. Here, the aim is to determine how sensitive a product or production procedure is to small changes in the factor settings. Such small changes usually correspond to fluctuations in the factors occurring during a "bad day" in production, or the customer not following the instructions for using the product. In the reporter gene assay example discussed in Appendix I, the point of departure for the robustness testing design was the optimal point identified by applying an optimization design. The ranges of the investigated factors corresponded to relevant variations that might be considered when running this assay on a routine basis.

The great advantage of using DOE is that it provides an organized approach, with which it is possible to address both simple and tricky experimental and production problems. The experimenter is encouraged to select an appropriate experimental objective, and is then guided through devising and performing an appropriate set of experiments for the selected objective. It does not take long to set up an experimental protocol, even for someone unfamiliar with DOE. What is perhaps counter-intuitive is that the user has to complete a whole set of experiments before any conclusions can be drawn.

The rewards of DOE are often immediate and substantial, for example higher product quality may be achieved at lower cost, and with a more environmentally-friendly process performance. However, there are also more long-term gains which might not be apparent at first glance. These include a better understanding of the investigated system or process, greater stability, and greater preparedness for facing new, hitherto unforeseen challenges.

22.2.4 The emergence of new concepts – QBD and the Design Space

The importance of DOE to the pharmaceutical industry is underlined by the growing interest in the Quality by Design (QBD) concept. QBD is a systematic way of addressing pharmaceutical development. It should be appreciated that it is both a drug product design strategy *and* a regulatory strategy for continuous improvement. QBD is based on developing a product that meets the needs of the patient and fulfils the stated performance requirements. QBD emphasizes the importance of understanding the influence of starting materials on product quality. Moreover, it also stresses the need to identify all critical sources of process variability and to ascertain that they are properly accounted for in the long run.

As mentioned above, the major advantage of DOE is that it enables all potential factors to be explored in a systematic fashion. Let us take, as an example, a formulator, who will be operating in a certain environment, with access to certain sets of equipment and raw materials that can be varied across certain ranges. With DOE, the formulator can investigate the effect of all factors that can be varied and their interactions. This means that the optimal formulation can be identified, consisting of the best mix of all the available excipients in just the right proportions. Furthermore, the manufacturing process itself can be enhanced and optimized in the same manner. When the formulation recipe and the manufacturing process have been screened, optimized and fine-tuned using a systematic approach based on DOE and MVDA, issues such as scale-up and process validation can be addressed very efficiently because of the comprehensive understanding that will have been acquired of the entire formulation development environment.

Thus, one objective of QBD is to encourage the use of DOE. The view of this approach is that, once appropriately implemented, DOE will aid in defining the so-called "design space" of the process. Inside the "design space", there is a region, or window, where the process meets all the specifications of the product. Outside this window of operation, there can be problems with one or more attributes of the product.

22.2.5 Structure of the remainder of the chapter

In the remaining part of this chapter we outline the benefits of DOE in PAT. To do this, it is first necessary to provide a brief overview of what PAT is, from the perspective of MVDA and DOE. This introduction is *not* an exhaustive account of the topic. A more detailed description of PAT can be found in the companion textbook: Eriksson, L., et al., Multi- and Megavariate Data Analysis, Part I, Chapter 16, Umetrics Academy.

22.3 Four levels of PAT

In this section a brief overview of PAT – as seen through the eyes of a multivariate analyst – will be presented. Basically, for PAT to work properly, as in any other inductive modelling, the data-set used for model training must be representative of the system or process when it is running under normal operating conditions. With access to such representative data it is possible to develop a multivariate process model through which the region of data space corresponding to an acceptable process can be defined. Conceptually, this smaller region, situated inside the total multivariate process data space, can be regarded as a window of operation for a stable and robust process. The limits of this window of operation are determined by the tolerance volume of the multivariate process model. A stringent definition of the size of the process tolerance volume is provided by the Hotelling's T^2 and DModX diagnostics.

From a chemometrics perspective, PAT can be regarded as having four levels of complexity:

- Level 1 – Multivariate (MV) off-line calibration
- Level 2 – At-line multivariate PAT for MV calibration or classification
- Level 3 – Multivariate on-line PAT for a single process step
- Level 4 – Multivariate on-line PAT for the entire process

It should be noted that each level, to a greater or lesser extent, relies on the principles of quality by design (QBD), for which DOE is one corner-stone. One of the main objectives of QBD is to minimize risk and establish the design space for the product and the production line. Risk minimization is discussed in Section 22.4, and the concept of the design space is further considered in Section 22.5.

22.3.1 Level 1 – Multivariate off-line calibration

At the first level of PAT, off-line chemical analyses in separate laboratories are usually substituted for at-line analysis, in which a combination of rapid measurements (e.g. using appropriate types of spectroscopy, fast chromatography, imaging and/or sensor arrays) and multivariate calibration is used. The latter converts PAT data into traditional representations, e.g. concentrations, disintegration rates, dissolution times, and so on. Typical applications include determinations of:

- The active pharmaceutical ingredient (API) concentration in the final product or its intermediates (Figure 22.1).
- Concentrations of impurities in the final product and in important intermediates.
- Moisture contents of samples after drying.
- Different crystal forms in API samples, tablets, etc.

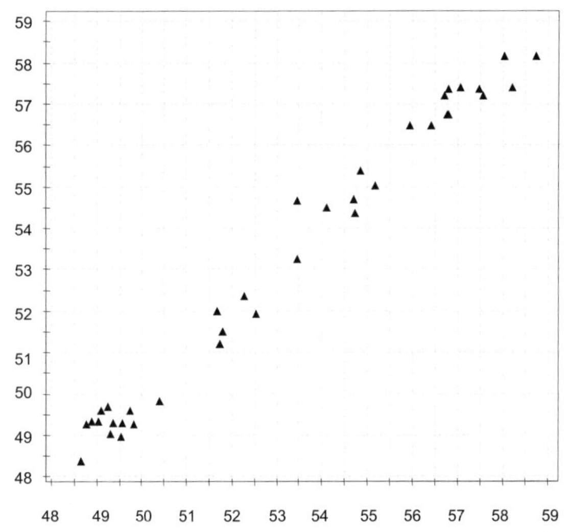

Figure 22.1: Relationship between observed percentage of active substance (abscissa) and predicted percentage of active substance (ordinate) in a pharmaceutical production situation.

There are many uses of DOE in multivariate calibration. One major application is for designing training sets that are representative of process variations that may occur over time. We can construct designs in both the X-space and the Y-space. Design in the Y-space is associated with introducing systematic changes in the levels of the main analytes, while design in the X-space is usually done to introduce variations in secondary factors such as ambient temperature and humidity, and raw material composition. The latter is accomplished using multivariate characterization of the raw material, PCA of the measured data, and coupling the PCA scores to a multivariate design. This is similar in scope to the example of raw material characterization discussed in Section 22.2.2. The principles of multivariate design were discussed in Chapter 19.

22.3.2 Level 2 – At-line multivariate PAT for classification and/or calibration

PAT level 2 includes PAT level 1. At this level, the at-line PAT data are used directly without conversion to traditional numerical representations. The objective is often to classify an intermediate or product (Figure 22.2), possibly adhering to a specification, with the same additional objectives as PAT level 1, i.e. to reduce production time and enhance accuracy. DOE is used in a similar manner as in level 1.

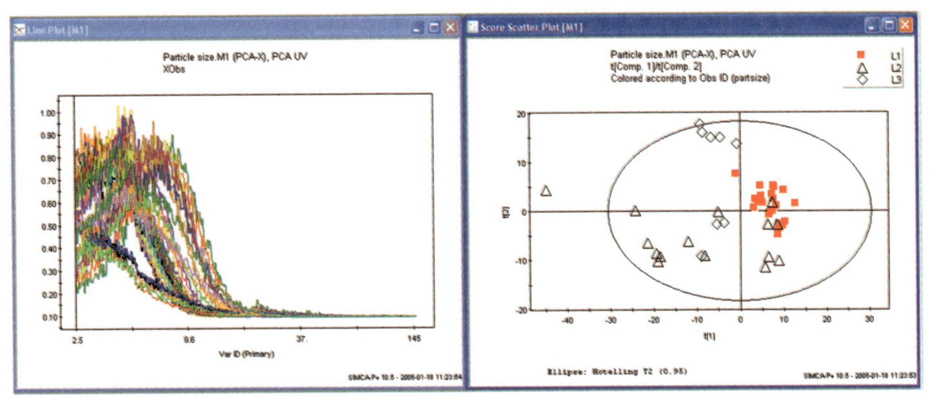

Figure 22.2: (left) An illustration of raw material characterization based on particle size distribution measurements. The data come from a pharmaceutical company and relate to all incoming batches during approximately two years of production. There is a total of 300+ variables corresponding to particle size intervals ranging from 2.5 to 150 µm. (right) PCA score plot from the analysis of the particle size distribution curves. There were three vendors of the excipient, and the plot is symbol- and color-coded according to vendor. The score plot reveals that supplier L1 provided the most homogeneous and consistent starting material, with little quality variation over time.

22.3.3 Level 3 – Multivariate on-line PAT on a single process step

Here, data are also measured during each production step, in order to form the basis for on-line MSPC and BSPC. At this level, which incorporates levels 1 and 2, the objective is real-time process monitoring, and all available data are used to determine whether the batch process is running normally, and whether it has reached normal completion (Figure 22.3). The latter is related to end-point detection, in which on-line PAT data are used to determine the degree of completion in, for instance, synthesis of the API, a granulator, a dryer, or a tablet-coating step.

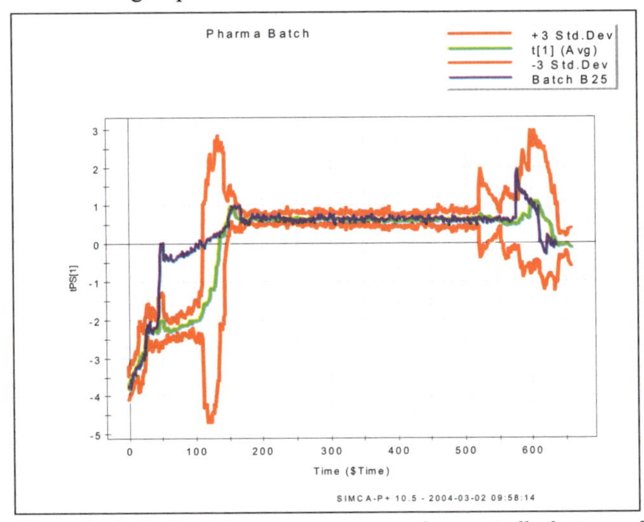

Figure 22.3: On-line PAT: Data are measured sequentially during each production step in on-line BSPC (Batch Statistical Process Control). Batch progress is conveniently monitored in terms of a BSPC control chart. Here we see a batch that has gone out of control (OOC) at around time point 50.

DOE is used here to establish a workable and reliable multivariate process model. The starting point is access to ample historical data about good batches, combined with process knowledge and experience. With this as a foundation, DOE is then used to set up a design in the important factors (raw material attributes, process variables and their trajectory shapes). This may lead to more than 30 additional batches being run, until completion, and classified into good or bad batches. The area or volume in the multivariate process space corresponding to the window of operation for the good batches can then be defined by multivariate data analysis of the data relating to the good batches.

22.3.4 Level 4 – Multivariate on-line PAT for the entire process

The final level, which incorporates levels 1–3, involves integrating data from all process steps and raw materials to obtain an overview of the total process; an overall process fingerprint (Figure 22.4). This integrated knowledge is used for developing models of the process variation and its influence on product quality. The objectives include using process understanding to drive process improvement and to discern, in real-time, whether or not the final product is within specifications, and thus can be shipped. This concept of parametric release is one major benefit of PAT as outlined by the FDA. The rationale is that if the process has remained within specification at all times, and all raw materials were within specifications, then the final product must also be within specification. This reasoning is, of course, based on the assumption that adequate data have been collected for all parts of the process sufficiently frequently, and thus that robust, appropriate validation schemes have been applied and measurements have been frequently obtained from all relevant instruments.

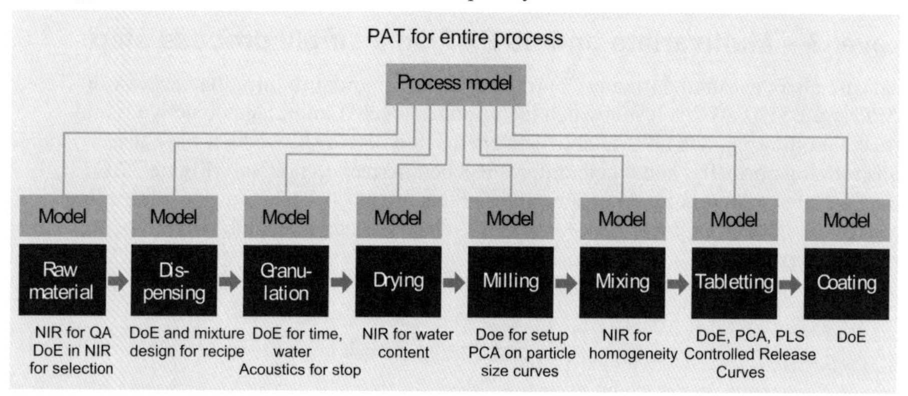

Figure 22.4: *Overall on-line PAT: Combining information from all unit operations and raw materials for a complete overview of the entire process. This incorporates levels 1-3.*

At this level, DOE has a number of uses, some of which are indicated in Figure 22.4. DOE is used in spectroscopic applications for selecting informative training sets. It is used in formulation, in both mixture design and process design, to identify optimal combinations of excipients and process conditions. DOE can be set up for almost any group of factors arising from the various unit operations, in order to define optimal and robust conditions. Last, but certainly not least, DOE can also be applied across many unit operations and raw materials simultaneously in order to map and optimize the entire process, and to investigate and ensure that the process is robust.

22.4 Risk minimization

Many of the PAT objectives can be formulated in terms of risk minimization, and the US FDA has re-defined its approach to "control" of the pharmaceutical industry as being one of "risk minimization" rather than rule-based. For example, high risks of product defects and waste are associated with highly variable processes. To identify approaches that minimize risk, DOE is required. One serious potential risk is that of overlooking influential factors in the experimental work. Fortunately, a tool that reduces the risk of forgetting important factors is available – the fishbone diagram – which is described in Section 22.4.2.

22.4.1 Types of risk

Several types of risk can be recognized in the present context:

- Total risk – patients may receive inadequate medicine (+ commercial risk for the company)
- Chemical risks – there may be incorrect concentrations of constituents and/or the wrong compounds in the formulation (due to mistakes or reactions)
- Physical risks – due for instance to wrong particle size distributions, crystal forms, disintegration and/or inappropriate solubility
- Biological risks – microbial contamination, and/or possible interactive effects with the microbial flora of the human body
- Engineering risks – the process may be run under the wrong conditions, leading to any of the above problems
- Analytical risk – inadequate data measured for the process / product / …
- Maths-Stats risk – data may be inadequately analyzed in the development phase, or the mathematical model may not be applicable in the on-line operation
- Interpretation risks – wrong decisions may be taken because data or results are not properly understood
- Communication risks – different parts of the organization may be aware of different aspects of the process, but fail to report anomalies

Some of the risk-types mentioned above may be investigated and understood through the use of QBD and DOE. With adequate data and DOE, physical, chemical, biological and engineering risks can be kept under control. In addition, a multitude of measured data, plus chemometric or similar models based on these data, provide diagnostics for levels of risk. Aspects addressed when evaluating the data include the precision of estimated results, confidence intervals, model error, diagnostics, etc. In addition it is necessary to put all this together into a manageable and consistent framework, incorporating chemometric methods into the overall risk evaluation, together with sampling, assessments of instrument and process equipment reliability, the training of scientists, engineers and other staff, and so on.

22.4.2 The fishbone diagram

The fishbone diagram can be used in risk minimization, and can help when setting up a DOE problem. It was introduced in Chapter 5 and further illustrated in Chapter 21 using the example of a formulator. One way to apply the fishbone diagram in this context is based on the 7M approach, drawn from total quality management (TQM) theory, for dividing causes of problems or undesired final outcomes into seven subgroups (Figure 22.5): Management, Man, Method, Measurement, Machine, Material & Milieu.

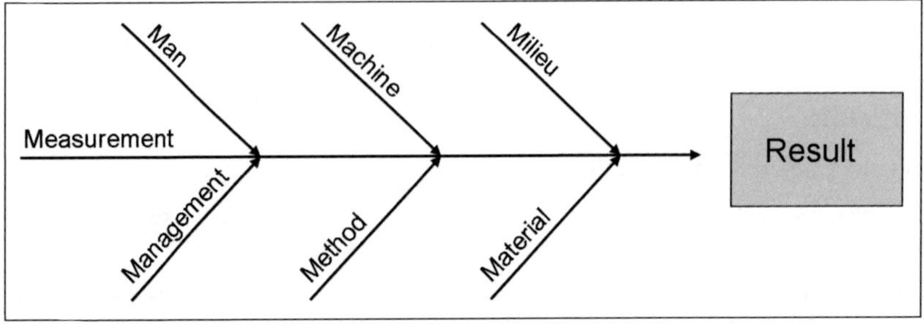

Figure 22.5: The fishbone diagram, or 7M-diagram, showing the seven main causes of a problem or bad result.

In summary, a DOE investigation should ideally start from a completed fishbone diagram with all factors and responses listed for all steps in the manufacturing chain. An example of a fishbone diagram for a formulation application was presented in Chapter 21 (Figure 21.18). For each selected factor, an investigation range must be defined and for each response the desired numerical result must be specified. The power of DOE is that it requires the user to state the purpose of the investigation (screening, optimization or robustness testing). In the next stage, the approach guides the user to select factors and responses that are relevant for the stated purpose and the questions to be addressed. In the final step, DOE then delivers answers to the questions being asked. In this respect, DOE is vital to QBD and risk minimization.

22.5 Establishing the Design Space of the process

In this section, we discuss how MVDA and DOE are used to estimate the "design space" of a process. However, the way this is done will depend on whether a process that is already running is being addressed in a PAT campaign, or a new process is being set up, in which PAT and QBD are being applied from the very start of the project.

In the first case, the starting point is the multivariate data analysis of historical data, in much the same way as discussed in Section 22.2.2 for the example on routine monitoring of a manufacturing process. The objective is to derive a good model for the tolerance intervals of the influential factors. Thus, in scope this is quite similar to a robustness test. Here, one typically identifies the most influential factors and performs DOE using them. Quite frequently, this is an iterative process and often several design cycles are needed before the result is satisfactory. One must also remember that important factors that are subject to tight process control may appear unimportant in the multivariate models since, historically, they have been kept almost constant.

In the second case, the investigator has to start with familiarization experiments in order to pinpoint the factors that are likely to be the most important, and determine their ranges. Only then can DOE be used. This is also an iterative process where ranges of the factors might have to be altered, new factors added, new responses considered, and new designs constructed. The work must be conducted at different scales, starting with laboratory and pilot plant experiments and culminating at the production scale.

Furthermore, since each process drifts and changes due to changes in environmental variables (notably temperature and humidity), wear to equipment, changes between batches of raw materials, etc., continuous monitoring and an understanding of such variations in operating conditions and their effects on product quality are required. In order to preserve good operating conditions, and keep the process close to an optimal point, "small" designs are needed on occasion to compensate for this undesired variability. Thus, DOE is also essential in process maintenance.

As discussed above, there are numerous important applications of DOE in PAT. However, regardless of the context and regardless of exactly which design has been used, once the trials in an experimental design have been performed, regression or PLS analysis can be used to elucidate the relationships between all the process inputs (factors related to materials and the process itself) and all the critical quality attributes (the responses). By interrogating the coefficients of the regression models, an understanding can be obtained of how changes in the factors will impact on the responses. The response contour plot provides one way of visualizing such changes. Another visualization tool is the Sweet Spot plot, which was introduced in Chapter 6. This plot is an overlay of many response contour plots; it is color coded according to the specifications for the responses, and a sweet spot exists if all the quality specifications are fulfilled.

Figure 22.6 shows the Sweet Spot plot established from a robustness testing design for a tablet formulation example. The data come from the article: Dick, C.F., et al., Determination of the sensitivity of a tablet formulation to variations in excipient levels and processing conditions using optimization techniques, Int. J. Pharm., 38, 23-31, 1987. In this study, the optimization criteria (numerical goals) were specified with respect to four responses: disintegration time, tablet hardness, weight variation and content uniformity. The graph shows how many of these criteria are fulfilled for various combinations of two factors (while keeping another three factors fixed). The arrow indicates the sweet spot region.

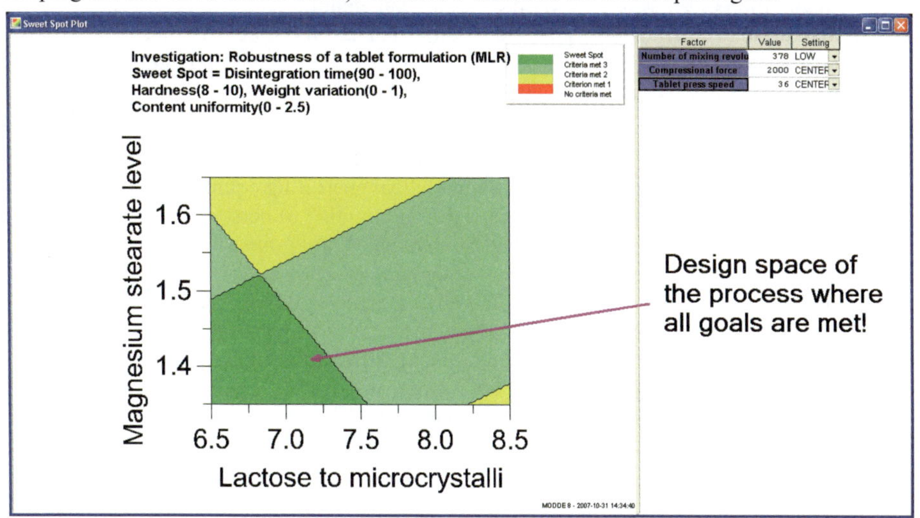

__Figure 22.6:__ Example of a Sweet Spot plot from a tablet formulation study. The arrow shows the location of the sweet spot. This is the region where all criteria are met. It can be said to reflect the "design space" of the process.

Once defined, the sweet spot reflects a volume in a sub-space, situated in the total multidimensional experimental design space, in which the combination and interactions of the process inputs (factors related to material attributes, excipient proportions and process settings) reliably deliver a product with the desired performance according to the profile of critical quality attributes defined for that product. According to the QBD jargon, this sweet spot area is the "design space" of the process. In conclusion, within the sweet spot or "design space", we have a window where the process meets all the specifications for the product. Outside this window of operation, there can be problems with one or more attributes of the product.

22.6 Questions for Chapter 22

- What is the objective of PAT?
- What are the different levels of PAT from a chemometric viewpoint?
- What goals are associated with the different levels?
- How can DOE enhance and hone PAT?
- What is QBD?
- In what context is the 7M approach useful?
- What is the design space of a process? How can it be established and visualized?

22.7 Summary and discussion

The PAT concept encompasses a wide range of analytical technologies for in-line, on-line or at-line measurements of pharmaceutical and biotechnological manufacturing processes. The basic objective is to determine whether the process being examined, its intermediates and final product are "within specifications". A successful PAT implementation is characterized by a smooth integration of process data sources and instrumentation with computers and pertinent software for data acquisition, data reduction, data analysis, storage of data and results in databases, and the presentation of results including the basis for the final batch record, and when necessary, alarms and diagnostics indicating causes of problems and their severity.

The implementation of PAT in a given process is not automatic and cannot be achieved overnight. However, with investment in adequate analytical measuring equipment, computer networks, databases, chemometric and other software, and training of personnel, the returns on these investments can be both substantial and immediate. Waiting times between sampling and analytical results tend to be dramatically reduced or even eliminated, the reliability of results improved, waste reduced, and process understanding improved. PAT not only provides more insight into the variability of raw material quality, but also allows users to combine spectral and wet chemistry data, and to model the relationship between raw material and final product quality. In addition, appropriate use of DOE combined with MVDA can augment the analysis and help to ensure that critical system parameters are varied together in a simultaneous and informational optimal manner.

In summary, setting up a "PAT-ified" process involves applying a combination of theory, experience, measurements, and experimentation. To make this efficient and reliable, DOE is needed. There are of course inherent difficulties, due to the intrinsic challenges involved in investigating and optimizing dynamic and non-linear batch processes. However, the rewards of applying DOE to PAT and QBD are substantial, immediate and provide improved knowledge and understanding of the process, increased process stability and robustness, and greater preparedness of the corporate organization to face and handle unforeseen changes in process behaviour and product quality.

Appendix I: The DOE work-flow

A1.1 Objective

DOE comprises three main experimental objectives, i.e., screening, optimization and robustness testing. Appendix I shows these stages put into practice on a real example dealing with the improvement of a reporter gene assay.

A1.2 Introduction

The data-set used in this exercise originates from Active Biotech AB in Lund, Sweden and we gratefully acknowledge Lena Schultz and Lisbeth Abramo for permitting us to use it. This study is unique in that it contains data related to the full spectrum of DOE applications, i.e. first a screening design was performed, then fold-over, then optimization and finally robustness testing. Our review of the application is structured accordingly in four phases:

- *Screening:* A 2^{6-2} fractional factorial design in 16 experiments + 3 center-points.
- *Fold-over:* The initial screening design was complemented by folding over.
- *Optimization:* A CCF design in 17 experiments to optimize three of the six factors.
- *Robustness Testing:* A 2^{5-1} fractional factorial design in 16 experiments + 3 center-points to investigate the sensitivity of the response to small changes in five of the factors.

This study deals with the luciferase reporter gene, one of a number of widely used reporter genes. Reporter gene assays are used in mechanistic studies of gene regulation. They also have great potential when applied to toxicology and drug development. A reporter gene has an easily measurable phenotype whose transcription is controlled by a promoter. Reporter gene assays provide important information of gene regulation relating to expression (i.e. number of copies) and when and where a particular protein is formed. A total of six factors were investigated using DOE and the objective was to increase and stabilize the signal-to-background ratio of the assay.

Factors:

- Cells – number of T-cells used in assay (number per well)
- PMA – agent added to stimulate T-cells (ng/ml)
- Ionomycin – agent added to stimulate T-cells (μg/ml)
- Stimulation time – duration of stimulation (hours)
- Lysing volume – volume of buffer needed to lyse T-cells (μl)
- Ratio – ratio of sample to substrate required to acquire a signal in the luciferase assay

Response:

- S/B – signal-to-background ratio computed as (signal-background)/background.

A1.3 Phase I – Screening

A1.3.1 Objective and configuration of the design

In screening the objective is to identify the most important factors and their ranges. Figure A1.1 shows the definition of the factors and Figure A1.2 the definition of the response variable.

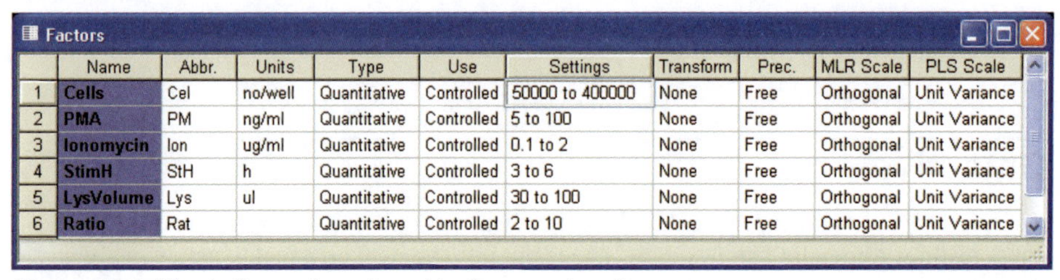

Factors

	Name	Abbr.	Units	Type	Use	Settings	Transform	Prec.	MLR Scale	PLS Scale
1	Cells	Cel	no/well	Quantitative	Controlled	50000 to 400000	None	Free	Orthogonal	Unit Variance
2	PMA	PM	ng/ml	Quantitative	Controlled	5 to 100	None	Free	Orthogonal	Unit Variance
3	Ionomycin	Ion	ug/ml	Quantitative	Controlled	0.1 to 2	None	Free	Orthogonal	Unit Variance
4	StimH	StH	h	Quantitative	Controlled	3 to 6	None	Free	Orthogonal	Unit Variance
5	LysVolume	Lys	ul	Quantitative	Controlled	30 to 100	None	Free	Orthogonal	Unit Variance
6	Ratio	Rat		Quantitative	Controlled	2 to 10	None	Free	Orthogonal	Unit Variance

Figure A1.1: *The six factors investigated in the screening phase.*

Responses

	Name	Abbr.	Units	Transform	MLR Scale	PLS Scale	Type
1	S/B	S/B		Log	None	Unit Variance	Regular

Figure A1.2: *The response variable measured in the screening phase.*

In order to vary the six factors, a 2^{6-2} fractional factorial design with three center-points was used (Figure A1.3).

Worksheet

	1	2	3	4	5	6	7	8	9	10	11
1	Exp No	Exp Name	Run Order	Incl/Excl	Cells	PMA	Ionomycin	StimH	LysVolume	Ratio	S/B
2	1	N1	1	Incl	50000	5	0.1	3	30	2	0
3	2	N2	12	Incl	400000	5	0.1	3	100	2	0
4	3	N3	9	Incl	50000	100	0.1	3	100	10	-0.2
5	4	N4	17	Incl	400000	100	0.1	3	30	10	0.1
6	5	N5	15	Incl	50000	5	2	3	100	10	0.6
7	6	N6	5	Incl	400000	5	2	3	30	10	6.3
8	7	N7	13	Incl	50000	100	2	3	30	2	1.5
9	8	N8	10	Incl	400000	100	2	3	100	2	6.7
10	9	N9	19	Incl	50000	5	0.1	6	30	10	0.2
11	10	N10	16	Incl	400000	5	0.1	6	100	10	0.4
12	11	N11	18	Incl	50000	100	0.1	6	100	2	-0.1
13	12	N12	11	Incl	400000	100	0.1	6	30	2	1.7
14	13	N13	4	Incl	50000	5	2	6	100	2	5.7
15	14	N14	8	Incl	400000	5	2	6	30	2	52.3
16	15	N15	7	Incl	50000	100	2	6	30	10	21.5
17	16	N16	3	Incl	400000	100	2	6	100	10	117
18	17	N17	6	Incl	225000	52.5	1.05	4.5	65	6	3.3
19	18	N18	14	Incl	225000	52.5	1.05	4.5	65	6	2
20	19	N19	2	Incl	225000	52.5	1.05	4.5	65	6	5.3

Figure A1.3: *The worksheet of the experimental design used in the screening phase.*

A1.3.2 Evaluation of raw data

The evaluation of the raw data indicates a few very large measurements and their histogram is highly skewed (Figure A1.4). A logarithmic transformation seems justified. However, since the response contains negative numbers, a small constant must be added before applying the transformation. The lowest response is –0.2. The second histogram (Figure A1.5) shows the effect of the log-transform using 1 as the constant. The response is still not approximately normally distributed. The third histogram (Figure A1.6) shows the effect of changing the constant to 0.21. Now, the histogram and replicate plot look much better. A plot of descriptive statistics (not shown here) confirms that the log-transform is appropriate.

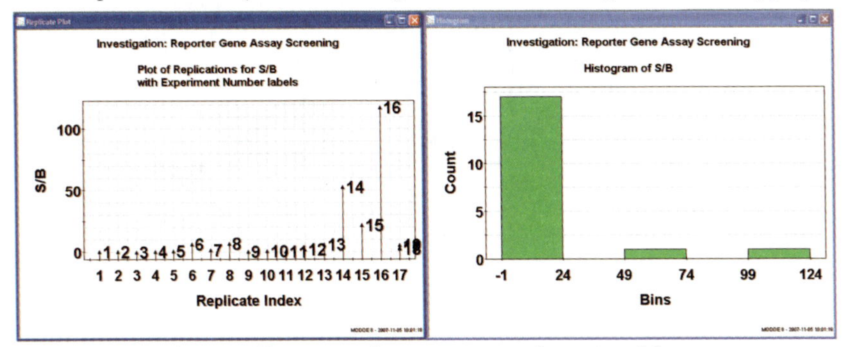

Figure A1.4: Replicate and histogram plots using no transformation of the response.

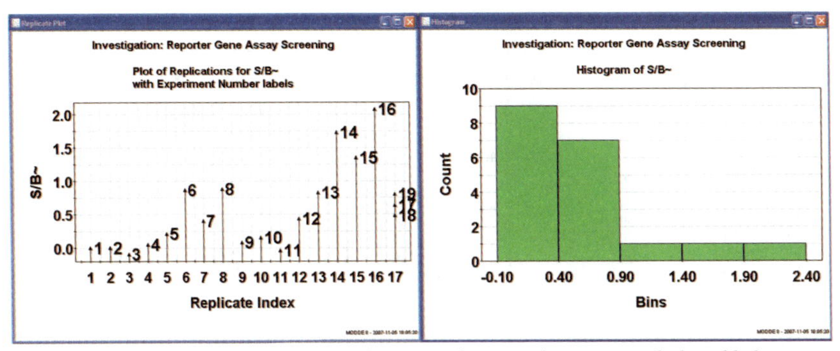

Figure A1.5: Replicate and histogram plots using log-transformation with the added constant of 1.

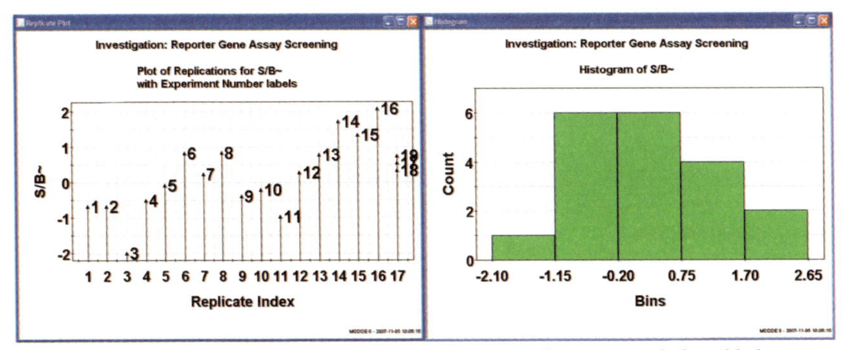

Figure A1.6: Replicate and histogram plots using log-transformation with the added constant of 0.21.

A1.3.3 Data analysis and model interpretation

The default linear model looks good with no evidence of lack of fit (Figure A1.7).

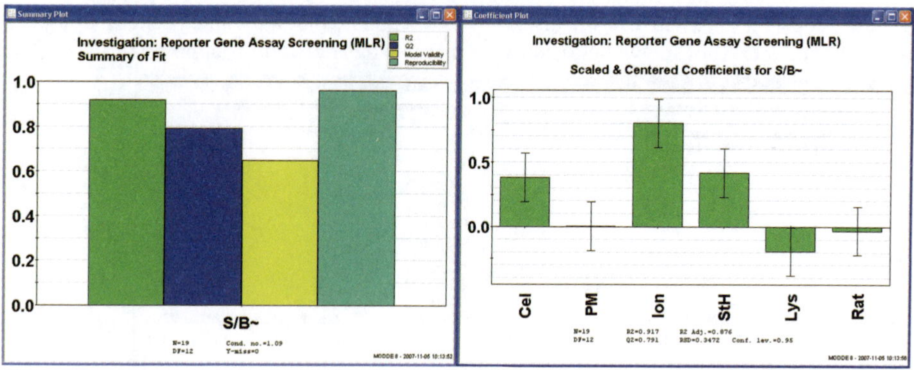

Figure A1.7: *The default linear model with* $R^2 = 0.92$ *and* $Q^2 = 0.79$.

To try and improve the model, PMA and Ratio were removed and the six two-factor interactions of the four remaining factors added, of which only three were worth keeping (Cel*Lys, Ion*StH, and Ion*Lys). The revised model is much better (Figure A1.8).

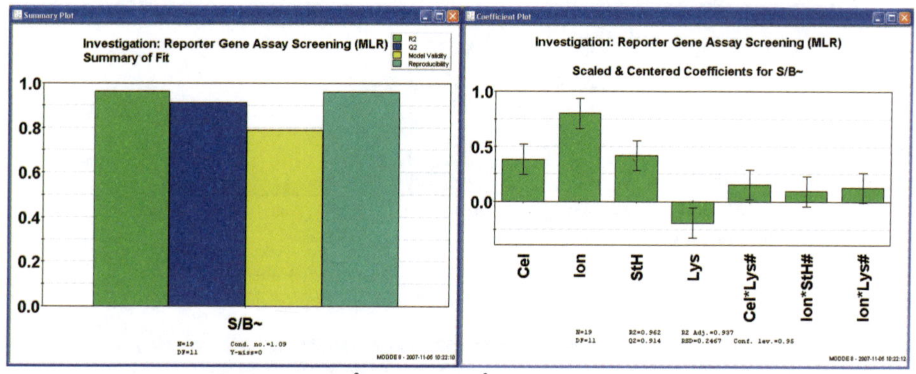

Figure A1.8: *The revised model with* $R^2 = 0.96$ *and* $Q^2 = 0.91$.

The revised model contains no outliers and the size of the residual is fairly independent of the predicted value, which is good (Figure A1.9).

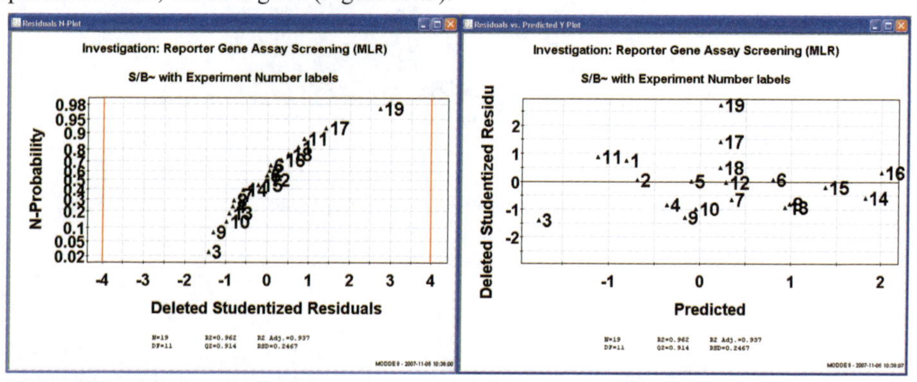

Figure AI.9: *Plot of residuals of the revised model. (left) N-plot. (right) Residual vs predicted response value.*

A1.3.4 Use of model for prediction

The contour plot below (Figure A1.10) shows how the signal-to-background ratio is predicted to change as a function of the factors Cells and LysVolume, while fixing the other factors at their maximum value. The combination of Cells and LysVolume was chosen to explore their borderline significant two-factor interaction.

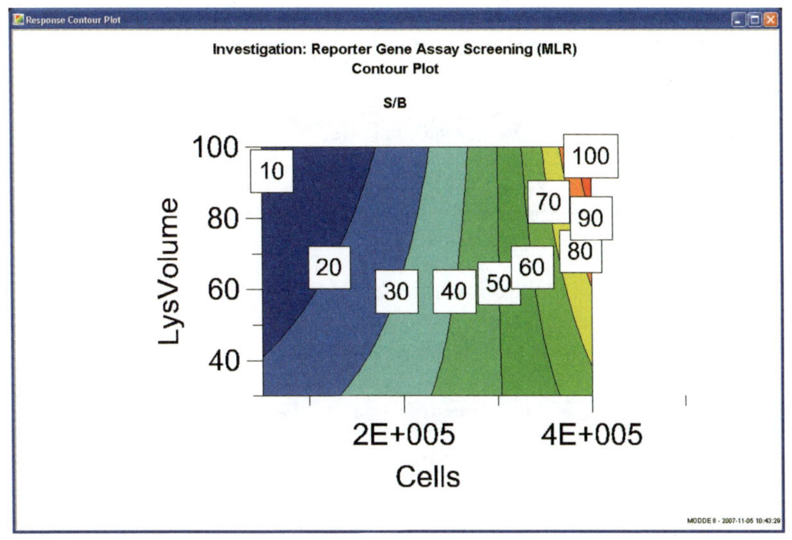

Figure A1.10: Response contour plot of the revised model.

A1.3.5 Conclusions of Phase 1

The three most important factors are Cells, Ionomycin and Stimulation Time. There are a few two-factor interactions which look interesting as they improve the predictive power of the model. However, these two-factor interactions are confounded with other two-factor interactions. Such confounding can be resolved using the Fold-over technique, see Phase 2 of the experimental work.

A1.4 Phase 2 – Fold-over

A1.4.1 Objective and configuration of the design

Fold-over is applied to screening designs to increase the number of experiments so that confounded terms may be resolved. When using Fold-over, MODDE will construct a new investigation including the existing data and the new runs. It will also add a block-factor ($Block) which is a precautionary measure to check whether the response has drifted with time. If $Block is non-significant, it will be removed from the model.

Due to the addition of the auto-generated block factor, the fold-over design together with the initial design comprises seven factors (Figure A1.11). The investigation still concerns the same single response variable (Figure A1.12).

Factors

	Name	Abbr.	Units	Type	Use	Settings	Transform	Prec.	MLR Scale	PLS Scale
1	Cells	Cel	no/well	Quantitative	Controlled	50000 to 400000	None	Free	Orthogonal	Unit Variance
2	PMA	PM	ng/ml	Quantitative	Controlled	5 to 100	None	Free	Orthogonal	Unit Variance
3	Ionomycin	Ion	ug/ml	Quantitative	Controlled	0.1 to 2	None	Free	Orthogonal	Unit Variance
4	StimH	StH	h	Quantitative	Controlled	3 to 6	None	Free	Orthogonal	Unit Variance
5	LysVolume	Lys	ul	Quantitative	Controlled	30 to 100	None	Free	Orthogonal	Unit Variance
6	Ratio	Rat		Quantitative	Controlled	2 to 10	None	Free	Orthogonal	Unit Variance
7	$Block	$Bl		Quantitative	Controlled	-1 to 1	None	Free	Orthogonal	Unit Variance

Figure A1.11: *The six factors plus the auto-generated block factor investigated in the fold-over design.*

Responses

	Name	Abbr.	Units	Transform	MLR Scale	PLS Scale	Type
1	S/B	S/B		Log	None	Unit Variance	Regular

Figure A1.12: *The response variable investigated in the fold-over design.*

When adding the fold-over complement to the original screening design, the number of factorial points in the worksheet is doubled. In this particular case, it was chosen to also double the number center-points. Hence, the resulting worksheet comprises 2 * 19 = 38 runs (Figure A1.13).

Worksheet

	1	2	3	4	5	6	7	8	9	10	11	12
1	Exp No	Exp Name	Run Order	Incl/Excl	Cells	PMA	Ionomycin	StimH	LysVolume	Ratio	$Block	S/B
2	1	N1	1	Incl	50000	5	0.1	3	30	2	-1	0
3	2	N2	12	Incl	400000	5	0.1	3	100	2	-1	0
4	3	N3	9	Incl	50000	100	0.1	3	100	10	-1	-0.2
5	4	N4	17	Incl	400000	100	0.1	3	30	10	-1	0.1
6	5	N5	15	Incl	50000	5	2	3	100	10	-1	0.6
7	6	N6	5	Incl	400000	5	2	3	30	10	-1	6.3
8	7	N7	13	Incl	50000	100	2	3	30	2	-1	1.5
9	8	N8	10	Incl	400000	100	2	3	100	2	-1	6.7
10	9	N9	19	Incl	50000	5	0.1	6	30	10	-1	0.2
11	10	N10	16	Incl	400000	5	0.1	6	100	10	-1	0.4
12	11	N11	18	Incl	50000	100	0.1	6	100	2	-1	-0.1
13	12	N12	11	Incl	400000	100	0.1	6	30	2	-1	1.7
14	13	N13	4	Incl	50000	5	2	6	100	2	-1	5.7
15	14	N14	8	Incl	400000	5	2	6	30	2	-1	52.3
16	15	N15	7	Incl	50000	100	2	6	30	10	-1	21.5
17	16	N16	3	Incl	400000	100	2	6	100	10	-1	117
18	17	N17	6	Incl	225000	52.5	1.05	4.5	65	6	-1	3.3
19	18	N18	14	Incl	225000	52.5	1.05	4.5	65	6	-1	2
20	19	N19	2	Incl	225000	52.5	1.05	4.5	65	6	-1	5.3
21	20	C20	27	Incl	50000	5	0.1	3	100	10	1	-0.2
22	21	C21	34	Incl	400000	5	0.1	3	30	10	1	0
23	22	C22	29	Incl	50000	100	0.1	3	30	2	1	-0.1
24	23	C23	28	Incl	400000	100	0.1	3	100	2	1	0
25	24	C24	32	Incl	50000	5	2	3	30	2	1	0.8
26	25	C25	20	Incl	400000	5	2	3	100	2	1	2.9
27	26	C26	31	Incl	50000	100	2	3	100	10	1	0.9
28	27	C27	22	Incl	400000	100	2	3	30	10	1	22.5
29	28	C28	21	Incl	50000	5	0.1	6	100	2	1	0.1
30	29	C29	26	Incl	400000	5	0.1	6	30	2	1	0.7
31	30	C30	24	Incl	50000	100	0.1	6	30	10	1	0.1
32	31	C31	37	Incl	400000	100	0.1	6	100	10	1	0.5
33	32	C32	30	Incl	50000	5	2	6	30	10	1	12.5
34	33	C33	38	Incl	400000	5	2	6	100	10	1	15.8
35	34	C34	36	Incl	50000	100	2	6	100	2	1	3.8
36	35	C35	33	Incl	400000	100	2	6	30	2	1	165
37	36	C36	25	Incl	225000	52.5	1.05	4.5	65	6	1	4.1
38	37	C37	35	Incl	225000	52.5	1.05	4.5	65	6	1	3.9
39	38	C38	23	Incl	225000	52.5	1.05	4.5	65	6	1	2.2

Figure A1.13: *The worksheet of the fold-over experimental design.*

A1.4.2 Evaluation of raw data

For ease of interpretation, the same response transformation was applied as before. This looks reasonable given the replicate and histogram plots below (Figure A1.14).

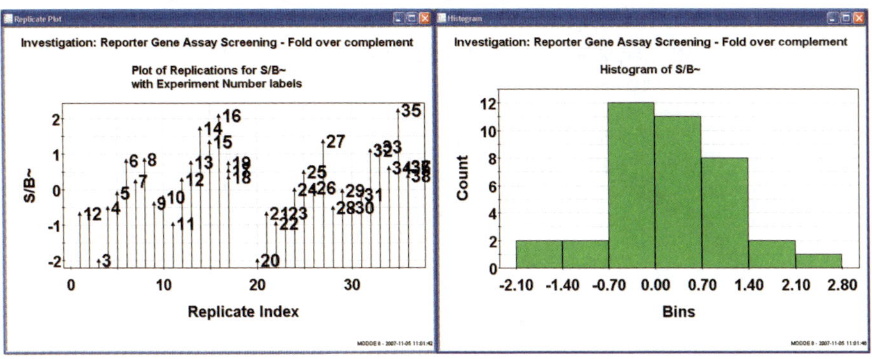

Figure A1.14: Replicate and histogram plots using log-transformation with the added constant of 0.21.

A1.4.3 Data analysis and model interpretation

The default linear model is very good (Figure A1.15). The Block factor is not significant so there is no evidence of a time drift between the two sets of experiments.

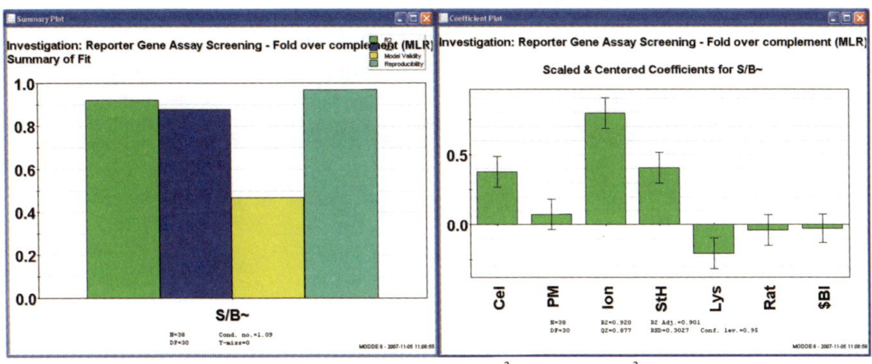

Figure A1.15: The default linear fold-over model with $R^2 = 0.92$ and $Q^2 = 0.88$.

To try and improve the model, PMA, Ratio and $Block were removed. Two-factor interactions were tested but they did not improve the model. The refined model is only marginally better (Figure A1.16).

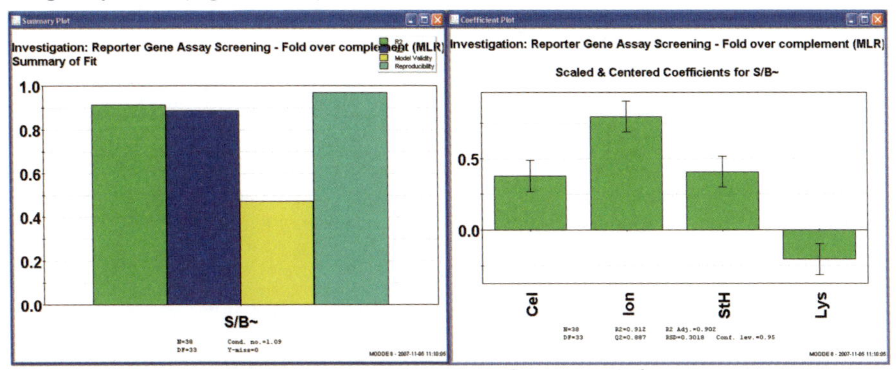

Figure A1.16: The revised linear fold-over model with $R^2 = 0.91$ and $Q^2 = 0.89$.

The revised model contains no outliers (Figure A1.17). However, the plot of the deleted studentized residuals versus the predicted value (Figure A1.17) indicates that some of the largest residuals correspond to the six center-points. A similar phenomenon was present also in the initial screening design (Figure A1.9). This hints at curvature problems. Curvature is easy to handle with a quadratic regression model but not with the linear model used here.

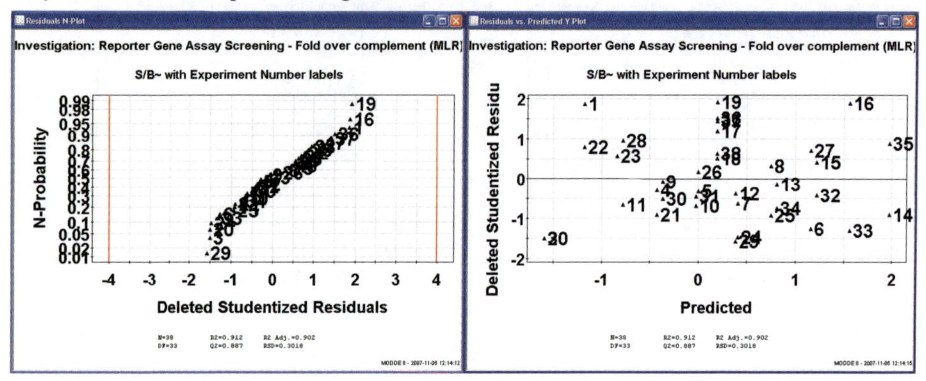

Figure A1.17: *Plot of residuals of the revised fold-over model. (left) N-plot. (right) Residual vs predicted response value.*

A1.4.4 Use of model for prediction

MODDE's optimizer was used to locate the factor combination which maximizes the response. PMA, Ratio and $Block were not included in the final model and are therefore grayed out in the optimizer factor spreadsheet (Figure A1.18).

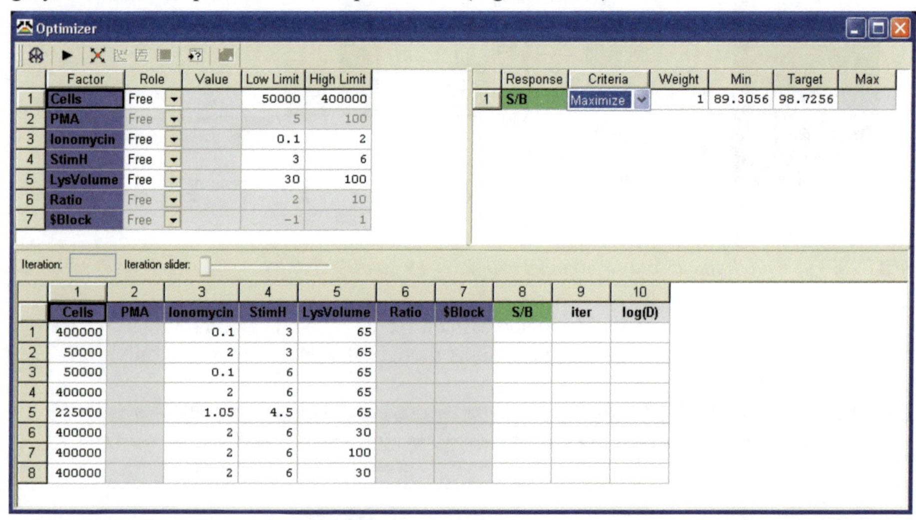

	Factor	Role	Value	Low Limit	High Limit
1	Cells	Free		50000	400000
2	PMA	Free		5	100
3	Ionomycin	Free		0.1	2
4	StimH	Free		3	6
5	LysVolume	Free		30	100
6	Ratio	Free		2	10
7	$Block	Free		-1	1

	Response	Criteria	Weight	Min	Target	Max
1	S/B	Maximize	1	89.3056	98.7256	

	Cells	PMA	Ionomycin	StimH	LysVolume	Ratio	$Block	S/B	iter	log(D)
1	400000		0.1	3	65					
2	50000		2	3	65					
3	50000		0.1	6	65					
4	400000		2	6	65					
5	225000		1.05	4.5	65					
6	400000		2	6	30					
7	400000		2	6	100					
8	400000		2	6	30					

Figure A1.18: *Start view of the MODDE optimizer.*

The results of running the optimizer are shown below (Figure A1.19). The optimum point corresponds to having three factors at their upper limit and one at its lower limit.

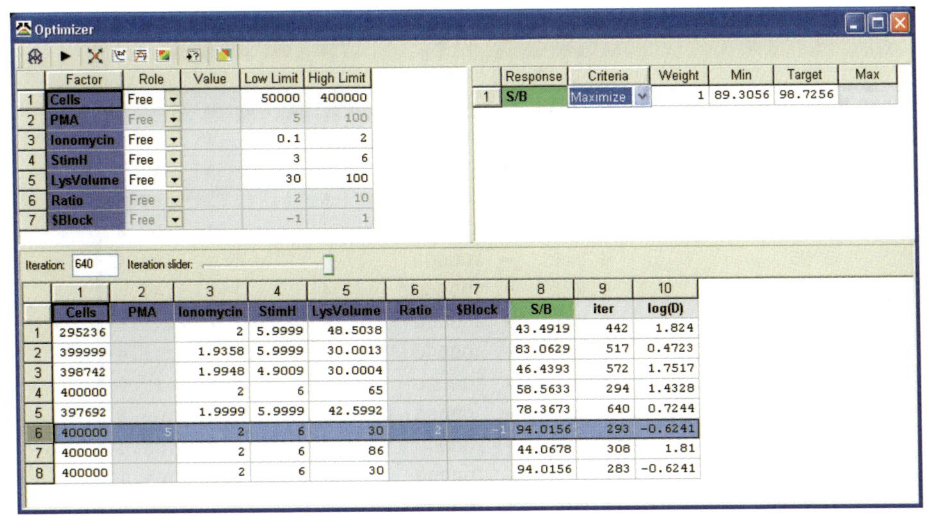

Optimizer

	Factor	Role	Value	Low Limit	High Limit
1	Cells	Free		50000	400000
2	PMA	Free		5	100
3	Ionomycin	Free		0.1	2
4	StimH	Free		3	6
5	LysVolume	Free		30	100
6	Ratio	Free		2	10
7	$Block	Free		-1	1

	Response	Criteria	Weight	Min	Target	Max
1	S/B	Maximize	1	89.3056	98.7256	

Iteration: 640 Iteration slider:

	1	2	3	4	5	6	7	8	9	10
	Cells	PMA	Ionomycin	StimH	LysVolume	Ratio	$Block	S/B	iter	log(D)
1	295236		2	5.9999	48.5038			43.4919	442	1.824
2	399999		1.9358	5.9999	30.0013			83.0629	517	0.4723
3	398742		1.9948	4.9009	30.0004			46.4393	572	1.7517
4	400000		2	6	65			58.5633	294	1.4328
5	397692		1.9999	5.9999	42.5992			78.3673	640	0.7244
6	400000	5	2	6	30	2	-1	94.0156	293	-0.6241
7	400000		2	6	86			44.0678	308	1.81
8	400000		2	6	30			94.0156	283	-0.6241

Figure A1.19: *Convergence view of the MODDE optimizer.*

A1.4.5 Conclusions of Phase 2

The Fold-over experiments did not indicate any large two-factor interactions. Instead, it confirmed that three of the factors dominate: Cells, Ionomycin and Stimulation Time. These three factors form the basis of the optimization design employed during Phase 3, which will be better suited to handling the non-linear behaviour noted above.

A1.5 Phase 3 - Optimization

A1.5.1 Objective and configuration of the design

In optimization, the objective is to locate an optimal factor combination which can be used as a future set point. When defining the RSM investigation with three factors (Figure A1.20) and one response (Figure A1.21), the factor ranges were modified according to the results of the screening phase. The new design defines a much smaller experimental domain.

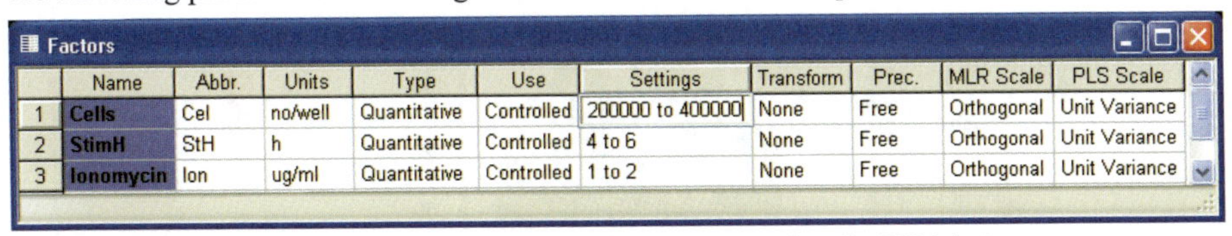

Factors

	Name	Abbr.	Units	Type	Use	Settings	Transform	Prec.	MLR Scale	PLS Scale
1	Cells	Cel	no/well	Quantitative	Controlled	200000 to 400000	None	Free	Orthogonal	Unit Variance
2	StimH	StH	h	Quantitative	Controlled	4 to 6	None	Free	Orthogonal	Unit Variance
3	Ionomycin	Ion	ug/ml	Quantitative	Controlled	1 to 2	None	Free	Orthogonal	Unit Variance

Figure A1.20: *The three factors with modified ranges investigated in the RSM design.*

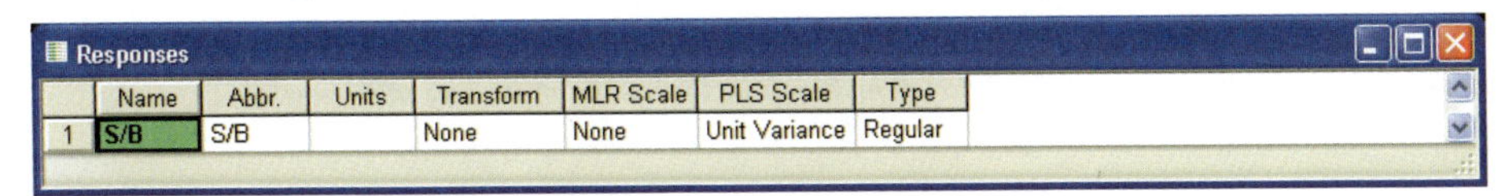

Responses

	Name	Abbr.	Units	Transform	MLR Scale	PLS Scale	Type
1	S/B	S/B		None	None	Unit Variance	Regular

Figure A1.21: *The response variable investigated in the RSM design.*

In order to vary the three factors, a CCF design augmented with four center-points was constructed (Figure A1.22). This design supports a quadratic regression model.

	1	2	3	4	5	6	7	8
1	Exp No	Exp Name	Run Order	Incl/Excl	Cells	StimH	Ionomycin	S/B
2	1	N1	10	Incl	200000	4	1	52.2
3	2	N2	16	Incl	400000	4	1	63.6
4	3	N3	13	Incl	200000	6	1	49.4
5	4	N4	7	Incl	400000	6	1	90.3
6	5	N5	4	Incl	200000	4	2	84.4
7	6	N6	9	Incl	400000	4	2	149.2
8	7	N7	6	Incl	200000	6	2	173.6
9	8	N8	11	Incl	400000	6	2	221.4
10	9	N9	2	Incl	200000	5	1.5	17.9
11	10	N10	5	Incl	400000	5	1.5	104.2
12	11	N11	14	Incl	300000	4	1.5	105.4
13	12	N12	17	Incl	300000	6	1.5	161.3
14	13	N13	1	Incl	300000	5	1	90.5
15	14	N14	8	Incl	300000	5	2	209.6
16	15	N15	3	Incl	300000	5	1.5	161
17	16	N16	12	Incl	300000	5	1.5	166.5
18	17	N17	15	Incl	300000	5	1.5	161.6
19	18	N18	18	Incl	300000	5	1.5	110.2

Figure A1.22: The worksheet of the experimental design used in the RSM phase.

A1.5.2 Evaluation of raw data

The replicate plot shows that the signal-to-background ratio is much higher than in the screening designs (Figure A1.23). The histogram and Box-Whisker plots (Figure A1.23) indicate that a response transformation is no longer required.

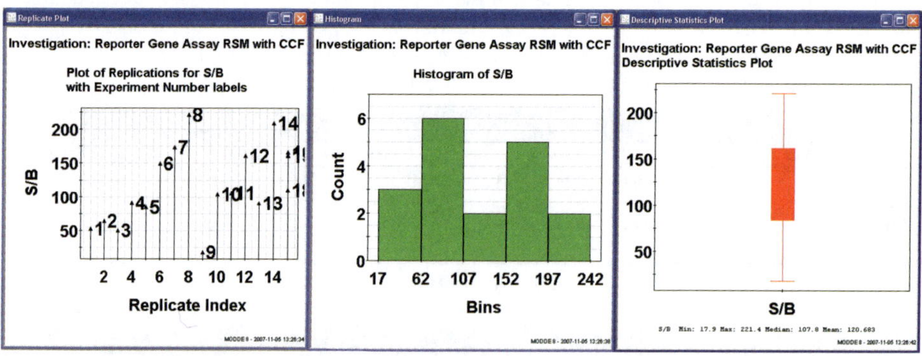

Figure A1.23: Replicate, histogram and Box-Whisker plots indicating that there is no need to transform the response variable.

A1.5.3 Data analysis and model interpretation

The default quadratic model has a relatively poor Q^2 ($R^2 = 0.91$, $Q^2 = 0.56$, Figure A1.24).

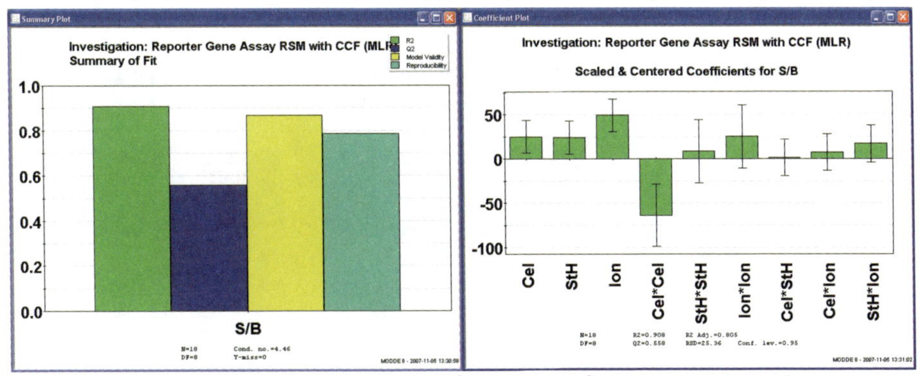

Figure A1.24: The default quadratic model with $R^2 = 0.91$ and $Q^2 = 0.56$.

The model was pruned, by removing small terms, which gave a better model (Figure A1.25).

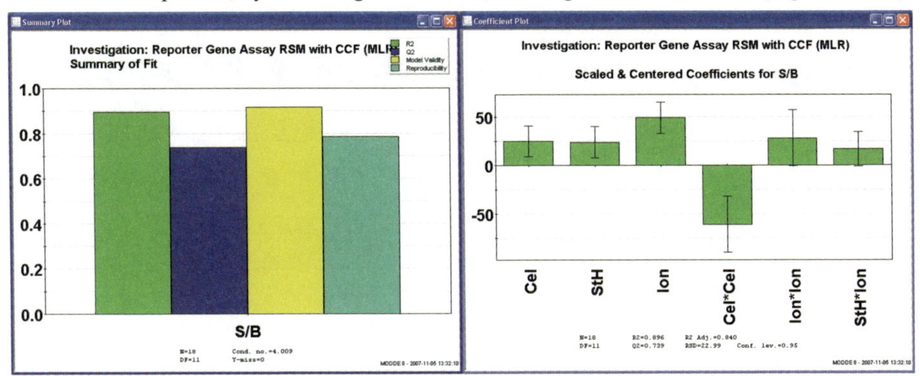

Figure A1.25: The revised quadratic model with $R^2 = 0.89$ and $Q^2 = 0.74$.

There are no outliers and the residuals are unrelated to the predicted value (Figure A1.26).

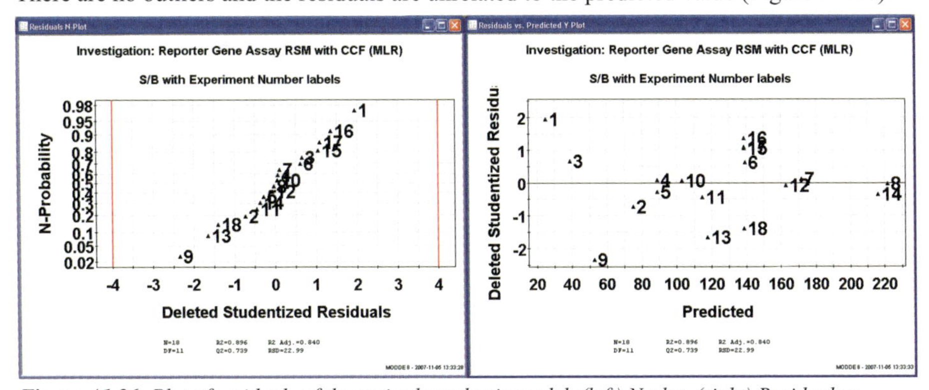

Figure A1.26: Plot of residuals of the revised quadratic model. (left) N-plot. (right) Residual vs predicted response value.

A1.5.4 Use of model for prediction

The contour plots below (Figure A1.27) show how the signal-to-background ratio varies in relation to the three factors. The optimum factor combination is high Stimulation time (6 hours), high Ionomycin (2) and intermediate Cells (around 320000).

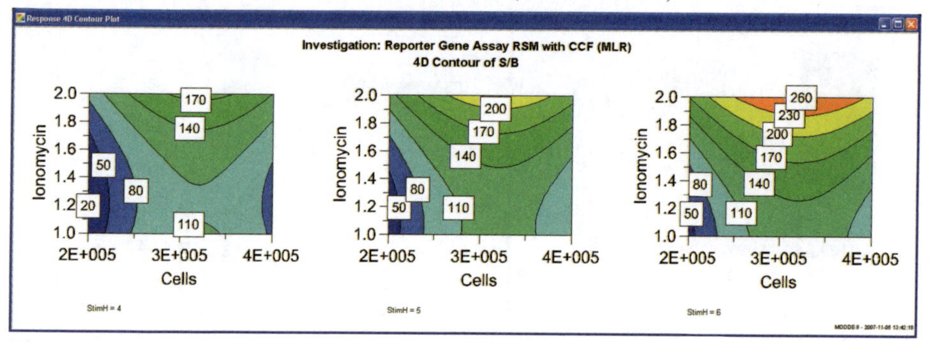

Figure A1.27: Response contour plots showing that maximum signal is obtained at Cells ≈ 320000, Stimulation Time = 6 and Ionomycin = 2.

The optimizer was used to obtain more exact co-ordinates of the optimum (Figure A1.28).

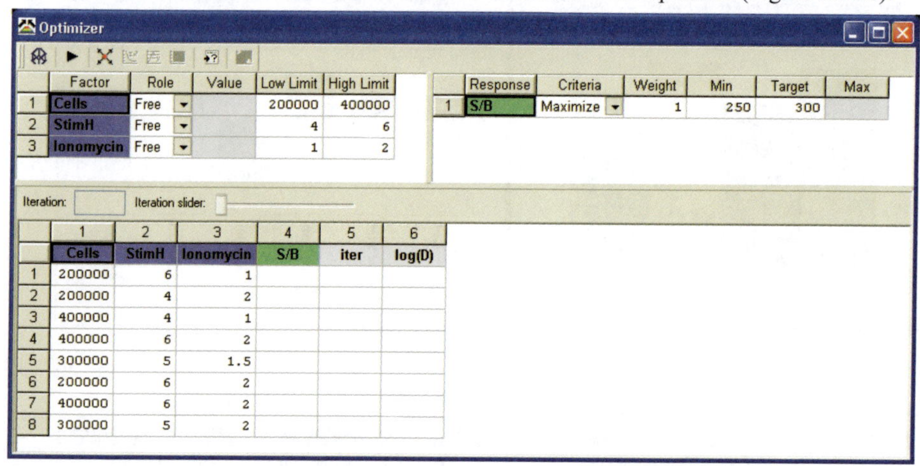

	Factor	Role	Value	Low Limit	High Limit
1	Cells	Free		200000	400000
2	StimH	Free		4	6
3	Ionomycin	Free		1	2

	Response	Criteria	Weight	Min	Target	Max
1	S/B	Maximize	1	250	300	

	1	2	3	4	5	6
	Cells	StimH	Ionomycin	S/B	iter	log(D)
1	200000	6	1			
2	200000	4	2			
3	400000	4	1			
4	400000	6	2			
5	300000	5	1.5			
6	200000	6	2			
7	400000	6	2			
8	300000	5	2			

Figure A1.28: Start view of the optimizer as applied to the RSM data and model.

After the first optimization round the 8th simplex was found to be best (Figure A1.29).

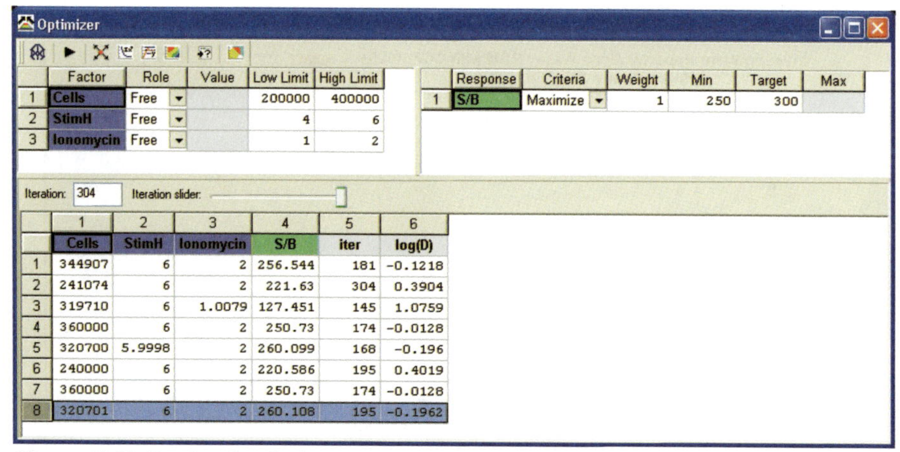

Figure A1.29: *Results after the first optimization cycle.*

During the second optimization round, new starting points were generated in the vicinity of the best simplex from the first round (Figure A1.30).

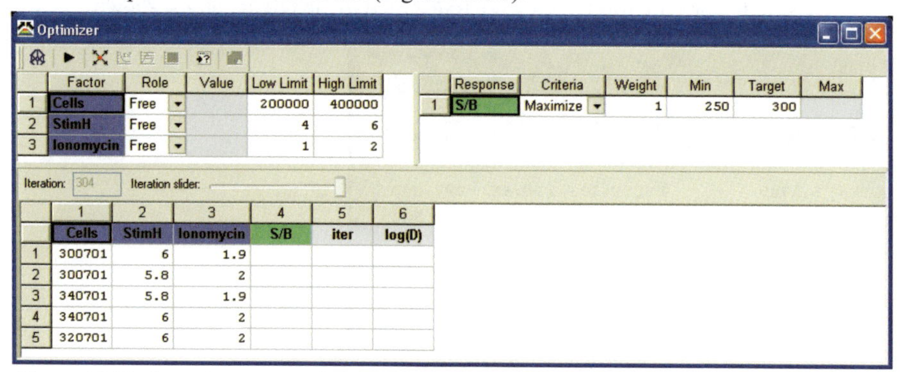

Figure A1.30: *Start view prior to the second optimization round.*

Four of the five new simplexes converge to the same point: Cells ≈ 320000, Stimulation Time = 6 and Ionomycin = 2 (Figure A1.31).

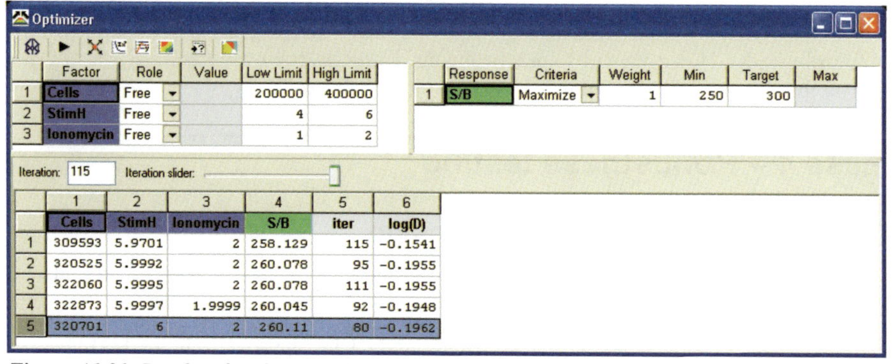

Figure A1.31: *Results after the second optimization cycle.*

The results for the best predicted simplex were transferred to the Sweet Spot plot, a plot which clearly shows the location of the optimal point (Figure A1.32).

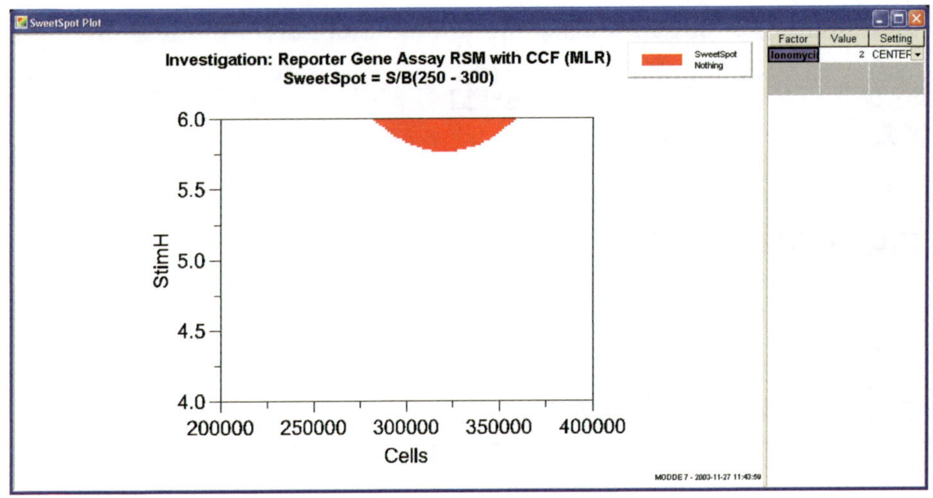

Figure A1.32: *Sweet Spot plot showing the location of the optimum.*

Further, the five simplex factor co-ordinates were transferred to the prediction list showing that the predicted optimal S/B value is 260 ± 40 (Figure A1.33).

Predictions List

	1	2	3	4	5	6
1	Cells	StimH	Ionomycin	S/B	Lower	Upper
2						
3	309086	5.9729	2	258.173	218.453	297.893
4	319842	5.9999	2	260.102	220.397	299.806
5	322361	5.9998	2	260.085	220.515	299.655
6	323608	5.9998	1.9999	260.026	220.536	299.516
7	321016	6	2	260.109	220.464	299.755

Figure A1.33: *Predicted values with their 95% confidence intervals.*

A1.5.5 Conclusions of Phase 3

The optimal factor combination within the investigated experimental domain is Cells ≈ 320000, Stimulation Time = 6 and Ionomycin = 2. In the final DOE stage, this point will be assessed for robustness. However, due to practical considerations, robustness testing was not performed on this precise point but rather one close to it (see Phase 4).

A1.6 Phase 4 – Robustness testing

A1.6.1 Objective and configuration of design

In robustness testing, the objective is to explore the robustness of an assay or method around its set point. The following set point was identified using the RSM results as basis:

- Cells = 300000 (320000 was optimal according to the CCF design but 300000 is more practical as it means less crowding of the sample volume.)
- PMA = 10 (Had virtually no effect in the screening phase, low level chosen.)

- Ionomycin = 1.5 (Two was optimal according to the CCF design but 1.5 is more practical. Too high a concentration creates an interference with the "real" signal which could reduce the signal-to-background ratio.)
- Stimulation time = 5.5 (Six hours was optimal according to the CCF design but 5.5h fits in better with an eight hour working day.)
- LysVolume = 30 (Low level which was found to be optimal during the screening phase.)

The specification of the response was that the signal-to-background ratio should exceed 50 regardless of the factor combination. The definition of the five factors (Figure A1.34) and the response (Figure A1.35) are seen below.

Factors

	Name	Abbr.	Units	Type	Use	Settings	Transform	Prec.	MLR Scale	PLS Scale
1	Cells	Cel	no/well	Quantitative	Controlled	295000 to 305000	None	Free	Orthogonal	Unit Variance
2	PMA	PM	ng/ml	Quantitative	Controlled	9.8 to 10.2	None	Free	Orthogonal	Unit Variance
3	Ionomycin	Ion	ug/ml	Quantitative	Controlled	1.47 to 1.53	None	Free	Orthogonal	Unit Variance
4	StimH	StH	h	Quantitative	Controlled	5.417 to 5.583	None	Free	Orthogonal	Unit Variance
5	LysVolume	Lys	ul	Quantitative	Controlled	29.5 to 30.5	None	Free	Orthogonal	Unit Variance

Figure A1.34: The five factors studied using the robustness testing design.

Responses

	Name	Abbr.	Units	Transform	MLR Scale	PLS Scale	Type
1	S/B	S/B		None	None	Unit Variance	Regular

Figure A1.35: The response variable studied using the robustness testing design.

In order to investigate the robustness of the reporter gene assay close to the set point, a 2^{5-1} fractional factorial design in resolution V, expanded with three center-points, was created. The worksheet of this design is given in Figure A1.36.

Worksheet

	1	2	3	4	5	6	7	8	9	10
1	Exp No	Exp Name	Run Order	Incl/Excl	Cells	PMA	Ionomycin	StimH	LysVolume	S/B
2	1	N1	6	Incl	295000	9.8	1.47	5.417	30.5	75.9
3	2	N2	12	Incl	305000	9.8	1.47	5.417	29.5	67.5
4	3	N3	7	Incl	295000	10.2	1.47	5.417	29.5	60.5
5	4	N4	13	Incl	305000	10.2	1.47	5.417	30.5	65.7
6	5	N5	18	Incl	295000	9.8	1.53	5.417	29.5	69.2
7	6	N6	2	Incl	305000	9.8	1.53	5.417	30.5	79.6
8	7	N7	11	Incl	295000	10.2	1.53	5.417	30.5	74.6
9	8	N8	3	Incl	305000	10.2	1.53	5.417	29.5	84.9
10	9	N9	17	Incl	295000	9.8	1.47	5.583	29.5	65
11	10	N10	1	Incl	305000	9.8	1.47	5.583	30.5	52.6
12	11	N11	19	Incl	295000	10.2	1.47	5.583	30.5	56
13	12	N12	16	Incl	305000	10.2	1.47	5.583	29.5	60.5
14	13	N13	9	Incl	295000	9.8	1.53	5.583	30.5	89.1
15	14	N14	10	Incl	305000	9.8	1.53	5.583	29.5	61.6
16	15	N15	14	Incl	295000	10.2	1.53	5.583	29.5	81.7
17	16	N16	5	Incl	305000	10.2	1.53	5.583	30.5	81.6
18	17	N17	4	Incl	300000	10	1.5	5.5	30	59.5
19	18	N18	8	Incl	300000	10	1.5	5.5	30	64.9
20	19	N19	15	Incl	300000	10	1.5	5.5	30	67.6

Figure A1.36: The worksheet of the robustness testing design.

A1.6.2 Evaluation of raw data

The histogram and replicate plots indicate that a transformation is unnecessary (Figure A1.37). The replicate plot also shows that all the response values are above 50, i.e. within specification.

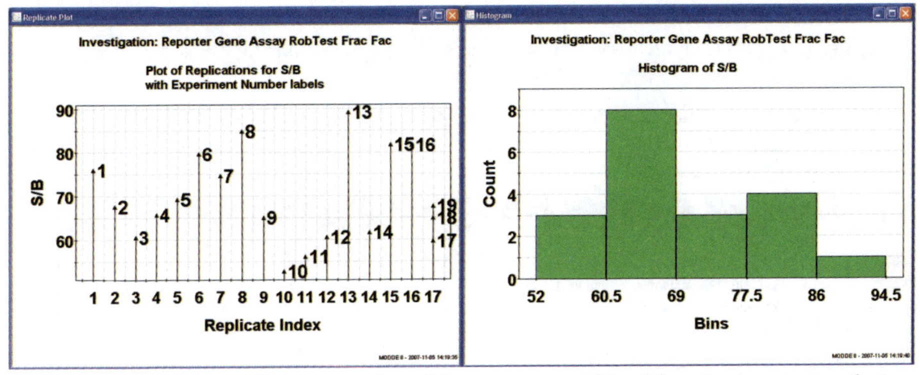

Figure A1.37: Replicate and histogram plots of the response variable of the robustness testing design.

A1.6.3 Data analysis and model interpretation

In the analysis of the data shown in Figure A1.37, the four limiting cases of robustness testing were considered in order to determine which applies here. These four limiting cases were illustrated in Chapter 7. Briefly, they are: Inside specification/Significant model (Limiting case 1); Inside specification/ Non-significant model (Limiting case 2); Outside specification/Significant model (Limiting case 3); Outside specification/Non-significant model (Limiting case 4).

In robustness testing model refinement is usually not performed and the ideal result is no model at all. Indeed, the model obtained is poor ($R^2 = 0.93$, $Q^2 =$ negative). However, the regression coefficient plot indicates that S/B *is* sensitive to changes in Ionomycin concentration (Figure A1.38).

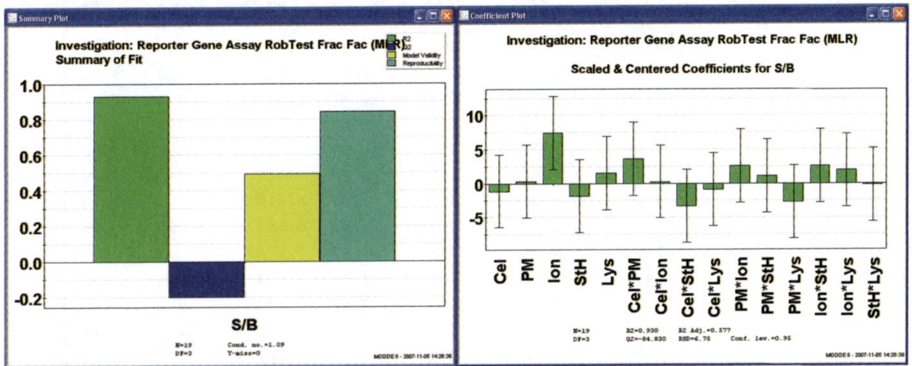

Figure AI.38: The default linear model is insignificant with $R^2 = 0.93$ and $Q^2 =$ negative. However, the response is sensitive to the amount of Ionomycin.

Above it was shown that S/B is not robust to changes in Ionomycin concentration. However, the response data themselves *are* robust given that they are within specification. The factor range of Ionomycin must be reduced by half in order to make S/B robust. Hence, the concentration range for Ionomycin within which robustness can be claimed is 1.485–1.515 µg/ml rather than 1.47–1.53 µg/ml.

A1.6.4 Conclusions of Phase 4

The final DOE phase illustrated the first limiting case of robustness testing, i.e., a significant model and inside specification. S/B was most sensitive to changes in Ionomycin concentration.

A1.7 Conclusions of example

Of the six factors originally investigated, three dominated the initial screening phase - Cells, Ionomycin and Stimulation Time (StimH). A few two-factor interactions were also used in the screening model as they increased the predictive power. However, the problem with two-factor interactions in medium-resolution screening designs is that they are often confounded with other two-factor interactions. Such confounding can be resolved using the fold-over technique.

The addition of the 19 fold-over experiments showed two things. First of all, there was no systematic shift in the response data between the two sets of experiments. Secondly, there were no important two-factor interactions. On the contrary, it confirmed the importance of the three key factors. There was also some evidence of curvature (non-linear behaviour) which can be investigated in more detail by using a central composite design.

In the RSM phase, the key factors Cells, Ionomycin and Stimulation Time were optimized using a CCF design in 17 runs. The factor ranges were adjusted in accordance with the findings of the screening phases. This design identified the optimal factor combination Cells ≈ 320000, Stimulation Time = 6 and Ionomycin = 2.

In the final robustness testing, the set point was defined as

- Cells = 300000
- PMA = 10
- Ionomycin = 1.5
- Stimulation time = 5.5
- LysVolume = 30

The signal-to-background ratio is most sensitive to changes in the Ionomycin concentration. However, the response may be regarded as robust given that all the values were within specification. The final conclusion is that the results of the four phases are both coherent and consistent. This indicates the high quality of the underlying experimental data.

A1.8 Concluding remarks

As demonstrated using the data of the reporter gene assay example, there are three main types of problem to which DOE is applicable. The first experimental objective is screening. Screening is used to uncover the most influential factors, and to determine in which ranges these should be investigated. This is a rather uncomplicated question, and therefore screening designs use few experiments in relation to the number of factors. Sometimes more than one screening design might be needed. Recall that in the current case a fold-over design was added to the initial screening design.

The second experimental objective is optimization. Now, the interest lies in defining which approved combination of the important factors will result in optimal operating conditions. Since optimization is more complex than screening, optimization designs demand more experiments per factor. With such an enriched body of data it is possible to compute quadratic regression models.

The third experimental objective is robustness testing. Here, one wants to determine how sensitive a product or production procedure is to small changes in the factor settings. Such small changes usually correspond to fluctuations in the factors occurring during ”a bad day” in the production, or the customer not following product usage instructions. In the reporter gene assay, the point of departure of the robustness testing design was the optimal point discovered through the optimization design. The ranges of the investigated factors corresponded to relevant variations that might be considered running this assay on a routine basis.

In summary, the great advantage of using DOE is that it provides an organized approach, with which it is possible to address both simple and tricky experimental problems. The experimenter is encouraged to select an appropriate experimental objective, and is then guided to devise and perform a set of experiments, which is adequate for the selected objective. It does not take long to set up an experimental protocol, not even for the person unfamiliar with DOE. What is perhaps counter-intuitive is that the user has to complete a whole set of experiments before any conclusions can be drawn.

The rewards of DOE are often immediate and substantial, such as a higher product quality obtained at a lower cost and with a more environmentally-friendly process performance. However, there are also more long-term gains which might not be thought of at a first quick glance. These involve better understanding of the investigated system or process, higher stability of the same, and a better preparedness for facing new, hitherto unseen challenges.

Appendix II: Statistical Notes

A2.1 Fit methods

MODDE supports Multiple Linear Regression (MLR) and PLS (Projections to Latent Structures) for fitting the model to the data.

A2.1.1 Multiple Linear Regression (MLR)

Multiple regression is extensively described in the literature, and this chapter will only identify the numerical algorithms used to compute the regression results, the measures of goodness of fit and diagnostics used by MODDE. For additional information on MLR, see Draper and Smith "Applied Regression Analysis", Second Edition, Wiley, New York.

MODDE uses the singular value decomposition (SVD) to solve the system of equations:

$Y = X*B+E$

Y is an n*m matrix of responses, and X (the extended design matrix) an n*p matrix, with p the number of terms in the model including the constant, and B the matrix of regression coefficients, E the matrix of residuals. See Golub and Van Loan (1983) for a description of the SVD and its use to obtain the regression results.

In case of missing data in a row (x or/and y), this row is eliminated before the MLR fitting.

A2.1.2 Partial Least Squares (PLS)

When several responses have been measured, it is useful to fit a model simultaneously representing the variation of all the responses to the variation of the factors. PLS deals with many responses simultaneously, taking their covariances into account. This provides an overview of the relationship between the responses and of how all the factors affect all the responses. This multivariate method of estimating the models for all the responses simultaneously is called PLS.

PLS contains the multiple linear regression solution as a special case, i.e. with one response and a certain number of PLS dimensions, the PLS regression coefficients are identical to those obtained by multiple linear regression.

PLS has been extensively described in the literature and only a brief description is given here. PLS finds the relationship between a matrix Y (response variables) and a matrix X (predictor or factor variables) expressed as:

$Y = XB+E$

The matrix Y refers to the characteristics of interest (responses). The matrix X refers to the predictor variables and to their square or/and cross terms if these have been added to the model.

PLS creates new variables (t_a) called X-scores as weighted combinations of the original X-variables: $t_a = Xw_a$, where w_a are the combination weights. These X-scores are few, often just two or three, and orthogonal. The X-scores are then used to model the responses.

With several responses, the Y-variables are similarly combined to a few Y-scores (u_a) using weights c_a, $u_a = Yc_a$. The PLS estimation is done in such a way that it maximizes the correlation, in each model dimension, between t_a and u_a. One PLS component (number a) consists of one vector of X-scores (t_a), and one of Y-scores (u_a), together with the X and Y-weights (w_a and c_a).

Hence the PLS model consists of a simultaneous projection of both the X- and Y-spaces on a low dimensional hyper-plane with the new coordinates T (summarizing X) and U (summarizing Y), and then relating U to T. This analysis has the following two objectives:

- To well approximate the X and Y spaces by the hyper-planes.
- To maximize the correlation between X and Y (t and u).

Mathematically the PLS model can be expressed as:

$X = TP' + E$

$Y = TC' + E$

Geometrically, we can see the matrices X and Y as n points in two spaces (Figure A2.1), the X-space with p axes, and the Y-space with m axes, p and m being the number of columns in X (terms in the model) and in Y (responses).

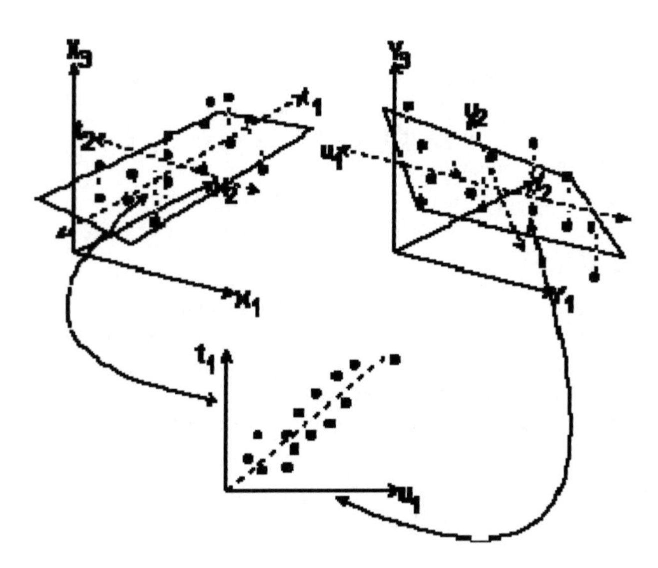

Figure A2.1: Schematic representation of PLS.

The model dimensionally, the number of significant PLS components, is determined by cross-validation (CV), where PRESS (see below) is computed for each model dimension. One selects the number of PLS dimensions that give the smallest PRESS.

A2.1.2.1 Model predictive power

The predictive power of an MLR or a PLS model is given by Q2, which is based on the Prediction Residual Sum of Squares, PRESS. This is a measure of how well the model will predict the responses for new experimental conditions. The computations are repeated several times with, each time, different observations kept out of the calculation of the model. PRESS is then computed as the squared difference between observed Y and predicted Y when the observations (rows in the tables X and Y) were kept out from the model estimation. Q^2 is computed as:

$Q^2 = (SS - PRESS)/SS$

Here SS = sum of squares of Y corrected for the mean. A Q^2 larger than zero indicates that the component is significant and predictive. An overall Q^2 is computed for all PLS components, for all the responses and for each individual response, and represent the percent variation of Y that is predictive. Large Q^2, 0.5 or larger, indicates that the model has good predictive ability and will have small prediction errors. Q^2 is the predictive measure corresponding to the measure of fit, R^2, the percent variation of the response explained by the model.

$R^2 = (SS - SS_{resid})/ SS$

Q^2 gives a lower, conservative estimate of how well the model predicts the outcome of new experiments, while R^2 gives an upper, optimistic estimate.

A2.1.2.2 Automatic cross-validation rules

A PLS component is cross-validated significant if:

Rule 1: PRESS for all Y's together < 1.2

Or

Rule 2: PRESS for at least $M^{1/2}$ Y's < 1.2

And

Rule 3: SS explained for all Y' s together > 1%

Or

Rule 4: SS explained for all separate Y's > 2%

MODDE computes a minimum of two PLS components (if they exist), even if not significant.

A2.2 Model

You may edit the model and add or delete terms. You may add up to third order terms, that is, cubic terms, or three-factor interactions. If your design is singular with respect to your model, MODDE will fit the model with PLS, and MLR will not be available.

A2.2.1 Hierarchy

MODDE enforces hierarchy of the model terms. You cannot delete the constant term. You can only delete a linear term if no higher order term containing the factor is still included in the model.

A2.3 Scaling

A2.3.1 Scaling X

When fitting the model with multiple linear regression, the design matrix X is scaled and centered as specified in the factor definition box: MLR scaling. If the choice is not orthogonal, the condition number will differ from the one displayed using Menu Analysis: Worksheet: Evaluate.

When fitting the model with PLS the X matrix is always scaled to unit variance. If warranted, the scaled X matrix is extended with squares and / or cross terms according to the selected model.

The choices of scaling are: (x denotes the original factor value and z the scaled one)

Orthogonal scaling:

$z_i = (x_i - M)/R$; where M = midrange, R = Range/2.

Midrange scaling:

$z_i = (x_i - M)$

Unit variance scaling:

$z_i = (x_i - M)/s$; where m = average, s = standard deviation computed from the worksheet.

Note that Orthogonal and Midrange scaling are only available with MLR. MODDE default scaling for MLR is the orthogonal scaling.

A2.3.2 Scaling Y

When fitting the model with PLS, the matrix of responses, Y, is scaled by default to unit variance. You can modify the unit variance scaling weight by using the PLS scaling box in the response definition. With MLR the Y's are not scaled.

A2.4 Condition number

The condition number of the orthogonally scaled and centered extended design matrix using the SVD (Singular value decomposition) is computed when clicking **Evaluate** on the **Analysis** menu. The X matrix is taken from the worksheet. The calculation depends on fit method (MLR, PLS) and which factors are involved. A message informs the user how the condition number is calculated.

A2.4.1 Condition number definition

The condition number is the ratio of the largest and the smallest singular values of X (eigenvalues of X'X). This condition number represents a measure of the sphericity of the design (orthogonality). All factorial designs, without center points have a condition number of 1 and the design points are situated on the surface of a sphere.

The condition number is calculated for the extended design matrix (X). The extended design matrix is created as follows:

- The factor values, taken from the worksheet, are centered and scaled according to the factor definition box, MLR scaling. With PLS the condition number is calculated with the factors scaled to unit variance.
- The design matrix is then extended according to the selected model and the condition number is computed.

If you select Midrange scaling, and your factors have different ranges, the condition number of the worksheet becomes very large. This is only a numerical artifact, but due to the fact that MODDE uses the SVD with MLR, the model should be fitted with PLS.

Note: If you selected a different scaling than orthogonal, the condition number will be different than the one computed in Analysis | Evaluate. In particular, if you select Midrange Scaling and your factors have different ranges, the condition number of the worksheet becomes very large. When fitting the model with PLS, the condition number refers to the X matrix, with unit variance scaling.

A2.4.2 Condition number with mixture factors

The condition number with mixture data depends on the method of fit and the type of model.

A2.4.3 PLS and the Cox reference mixture model

When the method of fit is PLS (Cox model) the data are scaled and centered. The condition number is computed from the worksheet, with the slack variable model (mixture factor with the largest variance removed) and all mixture factors scaled orthogonally.

A2.4.4 MLR and the Cox model

The condition number displayed is the condition number of the Cox model (with the Cox constraints), derived from the worksheet, with all mixture factors unscaled and uncentered.

A2.4.5 MLR and the Scheffé model

The condition number of the Scheffé model, derived from the worksheet, with all mixture factors unscaled and uncentered.

Note: With formulation factors when fitting with PLS, the condition number is computed by excluding the factor with the largest range, and scaling the remaining ones orthogonally. When fitting with MLR the condition number is computed without centering and scaling the factors.

A2.5 Missing data

A2.5.1 Missing data in X

Missing data in X are not allowed, and will disable the fit. This also applies to uncontrolled X-variables for MLR, while PLS can handle this. If the user still wants to analyze the data, an arbitrary value can be "filled in" at the place of the missing value, and the row then deleted from the calculations by setting it as "out" in the 4th worksheet column ("Incl/Excl").

A2.5.2 Missing data in Y with MLR

When fitting with MLR all rows with missing data in any Y are excluded from the analysis for all Y's, hence N, displayed on plots and lists, is the number of observations (experiments) without missing data in any Y.

A2.5.3 Missing data in Y with PLS

With PLS, missing data are handled differently. When all Y values are missing in a row, that row is excluded from the analysis. When there are some "present" Y-data in a row, the row is NOT excluded, but included in the projection estimation in PLS. This leads, however, to minor differences in the displayed N and DF at the bottom of plots and lists.

A2.5.4 Y-miss

At the bottom of all plots and lists, MODDE displays Y-miss = the number of missing elements in the total Y matrix (all responses). Y-miss is always equal to 0 when fitting with MLR or with PLS and there are no missing data.

A2.6 N-value

The N-value used in ANOVA, and for the computation of R^2 adjusted, is the actual number of non-missing observations (experiments) for each response-column. This N-value and DF = N-p are displayed at the bottom of the ANOVA plots and lists, and on all residual plots, including observed vs. predicted Y.

A2.7 Residual Standard Deviation (RSD)

The residual standard deviation displayed in the summary table and at the bottom of all plots and lists including ANOVA is computed with the total number of observations (experiments) without excluding the missing values. This is the RSD used in the computation of confidence interval, for coefficients and predictions.

A2.8 ANOVA

The analysis of variance (ANOVA) partitions the total variation of a selected response (SS, Sum of Squares corrected for the mean) into a part due to the regression model and a part due to the residuals.

$SS = SS_{regr} + SS_{resid}$

If there are replicated observations (experiments), the residual sum of squares is further partitioned into pure error (SS_{pe}) and Lack of fit (SS_{lof}).

$SS_{resid} = SS_{pe} + SS_{lof}$

$DF_{resid} = (n - p)$

$$SS_{pe} = \sum_{ki} (e_{ki} - e_k)^2$$

$$DF_{pe} = \sum_{k} (n_k - 1)^2$$

$$DF_{lof} = n - p - \sum_{k} (n_k - 1)^2$$

- n = number of experimental runs (excluding missing values)
- n_k = number of replicates in the kth set
- p = number of terms in the model, including the constant
- e_k = average of the n_k residuals in the kth set of replicates
- j = j^{th} residual in the kth set of replicates
- A goodness of fit test is performed by comparing the MS (mean square) of lack of fit to the MS of pure error.
- Two ANOVA plots are displayed:
  1. The regression goodness of fit test
  2. The LOF goodness of fit test

A2.8.1 Checking for replicates

MODDE checks the rows of the worksheet for replicates. Rows in the worksheet are considered replicates if they match all factor values plus or minus a 5% tolerance.

A2.9 Measures of goodness of fit

MODDE computes and displays some model performance indicators.

A2.9.1 Q2

$Q^2 = (SS - PRESS)/SS$

With

$$PRESS = \sum_i \frac{(Y_i - \hat{Y}_i)^2}{(1 - h_i)^2}$$

and h_i is the ith diagonal element of the Hat matrix:

$X(X'X)^{-1}X'$

A2.9.2 R2

$R^2 = (SS - SS_{resid})/SS$

R^2 adj = $(MS - MS_{resid})/MS$

$MS = SS / (n - 1)$

$MS_{resid} = SS_{resid} / (n - p)$

RSD = Residual standard deviation = $\sqrt{MSE_{resid}}$

A2.9.3 Degrees of freedom

MODDE always computes the real degrees of freedom RDF of the residuals:

$$RDF = n - p - \sum_i (n_i - 1)$$

n = number of experimental runs

n_i = number of replicates in the ith set

p = number of terms in the model, including the constant

A2.9.3.1 Saturated models

When RDF = 0 the model is saturated, and MODDE does not compute or display R^2, R^2 Adjusted or Q^2 when fitting the model with MLR. With PLS, only Q^2 is computed and displayed.

A2.9.3.2 Singular models

Singular models (condition number > 3000) are only fitted with PLS.

If $p > n - \sum_i (n_i - 1)$, the degrees of freedom of the residuals are computed as:

$DF_{resid} = 0$, if no replicates in the design.

A2.10 Coefficients

A2.10.1 Scaled and centered coefficients

The regression coefficients computed and displayed by MODDE refer to the centered and scaled data. You may also select to display the "unscaled and uncentered" coefficients.

A2.10.2 Normalized coefficients

In the overview plot, to make the coefficients comparable between responses, the "centered and scaled" coefficients are normalized with respect to the variation of Y. That is, they are divided by the standard deviation of their respective Y's.

A2.10.3 PLS orthogonal coefficients

The "centered and scaled" coefficients of PLS refer to factor values scaled to unit variance. The PLS orthogonal coefficients re-express the coefficients to correspond to factors centered and orthogonally scaled, i.e. using the Midrange and Low and High values from the Factor Definition. For matrices with condition number < 3000, MLR and PLS compute confidence intervals on coefficients as:

$\sqrt{((X'X)^{-1})} * RSD * t(\alpha/2, DF_{resid})$

For matrices with condition number > 3000, PLS does not compute confidence intervals on the coefficients.

A2.10.4 Confidence intervals

Confidence intervals of coefficients and predictions are computed using the total number of observations, regardless of missing values. This total number of observations is displayed as N at the bottom of all other plots and lists. This approximation is possible because the confidence intervals computed with the regression formulas are somewhat too large because the PLS solution is a shrunk estimator with smaller prediction errors than those of regression. Hence a small number of missing elements in Y does not make the PLS confidence intervals larger than those computed with the regression formulas and the total number of observations.

A2.11 Coding qualitative factors at more than two levels

If a term in the model comprises a qualitative factor, C, at k levels, there will be k-1 expanded terms associated with that term. For example, if the levels of the qualitative factor C are (a, b, c, d) the three expanded terms C(j) are as follows (Figure A2.2):

C	C(2)	C(3)	C(4)
a	-1	-1	-1
b	1	0	0
c	0	1	0
d	0	0	1

Figure A2.2: *Coding of qualitative factors at more than two levels.*

The coefficients of these expanded terms are given as the coefficients for level 2 (b), 3 (c), and 4 (d) of C, while the coefficient for level 1 (a) is computed as the negative sum of the three others. MODDE displays all the four coefficients in the coefficient table but notes that they are associated with only three degrees of freedom.

A2.12 Residuals

A2.12.1 Raw residuals

The raw residual is the difference between the observed and the fitted (predicted) value

$e_i = Y_i - \hat{Y}_i$

The raw residuals are displayed in the residual lists.

A2.12.2 Standardized residuals

The standardized residual is the raw residual divided by the residual standard deviation

$e_i / s \quad (s = RSD)$

These are MODDE default for PLS residual plots.

A2.12.3 Deleted studentized residuals

With MLR, for models with two or more degrees of freedom, deleted studentized residuals are MODDE default when plotting residuals. Deleted studentized residuals are not available with PLS. The deleted studentized residual is the raw residual e_i divided by its "deleted" standard deviation (s_i), which is the residual standard deviation (s_i) computed with observation (i) left out of the analysis, and corrected for leverage, i.e.:

$e_i = e_i /(s_i \sqrt{(1-h_i)})$

where s_i = is an estimate of the residual standard deviation with observation i left out of the model, and h_i is the ith diagonal element of the Hat matrix: $X(X'X)^{-1}X'$

For more information see Belsley, Kuh, and Welsch (1980).

Deleted studentized residuals require at least two degrees of freedom.

A2.13 Predictions

For X matrices with condition number < 3000, both MLR and PLS computes a confidence interval for the average predicted y:

$Y_i + \sqrt{h_i} * RSD * t(\alpha / 2, DF_{resid})$

where h_i is the ith diagonal element of the Hat matrix: $X(X'X)^{-1}X'$

For X matrices with condition number > 3000 and for all Cox mixture models, PLS computes only the standard error of the average predicted Y:

$SE(Y) = \sqrt{[1 / N) + t'_0 * (T'T)^{-1} * t'_0]} * RSD$

A2.14 PLS plots

Both scores and loading plots are available.

A2.14.1 Plot loadings

A2.14.1.1 WC plots (PLS)

Plots of the X- and Y-weights (w and c) of one PLS dimension against another, say, no.'s 1 and 2, show how the X-variables influence the Y-variables, and the correlation structure between the X's and the Y's. In particular, one better understands how the responses vary, their relation to each other, and which ones provide similar information.

A2.14.2 Plot scores

A2.14.2.1 TT, UU, and TU plots (PLS)

The t/t and u/u plots, of the X- and Y-scores of, say, dimensions 1 and 2 (i.e. t_1 vs. t_2, and u_1 vs. u_2), can be interpreted as windows into the X- and Y-spaces, respectively, showing how the design points (experimental conditions, X) and responses profile (Y) are situated with respect to each other. These plots show the possible presence of outliers, groups, and other patterns in the data.

The t/u plots (t_1 vs. u_1, t_2 vs. u_2, etc.) show the relation between X and Y, and display the degree of fit (good fit corresponds to small scatter around the straight line), indications of curvature, and outliers.

A2.15 PLS coefficients

PLS computes regression coefficients (B_m) for each response Y_m expressed as a function of the X's according to the assumed model (i.e. linear, linear plus, interactions or quadratic). These coefficients are (columns of B) computed as:

$B = W(P'W)^{-1} C'$

where W and C are (p*A) and (m*A) matrices whose columns are the vectors w_a and c_a, p = number of terms in the model, m = number of responses, and A = Number of PLS components.

A2.16 Box-Cox plot

A useful family of transformation on the necessarily positive Y's is given by the power transformation:

For not equal to zero $Z = Y^{\lambda}$

For equal to zero $Z = \ln Y$

MODDE, for values of between -2 and 2, computes L_{max} and plots it against the values of λ with a 95% confidence interval.

$L_{max}(\lambda) = ½ \ln(SS_{resid} / n) + (\lambda - 1)* \sum \ln Y$

SS_{resid} is the Residual Sum of Squares after fitting the model $Z = X * \beta + e$ for the selected value of λ.

The value of λ that maximizes $L_{max}(\lambda)$ is the maximum likelihood estimator of λ. The Box-Cox plots displays the values of lambda, λ, vs. the maximum likelihood.

The system uses your data to compute the best mathematical transformation of the response to achieve:

- A simple and parsimonious model.
- An approximately constant model error variance.
- An approximately normal model error distribution.

If the response values vary more than a magnitude of ten in the experimental domain, a transformation is often recommended. The maximum point on the Box-Cox plot gives the value of lambda (λ) for the response transformation Y that gives the best fit of the model. This is the maximum likelihood estimator for λ. For more information see Draper and Smith "Applied Regression Analysis, Second Edition" Wiley, New York or Box and Draper "Response Surface Modeling".

The Box-Cox plot is not available for PLS.

A2.17 Mixture data in MODDE

A2.17.1 Mixture factors only

A2.17.1.1 Model forms

When the investigation includes mixture factors only there are three model types available: Slack variable model, Cox reference model, and Scheffé model.

A2.17.1.2 Slack variable model

When you define a mixture factor as filler, MODDE generates the slack variable model by omitting the filler factor from the model. The model is generated according to the selected objective and is treated as a non-mixture model. You may select MLR or PLS to fit the model as with ordinary process factors. With MLR the factors will be orthogonally scaled.

A2.17.1.3 Cox reference model

When all mixture factors are formulation factors, MODDE generates, by default, the Cox reference model, and the complete polynomial model (linear or quadratic). MODDE supports also a special cubic and a full cubic model.

A2.17.1.4 Scheffé model

You may select to fit a Scheffé model, by selecting **Scheffé MLR** as fit method under the **Analysis** menu. MODDE then expresses the mixture model in the Scheffé form. The full cubic model is not supported as a Scheffé model.

A2.17.1.5 Analysis and fit method

In MODDE, the default fit method with mixture factors is PLS, and the model form is the Cox reference mixture model. All factors, including mixture factors, are scaled to unit variance, by default, prior to fitting. This is also done with mixture factors that have been transformed to pseudo components.

A2.17.1.6 The Cox reference model

The Cox reference model can be fitted by MLR (when obeying mixture hierarchy) and in all cases by PLS. The coefficients in the Cox model are meaningful and easy to interpret. They represent the change in the response when going from a standard reference mixture (with coordinates s_k) to the vertex k of the simplex. In other words when component x_k changes by Δ_k, the change in the response is proportional to b_k. Terms of second or higher degree are interpreted as with regular process variable models. The presence of square terms, though they are not independent, facilitates the interpretation of quadratic behaviour, or departure from non-linear blending. The constant term is the value of the response at the standard reference mixture.

A2.17.1.7 Changing the Standard Reference Mixture

Click **Model / Reference Mixture** from the **Edit** menu to change the coordinates s_k of the standard reference mixture. By default MODDE selects as reference mixture the centroid of the constrained region.

A2.17.1.8 Mixture Hierarchy with the Cox reference model

By default all Cox reference models, linear and quadratic, obey "mixture hierarchy". That is the group of terms constrained by:

$\Sigma b_k s_k = 0$

$\Sigma c_{kj} b_{kj} s_k = 0$ for k = 1,,,,q and for j = 1,,,,q

(c_{kj} = 1 when $j \neq k$ and c_{kj} = 2 when $k = j$.)
are treated as a unit, and terms cannot be removed individually.

If you want to remove terms individually, as with regular process models, clear the **Enforce the mixture model hierarchy** in the **Edit | Model / Reference Mixture** dialog. When the mixture hierarchy is not enforced (this includes cubic models), the Cox reference model can only be fitted by PLS. The coefficients are the regular PLS coefficients computed from the projection and not re-expressed relative to a stated standard reference mixture. Note that in all cases model hierarchy is enforced a term cannot be removed, if a higher order term containing that factor is still in the model.

A2.17.1.9 ANOVA with the Cox model

In the ANOVA table, the degrees of freedom for regression are the real degrees of freedom, taking into account the mixture constraint. These are the same as for the equivalent slack variable model.

A2.17.1.10 Screening plots

When the objective is to find the component effects on the response, the coefficients of the Cox reference linear model are directly proportional to the Cox effects. The Cox effect is the change in the response when component k varies from 0 to 1 along the Cox axis. That is the axis joining the reference point to the $\mathbf{k^{th}}$ vertex.

A2.17.1.11 Effect plot

The effect plot displays the adjusted Cox effects. The adjusted effect of component **k** is:

$k = r_k{*}t_k$; $r_k = U_k\text{-}L_k$; $t_k = b_k/(T\text{-}s_k)$

where r_k is the range of factor k; t_k is the total Cox effect, T is the mixture total (in most cases T=1); b_k is the unscaled uncentered coefficient and s_k is the value of the factor at the reference mixture.

The Effect plot is only available for screening designs using the Cox model.

A2.17.1.12 Main effect plot

For a selected mixture factor X_k, this plot displays the predicted change in the response when X_k varies from its **low** to its **high** level, adjusted for all other mixture factors, that is, by default, the relative amounts of all other mixture factors are kept in the same proportion as in the standard reference mixture (MODDE does not check if the other mixture factors are kept within their ranges). For example, if the Main effect of the mixture factor X_1 is being displayed, when X_1 takes the value x1, the other mixture factors are assigned the values: $x_j = (T - x_1)^* (s_j / T - s_1)$. S_k are the coordinates of the standard reference mixture. The standard reference mixture is the one used in the model. You can change this default and select to have all other mixture factors kept in the same proportion as their ranges (this ensures no extrapolation).

A2.17.1.13 Interaction Plot

The interaction plot is not available when you only have mixture factors.

A2.17.1.14 MLR and the Cox reference model

MODDE you can fit the Cox reference mixture model (linear or quadratic) with MLR, only when they obey mixture hierarchy. When fitting the model with MLR the mixture factors are not scaled and are only transformed to pseudo components when the region is regular. The model is fitted by imposing the following constraints on the coefficients:

Linear models: $\Sigma b_k s_k = 0$ (1)

Quadratic models: $\Sigma b_k s_k = 0$ (1)

$$\Sigma c_{kj} b_{kj} s_k = 0 \text{ for } k = 1,,,,q \text{ and for } j = 1,,,,q \quad (2)$$

Here $c_{kj} = 1$ when $j \neq k$ and $c_{kj} = 2$ when $k = j$, and s_k are the coordinates of the standard reference mixture.

A2.17.1.15 PLS and the Cox reference mixture

With PLS the standard reference mixture is not stated a priori as with multiple linear regression, and no constraints on the coefficients are explicitly imposed. PLS fits the mixture models, and deals with all collinearities by projecting on a lower dimensional subspace. The PLS coefficients can be interpreted as in the Cox model, relative to a reference mixture resulting from the projection, but not explicitly stated.

A2.17.1.16 Expressing PLS coefficients relative to an explicit standard reference mixture

With linear and quadratic models obeying hierarchy it is possible to re-express the PLS coefficients relative to a stated reference mixture with coordinates s_k (s_k expressed in pseudo component, if pseudo component transformation was used). On the fitted PLS model one imposes the following constraints, on the uncentered, unscaled coefficients:

Linear models: $\Sigma b_k s_k = 0$ (1)

Quadratic models: $\Sigma b_k s_k = 0$ (1)

$\Sigma c_{kj} b_{kj} s_k = 0$ for $k = 1,,,,q$ and for $j = 1,,,,q$ (2)

Here $c_{kj} = 1$ when $j \neq k$ and $c_{kj} = 2$ when $k = j$.

The scaled and centered coefficients are recomputed afterwards.

Note: In MODDE, with linear and quadratic models obeying the mixture hierarchy, (i.e. terms constrained by (1) or (2) can only be removed as a group, and not individually), by default the PLS coefficients are always expressed relative to a stated standard reference mixture.

With models containing terms of the third order, or disobeying mixture hierarchy, no constraints are imposed on the PLS coefficients. The coefficients are in this case the regular PLS coefficients and the reference mixture is implicit and results from the projection.

A2.17.1.17 Scheffé models derived from the Cox model

With the linear or quadratic Cox reference model, one can re-express the unscaled coefficients as those of a Scheffé model. The following relationship holds:

Linear: Scheffé, b_k = Cox (PLS) $b_0 + b_k$

Quadratic: Scheffé, b_k = Cox (PLS) $b_0 + b_k + b_{kk}$

Scheffé, b_{kj} = Cox (PLS) $b_{kj} - b_{kk} - b_{jj}$

A2.17.1.18 MLR solution derived from PLS

Because PLS contains multiple regression as a special case, you can with PLS derive the same solution as when you fit the Cox model with multiple linear regression. When you extract as many PLS components as available you get the same solution as MLR.

With MODDE you do the following: First fit the model. MODDE extracts only the significant PLS components. This is the PLS solution. Then click **Next Component** menu and continue extracting PLS components until no correlation between X and Y remains. This is the MLR solution.

A2.17.1.19 Scheffé models

The Scheffé models are only fitted with MLR and only the main effect plot is available. Scheffé models are only available for investigations with all factors being mixture factors.

A2.17.1.20 ANOVA

As described by Snee in "Test Statistics for Mixture Models" (Technometrics, Nov. 1974), the degrees of freedom in the ANOVA table are computed in the same way as with the slack variable model.

A2.17.1.21 Using the model

Prediction plot: This plot is available for all objectives and all model forms. As with process factors, this plot displays a spline representing the variation of the fitted function, when the selected mixture factor varies over its range, adjusted for the other factors. As with the main effect plot, this means that the relative amounts of all other mixture factors are kept in the same proportion as in the standard reference mixture. If no standard reference mixture is specified, the centroid of the constrained region is used as default.

Mixture contour plot: Trilinear contour plots are available with mixture factors and but no 3D plots.

Process and mixture factors together: When you have both process and mixture factors, you can select to treat them as one model, or to specify separate models for the mixture factors, and the process factors. With both mixture and process factors, the only model form available is the Cox reference mixture model.

When the model obeys mixture hierarchy, the PLS coefficients are expressed relative to a stated standard reference mixture. The following constraints are imposed on the coefficients:

Linear models: $\Sigma b_k s_k = 0.$ (1)

Quadratic models: $\Sigma b_k s_k = 0$ (1)

$\Sigma c_{kj} b_{kj} s_k = 0$ for $k = 1,,,,q$ and for $j = 1,,,,q$ (2)

Here $c_{kj} = 1$ when $j \neq k$ and $c_{kj} = 2$ when $k = j$.

If γ (gamma) are the coefficients of the interactions between the process and mixture factors: $\Sigma \gamma_k s_k = 0$.

Note: When the model contains terms of order three, or contains qualitative and formulation factors, the PLS coefficients are not adjusted relative to a stated standard mixture.

A2.17.2 MODDE plots

All MODDE plots are available when you have both mixture and process factors. For both the **Main Effect** and **Prediction plots**, when you select to vary a process factor, all of the mixture factors are set to the values of the standard reference mixture. When you select to vary a mixture factor, process factors are set on their average and the other mixture factors are kept in the same proportion as in the standard reference mixture or their ranges.

A2.18 Optimizer

The **Optimizer** uses a Nelder-Mead simplex method with the fitted response functions, to optimize an overall desirability function combining the individual desirability of each response.

A2.18.1 Desirability when a response is to be Minimum or Maximum

When the response is to be maximized or minimized we construct a one sided continuous desirability function:

$d_i = \exp[-\exp -(c_0 + c_1 Y_i)]$

Y_i = response to be maximized or minimized; c_0 and c_1 are scaling parameters determined by assigning values to the desirability d_i for different levels of Y_i using:

$\ln(\ln(1/d_i)) = c_0 + c_1 Y_i$

MODDE selects $d_i = 1$ for the maximum or the minimum specified values of Y_i, and $d_i = 0.1$ for the worst point (furthest from Max or Min) among the set of points calculated from the five starting simplexes.

A2.18.2 Desirability when a response aims for a Target

When the response is to be on target we construct a two sided desirability function:

$d_i\ (hi) = \exp[-\exp -(b_0 + b_1\ Y_{hi})]$ for values of Y_i above target

and

$d_i\ (lo) = \exp[-\exp -(c_0 + c_1\ Y_{lo})]$ for values of Y_i below target.

In both cases the scaling parameters b_0, c_0 and b_1, c_i are determined by assigning the following values to the desirability d_i for different levels of Y_i:

$\ln(\ln(1/d_i)) = c_0 + c_1\ Y_i$

With $d_i = 1$ when Y_i is equal to target and $d_i = 0.1$ for the worst point (furthest from Target) among the set of points calculated from the 5 starting simplexes

and

$\ln(\ln(1/d_i)) = b_0 + b_1\ Y_i$

With $d_i = 1$ when Y_i is equal to target and $d_i = 0.1$ for the worst point (furthest from Target) among the set of points calculated from the 5 starting simplexes

A2.18.3 Overall desirability

We finally form the overall desirability D as

$\log(D) = \Sigma w_i \log(d_i) / \Sigma w_i$

w_i are the assigned weights representing the importance of the response.

log(D) is the function to be optimized by using the Nelder-Mead Simplex method.

A2.18.4 Starting simplexes

The optimizer starts eight simplexes from eight starting runs, selected from four corners of the experimental region, the overall center, plus the 3 "best" runs from the worksheet. The user can modify these runs or add other preferred factorial combinations.

Each simplex is generated from the starting run by adding an additional run for each factor with an offset of 20% of the distance from the center to the maximum value, the other factors being kept at the same values. A check is made that all runs are within the defined experimental region.

A2.19 Orthogonal blocking

For example, suppose you are running a full factorial with five factors and 32 runs, and the batch size of raw material allow you to perform only eight runs per batch. You may want to run your experiment in 4 blocks, each composed of eight runs using homogeneous material.

The method of dividing 32 runs in four blocks of eight runs, each such as the difference between the blocks (the raw material) does not affect the estimate of the factors, is called *Orthogonal Blocking.*

MODDE supports Orthogonal Blocking for the two-level factorial, fractional factorial, and Plackett-Burman design, CCC, and Box-Behnken designs.

MODDE can also block D-Optimal designs. These designs are more flexible with respect to the number of blocks and the block size, but the blocks in D-Optimal design are not usually orthogonal to the main factors.

The only restriction with D-Optimal designs is that the number of runs must be a multiple of the block size.

Note: Blocking introduces extra factors in the design, and hence reduces the degrees of freedom of the residuals, and the resolution of the design. You should only block when the extraneous source of variability is high and cannot be dealt with by randomizing the run order.

A2.19.1 Blocking full and fractional factorial designs

The block size and the number of blocks of the two-level factorial designs, are always powers of two. The maximum number of blocks supported by MODDE is eight, with a minimum block size of four. The designs are blocked by introducing blocking factors called $BlockV. There is one blocking factor for two blocks, two for four blocks and three for eight blocks. The block effects consist of the effects of the blocking factors and all their interactions. Hence with eight blocks, there are seven block effects using seven degrees of freedom.

MODDE selects the generators of these blocking factors to achieve the highest possible pseudo-resolution of the design. The pseudo-resolution of the design is the resolution of the design when all the block effects (blocking factors and all their interactions) are treated as main effects under the assumption that there are no interactions between blocks and main effects, or blocks and main effects interactions.

A2.19.1.1 Block interaction

An interaction between a main effect and a block effect is called a block interaction. When the design supports the interactions between the block effects and the main effects, the check box, Block interactions, in the Objective dialogue is active. You can mark the check box if you want to add the block interactions to your model.

A2.19.2 Recoding the blocking factors

When blocking factors are generated, the blocks are assigned according to the combination of signs of the blocking factors. For example to generate four blocks, the following scheme of signs of the blocking factors is used:

$B1	$B2	
–	–	Block 1
+	–	Block 2
–	+	Block 3
+	+	Block 4

When you select Show | Design Matrix, the design is displayed in coded unit including the blocking factors. When the worksheet is generated, the blocking factors are recoded and the model is reparameterized. Rather than keeping the d blocking factors, such as $2^d = k$ (the number of blocks), MODDE generates one qualitative variable called $BlockV, with k levels called B1, B2,..Bk.

A2.19.3 Plackett Burman designs

These designs can only be split into two blocks by introducing one block variable, and using its signs to split the design.

A2.19.4 Inclusions and blocks

Adding inclusions to a blocked design is not supported, unless the inclusions belong to one of the blocks present.

A2.19.5 Blocking RSM designs

RSM designs are orthogonally blocked when they fulfill the following two conditions:

1. Each block must be a first order orthogonal block.

2. The fraction of the total sum of squares of each variable contributed by every block must equal the fraction of the total observations (experiments) allotted to the block.

A2.19.5.1 Central Composite Designs

The Central Composite design can be split into two blocks, the cube portion and the star portion, and satisfying the two above conditions when α (the distance of the star points to the center) is equal to

$\alpha = [k\ (1+p_s) / (1 + p_c)\ ]^{1/2}$

where **k** = number of factors, **ns** = number of star point runs, **nc** = number of runs from the cube portion, $\mathbf{p_s}$ = nso/ns proportion of center-points in the star portion, $\mathbf{p_c}$ = nco/nc proportion of center-points in the cube portion.

This is the value of α implemented in MODDE when you select blocking a CCC design. The cube portion of the Central Composite design (CCC) can be split into further blocks if (a) the factorial or the fractional factorial part of the design can be split into orthogonal blocks of pseudo resolution 5 and (b) each one of these blocks have the same number of center-points.

A2.19.5.2 Box Behnken designs

These designs can be orthogonally blocked, as specified by Box and Behnken (1960) and Box and Draper (1987).

A2.19.5.3 Central Composite Face designs

These designs can not be blocked.

A2.19.6 D-Optimal designs

MODDE can block D-Optimal designs, but usually the blocks are not orthogonal to the main factors. The following restrictions apply to blocking D-optimal designs:

1. The blocks must be of equal size.
2. You cannot have interactions between the block factor and the other factors in the model.
3. The selected number of runs of the design must be a multiple of the number of blocks.

MODDE blocks the D-Optimal design by generating a qualitative factor, $BlockV, with as many levels as the selected number of blocks. By default, it then selects only balanced designs with respect to the blocking factor. It is not always possible to generate a balanced D-optimal design with respect to the blocking factor. In this case you may want to change the model, the number of blocks, or generate an unbalanced design.

A2.19.6.1 Random versus fixed factors

You can select to have the block factor treated as fixed or random effect and the predictions computed accordingly. Select the block factor as **Fixed** when the external variability can be **set at will** (it is *controlled*) and the primary objective for blocking is to eliminate that source of variability. A fixed block can be modeled as a controlled qualitative factor with a limited number of levels. All predictions of the responses and contour plots will be made for a selected block level. The block is a fixed effect, for example, if you are making 32 experiments, and each eight runs are done on one of four different machines. There is no other machine than the four available, not now, nor in the future. You may want to have four blocks to eliminate the variability introduced by the machines, but all predictions of the response(s) are made for one of the specific machines.

Select the block factor as **Random** effect when the external variability **cannot** be controlled and set at will, and the primary objective is to make predictions without specifying the block level, and taking into account the external variability. Since the block level of future experiments is unknown, all predictions of the responses for random block effects are made without specifying the block level. The confidence intervals for the responses are increased to account for the uncontrolled external block variability. For example, the block is a random effect if you are making 32 experiments, and each block of eight runs are made with a different batch of raw material. Your primary objective is to make predictions for the next unknown batch of raw material.

A2.19.6.2 Analysis with random effects

When you treat the block factor as **Random** effect it is often desirable to investigate the consistency of the factor effects by including in the model all the interactions of the block factor with the main factors, if possible. In MODDE, the model is always fitted as a fixed effect model, that is with the block factor treated as a controlled qualitative variable, even when the blocks are specified as random.

If the random block interaction effects with the main factors are large and significant, the effect of the main factors varies from block to block, and the confidence intervals on the prediction will be large due to this uncontrolled variability. If the random block interaction effects are small and insignificant, the effects of the main factors are consistent from block to block and the uncertainty of the predictions is greatly reduced. To have a realistic size of the confidence intervals, trim the model and remove all insignificant block interactions effects. If the block factor is specified as fixed effect, the interactions of the main factors with the block factor are of less interest.

A2.19.6.3 Predictions with random effect

The prediction of the responses, when the block effect is specified as random, are computed without specifying the block level. MODDE uses the average block level to predict the

response but the confidence interval is increased to take into account the variability of the response due to the different blocks, plus the variability of the response due to uncertainty on the coefficients of the model including all the terms with the block factor.

Appendix III: Explaining DOE to children

A3.1 Objective

The example described in this appendix is a personal account from real life, related by one of the authors (Lennart Eriksson). It is included in this book because it shows how you can use DOE to solve an everyday problem. Informally, we could say that the objective of this appendix is to explain how you can make DOE understandable to children.

A3.2 Introduction

How do you explain to your kids what you do at work? I told my two sons, aged 7 and 10, that we do 'intelligent experiments'. Apparently, this was not sufficient to satisfy their curiosity, so I was asked several times during the recent summer break to reveal the secrets. In particular, I remember one situation that occurred at the famous Legoland (in Billund, Denmark) where all this started (Figure A3.1). We were having some popcorn and cotton candy. The cotton candy was quickly eaten by my two sons, but the popcorn did not taste good.

I realized that this was a good opportunity to introduce DOE as a way of "optimizing" the taste of popcorn. My two sons and I agreed that when we were back at home we should immediately set up an optimization design to study how microwave popcorn would be affected by changing the cooking time and the power of the microwave oven. And, as suggested by the ten-year-old, the name of the investigation should be 'Mission Popcorn'. The kids were very satisfied with this idea. It meant to them – I realized much later – that their dad had actually sanctioned an excessive consumption of popcorn within a short time-frame.

Figure A3.1: *Where it all started, at Legoland, Denmark. To the left and right are Andreas (aged 10) and Mattias (aged 7). In the middle is a somewhat older boy, who enjoys being with, and loves, his two sons enormously.*

A3.3 Example: Mission Popcorn

The objective of this experiment was to explain to children what DOE means, using an everyday problem (i.e., how to get good popcorn from the microwave) as an illustration. The data-set contains two factors, Time and Power, both adjustable on a continuous scale (Figure A3.2).

- Time (seconds), low level 170 seconds, high level 210 seconds.
- Power (watts), low level 600 watts, high level 800 watts.

Factors

	Name	Abbr.	Units	Type	Use	Settings	Transform	Prec.	MLR Scale	PLS Scale
1	Time	Tim	seconds	Quantitative	Controlled	170 to 210	None	Free	Orthogonal	Unit Variance
2	Power	Pow	W	Quantitative	Controlled	600 to 800	None	Free	Orthogonal	Unit Variance

Figure A3.2: *The two factors of the popcorn experiment.*

The dataset also contains two responses, Kernels and Taste (Figure A3.3).

- Kernels, this is simply the number of unpopped kernels.
- Taste, each person expressed his liking on a five-point scale (1=bad taste, …., 5=optimal taste). The response value is the sum across three people (we could not use the average as this was too complex for the younger brother).

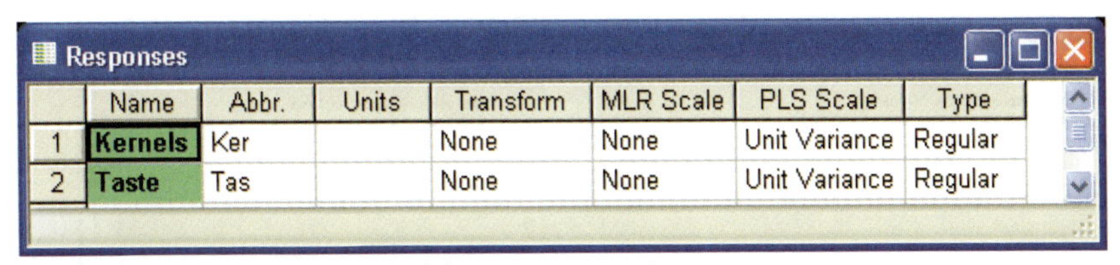

	Name	Abbr.	Units	Transform	MLR Scale	PLS Scale	Type
1	Kernels	Ker		None	None	Unit Variance	Regular
2	Taste	Tas		None	None	Unit Variance	Regular

Figure A3.3: The two responses of the popcorn experiment.

The design used was a CCF design, by default encoding 8+3 experiments in MODDE 8. One center-point was omitted since we bought a ten-pack of microwave popcorn (Figure A3.4).

	1	2	3	4	5	6	7	8
1	Exp No	Exp Name	Run Order	Incl/Excl	Time	Power	Kernels	Taste
2	1	N1	9	Incl	170	600	142	12
3	2	N2	2	Incl	210	600	53	11
4	3	N3	10	Incl	170	800	11	10
5	4	N4	3	Incl	210	800	0	3
6	5	N5	1	Incl	170	700	48	12
7	6	N6	6	Incl	210	700	0	8
8	7	N7	8	Incl	190	600	121	12
9	8	N8	4	Incl	190	800	1	7
10	9	N9	5	Incl	190	700	22	14
11	10	N10	7	Incl	190	700	18	13

Figure A3.4: The worksheet for the CCF design.

A3.4 Stage 1: Evaluation of raw data

In the analysis of the measured data, we first had to visually display the design used; a CCF design in 10 runs (Figure A3.5).

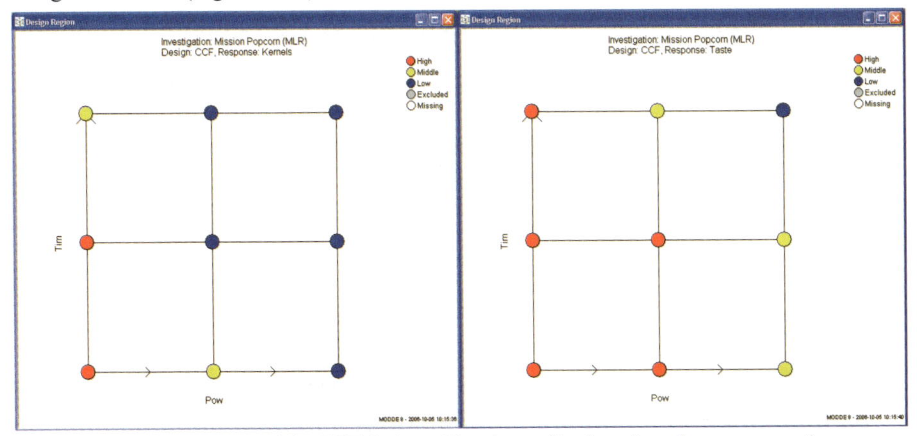

Figure A3.5: The geometry of the CCF design with color coding based on the response values.

The replicate plots indicate small variability among the replicates (Figure A3.6). The results for Kernels are not close to being normally distributed.

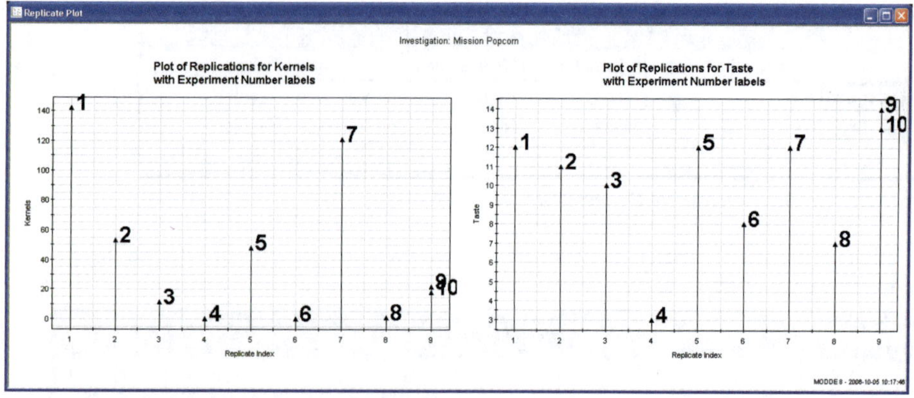

Figure A3.6: Replicate plots of the two responses.

However, the histogram plots still suggest we should proceed with the data analysis without transformations (Figure A3.7, very few data points make the shapes of the data distributions uncertain).

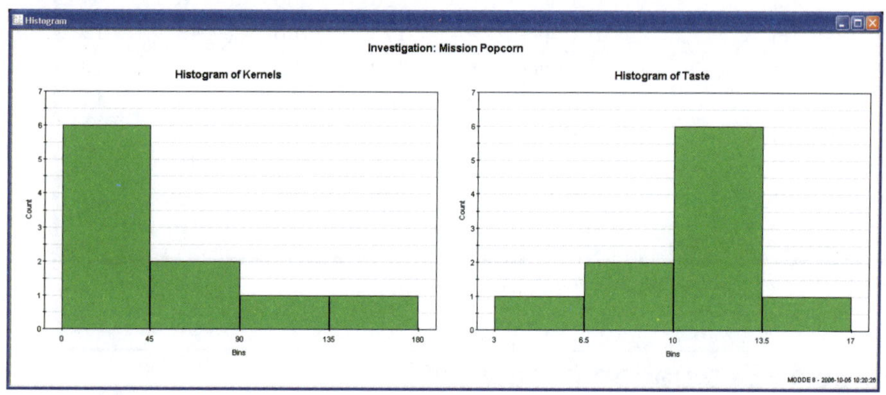

Figure A3.7: Histogram plots of the two responses.

A3.5 Stage 2: Data analysis and model interpretation

When fitting the default quadratic model to the data, we obtained surprisingly strong models (Figure A3.8).

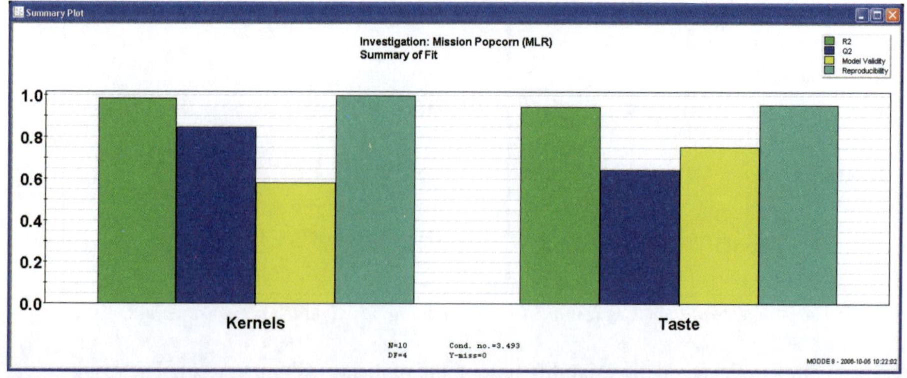

Figure A3.8: Summary of fit plots for the models of the two responses.

The regression coefficients show that Power has a stronger influence than Time on Kernels (Figure A3.9). To minimize the number of Kernels, both factors should be set high. Time and Power seem to have a similar impact on Taste. Adjusting both factors at a lower value corresponds to increasing the Taste score. Furthermore, because the two models are so good, no attempts at model refinement seem to be necessary.

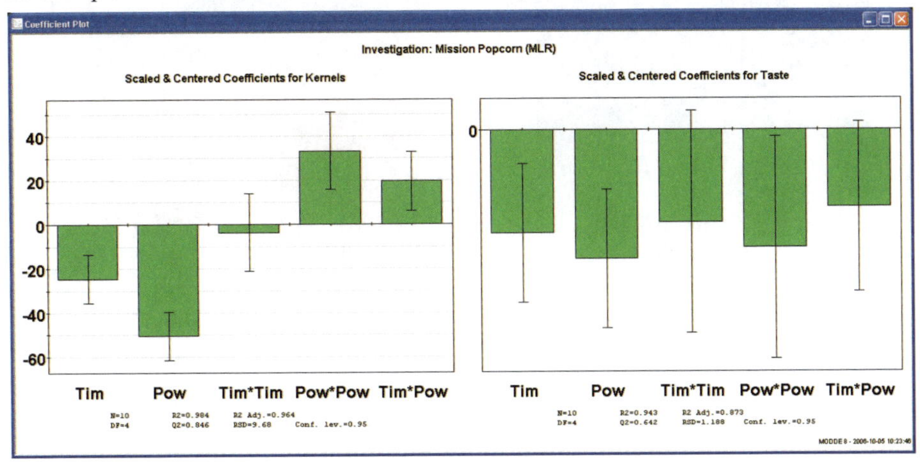

Figure A3.9: *Regression coefficient plots for the models of the two responses.*

A3.6 Stage 3: Use of the model for prediction

The third task was to use the model and, by means of the software optimizer, to try to pinpoint an optimal factor combination. To this end the following desirabilities were used (Figure A3.10):

	Response	Criteria	Weight	Min	Target	Max
1	**Kernels**	Minimize	1		0	30
2	**Taste**	Maximize	1	12	15	

Figure A3.10: *Desirabilities for the two responses.*

These desirabilities should be interpreted in the following ways. The number of Kernels should, ideally, be at the target value of 0 (zero). If the target (zero) is not achieved, the number of Kernels should definitely stay below the maximum of 30. Thirty was considered to be reasonable, since this would mean 10 unpopped kernels per bowl and consumer. Analogously, the Taste should ideally be at the target value of 15. If this could not be achieved, it was argued that we should at least exceed a minimum of 12.

Using these desirabilities and the MODDE optimizer an optimization was undertaken. The optimizer was run without changing the factor settings and convergence was instantaneous; the results are plotted in Figure A3.11. The contour plot for Taste indicates there is a point that encodes the highest taste score and that this is located within the investigated region, i.e., around Time ≈ 182 seconds and Power ≈ 657 watt. This point does not, however, correspond to the lowest number of Kernels.

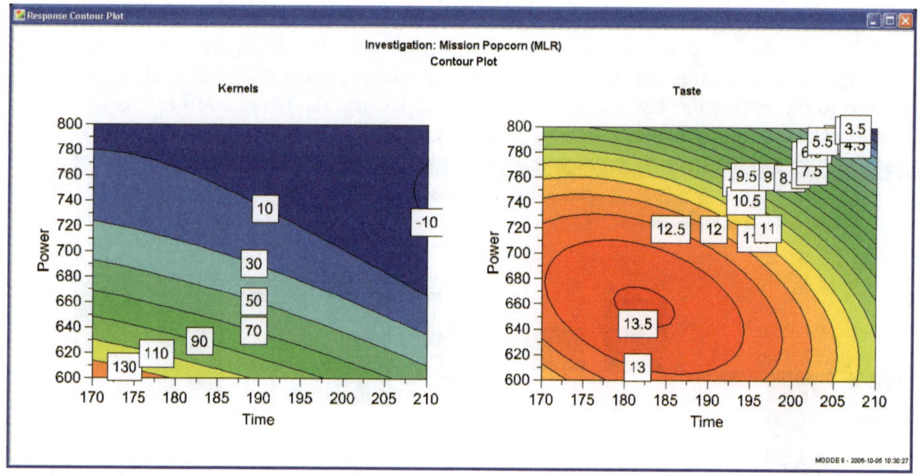

Figure A3.11: Response contour plots for the two responses.

Subsequently, we created a Sweet Spot plot to overview the system (Figure A3.12). This plot can be thought of as an overlay of several response contour plots, and colored according to the number of criteria (desirabilities) that are met. The sweet spot exists where all criteria are fulfilled. As shown by Figure A3.12 below, there is an optimum region within the searched space.

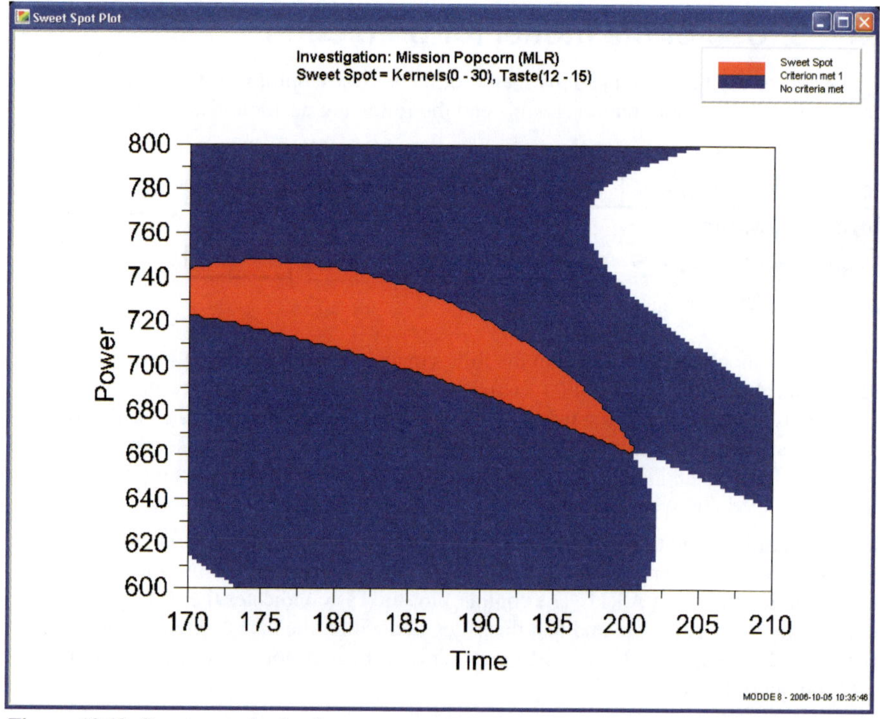

Figure A3.12: Sweet spot plot for the two responses.

A3.7 Conclusions

Based on our joint efforts we were able to discover a suitable combination of Time (= 190 seconds) and Power (= 700 watt). We are currently using this combination to great satisfaction. It produces popcorn that tastes good, but without undesirable side effects such as burning, unpleasant odor or large frequencies of unpopped kernels. One resulting bag is pictured below (Figure A3.13).

Figure A3.13: *A bag of popcorn containing the product with the optimum combination of taste, numbers of popped kernels and smell.*

The final result (apart from the popcorn) for the two end users (i.e., the two boys) was a better understanding of dad's work, plus having a lot of fun together with their father!

References

A foreword

This is a list covering a small selection of useful references (books and articles) in the fields of design of experiments (DOE) and multivariate data analysis (MVDA). It is emphasized that this is by no means an exhaustive account of the available literature. Rather, this compilation highlights references which may guide the reader for further studies.

References for DOE

Books

- Box, G.E.P., Hunter, W.G., and Hunter, J.S., Statistics for Experimenters, John Wiley & Sons, Inc., New York, 1978.
- Cornell, J.A., Experiments with mixtures, John Wiley & Sons, Inc., New York, 1981.
- Bayne, C. K., and Rubin, I.B., Practical Experimental Designs and Optimization Methods for Chemists, VCH Publishers, Inc., Deerfield Beach, Florida, 1986.
- Box, G.E.P., and Draper, N.R., Empirical Model-Building and Response Surfaces, John Wiley & Sons, Inc., New York, 1987.
- Haaland, P.D., Experimental Designs in Biotechnology, Marcel Dekker, Inc., New York, Basel, 1989).
- Montgomery, D.C., Design and Analysis of Experiments, John Wiley & Sons, New York, 1991, ISBN 0-471-52994-X.
- Morgan, E., Chemometrics: Experimental Design, John Wiley & Sons, N.Y., 1991.
- Nortvedt, R., et al., Anvendelse av kjemometri innen forskning og industri, Tidskriftsforlaget Kjemi AS, 1996, ISBN 82-91294-01-1.
- Carlson, R., and Carlson, J., Desigh and Optimization in Organic Synthesis, Volume 24: Second Revised and Enlarged Edition, Data Handling in Science and Technology, Elsevier, 2005.

Articles

- Rechtschaffner, R.L., Saturated fractions of 2^n and 3^n factorial designs, Technometrics, 9, 1967, 569-575.
- Doehlert, D.H., Uniform shell designs, Journal of the Royal Statistical Society, Series C, 19, 1970, 231-239.
- Cox, D.R., A Note on Polynomial Response Functions for Mixtures, Biometrica, 58, 1971, 155-159.

- Marquardt, D.W., and Snee, R.D., Test Statistics for Mixture Models, Technometrics, 16, 1974, 553-537.
- Hendrix, C., What Every Technologist Should Know About Experimental Design, Chemtech, 9, 1979, 167-174.
- Crosier, R.B., Mixture Experiments: Geometry and Pseudo-components, Technometrics, 26, 1984, 209-216.
- Hunter, J.S., Applying Statistics to Solving Chemical Problems, Chemtech, 17, 1987, 167-169.
- Dick, C.F., et al., Determination of the sensitivity of a tablet formulation to variations in excipient levels and processing conditions using optimization techniques, Int. J. Pharm., 38, 1987, 23-31.
- Kettaneh-Wold, N., Analysis of mixture data with partial least squares, Chemometrics and Intelligent laboratory Systems, 14, 1992, 57-69.
- Grize, Y.L., A Review of Robust Process Design Approaches, Journal of Chemometrics 9, 1995, 239-262.
- Ahlinder, S., et al., Smart Testing – Reaping the Benefits of DOE, Volvo Technology Report No 2 1997, www.volvo.se/rt/trmag/index.html.
- Nyström, A., and Karlsson, A., Enantiomeric Resolution on Chiral-AGP with the aid of Experimental Design. Unusual Effects of Mobile Phase pH and Column Temperature, Journal of Chromatography A, 763, 1997, 105-113.
- Eriksson, L., Johansson, E., Wikström, C., Mixture Design – Design Generation, PLS Analysis and Model Usage, Chemometrics and Intelligent Laboratory Systems, 43, 1998, 1-24.
- Lundstedt, T., et al., Experimental Design and Optimization, Chemometrics and Intelligent Laboratory Systems, 42, 1998, 3-40.
- Rappaport, K.D., et al., Perspectives on Implementing Statistical Modeling and Design in an Industrial/Chemical Environment, The American Statistician, 52, 1998, 152-159.
- Moberg, L., and Karlberg, B., Validation of a multivariate calibration method for the determination of chlorophyll a, b, and c and their corresponding pheopigments, Analytica Chimica Acta, 450, 2001, 143-153.
- Rambali, B., Baert, L., Thone, D., and MAssart, D.L., Using Experimental Design to Optimize the Process Parameters in Fluidized Bed Granulation, Drug Development and Industrial Pharmacy, 27, 2001, 47-55.
- Lindberg, N.O., and Gabrielsson, J., Use of software to facilitate pharmaceutical formulation – experiences from a tablet formulation, Journal of Chemometrics, 18, 2004, 133-138.
- Wikström, M., and Sjöström, M., Identifying cause of quality defect in cheese using qualitative variables in a statistical experimental design, Journal of Chemometrics, 18, 2004, 139-145.
- Flåten, G.R., and Walmsley, A.D., A design of experiment approach incorporating layered designs for choosing the right calibration model, Chemometrics and Intelligent Laboratory Systems, 73, 2004, 55-66.
- Katkevica, D., Trapencieris, P., Boman, A., Kalvins, I., and Lundstedt, T., The Nenitzescu reaction. An initial screening of experimental conditions for improvement of the yield of a model reaction., Journal of Chemometrics, 18, 2004, 183-187.

- Wormbs, G., Larsson, A., Alm, J., Tunklint-Aspelin, C., Strinning, O., Danielsson, E., and Larsson, H., The use of design of experiment and sensory analysis as tools for the evaluation of production methods for milk, Chemometrics and Intelligent Laboratory Systems, 73, 2004, 67-71.
- Gabrielsson, J., Multivariate Methods in Tablet Formulation, Ph.D. Thesis, Umeå University, Sweden, 2004.
- Nordfelth, R., et al., Small-Molecule Inhibitors Specifically Targeting Type III Secretion, Infection and Immunity, 73, 2005, 3104-3114.
- Goupy, J., Factorial experimental design: Detecting an outlier with the dynamic variable and the Daniel´s diagram, Chemometrics and Intelligent Laboratory Systems, 80, 2006, 156-166.
- Cau Dit Coumes, C., and Courtois, S., Design of experiments to investigate main effects and all two-factor interactions when one of the factors has more than two levels - application to nuclear waste cementation, Chemometrics and Intelligent Laboratory Systems, 80, 2006, 167-175.
- Baati, R., Kamoun, A., Chaabouni, M., Sergent, M., and Phan Tan Luu, R., Screening and optimization of the factors of a detergent admixture preparation, Chemometrics and Intelligent Laboratory Systems, 80, 2006, 198-208.
- Rosén, P., Experimental Design in Assay Development, Masters Thesis, Dept. of Chemical and Biological Engineering, Chalmers University of Technology, Gothenburg, Sweden, 2006.
- Qu, W., Statistical properties of Rechtschaffner designs, Journal of Statistical Planning and Inference, 37, 2007, 2156-2164.
- Andersson, M., Ringberg, A., and Gustafsson, C., Multivariate methods in tablet formulation for early drug development: Predictive models from a screening design of several linked responses, Chemometrics and Intelligent Laboratory Systems, 87, 2007, 151-156.
- Lundgren, J., Salomonsson, J., Gyllenhaal, O., and Johansson, E., Supercriticial fluid chromatography of metoprolol and analogues on aminopropyl and ethylpyridine silica without any additives, Journal of Chromatography A, 1154, 2007, 360-367.
- Andersson, M., Ringberg, A., and Gustafsson, C., Multivariate methods in tablet formulation suitable for early drug development: Predictive models from a screening design of several linked responses, Chemometrics and Intelligent Laboratory Systems, 87, 2007, 151-156.
- Johnson, R., et al., Applications of Multivariate Data Analysis in Biotech Processing, BioPharm International, October 2007, 130-144.

References for MVDA

Books

- Jollife, I.T., Principal component analysis, Springer-Verlag, New York, 1986, ISBN 0-387-96269-7.
- Draper, N.R., and Smith, H., Applied Regression Analysis, Second Edition, Wiley, New York, 1987.
- Martens, H., and Naes, T. Multivariate calibration, John Wiley, New York, 1989.

- Jackson, J.E., A user's guide to principal components, John Wiley, New York, 1991, ISBN 0-471-62267-2.
- Barnett, V., and Lewis, T., Outliers in Statistical Data, 3rd edition, John Wiley & Sons, New York, 1994, ISBN 0-471-93094-6.
- Anthology, Anvendelse av Kjemometri innen forskning og industri, Tidsskriftforlaget Kjemi AS, Bergen, Norway, 1996, ISBN 82-91294-01-1.
- Höskuldsson, A., Prediction Methods in Science and Technology, Thor Publishing, Copenhagen, Denmark, 1996, ISBN 87-985941-0-9.
- Massart, D.L., et al., Handbook of Chemometrics and Qualimetrics. Part A and B. Elsevier, Amsterdam, 1998.
- Eriksson, L., Johansson, E., Kettaneh-Wold, N., Trygg, J., Wikström, C., and Wold, S., Multi- and Megavariate Data Analysis – Part I: Basic Principles and Applications, Umetrics Academy, 2006, ISBN 91-973730-2-8.
- Eriksson, L., Johansson, E., Kettaneh-Wold, N., Trygg, J., Wikström, C., and Wold, S., Multi- and Megavariate Data Analysis – Part II: Advanced Applications and Method Extension, Umetrics Academy, 2006, ISBN 91-973730-3-6.

Articles, general

- Wold, S., Esbensen, K., and Geladi, P., Principal Component Analysis, Chemometrics and Intelligent Laboratory Systems, 2, 1987, 37-52.
- Höskuldsson, A., PLS Regression Methods, Journal of Chemometrics, 2, 1998, 211-228.
- Ståhle, L., and Wold, S., Multivariate Data Analysis and Experimental Design in Biomedical Research, In: Ellis, G.P., and West, G.B. (Eds) Progress in Medical Chemistry, Elsevier Science Publishers, 1998, 291-338.
- Wold, S., Albano, C., and Dunn W.J., et al., Multivariate Data Analysis: Converting Chemical Data tables to plots, In: Computer Applications in Chemical Research and Education, Heidelberg, Dr. Alfred Hütig Verlag, 1989.
- Stone M., and Brooks R.J., Continuum regression: Cross-validated sequentially constructed prediction embracing ordinary least squares, partial least squares and principal components regression, Journal of the Royal Statistical Society, Ser. B, 52,, 1990, 237-269.
- Frank, I.E., and Friedman, J.H., A Statistical View of Some Chemometrics Regression Tools, Technometrics, 35, 1993, 109-148.
- Wold, S., Exponentially Weighted Moving Principal Components Analysis and Projections to Latent Structures, Chemometrics and Intelligent Laboratory Systems, 23, 1994, 149-161.
- Wold, S., Eriksson, L., and Sjöström, M., PLS in Chemistry, In: The Encyclopedia of Computational Chemistry, Schleyer, P. v. R.; Allinger, N. L.; Clark, T.; Gasteiger, J.; Kollman, P. A.; Schaefer III, H. F.; Schreiner, P. R., Eds., John Wiley & Sons, Chichester, 1999, pp. 2006-2020.
- Wold, S., Sjöström, M., and Eriksson, L., PLS-Regression: A Basic Tool of Chemometrics, Chemometrics and Intelligent Laboratory Systems 58, 2001, 109-130.
- Eriksson, L., Antti, H., Holmes, E., Johansson, E., Lundstedt, T., Shockcor, J., and Wold, S., PLS in Cheminformatics. In: Gasteiger, J., (Ed.), Handbook of Chemoinformatics – From Data to Knowledge. Wiley-VCH, 2003, pp 1134-1166.

- Eriksson, L., Antti, H., Gottfries, J., Holmes, E., Johansson, E., Long, I., Lindgren, F., Lundstedt, T., Trygg, J., and Wold, S., Using Chemometrics for Navigating in the Large Data Sets of Genomics, Proteomics & Metabonomics (GPM). Analytical and Bioanalytical Chemistry, 380, 2004, 419-429.

Articles, process

- Kresta, J.V., MacGregor J.F., and Marlin T.E., Multivariate Statistical Monitoring of Process Operating Performance, The Canadian Journal of Chemical Engineering, 69, 1991, 35-47.
- Kourti, T., and MacGregor, J.F., Process Analysis, Monitoring and Diagnosis, Using Multivariate Projection Methods, Chemometrics and Intelligent Laboratory Systems, 28, 1995, 3-21.
- MacGregor, J.F., Using, On-line Process Data to Improve Quality, ASQC Statistics Division Newsletter, Vol. 16., No. 2., 1996, 6-13.
- Nijhuis, A., de Jong, S., and Vandeginste, B.G.M., Multivariate Statistical Process Control in Chromatography, Chemometrics and Intelligent Laboratory Systems, 38, 1997, 51-61.
- Rännar, S., McGregor, J.F., and Wold, S., Adaptive Batch Monitoring Using Hierarchical PCA, Chemometrics and Intelligent Laboratory Systems, 41, 1998, 73-81.
- Wikström, C., et al., Multivariate Process and Quality Monitoring Applied to an Electrolysis Process – Part I. Process Supervision with Multivariate Control Charts, Chemometrics and Intelligent Laboratory Systems, 42, 1998, 221-231.
- Wikström, C., et al., Multivariate Process and Quality Monitoring Applied to an Electrolysis Process – Part II. Multivariate Time-series Analysis of Lagged Latent Variables, Chemometrics and Intelligent Laboratory Systems, 42, 1998, 233-240.
- Wold, S., et al., Modelling and Diagnostics of Batch Processes and Analogous Kinetic Experiments, Chemometrics and Intelligent Laboratory Systems, 44, 1998, 331-340.
- Jaeckle, C.M., and MacGregor, J.F., Industrial Applications of Product Design Through Inversion of Latent Variable Models, Chemometrics and Intelligent Laboratory Systems, 50, 2000, 199-210.
- Eriksson, L., Hagberg, P., Johansson, E., Rännar, S., Whelehan, O., Åström, A., and Lindgren, T., Multivariate Process Monitoring of a Newsprint Mill. Application to Modelling and Predicting COD Load Resulting from De-Inking of Recycled Paper, Journal of Chemometrics, 15, 2001, 337-352.
- Jonsson, P., Sjöström, M., Wallbäcks, L., and Antti, H., Strategies for implementation and validation of on-line models for multivariate monitoring and control of wood chips properties, Journal of Chemometrics, 18, 2004, 203-207.

Articles, multivariate calibration

- Brown, P.J., Multivariate Calibration, Journal of the Royal Statistical Society, B44, 1982, 287-321.
- Beebe, K.R. and Kowalski, B.R., An Introduction to Multivariate Calibration and Analysis, Analytical Chemistry, 57, 1987, 1007-1017.
- Bro, R., Håndbog i Multivariabel Kalibrering, KVL, Copenhagen, Denmark, 1996.
- Trygg, J., and Wold, S., PLS Regression on Wavelet Compressed NIR Spectra, Chemometrics and Intelligent Laboratory Systems, 42, 1998, 209-220.

- Swierenga, H., et al., Improvement of PLS Model Transferability by Robust Wavelength Selection, Chemometrics and Intelligent Laboratory Systems, 41, 1998, 237-248.
- Wold, S., Antti, H., et al., Orthogonal Signal Correction of Near-Infrared Spectra, Chemometrics and Intelligent Laboratory Systems, 43, 1998, 123-134.
- Svensson, O., Josefsson, M., and Langkilde, F.W., Reaction Monitoring using Raman Spectroscopy and Chemometrics, Chemometrics and Intelligent Laboratory Systems, 49, 1999, 49-66.
- Janné, K., Pettersen J., N.-O. Lindberg and Lundstedt, T., Hierarchical principal component analysis (PCA) and projection to latent structure (PLS) technique on spectroscopic data as a data pretreatment for calibration, Journal of Chemometrics, 15, 2001, 203-213.
- Wiberg, K., Hagman, A., Burén, P., and Jacobsson, S.P., Determination of the content and identity of lidocaine solutions with UV-visible spectroscopy and multivariate calibration, Analyst, 126, 2001, 1142-1148.
- Dyrby, M., Engelsen, S.B., Nörgaard, L., Bruhn, M., and Lundsberg-Nielsen, L., Chemometric Quantitation of the Active Substance (Containing CN) in a Pharmaceutical Tablet Using Near-infrared (NIR) Transmittance and NIR FT-Raman Spectra, Applied Spectroscopy, 56, 2002, 579-585.
- Blanco, M., Peinado, A.C., and Mas, J., Analytical monitoring of alcoholic fermentation using NIR spectroscopy, Biotechnology and Bioengineering, 88, 2004, 536-542.
- Trygg, J., Prediction and spectral profile estimation in multivariate calibration, Journal of Chemometrics, 18, 2004, 166-172.
- Andersson, M., Svensson, O., Folestad, S., Josefson, M., and Wahlund, K.G., NIR spectroscopy on moving solids using a scanning grating spectrometer – impact on multivariate process analysis, Chemometrics and Intelligent Laboratory Systems, 75, 2005, 1-11.
- Wallbäcks, L., Multivariate data analysis of multivariate populations, Chemometrics and Intelligent Laboratory Systems, 86, 2007, 10-16.
- Christensen, J., Nörgaard, L., Bro, R., and Balling Engelsen, S., Multivariate autoflouresence of intact food systems, Chemical Reviews, 2007.

Articles, multivariate characterization

- Carlson, R., Lundstedt, T., and Albano, C., Screening of Suitable Solvents in Organic Synthesis – Strategies for Solvent Selection, Acta Chemica Scandinavica, B39, 1985, 79-91.
- Wallbäcks, L., Edlund, U. and Nordén, B., Multivariate Characterization of Pulp Using Solid-State ^{13}C NMR, FTIR and NIR, Tappi Journal, 74, 1991, 201-206.
- Cocchi, M., et.al., Theoretical versus Empirical Molecular Descriptors in Monosubstituted Benzenes – A Chemometric Study, Chemometrics and Intelligent Laboratory Systems, 12, 1992, 209-224.
- Eriksson, L., Verhaar, H.J.M., and Hermens, J.L.M., Multivariate Characterization and Modelling of the Chemical Reactivity of Epoxides, Environmental Toxicology and Chemistry, 13, 1994, 683-691.

- Lindgren, Å., and Sjöström, M., Multivariate Physico-Chemical Characterization of Some Technical Non-Ionic Surfactants, Chemometrics and Intelligent Laboratory Systems, 23, 1994, 179-189.
- Andersson, P., Haglund, P., and Tysklind, M., Ultraviolet Absorption Spectra of all 209 Polychlorinated Biphenyls Evaluated by Principal Component Analysis, Fresenius Journal of Analytical Chemistry, 357, 1997, 1088-1092.
- Marvanova, S., Nagata, Y., Wimmerova, M., Sykorova, J., Hynkova, K., and Damborsky, J., Biochemical characterization of broad-specificity enzymes using multivariate experimental design and a colorimetrics microplate assay: characterization of the haloalkane dehalogenase mutants, Journal of Microbiological Methods, 44, 2001, 149-157.
- Duarte, I., Barros, A., Belton, P.S., Righelato, R., Spraul, M., Humper, E., and Gil, A.M., High-resolution nuclear magnetic resonance spectroscopy and multivariate analysis for the characterization of beer, Journal of Agricultural and food chemistry, 50, 2002, 2475-2481.

Articles, QSAR

- Eriksson, L., Hermens, J.L.M., et al., Multivariate Analysis of Aquatic Toxicity Data with PLS, Aquatic Sciences, 57, 1995, 217-241.
- Eriksson, L., and Johansson, E., Multivariate Design and Modeling in QSAR, Chemometrics and Intelligent Laboratory Systems, 34, 1996, 1-19.
- Verhaar, H.J.M., Hermens, J.L.M., et al., Classifying Environmental Pollutants. Separation of Class 1 and Class 2 Type Compounds Based on Chemical Descriptors, Journal of Chemometrics, 10, 1996, 149-162.
- Goodford, P., Multivariate Characterization of Molecules for QSAR Analysis, Journal of Chemometrics, 10, 1996, 107-117.
- Lindgren, Å., et al., Quantitative Structure-Effect Relationships for Some Technical Non-ionic Surfactants, Journal of the American Oil Companies Society, 73, 1996, 863-875.
- Sandberg, M., et al., New Chemical Descriptors Relevant for the Design of Biologically Active Peptides. A multivariate Characterization of 87 Amino Acids, Journal of Medicinal Chemistry, 41, 1998, 2481-2491.
- Giraud, E., Luttmann, C., Lavelle, F., Riou, J.F., Mailliet, P., and Laoui, A., Multivariate data analysis using D-optimal designs, partial least squares, and response surface modelling, A directional approach for the analysis of farnesyltransferase inhibitors, Journal of Medicinal Chemistry, 43, 2000, 1807-1816.
- Cruciani, G., Baroni, M., Carosati, E., Clementi, M., Valigi, R., and Clementi, S., Peptide studies by means of principal properties of amino acids derived from MIF descriptors, Journal of Chemometrics, 18, 2004, 146-155.
- Wold, S., Josefson, M., Gottfries, J., and Linusson, A., The utility of multivariate design in PLS modeling, Journal of Chemometrics, 18, 2004, 156-165.
- Eriksson, L., Arnhold, T., Beck, B., Fox, T., Johansson, E., and Kriegl, J.M., Onion design and its application to a pharmaceutical QSAR problem, Journal of Chemometrics, 18, 2004, 188-202.
- Gilbert, A.M., Kirisits, M., Toy, P., Nunn, D.S., Failli, A., Dushlin, E.G., Novikova, E., Petersen, P.J., Joseph-McCarthy, D., McFadyen, I., and Fritz, C.C., Anthranilate

4H-oxazol-5-ones: Novel small molecule antibacterial acyl carrier protein synthase inhibitors, Bioorganic and Medicinal Chemistry Letters, 14, 2004, 37-42.

- Eriksson, L., Gottfries, J., Johansson, E., and Wold, S., Time-resolved QSAR: an approach to PLS modelling of three-way biological data, Chemometrics and Intelligent Laboratory Systems, 73, 2004, 73-84.
- Wu, J., Claesson, O., Fängmark, I.E., and Hammarström, L.G., A systematic invstigation of the overall rate coefficient in the Wheeler-Jonas equation for adsorption on dry activated carbons, Carbon, 43, 481-490, 2005.
- Larsson, A., Johansson, S.M.C., Pinkner, J.S., Hultgren, S.J., Almqvist, F., Kihlberg, J., and Linusson, A., Multivariate design, synthesis, and biological evaluation of peptide inhibitors of FimC/FimH protein-protein interactions in uropathonegic Escherichia Coli, Journal of Medicinal Chemistry, 2005.
- Eriksson, L., Andersson, P.M., Johansson, E., and Tysklind, M., Megavariate analysis of environmental QSAR data. Part I - A basic framework founded on principal component analysis (PCA), partial least squares (PLS), and statistical molecular design (SMD). Molecular Diversity 10, 2006, 169-186.
- Eriksson, L., Andersson, P.M., Johansson, E., and Tysklind, M., Megavariate analysis of environmental QSAR data. Part II – Investigating very complex problem formulations using hierarchical, non-linear and batch-wise extensions of PCA and PLS. Molecular Diversity 10, 2006, 187-205.

Index

2

7

A

B

C

D

E

F

G

H

I

L

Q

R

S

T

U

V

W